PRENTICE HALL
Life
Science

Prentice Hall
Life Science

Jill Wright

Associate Professor of Science Education
University of Tennessee
Knoxville, Tennessee

Charles R. Coble

Dean of School of Education
East Carolina University
Greenville, North Carolina

Jean Hopkins

Science Department Chairperson
John H. Wood Middle School
San Antonio, Texas

Susan Johnson

Professor of Biology
Ball State University
Muncie, Indiana

David LaHart

Senior Instructor
Florida Solar Energy Center
Cape Canaveral, Florida

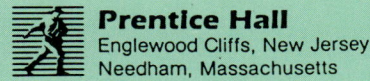 **Prentice Hall**
Englewood Cliffs, New Jersey
Needham, Massachusetts

Prentice Hall Life Science Program

Student Text and Annotated Teacher's Edition

Laboratory Manual with Annotated Teacher's Edition

Teacher's Resource Book

Test Bank with Software and DIAL-A-TEST™ Service

Life Science Critical Thinking Skills Transparencies

Life Science Courseware

Other programs in this series

Prentice Hall Physical Science © 1991

Prentice Hall Earth Science © 1991

Prentice Hall Life Science presents science as a process that involves research, experimentation, and the development of theories that can hold great explanatory and predictive power. This approach reflects the challenges and intellectual rewards available to students in the ever-changing discipline of science.

Photo credits begin on p. 630

The photograph on the cover shows an adult cheetah and her cub resting on the plains of Africa. (DPI)

Life Science Reviewers:

James Beau Horn
Science Instructor
Sellwood Middle School
Portland, Oregon

Marilyn R. Sinclair
Science Coordinator
Columbia Middle School
Champaign, Illinois

Gail Holcomb Sones
Life Science Instructor
Memorial Parkway Junior High
Katy, Texas

John F. Westwater
Science Instructor
Saint Joseph Junior High
Medford, Massachusetts

Reading Consultant

Patricia N. Schwab
Director of Undergraduate
    Advisement and Lecturer
University of South Carolina
Columbia, South Carolina

SECOND EDITION

© 1991, 1988 by Prentice-Hall, Inc., Englewood Cliffs, New Jersey 07632. All rights reserved. No part of this book may be reproduced in any form or by any means without permission in writing from the publisher. Printed in the United States of America.

ISBN 0-13-713991-8

10 9 8 7 6 5 4 3 2

Prentice Hall
A Division of Simon & Schuster
Englewood Cliffs, New Jersey 07632

Contents

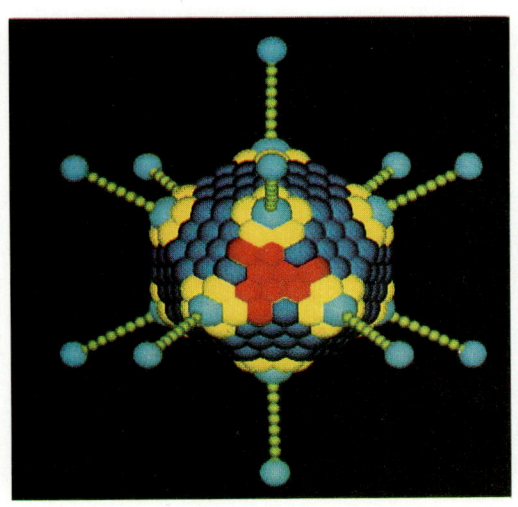

UNIT FIVE Human Biology 270–445

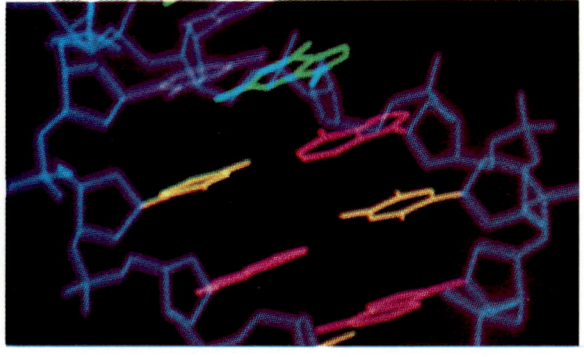

Reference Section

Characteristics of Living Things

Along most coastlines, there is an area where ocean waves pound against the rocky shore. Here, where the sea meets the land, you might think that it would be difficult to find living things. But a close look just might reveal hundreds of different plants and animals. Each of these living things has found a way to survive amid the shifting tides and crashing waves.

Seaweed, for example, is flattened against the rocks with each passing wave. The rocks, which seem white at first glance, are actually covered with living barnacles. Clusters of blue-black mussels live on the rocks as well.

Snails, called whelks, eat the mussels by drilling holes in the mussel shells and sucking out their flesh. Nearby, a tiny hermit crab is looking for a discarded shell to use as a home. A seagull circling overhead searches for food in the water below. And beneath the water, a purple starfish crawls about.

Each of these living things has its own characteristics that help it survive in this ever-changing environment. In this unit, you will learn what it means to be a living organism. You will also learn how the different body parts of an organism work in harmony. And you will learn how organisms are classified on the basis of their characteristics and body parts.

CHAPTERS

Along this California coastline, seaweed is flattened against the rocks with each passing wave.

Exploring Living Things

CHAPTER OBJECTIVES

After completing this chapter, you will be able to

1–1 Describe the steps in the scientific method.

1–1 Identify various branches of life science.

1–2 Construct and analyze a graph from experimental data.

1–2 Make calculations using the metric system.

1–3 Describe the importance of various life science tools.

1–3 Compare different types of microscopes.

1–4 Apply safety procedures in the classroom laboratory.

You probably would not be surprised to see a toad catch a fly and gobble it up. But what if you saw just the opposite? On the night of August 27, 1982, in a desert near Portal, Arizona, a biologist discovered a fly that captures and eats live toads!

The biologist Thomas Eisner was out collecting insects with a photographer and a graduate student. It was the photographer who first noticed something peculiar.

Thousands of spadefoot toads had just emerged from their tadpole stage in a small pond. The toads, which were only about 1.5 to 2 centimeters long, covered the muddy shores around the pond. But some of the tiny toads were dying and sinking into the mud! When the biologist tried to pull the dying toads out of the mud, he felt a strong pull in the opposite direction.

After some investigation, Eisner discovered that the culprit was the larvae, or young, of a type of horsefly. These large, wormlike organisms bury themselves in the mud, leaving only their heads and hooklike mouthparts above the surface. When a toad comes to rest above them, the horsefly larvae use their pointed mouthparts to grab the toad, inject it with a poison, and then feed on its body fluids.

Before this surprising finding, horsefly larvae had never been known to feed on such large animals. The diet of the larvae had been thought to consist only of other insects. As you read this book, you will come to understand how accidental observations, curious minds, and smart detective work can lead to a scientific discovery, such as a toad-catching fly.

A spadefoot toad peers out from its burrow.

Section Objective

To describe the steps in the scientific method

Fact or fiction? Turtle eggs develop into males in cold temperatures and females in warmer temperatures. Insect-eating bats can eat as many as 3000 insects in one night. The average American teenager eats about 825 kilograms of food a year. Believe it or not, all of these statements are true. They are all facts. The study of science involves the discovery of facts about nature.

The word *science* comes from the Latin word *scire,* which means "to know." Scientists seek to know more about nature by uncovering facts. However, science is more than a simple list of facts. Science involves a constant search for information about the universe. A famous nineteenth-century French scientist, Jules Henri Poincare, said, "Science is built up with facts as a house is with stones. But a collection of facts is no more a science than a heap of stones is a house,"

Not only do scientists find facts, they try to tie the facts together to explain mysteries of nature. For example, scientists may try to determine how a particular disease is caused. After they determine what causes the disease, they will try to determine how the disease can be cured.

Figure 1–1 *Life scientists seek to discover facts about living things, such as the greater horseshoe bat (left) and the painted turtle (right). What information must be known to determine whether this turtle is a male or a female?*

Figure 1–2 *According to one scientific theory, the extinction of dinosaurs was caused when an asteroid struck the earth. What is the difference between a theory and a law?*

To gather knowledge about nature, scientists make observations about the world around them. Then, the scientists come up with ideas to explain what they have observed. And they perform experiments to test their explanations. After a study of facts, observations, and experiments, scientists may develop a **theory.** A theory is the most logical explanation of events that happen in nature.

Once a scientific theory has been developed, it must be tested over and over again. If test results do not agree, the theory is changed or rejected. When a scientific theory has been tested many times and is generally accepted as true, scientists may call it a **law.** However, even scientific laws can be changed as a result of future observations and experiments.

Scientific Method

When scientists search for an answer to a problem, they usually take an orderly and systematic approach. The **scientific method** is the systematic

Sharpen Your Skills

Theories and Laws

Scientists sometimes use observations and experiments to develop a theory. Use books and reference materials in the library to find out about a theory of life science that became a law. Present your findings to the class in an oral report.

Figure 1–3 *Poisonous snakes, such as the rattlesnake, have fangs. What are the fangs used for?*

Sharpen Your Skills

Making Observations

1. Collect ten similar objects, such as maple leaves, string beans, or blades of grass.

2. Choose one object from your collection and observe it carefully.

3. Write down all of your observations and draw a diagram of the object.

4. Put all ten objects together and mix them up.

5. Now see if a classmate can pick out your object using your observations.

Which observations were helpful in identifying your object?

approach to problem solving used by all scientists. **The basic steps of any scientific method are**

> **Stating the problem**
>
> **Gathering information on the problem**
>
> **Forming a hypothesis**
>
> **Experimenting**
>
> **Recording and analyzing data**
>
> **Stating a conclusion**

The following example shows how a life scientist used the scientific method to solve a problem.

Stating the Problem

Most people know enough to walk the other way if they should run into a rattlesnake. However, if you could safely observe a rattler, you would discover a rather curious kind of behavior.

With fangs flashing and body arching, the deadly rattler strikes. The snake's fangs quickly inject poisonous venom into its victim. Then, in a surprise move, the rattler allows the wounded animal to run away! But the rattlesnake will not miss its intended meal. After waiting for its poison to take effect, the rattler follows the trail of the injured animal.

Although the rattler cannot see well, somehow it manages to find its victim on the dense, dark, forest floor. Clearly, something leads the snake to its prey. What invisible trail does the snake follow in tracking down its bitten prey? This is a *problem* that scientists recently tried to solve.

Gathering Information on the Problem

The first step in solving a scientific problem is to find out or review everything important related to it. For example, the scientists trying to solve the rattlesnake mystery knew that a rattlesnake's eyes are only sensitive to visible light. However, they also knew that a pair of organs located under the animal's eyes detects invisible light in the form of heat. These heat-sensing pits pick up signals from warm-blooded animals. The signals help the snake locate its intended prey. But the heat-sensing pits cannot help the snake find a wounded victim that has run many meters away. Some other process must be responsible for that.

The scientists knew that a rattler's tongue "smells" certain odors in the air. The rattler's tongue picks up these odors on an outward flick. The odors enter the snake's mouth on an inward flick. The scientists also knew that the sight or smell of an unbitten animal did not trigger the rattler's tracking action. Using all this information, the scientists were able to suggest a solution to the problem.

Forming a Hypothesis

A suggested solution is called a **hypothesis** (high-PAH-thuh-sihs). A hypothesis is almost always formed after the information related to the problem has been carefully studied. But sometimes a hypothesis is the result of creative thinking that often involves bold, original guesses about the problem. In this regard, forming a hypothesis is like good detective work, which involves not only logic, but hunches, intuition, and the taking of chances.

To the problem, "What invisible trail does a rattler follow in tracking down its prey?" the scientists suggested a hypothesis. The scientists suggested that *after the snake wounds its victim, the snake follows the smell of its own venom to locate the animal.*

Figure 1–4 *The timber rattlesnake was able to track down its wounded prey after the prey had run away. What hypothesis could explain the snake's behavior?*

Your friend tells you that plants grow slower when salt is added to their soil.

1. Design and conduct an experiment to find out if this is true. Make sure your experiment has a control and a variable.

2. Record all data.

3. Then make a graph of your results.

What conclusion can you draw from the graph?

Experimenting

The scientists next had to test their hypothesis by performing certain activities and recording the results. These activities are called *experiments.* Whenever scientists test a hypothesis using an experiment, they must make sure that the results of the experiment clearly support or do not support the hypothesis. That is, they must make sure that one, and only one, factor affects the results of the experiment. The factor being tested in an experiment is called the **variable.** In any experiment, only one variable is tested at a time. Otherwise it would not be clear which variable had caused the results of the experiment.

In the rattlesnake experiments, the scientists tested whether the snake's venom formed an invisible trail that the snake followed. The venom was the variable, or single factor, that the scientists wanted to test. The scientists performed the experiment to test this variable.

First, the scientists dragged a dead mouse that had been struck and poisoned by a rattlesnake along a curving path on the bottom of the snake's empty cage. When the snake was placed in its cage, its tongue flicked rapidly, its head moved slowly from side to side, and it followed the exact trail the scientists had laid out. The results seemed clear, but the scientists had one more experiment to perform.

Figure 1–5 *Snakes, such as the rattlesnake, are able to swallow their prey whole.*

To be sure it was the scent of the venom and no other odor that the snake followed, the scientists ran a **control** experiment. A control experiment is run in exactly the same way as the experiment with the variable, but the variable is left out. So the scientists dragged an unbitten dead mouse along a path in the cage. The experiment was exactly the same, except this mouse had not been poisoned. This time the snake seemed disinterested. Its tongue flicked very slowly and it did not follow the path.

Recording and Analyzing Data

The rattlesnake experiments were repeated many times, and the scientists carefully recorded the **data** from the experiments. Data include observations such as measurements. Then the scientists analyzed the recorded data.

Stating a Conclusion

After analyzing the recorded data, the scientists came to a *conclusion*. They concluded that the scent of venom was the only factor that could cause a rattlesnake to follow its bitten victim. Rattlesnake venom is made up of many different substances. Exactly which ones are responsible for the snake's behavior are as yet unknown. As is often the case in science, a solution to one mystery brings to light another mystery. Using scientific methods similar to those described here, scientists hope to follow a path that leads to the solution to this new mystery.

Branches of Life Science

Life science is one major branch of science. It deals with living things and their relationship to one another and to their environment. The major branches of science have become more specialized as scientific knowledge has expanded.

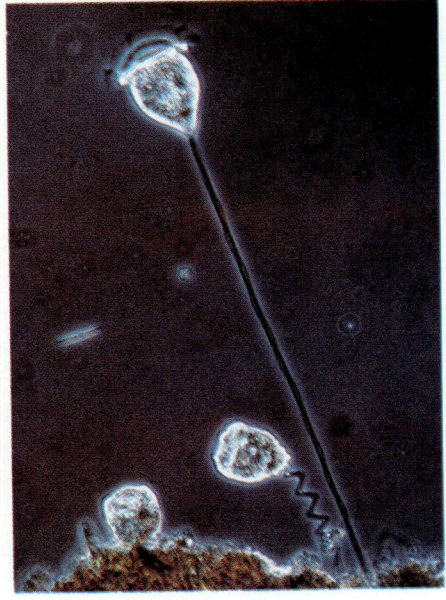

Figure 1–6 *Life science can be divided into many specialized branches. Ecology is the study of the relationships between organisms and their environment (top). Botany is the study of plants (center). And, microbiology is the study of microscopic organisms such as* Vorticella *(bottom).*

Sharpen Your Skills

What's the Word?

Use the prefixes and suffixes in Figure 1–7 to determine the meanings of the following words.

biology	exostosis
gastropod	exoskeleton
hemostasis	epidermis
homeostasis	cytology
photometer	

Life science can be divided into a number of specialized branches. One branch of life science is **zoology** (zoh-AHL-uh-jee). Zoology is the study of animals. Another branch of life science is **botany** (BAHT-uh-nee), which is the study of plants. A third branch of life science is **ecology,** or the study of the relationships between living organisms and their environment. Still another branch of life science is **microbiology.** Microbiology is the study of **microorganisms,** or microscopic organisms.

Just like any branch of science, life science has many special terms that may be unfamiliar to you. The chart in Figure 1–7 gives the meanings of some prefixes and suffixes that are commonly used in science vocabulary. If you learn the meanings of these prefixes and suffixes, learning new science terms will be easier. For example, suppose you are reading a magazine article and you come across the term *osteology.* You know that the prefix *osteo-* means bone and the suffix *-logy* means study. By joining the meanings of the prefix and suffix, you learn that osteology is the study of bones.

Figure 1–7 *Learning the meanings of these prefixes and suffixes will make learning new science terms easier. What is the meaning of the word* arthropod?

Prefix	Meaning	Prefix	Meaning	Suffix	Meaning
anti-	against	herb-	pertaining to plants	-cyst	pouch
arth-	joint, jointed	hetero-	different	-derm	skin, layer
auto-	self	homeo-	same	-gen	producing
bio-	related to life	macro-	large	-itis	inflammation
chloro-	green	micro-	small	-logy	study
cyto-	cell	multi-	consisting of many units	-meter	measurement
di-	double	osteo-	bone	-osis	condition, disease
epi-	above	photo-	pertaining to light	-phase	stage
exo-	outer, external	plasm-	forming substance	-phage	eater
gastro-	stomach	proto-	first	-pod	foot
hemo-	blood	syn-	together	-stasis	stationary condition

1. Identify the steps in the scientific method.
2. Explain why an experiment has only one variable.
3. Using Figure 1–7, determine what the terms *arthritis* and *hematology* mean.
4. Design an experiment that tests the hypothesis that a green plant needs light to live. Identify the variable and the control in the experiment.

1–2 Scientific Measurements

To compare metric units used to measure length, volume, mass, and temperature

As part of the process of experimenting and gathering information, scientists must make accurate measurements. The scientists may need to know the size of a cell or the temperature of a bird's body. They must also be able to share their information with other scientists. To do this, scientists must speak the same measurement "language." **The common language of measurement in science used all over the world is the metric system.**

The **metric system** is a decimal system, or a system based on ten. That is, each unit in the metric system is ten times larger or ten times smaller than the next unit. Metric calculations are easy to do because they involve multiplying or dividing by ten. Some frequently used metric units and their abbreviations are listed in Figure 1–8 on page 14.

Length

The **meter (m)** is the basic unit of length in the metric system. One meter is equal to 39.4 inches. Sometimes scientists must measure distances much longer than a meter, such as the distance a bird may fly across a continent. To do this, scientists use a unit called a kilometer (km). The prefix *kilo-* means 1000. So a kilometer is 1000 meters, or about the length of five city blocks.

A hummingbird, on the other hand, is too small to be measured in meters or kilometers. So scientists use the centimeter (cm). The prefix *centi-* means that 100 of these units make a meter. Thus there are 100 centimeters in a meter. The hummingbird is only about 5 centimeters long. For objects even

Using Metric Measurements

Use the appropriate scientific tools and the metric system to measure the following. Construct a chart of your results.

length of the textbook
length of your arm
temperature indoors
temperature outdoors
volume of a glass of water
volume of a bucket of water

COMMONLY USED METRIC UNITS

Length

Length is the distance from one point to another.
A meter is slightly longer than a yard.
1 meter (m) = 100 centimeters (cm)
1 meter = 1000 millimeters (mm)
1 meter = 1,000,000 micrometers (μm)
1 meter = 1,000,000,000 nanometers (nm)
1 meter = 10,000,000,000 angstroms (Å)
1000 meters = 1 kilometer (km)

Mass

Mass is the amount of matter in an object.
A gram has a mass equal to about one paper clip.
1 kilogram (kg) = 1000 grams (g)
1 gram = 1000 milligrams (mg)
1000 kilograms = 1 metric ton (t)

Volume

Volume is the amount of space an object takes up.
A liter is slightly larger than a quart.
1 liter (L) = 1000 milliliters (mL) or 1000 cubic centimeters (cm^3)

Temperature

Temperature is the measure of hotness or coldness in degrees Celsius (°C)
0°C = freezing point of water
100°C = boiling point of water

Figure 1–8 *The metric system is easy to use because it is based on units of ten. What is the basic unit of length in the metric system?*

smaller, another division of the meter is used. The millimeter (mm) is one thousandth of a meter. The prefix *milli-* means ¹/₁₀₀₀. One meter equals 1000 millimeters. In bright light, the diameter of the pupil of your eye is about 1 millimeter. Even this unit is too large to use when describing the sizes of the smallest germs, such as the bacteria that cause sore throats. These germs are measured in micrometers, or millionths of a meter, and nanometers, or billionths of a meter.

Volume

The amount of space an object takes up is called its volume. The **liter (L)** is the basic unit of volume in the metric system. The liter is slightly larger than the quart. Here again scientists use divisions of the liter to measure smaller volumes. The milliliter (mL), or cubic centimeter (cm^3), is ¹/₁₀₀₀ of a liter. There are 1000 milliliters, or cubic centimeters, in a liter. The volumes of both liquids and gases are measured in liters and milliliters. For example, a scientist may remove a few milliliters of blood from an animal in order to study the types of substances the blood contains.

Mass and Weight

The **kilogram (kg)** is the basic unit of mass in the metric system. **Mass** is a measure of the amount of matter in an object. There is more matter in a tree trunk than in a leaf. Therefore, a tree trunk has more mass than a leaf.

One kilogram is slightly more than two pounds. For smaller units of mass, the gram (g) is used. Remember that the prefix *kilo-* means 1000. There are 1000 grams in a kilogram. One milligram (mg) measures an even smaller mass, $\frac{1}{1000}$ of a gram. How many milligrams are there in one kilogram?

Mass is not the same as **weight.** Weight is a measure of the force of attraction between objects due to **gravity.** All objects exert a force of gravity on other objects. The strength of the force depends on

Figure 1–9 *Scientists use many tools to study their world. What is being measured by the triple-beam balance (top left) and the Celsius thermometer (top right)? For what measurements are the metric ruler (bottom left) and the graduated cylinder (bottom right) used?*

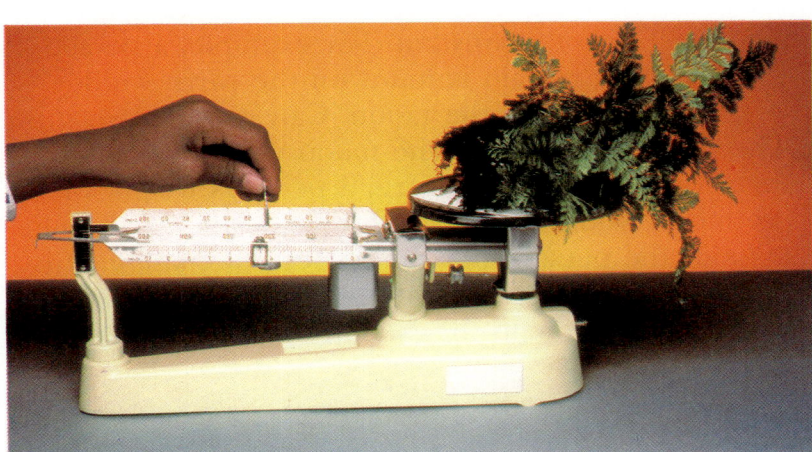

Figure 1–10 *The kilogram is the basic unit of mass in the metric system. Which has more mass, an elephant or a snail?*

the mass of the objects and the distance between them. You may not be aware of it, but the gravity of your body pulls the earth toward you. At the same time, the gravity of the earth pulls you toward its center. The mass of the earth is, however, much greater than your mass. As a result, the force of gravity you exert is very small compared to that exerted by the earth.

You remain on the surface of the earth because of the force of the earth's gravity. And it is the size of this force that is measured by your weight. As the distance between objects decreases, the force of gravity between them increases. Therefore, you would weigh a little more at sea level than on the top of a mountain. At sea level, you are closer to the center of the earth.

It may be apparent to you now that mass is a constant and weight is not. Because weight can change, it is not a constant. For example, a person who weighs a certain amount on the earth would weigh much less on the moon. Can you explain this? The force of gravity of the moon is about one-sixth that of the earth. Therefore, a person's weight on the moon would be one-sixth that on the earth. The mass of the person, however, does not change. It is the same on the moon or the earth because the amount of matter in the person does not change.

Figure 1–11 *As astronauts move farther from the center of the earth, the pull of the earth's gravity decreases and they appear to be "weightless." Does the mass of the astronaut change as well?*

Density

The measurement of how much mass is contained in a given volume of an object is called its **density.** Density can be defined as the mass per unit volume of a substance. The following formula shows the relationship between density, mass, and volume.

$$\text{density} = \frac{\text{mass}}{\text{volume}}$$

For example, the mass of 100 cubic centimeters of corn oil is 92.2 grams. The density of corn oil would be

$$\text{density} = \frac{\text{92.2 g (mass)}}{\text{100 cm}^3 \text{ (volume)}} = 0.922 \text{ g/cm}^3$$

Density is an important concept because it allows scientists to identify and compare substances. Each substance has its own characteristic density. For example, the density of water is 1 g/cm^3 while that of butterfat is 0.91 g/cm^3. An object with a density less than 1 g/cm^3 will float on water. Based on this data, will butterfat sink or float in water?

Temperature

Scientists measure temperature according to the **Celsius** scale, in degrees Celsius (°C). The fixed points on the scale are the freezing point of water at sea level, 0°C, and the boiling point of water, 100°C. The range between these points is 100 degrees, and each degree is $\frac{1}{100}$ of the difference between the freezing point and boiling point of water. Normal human body temperature is 37°C, while some birds maintain a body temperature of 41°C.

Analyzing Measurements

As you know, when scientists perform experiments, they must record data. Important data is often recorded as scientific measurements. Such measurements are easily analyzed when they are organized in data tables.

The tables in Figure 1–13 on page 18 show data from an experiment on a type of yeast normally grown at 25°C. Yeast is a one-celled microscopic

The tables in Figure 1–13 on page 18 show data

Figure 1–12 *Steam rises from a hot spring in Yellowstone National Park, Wyoming. What tool would you use to measure the temperature of the hot spring?*

GROWTH OF YEAST CELLS

Temperature at 15°C *(experimental setup)*

Time *(hours)*	0	2	4	6	8	10	12	14	16	18
Number of Yeast Cells	1	4	20	38	90	150	220	350	430	550

Temperature at 25°C *(control setup)*

Time *(hours)*	0	2	4	6	8	10	12	14	16	18
Number of Yeast Cells	1	26	74	170	330	550	600	650	665	665

Figure 1–13 *Scientists often organize important information in data tables. What are the time intervals in these data tables?*

organism. The hypothesis tested in the experiment was that the rate of yeast growth decreases when the temperature of their environment is lowered to 15°C. What is the variable in the experiment?

To visually compare data from an experimental setup with data from a control setup, a scientist constructs a graph on which to plot the data. The data tables in Figure 1–13 have two types of measurements. A graph of the data then would have two axes. See Figure 1–14.

The horizontal axis of the graph represents time measurements in the data tables. Notice that the horizontal axis is clearly labeled. Because time measurements were made every two hours, the horizontal axis is marked with intervals of two hours. In any graph, the space between equal intervals has to be equal. For instance, the space between two hours and four hours must be the same as the space between four hours and six hours.

The vertical axis of the graph represents the measurements of cell number in the data tables. The vertical axis also must be clearly labeled. Because the experiment began with 1 cell and ended with 665 cells, the vertical axis begins at 1 and ends at 665. Again, the space between equal intervals must be equal.

After the axes of the graph were set up, the scientist first plotted the data from the experimental

Sharpen Your Skills

Metric Conversion

Complete the following metric conversions.

a. 21,537 millimeters = __________ meters

b. 425 kilometers = __________ centimeters

c. 6.87 grams = __________ kilograms

d. 96.3 milliliters = __________ liters

e. 11 milliliters = _____ cubic centimeters

setup. Each pair of points in the data table was plotted. At zero hours the number of yeast cells was one. So the scientist placed a dot where zero hours and one yeast cell intersect—the lower left corner of the graph. The next pair of data points was two hours and four yeast cells. So the scientist lightly drew a vertical line from the two-hour interval of the horizontal axis and then a horizontal line from the four-cell interval of the vertical axis. The scientist then placed a dot at the place where the two lines intersected. This dot represents the data point two hours and four yeast cells. The scientist plotted the rest of the data pairs from the data table in a similar manner.

After all of the data pairs were plotted, the scientist drew a line connecting all the dots on the

Figure 1–14 *The information in a data table can be placed on a graph. What conclusion about the effect of temperature on the rate of yeast growth can you draw from this graph?*

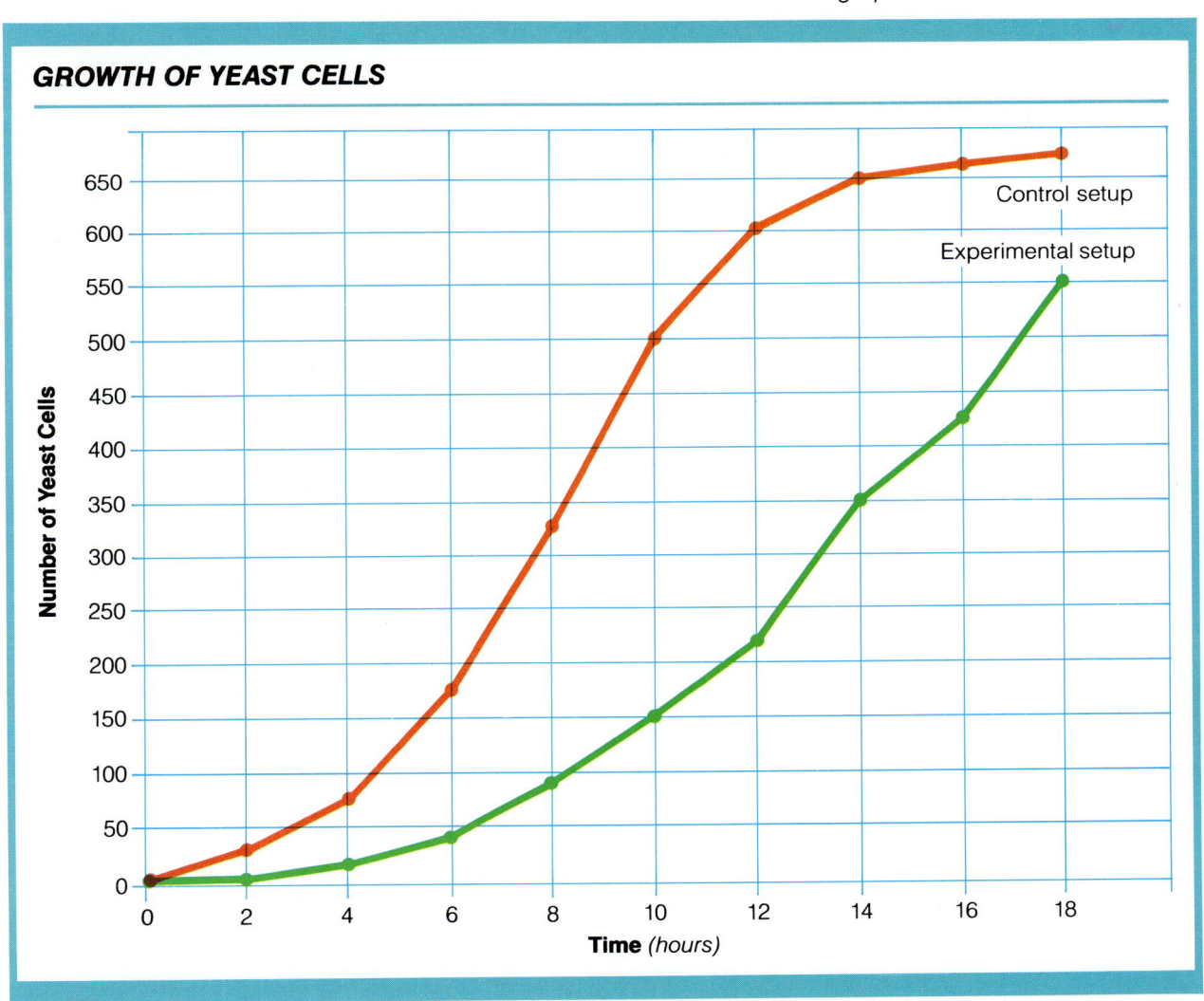

GROWTH OF YEAST CELLS

graph. This line represents the data from the experimental setup. Then the scientist graphed all the data pairs from the control setup. Figure 1–14 shows what the two lines look like. After analyzing the lines on the graph, the scientist concluded that the rate of yeast growth decreases when the temperature of their environment is lowered to 15°C.

SECTION REVIEW

1. What are the basic units of mass, length, volume, and temperature in the metric system?
2. What is the difference between mass and weight?
3. Why would a scientist graph data from a control setup on the same graph with data from an experimental setup?
4. To help identify an unknown substance, would a scientist measure its mass or its density? Explain.

1–3 Tools of a Life Scientist

Scientists use a variety of tools to explore the world around them ranging from simple microscopes to complex computers. The kind of tools scientists use depends on the problems they want to solve. A scientist may use a metric ruler to measure the length of a leaf. At another time, the same scientist may use a computer to analyze large amounts of data concerning hundreds of leaves.

Tools are very important in the advancement of science. In fact, some great discoveries were not made until the appropriate tools were developed. An example is the **microscope,** which is an instrument that produces an enlarged image of an object. Simple microscopes have been used by scientists for many years to study the structure of living things. But it was not until several hundred years after the first microscopes that scientists were able to discover the smallest structures that make up living things!

Microscopes

Have you ever looked at a ladybug through a magnifying glass? If you have, then you used a type of microscope. A magnifying glass is a simple micro-

HELP WANTED: ELECTRON MICROS-COPIST Bachelor's degree in histo-technology preferred. Training in the operation of the electron microscope and the interpretation of electron micrographs required.

Until the invention of the electron microscope in the 1930s, scientists examined the inner structures of living things by viewing slices of plant and animal tissues through light microscopes. Light microscopes magnify only large cell structures, not tiny structures. The tiny cell structures can be seen only when viewed through an electron microscope. Users of electron microscopes need special training and extensive practice before they become **electron microscopists.** Once their skill is developed, they can use the electron microscope to bring objects as small as bacteria, viruses, or even atoms into focus.

Electron micrographs are photographs of specimens seen through the electron microscope. By studying these photographs, research scientists learn more about the normal activities of cells. Human tissue micrographs help doctors diagnose patients.

Electron microscopists work in hospitals, universities, and research laboratories. If you wish to know more about a career in electron microscopy, write to the National Society for Histotechnology, P.O. Box 36, Lanham, MD 20706.

scope because it has only one **lens.** A lens is a curved piece of glass. As light rays pass through the glass, they bend. In some kinds of lenses, the bending of the light rays increases the size of an object's image.

THE COMPOUND LIGHT MICROSCOPE **Compound light microscopes** are microscopes that have more than one lens. Like magnifying glasses, compound microscopes use light to make objects appear larger. Magnifying glasses produce an image only a few times larger than the object. But a compound light microscope can make an object appear up to 2000 times larger than it actually is! This amount of magnification allows you to see inside the cells of a leaf. Appendix B on pages 609 to 610 describes the various parts of a compound light microscope.

THE ELECTRON MICROSCOPE Instead of using light to magnify the image of an object, an **electron microscope** uses a beam of tiny particles called electrons. The beam produces pictures, which are focused on photographic film or a television screen.

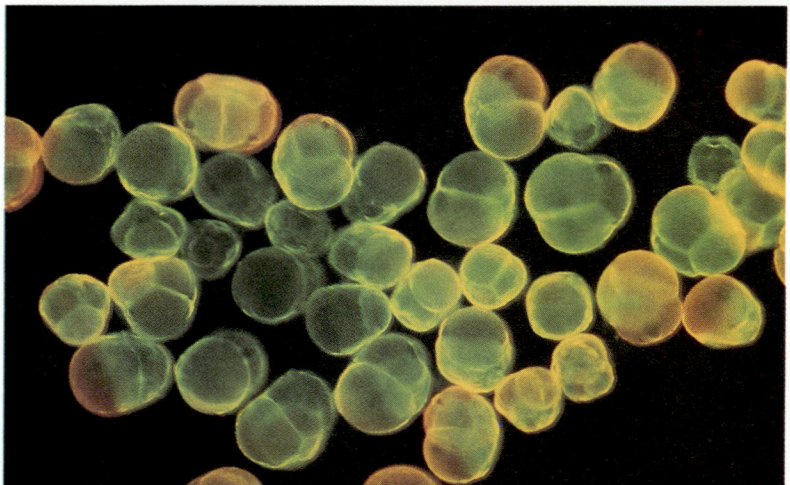

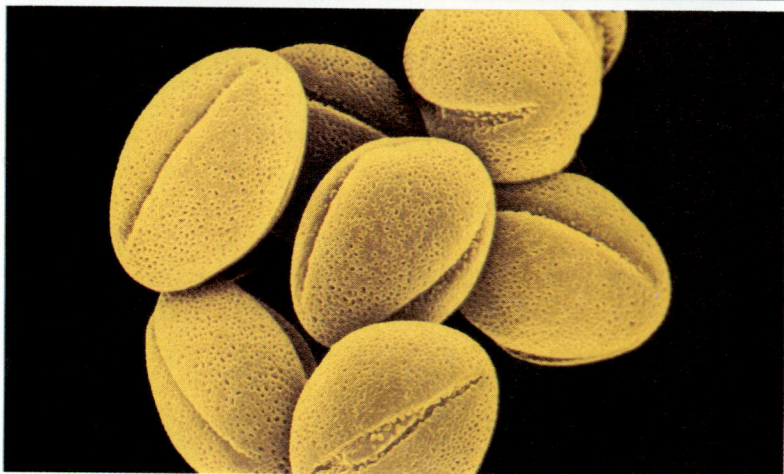

Figure 1–15 *Notice the pollen grains visible on the flower (top). Then look at pollen grains that have been magnified 60 times through a microscope (center). This three-dimensional image of pollen grains (bottom) was produced by a scanning electron microscope. The grains have been magnified to 378 times their normal size.*

Electron microscopes can magnify objects hundreds of thousands of times. The scanning electron microscope, or SEM, is a type of electron microscope that can produce three-dimensional images. With electron microscopes, scientists can study the smallest parts of plant and animal cells.

Looking Through Barriers

To learn more about living things, scientists must be able to see the inside of an organism *from the outside*. To do this, scientists use certain tools.

X-RAY For almost 100 years, scientists have been using invisible radiations known as X-rays. X-rays are blocked by dense materials such as bone, but pass easily through skin and muscle. For this reason, X-rays are often used for taking pictures of bones inside an organism.

CAT Computerized Axial Tomography, or CAT scan, is a relatively new technique that produces cross-sectional pictures of an object. An X-ray machine in the CAT-scanner is used to take up to 720 different exposures of an object. Each picture shows a "slice" of the object. Then a computer analyzes

Figure 1–16 *Because this 30-million-year-old skull of a small mammal (left) was a rare find, scientists did not want to break it open to see what it looked like inside. But by using a CAT-scanner, scientists were able to get a three-dimensional view of the inside of the skull (right).*

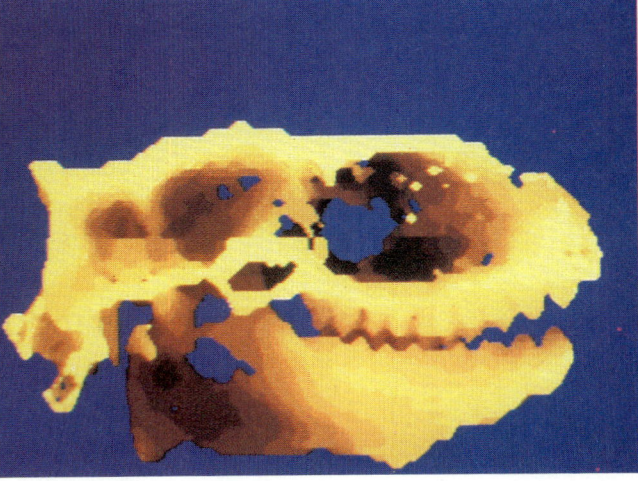

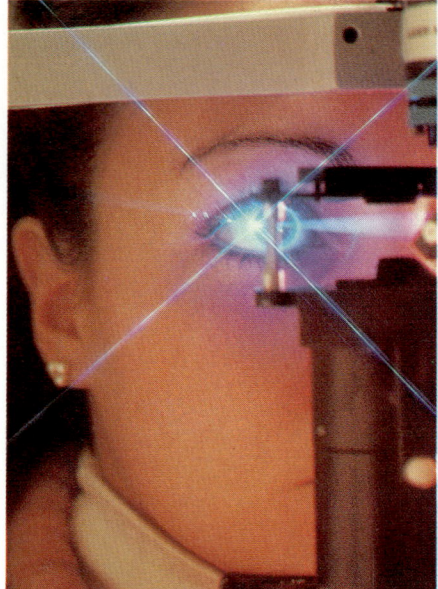

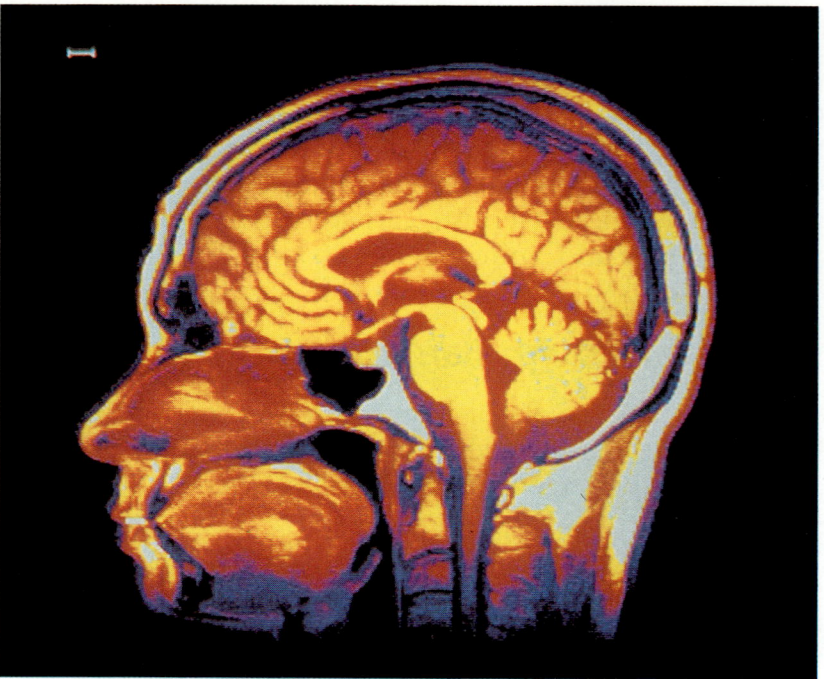

Figure 1–17 *Laser beams (left) can be used in certain types of eye surgery. NMR images (right) help scientists study the inside of the body. What important part of the body can be studied from this NMR image?*

and combines each exposure to construct one picture. Doctors use CAT scans to examine the brain and other parts of the body.

NMR Nuclear Magnetic Resonance, or NMR, is another tool that helps scientists see the inside of objects. NMR uses magnetism and radio waves to produce images. Scientists can use NMR to study the structure and function of body cells without harming the living tissue.

Lasers

Lasers are another tool used by scientists. A laser is an instrument that produces a narrow, intense beam of light. Unlike a microscope, the laser is not used to magnify images. But it can be used as a kind of scalpel. For example, a surgeon may use a laser beam to remove unhealthy cells, such as cancer cells, from the body. Lasers are also used in certain types of eye surgery.

Computers

A computer is another useful tool for scientists. Computers are electronic devices that collect, analyze, display, and store data. Computers help doctors diagnose diseases and prescribe treatments. In some branches of life science, a computer might be used to help researchers rapidly analyze information about the structure and functions of a living thing.

As you can see, modern life scientists have a variety of tools that they can use in their search for information. And perhaps future scientists will be using even better tools that have yet to be developed. Science, it seems, needs more than keen minds. It needs the kinds of data that very often only special tools can provide.

SECTION REVIEW

1. List five tools used by life scientists and describe the importance of each tool.
2. Explain the difference between the compound light microscope and the electron microscope.
3. What does NMR use to produce an image?
4. Explain why a CAT scan might be more useful to a doctor than a simple X-ray.

1–4 Science Safety in the Laboratory

Section Objective

To apply safety rules in the laboratory

The scientific laboratory is a place of adventure and discovery. Some of the most exciting events in scientific history have happened in laboratories. The structure of DNA, the blueprint of life, was discovered by scientists in laboratories. The artificial skin that is used to replace skin destroyed by burns was first made by scientists in a laboratory. The list goes on and on.

To better understand the facts and concepts you will read about in life science, you may work in the laboratory this year. If you follow instructions and are as careful as a scientist would be, the laboratory will turn out to be an exciting experience for you.

When working in the laboratory, scientists know that it is very important to follow safety procedures. Therefore, scientists take as many precautions as possible to protect themselves and their fellow workers.

All of the work you will do in the laboratory this year will include experiments that have been done over and over again. When done properly, the experiments are not only interesting but also perfectly safe. But if they are done improperly, accidents can occur. How can you avoid such problems?

The most important safety rule is to always follow your teacher's directions or the directions in your textbook exactly as stated. Never try anything on your own without asking your teacher first. And when you are not sure what you should do, always ask first. As you read the laboratory investigations in the textbook, you will see safety alert symbols. Look at Figure 1–19 to learn the meanings of the safety symbols and all the important safety precautions you should take.

In addition to the safety procedures listed in Figure 1–19, there is a more detailed list of safety procedures in Appendix C on pages 611 to 612 of this textbook. Before you enter the laboratory for the first time, make sure that you have read each rule carefully. Then read them over again. Make sure that you understand each rule. If you do not understand a rule, ask your teacher to explain it. If you wish, you may even want to suggest further rules that apply to your particular classroom.

Figure 1–18 *Taking accurate measurements is important in laboratory work. Following safety rules is also important. Are these students following all necessary safety rules?*

LABORATORY SAFETY: RULES AND SYMBOLS

Glassware Safety

1. Whenever you see this symbol, you will know that you are working with glassware that can be easily broken. Take particular care to handle such glassware safely. And never use broken glassware.
2. Never heat glassware that is not thoroughly dry. Never pick up any glassware unless you are sure it is not hot. If it is hot, use heat-resistant gloves.
3. Always clean glassware thoroughly before putting it away.

Fire Safety

1. Whenever you see this symbol, you will know that you are working with fire. Never use any source of fire without wearing safety goggles.
2. Never heat anything—particularly chemicals—unless instructed to do so.
3. Never heat anything in a closed container.
4. Never reach across a flame.
5. Always use a clamp, tongs, or heat-resistant gloves to handle hot objects.
6. Always maintain a clean work area, particularly when using a flame.

Heat Safety

Whenever you see this symbol, you will know that you should put on heat-resistant gloves to avoid burning your hands.

Chemical Safety

1. Whenever you see this symbol, you will know that you are working with chemicals that could be hazardous.
2. Never smell any chemical directly from its container. Always use your hand to waft some of the odors from the top of the container towards your nose—and only when instructed to do so.
3. Never mix chemicals unless instructed to do so.
4. Never touch or taste any chemical unless instructed to do so.
5. Keep all lids closed when chemicals are not in use. Dispose of all chemicals as instructed by your teacher.

6. Rinse any chemicals, particularly acids, off your skin and clothes with water immediately. Then notify your teacher.

Eye and Face Safety

1. Whenever you see this symbol, you will know that you are performing an experiment in which you must take precautions to protect your eyes and face by wearing safety goggles.
2. Always point a test tube or bottle that is being heated away from you and others. Chemicals can splash or boil out of the heated test tube.

Sharp Instrument Safety

1. Whenever you see this symbol, you will know that you are working with a sharp instrument.
2. Always use single-edged razors; double-edged razors are too dangerous.
3. Handle any sharp instrument with extreme care. Never cut any material towards you; always cut away from you.
4. Notify your teacher immediately if you are cut in the lab.

Electrical Safety

1. Whenever you see this symbol, you will know that you are using electricity in the laboratory.
2. Never use long extension cords to plug in an electrical device. Do not plug too many different appliances into one socket or you may overload the socket and cause a fire.
3. Never touch an electrical appliance or outlet with wet hands.

Animal Safety

1. Whenever you see this symbol, you will know that you are working with live animals.
2. Do not cause pain, discomfort, or injury to an animal.
3. Follow your teacher's directions when handling animals. Wash your hands thoroughly after handling animals or their cages.

SECTION REVIEW

1. What is the most important general rule to keep in mind when working in the laboratory this school year?
2. Explain why the laboratory is important in scientific research.
3. Suppose your teacher asked you to boil some water in a test tube. What precautions would you take so that this activity would be done safely?

Figure 1–19 *You should become familiar with these safety symbols because you will see them in the laboratory investigations in the textbook. What is the symbol for special safety precautions with sharp instruments?*

Determining Densities

Problem

How can you determine the density of a regular object, an irregular object, and a liquid?

> **Materials** *(per group)*
>
> small block of wood   lead pencil
> small block of plastic   metric ruler
> small block of metal
> small ball of aluminum foil
> 100 mL of water
> triple-beam balance
> 100-mL graduated cylinder

Procedure

1. Using the triple-beam balance, find the mass in grams of the wood, plastic, metal, aluminum, and pencil. Record your results in a chart.
2. To determine the mass of water, first use the balance to find the mass of the empty graduated cylinder. Then take the cylinder off the balance and fill it with 100 mL of water. Determine the mass of the filled cylinder. Subtract the mass of the empty cylinder from the mass of the filled cylinder to find the mass of the water.

Substance	Mass (g)	Volume (cm³)	Density (g/cm³)
Wood			
Plastic			
Metal			
Aluminum foil			
Pencil			
Water			

3. One mL of water is equivalent to 1 cm³ of water. So 100 mL of water has a volume of 100 cm³. Record this information.
4. The volume of a regular object is found by multiplying the length times the width times the height of the object. Use the metric ruler to obtain these measurements for the blocks of wood, metal, and plastic. Then calculate the volume of each object. Record this information.
5. The volume of an irregular object, such as the aluminum foil or the pencil, is found by filling the graduated cylinder with 75 mL of water and then placing the object into the cylinder. The volume of the object is equal to the change in water level in the cylinder. (Remember 1 mL = 1 cm³.) Calculate the volume of the aluminum foil and the pencil using this method. Record this information in your chart.
6. Find the density of the wood, plastic, metal, aluminum foil, pencil, and water by dividing the mass by the volume. The formula is

$$\text{density} = \frac{\text{mass in grams}}{\text{volume in cubic centimeters}}$$

Observations

1. Which object has the greatest density?
2. Which object has the lowest density?

Conclusions

1. Based on your observations, which of the objects you tested will float in water? Explain your answer.
2. Suppose you doubled the size of the wooden block. How would its density be affected? What about the metal block?
3. How does changing an object's size affect its density?
4. If an object has a mass of 56 g and a volume of 22 cm³, will the object float in water? Explain your answer.

CHAPTER REVIEW

1–1 What Is Science?

❏ The basic steps of any scientific method are stating the problem, gathering information, forming a hypothesis, experimenting, recording, and analyzing data, and stating a conclusion.

❏ An experiment should have only one variable, or factor being tested, and a control, or the same experiment without the variable.

1–2 Scientific Measurements

❏ In the metric system, the meter is the basic unit of length, the liter is the basic unit of volume, and the kilogram is the basic unit of mass. The Celsius scale is used to measure temperature.

❏ Mass is a measure of the amount of matter in an object. Weight is a measure of the attraction between objects due to gravity.

❏ Density is the measurement of the amount of mass in a given volume of an object.

1–3 Tools of a Life Scientist

❏ Microscopes magnify small objects and produce enlarged images of them.

❏ X-rays, CAT, and NMR can provide pictures of internal body structures.

❏ Lasers and computers have important applications in life science.

1–4 Science Safety in the Laboratory

❏ When working in the laboratory, it is important to follow safety procedures.

❏ The laboratory can be a safe place to work if proper safety rules are followed at all times.

VOCABULARY

Define each term in a complete sentence.

botany	ecology	liter	scientific method
Celsius	electron microscope	mass	theory
compound light microscope	gravity	meter	variable
control	hypothesis	metric system	weight
data	kilogram	microbiology	zoology
density	law	microorganism	
	lens	microscope	

CONTENT REVIEW: MULTIPLE CHOICE

On a separate sheet of paper, write the letter of the answer that best completes each statement.

1. A hypothesis is a (an)
   a. question.   b. answer.   c. suggested solution.   d. conclusion.
2. The factor being tested in an experiment is the
   a. variable.   b. hypothesis.   c. conclusion.   d. control.
3. Recorded observations and measurements are
   a. hypotheses.   b. data.   c. variables.   d. conclusions.

4. Scientists must analyze the results of an experiment before they form a (an)
   a. hypothesis.   b. experiment.   c. control.   d. conclusion.
5. The branch of life science that deals with the study of plants is
   a. ecology.   b. botany.   c. microbiology.   d. zoology.
6. The basic unit of volume in the metric system is the
   a. meter.   b. gram.   c. liter.   d. degree Celsius.
7. The meter is the basic unit of
   a. volume.   b. mass.   c. weight.   d. length.
8. One kilogram is equal to
   a. 100 grams.   b. 0.1 gram.   c. 1000 grams.   d. 10 grams.
9. The amount of mass contained in a given volume of an object is called
   a. gravity.   b. area.   c. temperature.   d. density.
10. The most important laboratory safety rule is to
   a. use the metric system.   b. wear a lab coat.
   c. have a partner.   d. always follow directions.

CONTENT REVIEW: COMPLETION

On a separate sheet of paper, write the word or words that best complete each statement.

1. When a scientific theory has been tested many times and is generally accepted as true, it may be called a (an) _____.
2. An orderly and systematic approach to problem solving is a (an) _____.
3. To test their hypotheses, scientists perform _____.
4. A control experiment does not have a (an) _____.
5. The branch of science that deals with the study of microorganisms is _____.
6. The system of measurement that is used by all scientists today is the _____.
7. The prefix _____ means one-hundredth.
8. The _____ is the basic unit of mass in the metric system.
9. An instrument that produces an enlarged image of an object is known as a (an) _____.
10. A (An) _____ produces a narrow, intense beam of light.

CONTENT REVIEW: TRUE OR FALSE

Determine whether each statement is true or false. Then on a separate sheet of paper, write "true" if it is true. If it is false, change the underlined word or words to make the statement true.

1. The <u>last</u> step in solving a scientific problem is to review important information related to the problem.
2. After scientists analyze the data from an experiment, they come to a <u>hypothesis</u>.
3. <u>Microbiology</u> is the study of microorganisms.
4. The metric system is a <u>decimal</u> system.
5. The prefix *milli-* means 1000.
6. The <u>gram</u> is the basic unit of length.
7. <u>Gravity</u> is the mass per unit volume of a substance.
8. Your <u>weight</u> on the moon would be the same as it is on the earth.
9. A <u>magnifying glass</u> is an example of a simple microscope.
10. Never heat anything in an <u>open</u> container.

CONCEPT REVIEW: SKILL BUILDING

Use the skills you have developed in the chapter to complete each activity.

1. **Classifying metric units** Which metric units would you use to measure each of the following?

   a. the length of your pencil
   b. the amount of milk you had for breakfast
   c. the temperature of ocean water
   d. the distance from school to your home
   e. the mass of a caterpillar
   f. the length of a humpback whale
   g. the amount of water it would take to fill your bathtub
   h. the mass of your life science textbook

2. **Making calculations** Complete the following metric conversions.

   a. 32 km = _____ m
   b. 65 cm = ___ mm
   c. 556 g = _____ kg
   d. 13,026 mg = _ g
   e. 45 mL = ___ cm^3
   f. 15 L = _____ mL

3. **Applying concepts** Explain the following statement: Every substance has a characteristic density, but no substance has a characteristic mass.

4. **Designing an experiment** Design an experiment to test the following hypothesis: Turtle eggs develop into males in cold temperatures and females in warmer temperatures. Include a control and a variable.

5. **Making graphs** A scientist grew a population of yeast cells in a glass dish. Every two hours she counted the number of cells in the dish. She organized her data in the following chart. Use these data to construct a graph of her results. Plot the number of hours on the horizontal axis and the

Hours	Number of cells
0	1
2	26
4	74
6	120
8	340
10	500

number of cells on the vertical axis. What conclusions can you draw from the graph about the growth of yeast cells?

6. **Making predictions** What do you think might happen to the yeast cells in question 5 after a few days? Explain your answer.

7. **Making calculations** The density of water is equal to 1 g/cm^3. Therefore, objects with densities greater than 1 g/cm^3 sink in water. Objects with densities of less than 1 g/cm^3 float in water.

   a. Is the density of air greater or less than 1?   b. Will an object with a mass of 53 g and a volume of 18 cm^3 float in water?   c. Sample Y sinks in water. Would a 30 cm^3 sample of Y have a mass of 60 g or a mass of 15 g?

CONCEPT REVIEW: ESSAY

Discuss each of the following in a brief paragraph.

1. Define five branches of life science.
2. Explain the difference between an experimental setup and a control setup.
3. Explain how chance sometimes plays a role in scientific experiments.
4. Compare a theory and a law.
5. Explain each step in a scientific method.
6. Compare a simple, a compound, and an electron microscope.

The Processes of Life

2

CHAPTER OBJECTIVES

After completing this chapter, you will be able to

2–1 List the characteristics of living things.

2–1 Describe metabolism and the activities involved in metabolism.

2–2 Identify the needs of living things.

2–2 Explain how the sun is the primary source of energy for all living things.

2–3 Describe the composition of matter.

2–3 Identify the particles that make up an atom.

2–3 Define element, compound, and molecule.

Imagine a scientist turning a flask full of colorless gases into the "stuff" of which life is made. Magic? It may seem to be, and at times the scientist may seem to be a magician. But this exciting experiment was actually performed. And its results are being used by scientists today to solve a key mystery of life science: How did the chemical building blocks that make up all living things first form on the earth?

In 1952, the American scientist Stanley Miller set out to answer this question. Miller filled a clear glass flask with four gases: hydrogen, methane, ammonia, and gaseous water. Miller knew that these gases made up much of the atmosphere on the earth some 4 billion years ago. He sealed the flask and let it boil for a week. But the mixture needed a spark to produce the necessary change. So while the gases were boiling, Miller sent a surge of electricity through the flask again and again. The electricity recreated lightning bolts that had streaked through the atmosphere of primitive earth.

By the end of the week, a sticky brown coating had formed on the walls of the flask. Miller analyzed this tarlike substance and found that it contained several of the chemical building blocks that make up all living things. From four common gases and electricity, Stanley Miller had created some of the building blocks of life!

But life is more than just chemical building blocks. In this chapter, you will discover exactly what it is that makes a thing alive.

In this photograph, Stanley Miller recreates some of the building blocks of life.

2–1 Characteristics of Living Things

Figure 2–1 *The katydid (bottom left) may not be as familiar to you as the walrus (bottom right). The tree cactus (top left) and the flower of an Acacia plant (top right) may also be unfamiliar. Yet all are made up of the same four basic elements. What are these elements?*

Whether you live in the city or in the country, even the shortest walk will show that there is an enormous variety of living things on the earth. In fact, scientists estimate that there are more than five million different types of living things, ranging from one-celled bacteria to huge blue whales. Yet all living things are composed almost entirely of the same basic elements: carbon, hydrogen, nitrogen, and oxygen. These elements make up the gases Miller placed in his glass flask. These four elements along with iron, calcium, phosphorus, and sulfur all link together in chains, rings, and loops to form the stuff of life.

Well-known chemical rules govern the way these elements combine and interact. But less well understood is what gives this collection of chemicals a very special property—the property of life.

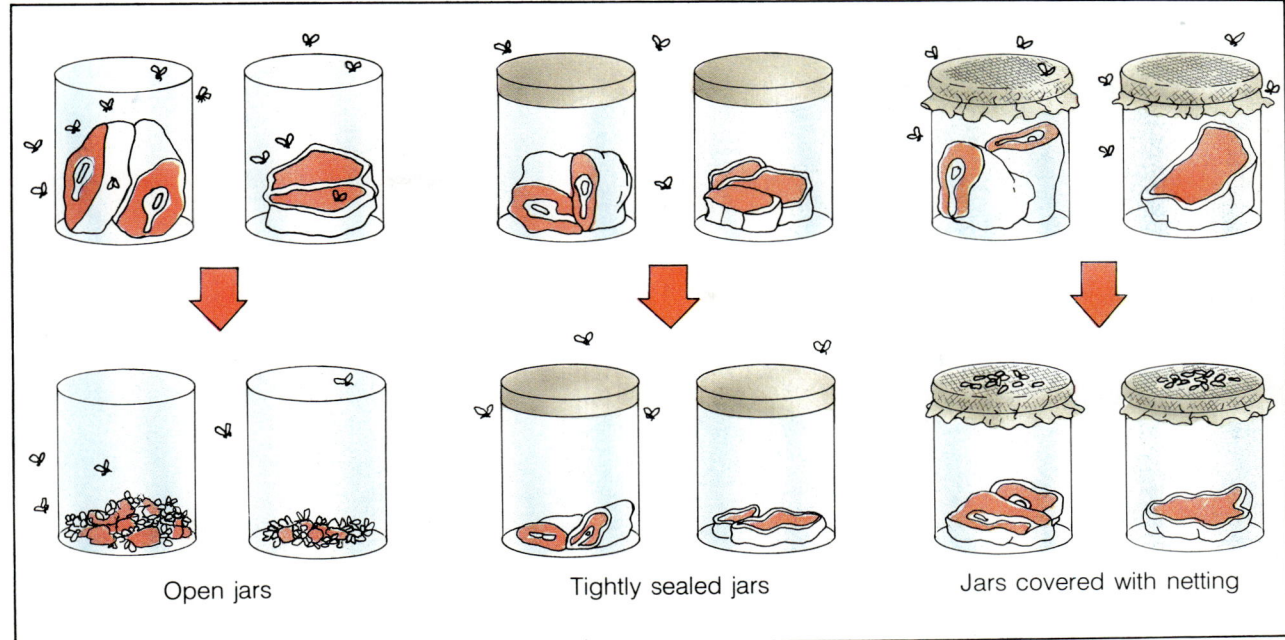

Open jars　　　　Tightly sealed jars　　　　Jars covered with netting

Spontaneous Generation

People did not always understand that living matter is so special. Until the 1600s, many people believed in the theory of **spontaneous generation.** According to this theory, life could spring from nonliving matter. For example, people believed that mice came from straw and frogs and turtles developed from rotting wood and mud at the bottom of a pond.

In 1668, an Italian doctor named Francesco Redi disproved this theory. In those days, maggots, a wormlike stage in the life cycle of a fly, often appeared on decaying meat. People believed that the rotten meat had actually turned into maggots, and that flies formed from dead animals. In a series of experiments, Redi proved that the maggots hatched from eggs laid by flies. See Figure 2–2. Today there is no doubt that living things can arise *only* from other living things.

But what are the characteristics that make living things special? That is, what distinguishes even the smallest organism from a lifeless streak of brown tar on a laboratory flask? **Living things are able to move, grow, reproduce, respond to a stimulus, and perform certain chemical activities.**

Figure 2–2 *Redi's experiments helped disprove spontaneous generation. No maggots were found on the meat in jars that were covered with netting or were tightly sealed. Maggots appeared only when adult flies were able to enter the jars.*

Sharpen Your Skills

Life From Life

Redi was not the only person involved with the theory of spontaneous generation. Using reference materials in the library, find out about each of the following scientists: John Needham, Lazzaro Spallanzani, Louis Pasteur.

In a written report, describe how each scientist used the scientific method to prove or disprove the theory. Describe how each scientist contributed to a better understanding of living things.

Movement

An ability to move through the environment is an important characteristic of many living things. The arctic tern, for example, is a small, gull-like bird that at first glance may not appear to be a record breaker. See Figure 2–3. Yet for seven months of every year, the arctic tern is in flight, covering a distance of nearly 32,000 kilometers. The tern begins its journey near the Arctic, travels south to its winter quarters in the Antarctic, and then flies back north again. Its yearly trip gives the tern the "long-distance record" for birds in flight.

Animals must be able to move in order to find food and shelter. In times of danger, swift movement can be the difference between safety and death. Of course, animals move in a great many ways. Fins enable fish such as salmon to swim hundreds of kilometers in search of a place to mate. The kangaroo uses its entire body as a giant pogo stick to gracefully bounce along the Australian plains looking for scarce patches of grass upon which to graze.

Most plants do not move in the same way animals do. Only parts of the plants move. The stems of the common houseplant *Pothos* bend toward the sunlight coming through windows. Its leaves turn to catch the sun's rays.

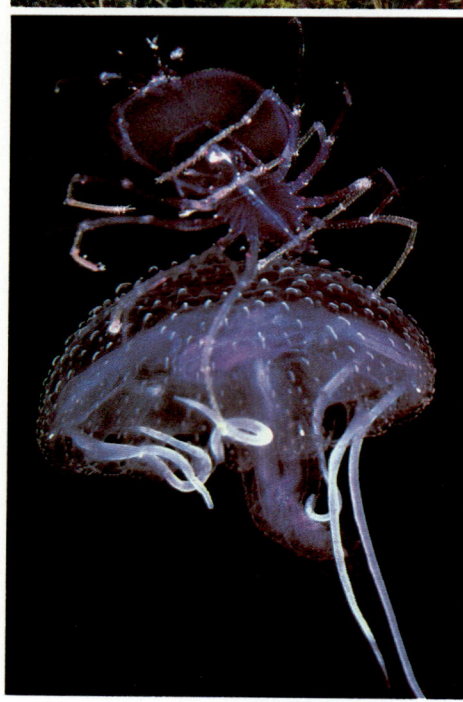

Figure 2–3 *In order to find food and shelter, animals must be able to move. The arctic tern (top) survives by traveling between the Arctic and the Antarctic each year. The larval crab gets around by hitching a ride on a jellyfish (bottom). How do the movements of plants and animals differ?*

Metabolism

Building up and breaking down is a good way to describe the chemical activities that are essential to life. During some of these activities, simple substances combine to form complex substances. These substances are needed for growth, to store energy, and to repair or to replace living materials. During other activities, complex substances are broken down, releasing energy and usable food substances. Together, these chemical activities in an organism are called **metabolism** (muh-TAB-uh-lihz-uhm). Metabolism, as you can see, is another characteristic of living things.

Metabolism includes many chemical reactions that go on in the body of a living thing. However, metabolism usually begins with eating.

Figure 2–4 *These brown bears are performing the first step in metabolism—ingestion, or eating. What is metabolism?*

INGESTION   All living things must either take in food or produce their own food. For most animals, **ingestion,** or eating, is as simple as putting food into their mouths. But some organisms, such as worms that live inside animals, absorb food directly through their skin.

Green plants do not have to ingest food because they make it. Using their roots, green plants absorb water and minerals from the soil. Tiny openings in the underside of their leaves allow carbon dioxide to enter. The green plants use the water and carbon dioxide, along with energy from the sun, to make the food they need.

DIGESTION   Getting food into the body is the first step in metabolism, but there is a lot more to metabolism than just eating. The food must be digested in order to be used. **Digestion** is the process by which food is broken down into simpler substances. Later, some of these simpler substances are reassembled into more complex materials for use in the growth and repair of the living thing. Other simple substances store energy that the organism will use for its many activities.

RESPIRATION   Organisms combine oxygen with the products of digestion to produce energy. The energy is used to do all the work of the organism. The process by which living things take in oxygen and use it to produce energy is called **respiration.**

Land animals like yourself have lungs that remove oxygen from the air. Most sea animals have

Figure 2–5 *A cheetah (left) could not run very fast without the energy produced during respiration. A killer whale (right) also needs energy so that it can swim through the water. What is respiration?*

gills that absorb oxygen dissolved in water. Some sea animals, however, come to the surface to breathe with their lungs. Whales, porpoises, and dolphins are examples of air-breathing sea animals. Some of these animals can remain under water for as long as 120 minutes!

Plants, too, need oxygen to stay alive. Most plants absorb oxygen through tiny pores in their leaves. Plants use oxygen for respiration, as do almost all living things. Respiration is the main process that provides energy necessary to living things. You get this energy by combining the foods you eat with the oxygen you breathe.

EXCRETION Not all the products of digestion and respiration can be used by the organism. Some products are waste materials that must be released. The process of getting rid of waste materials is called **excretion.** If waste products are not removed, they will eventually poison the organism.

Growth and Development

Metabolism supplies living things with the energy they need to grow. And growth is one of the characteristics of living things. But growing involves more than just increasing in size. Growing things also develop and become more complex. Sometimes this development results in dramatic changes. A tadpole, for example, swims for weeks in a summer pond. However, one day that tadpole becomes the frog that sits near the water's edge. And surely the

Figure 2–6 *Each of these monarch caterpillars (left) will grow and develop into an adult monarch butterfly (right).*

caterpillar creeping through a garden gives little hint of the beautiful butterfly it will soon become. Certainly development must be added to your list of the characteristics of living things.

Different organisms grow at different rates. A person will grow from a newborn baby to an adult in about 18 years. But puppies become adult dogs in only a few years. And a lima bean seed becomes a bean plant in just a few weeks. Some living things, such as certain insects, can change from an egg to an adult within a few days.

Each type of organism has a different **life span.** An organism's life span is the maximum length of

Figure 2–7 *As you can see from this chart of maximum life spans, the life span of organisms varies greatly. What is the maximum life span of a blue whale?*

MAXIMUM LIFE SPANS

Organism	Life Span	Organism	Life Span
Adult mayfly	1 day	Asiatic elephant	78 years
Marigold	8 months	Blue whale	100 years
Mouse	1–2 years	Human	117 years
Dog	20 years	Tortoise	152 years
Horse	20–30 years	Bristlecone pine	5500 years

Figure 2–8 *The adult mayfly (left) lives for only a day. The bristlecone pine (right) has a life span of 5500 years. What is a life span?*

time it can be expected to live. An elephant, for example, has a life span of about 60 or 70 years. A bristlecone pine tree has a life span of close to 5500 years! What is the difference between life span and life expectancy?

Response

When the alarm clock rings in the morning, you wake up. The smell of eggs and toast makes your mouth water as you hurry to breakfast. On your way to school, you stop at a red light. In just a matter of several hours, you have reacted to signals in your surroundings that determine much of your behavior.

Scientists call each of the signals to which an organism reacts a **stimulus.** A stimulus is any change in the environment, or surroundings, of an organism that produces a **response.** A response is some action or movement of the organism.

Some plants have special responses that protect them. For example, when a gypsy moth caterpillar chews on the leaf of an oak tree, the tree, in time, responds by producing bad-tasting chemicals in its other leaves. The chemicals discourage all but the hungriest caterpillars from eating these leaves. Can you think of responses that help you protect yourself?

Figure 2–9 *When a fly touches the tiny hairs lining a leaf of a Venus' flytrap plant (left), the leaf responds by closing and trapping the fly (right). What is the stimulus in this situation?*

Reproduction

You probably know that dinosaurs that lived 230 million years ago are now extinct. Yet crocodiles, which appeared before dinosaurs, are still living today. An organism becomes extinct when it no longer produces other organisms of the same kind.

The process by which living things give rise to the same type of living thing is **reproduction.** Crocodiles, for example, do not produce dinosaurs. Crocodiles only produce more crocodiles. You are a human and not a water buffalo, a duck, or a tomato plant because your parents are human. An easy way to remember this is *like produces like.*

There are two different types of reproduction: **sexual reproduction** and **asexual reproduction.** Sexual reproduction usually requires two parents. Most higher forms of plants and animals reproduce by sexual reproduction.

Some living things reproduce from only one parent. This is asexual reproduction. This type of reproduction can be as simple as an organism dividing into two parts. Bacteria reproduce this way. Yeast form growths called buds, which break off and then form new yeast plants. Geraniums and African violets can grow new plants from part of a stem, root, or leaf of the parent plant. All of these examples demonstrate asexual reproduction.

Sexual and asexual reproduction have an important function in common. In each case, the offspring receive a set of very special chemical "blueprints," or plans. These blueprints determine the characteristics of that type of living thing and are passed from one generation to the next.

Figure 2–10 *Notice that this female crocodile is carrying her newly hatched baby in her mouth. Do crocodiles reproduce asexually or sexually?*

SECTION REVIEW

1. List five characteristics of living things.
2. Explain how Redi showed that the theory of spontaneous generation was incorrect.
3. Define metabolism and give four examples of metabolism in living things.
4. What is the difference between sexual reproduction and asexual reproduction?
5. How does responding to stimuli help an organism adapt to its environment?

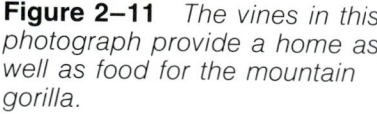
To identify the basic needs of all living things

2–2 Needs of Living Things

Living things do not live in a vacuum. They interact with one another and with their environment. Of course, these interactions are as varied as the living things themselves. Birds, for example, use dead twigs to build nests, but they eat live worms and insects. Crayfish build their homes in the sand or mud of streams and swamps. They absorb a chemical called lime from these waters to build a hard body covering. But crayfish rely on living snails and tadpoles for their food. And some people rely on crayfish for a tasty meal!

It seems clear, then, that living things depend on both the living and nonliving parts of their environment. **In order for a living organism to survive, it needs energy, food, water, oxygen, living space, and the ability to maintain a fairly constant body temperature.**

Figure 2–11 *The vines in this photograph provide a home as well as food for the mountain gorilla.*

Energy

All living things need energy. The energy can be used in different ways. A lion uses energy to chase and capture its prey. The electric eel defends itself by shocking its attackers with electric energy.

What is the source of this energy so necessary to living things? The primary source of energy for most living things is the sun. Plants use the sun's light energy to make food. Some animals feed on plants and in that way obtain the energy stored in the plants. Other animals may then eat the plant eaters. In this way, the energy from the sun is passed on from one living thing to another.

Food and Water

Have you ever heard the saying "You are what you eat"? Certainly this saying does not mean you will become a carrot or a hamburger if you eat these foods. But it does tell you that what you eat is very important. Food is a need of living things.

FOOD The kind of food organisms eat varies considerably. You would probably not want to eat eucalyptus leaves, yet that is the only food a koala eats. A diet of wood may not seem tempting, but for the termite it is a source of energy and necessary chemical substances. And surely a green plant's diet of simple sugar would not suit your taste.

Although all plants and animals must obtain food to live, some do it in very unusual ways. For example, the butterwort plant acts like fly paper to trap insects. After an insect lands on a leaf and gets

Figure 2–12 *The electric eel (top) uses energy to produce an electric shock. Sunflowers (bottom) use energy to produce food. What is the primary source of energy for all living things?*

Figure 2–13 *Many animals have unusual ways of obtaining food. The chameleon uses its tongue to capture its next meal. Why do living things need food?*

Sharpen Your Skills

You're All Wet

About 65% of your body mass is water.

1. Find your mass in kilograms.

2. Use that number to determine how many kilograms of water you contain.

stuck, the leaf curls around the victim. Then the plant produces fluids to digest its meal.

WATER Although you would probably not enjoy it, you could live a week or more without food. But you would die in a few days without water. It may surprise you to learn that 55 to 75 percent of your body is water. Other living things are also made up mainly of water.

Aside from making up much of your body, water serves many other purposes. Most substances dissolve in water. In this way, important chemicals can be transported easily throughout an organism. The blood of animals and the sap of trees, for example, are mainly water.

Most chemical reactions in living things cannot take place without water. Metabolism would come to a grinding halt without water. And it is water that carries away many of the metabolic waste products produced by living things. For green plants, water is also a raw material needed to make food. What else do green plants need to make food?

CAREER

Science Teacher

HELP WANTED: SCIENCE TEACHER
To teach life science to junior high school students. Science program includes laboratory and field activities. Teaching certification in science required. Must be available to guide students in after-school activities such as science club.

Saturday afternoon provided ideal weather for some backyard gardening. But the teacher's mind was back in the classroom rather than on the sprouting bean plants. This week the unit on life science was to begin. Many students would have trouble understanding what they could not actually see. The teacher thought about the best way to begin the lesson.

Science teachers spend much of their time developing lesson plans for different topics. They create interesting ways to present information to their students. Activities such as lectures, demonstrations, laboratory work, and field trips are often used.

Science teachers have other duties as well. They must correct homework and test papers, make up tests and quizzes, record pupil attendance and progress, and issue reports to parents. They also attend meetings, conferences, and workshops. If you are interested in this career, you can learn more by writing to the National Science Teachers Association, 1742 Connecticut Ave. N.W., Washington, DC 20009.

Air

You already know that oxygen is necessary for the process of respiration. Most living things get oxygen directly from the air or from air that has dissolved in water. When you breathe out, you release the waste gas called carbon dioxide. However, this gas is not wasted. The green plants around you use carbon dioxide in the air to make food.

Living Space

Do you enjoy hearing the chirping of birds on a lovely spring morning? Surprisingly, the birds are not simply singing. Rather, they are staking out their territory and warning intruders to stay away.

Often there is only a limited amount of food, water, and energy in an environment. As a result, only a limited number of the same kind of living thing can survive there. That is why many animals defend a certain area that they consider to be their living space. The male sunfish, for example, defends its territory in ponds by flashing its colorful fins at other sunfish and darting toward any sunfish that comes too close. You might think of this behavior as a kind of **competition** for living space.

Competition is the struggle among living things to get the proper amount of food, water, and energy. Animals are not the only competitors. Plants compete for sunlight and water. Smaller, weaker plants often die in the shadow of larger plants.

Proper Temperature

During the summer, temperatures as high as 58°C have been recorded on the earth. Winter temperatures can dip as low as −89°C. Most organisms cannot survive at such temperature extremes because many metabolic activities cannot occur at these temperatures. Without metabolism, an organism cannot function and dies.

Actually, most organisms would quickly die at far less severe temperature extremes if it were not for **homeostasis** (hoh-mee-oh-STAY-sihs). Homeostasis is the ability of an organism to keep conditions inside its body the same even though conditions in its external environment change. Maintaining a constant

Figure 2–14 *These kissing gouramis are actually engaged in battle. What is this behavior called?*

Figure 2–15 *A coldblooded tortoise (left) must absorb the sun's heat to keep warm. Flowers such as these coltsfoots (right) can grow in extremely cold conditions.*

body temperature, no matter what the temperature of the surroundings, is part of homeostasis.

Birds and certain other animals, such as dogs and horses, produce enough heat to keep themselves warm at low temperatures. Trapped air in the feathers of birds keeps them cool when temperatures get too high. Panting and sweating do the same for dogs and horses. Animals that maintain a constant body temperature are called **warmblooded** animals. Warmblooded animals can be active during both day and night, in hot weather and in cold.

Animals such as reptiles and fish have body temperatures that can change somewhat with changes in the temperature of the environment. These animals are called **coldblooded** animals. To keep warm, a coldblooded reptile, such as a crocodile, must spend part of each day lying in the sun. At night, when the air temperature drops, so does the crocodile's body temperature. The crocodile becomes lazy and inactive. Coldblooded animals do not move around much at relatively high or low temperatures. In what parts of the world would you *not* expect to see coldblooded animals?

SECTION REVIEW

1. List six things that all organisms need.
2. What is the primary source of energy for all living things?
3. List three reasons why water is important to all organisms.
4. What is homeostasis?
5. Why is it more advantageous for an animal to be warmblooded than it is for it to be coldblooded?

2–3 Chemistry of Living Things

All the world is made of **matter.** Matter is anything that takes up space and has mass. Every living and nonliving thing is made of matter.

As you read through the chapters in this book, you will be introduced to many different substances that make up the matter in living things. For example, in Chapter 8 you will discover the chemical substances formed when plants use energy from the sun to produce food. In Chapter 13, you will learn about the chemicals in the matter that makes up the foods you eat. In order to better understand the chemistry of living things, you will now look at some of the particles that make up matter.

Atoms

The basic building blocks of matter are atoms. Atoms are composed of three main particles: protons, neutrons, and electrons.

The center of the **atom** is called the nucleus. Two different kinds of particles are found in the nucleus. One of these is the **proton.** The proton is a positively charged particle. The other particle that makes up the nucleus is the **neutron.** A neutron is a neutral particle. It has no charge.

Figure 2–16 *Living things, such as these Bengal tigers (left), and nonliving things, such as these rocky cliffs (right), are made up of matter. What is matter?*

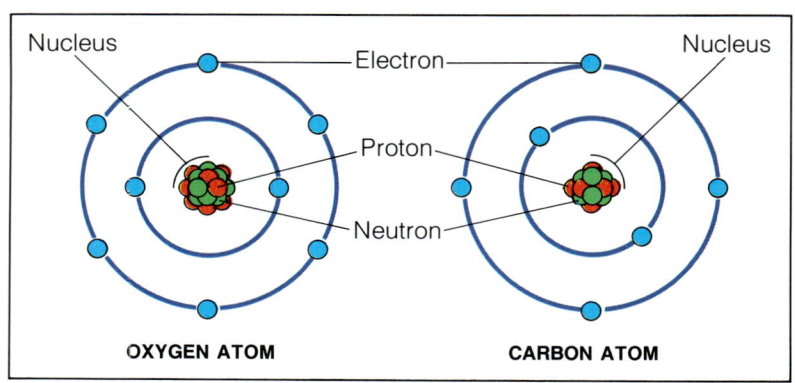

Figure 2–17 *The basic building blocks of matter are atoms. Most atoms are composed of protons, neutrons, and electrons. How many electrons does an atom of oxygen (left) contain? An atom of carbon (right)?*

OXYGEN ATOM   **CARBON ATOM**

Whirling around outside the nucleus are particles called **electrons.** An electron has a negative charge. In a neutral atom, the number of negatively charged electrons is equal to the number of positively charged protons. What, then, is the total charge on a neutral atom?

Figure 2–18 *The names and symbols of some common elements are listed in this table. What is the symbol for calcium? For sodium?*

Elements

Pure substances that cannot be separated into simpler substances by ordinary chemical processes are called **elements.** The particles making up an element are in their simplest form. All elements are made of atoms. An atom is the smallest particle of an element that still has the properties of that element.

Each element is represented by a chemical symbol. Chemical symbols are a shorthand way of representing the elements. Each symbol consists of one or two letters, usually taken from the element's name. The symbol for the element oxygen is O. The symbol for carbon is C and the symbol for hydrogen is H. When the first letter of an element's name has already been used for another element, two letters are needed for that element's symbol. For example, the symbol for carbon is C and the symbol for calcium is Ca. Sometimes scientists use the Latin name of an element to create its symbol. The symbol for iron is Fe. The Latin word for iron is *ferrum.*

COMMON ELEMENTS

Name	Symbol
Calcium	Ca
Carbon	C
Chlorine	Cl
Fluorine	F
Hydrogen	H
Iron	Fe
Magnesium	Mg
Nitrogen	N
Oxygen	O
Potassium	K
Silicon	Si
Sodium	Na
Sulfur	S

Compounds

A **compound** is two or more elements chemically joined, or bonded. Water is a compound that is composed of the elements hydrogen and oxygen.

Sugar is a compound made of the elements carbon, hydrogen, and oxygen. Can you name any other compounds?

A **molecule** (MAHL-uh-kyool) is the smallest particle of a compound that has all the properties of that compound. A molecule is made of two or more atoms chemically bonded together. Just as all atoms of a certain element are alike, all molecules of a compound are alike.

Chemical Formulas

You learned that elements are represented by chemical symbols. Because compounds are made of two or more elements, chemical symbols can be combined to represent compounds. Combinations of chemical symbols are called chemical **formulas.**

A chemical formula shows the elements that make up a compound. A chemical formula also shows the number of atoms of each element in a molecule of that compound. For example, the chemical formula for water is H_2O. This indicates that every molecule of water is made of two hydrogen atoms and one oxygen atom.

When writing a chemical formula, you use the symbol of each element in the compound. You also use small numbers called subscripts. Subscripts are placed to the lower right of the symbols. A subscript gives the number of atoms of the element in the compound. When there is only one atom of an element, the subscript 1 is not written. It is understood to be 1.

Carbon dioxide is a compound of the elements carbon and oxygen. Its formula is CO_2. By looking at its formula, you can tell that every molecule is made up of one atom of carbon, C, and two atoms of oxygen, O. The formula for glucose, a simple sugar, is $C_6H_{12}O_6$.

SECTION REVIEW

1. Classify the particles that make up an atom.
2. What is an element?
3. Compare chemical symbols to chemical formulas.
4. What elements and how many atoms of each element are in a molecule of glucose?

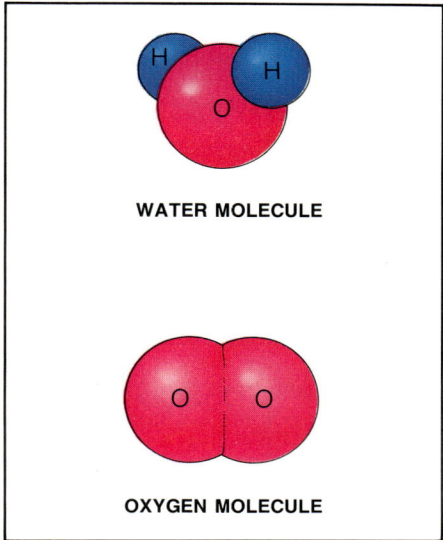

Figure 2–19 *A water molecule (top) is made up of two atoms of hydrogen and one atom of oxygen joined together. An oxygen molecule (bottom) is made up of two atoms of oxygen. Which molecule is an element? Which is a compound?*

Sharpen Your Skills

Chemical Formulas

What elements and how many atoms of each are there in the following compounds?

CO
$C_{14}H_{18}O_{10}$
HCl
NaCl
H_2SO_4
$CaCl_2$

Investigating Variables that Affect the Growth of Bread Mold

Problem

Do hidden variables affect the results of an experiment?

Materials (per group)

slice of bread
2 jars with lids
medicine dropper

Procedure

1. Tear a slice of bread in half.
2. Place each half into a separate jar.
3. Use the medicine dropper to moisten each half with 10 drops of water. Cover the jars.
4. Place one jar in sunlight and the other jar in a dark closet.
5. Observe the jars every few days for about two weeks. Record your observations in a data table.

Observations

1. Did you observe mold growth in either jar?
2. If so, describe any differences or similarities in mold growth in the two jars.
3. Do you think that light affects the growth of bread mold? If so, how?

Conclusions

1. What variable is being tested in this experiment? Which part is the control?
2. What conclusions can you draw from this experiment?
3. Often hidden variables affect the outcome of an experiment. Scientists know that light *does not* affect the growth of bread mold. What other variables could have affected mold growth in this investigation?
4. Design an experiment to test one of these other variables. Include a hypothesis, control, and variable in your design.
5. Cathy set up the following experiment: She placed a piece of orange peel in each of two jars. She added 3 mL of water to jar 1 and placed it in the refrigerator. She added no water to jar 2 and placed it on a windowsill in the kitchen. At the end of a week, she noticed more mold growth in jar 2. Cathy concluded that light, a warm temperature, and no moisture are ideal conditions for mold growth. Discuss the accuracy of Cathy's conclusion.

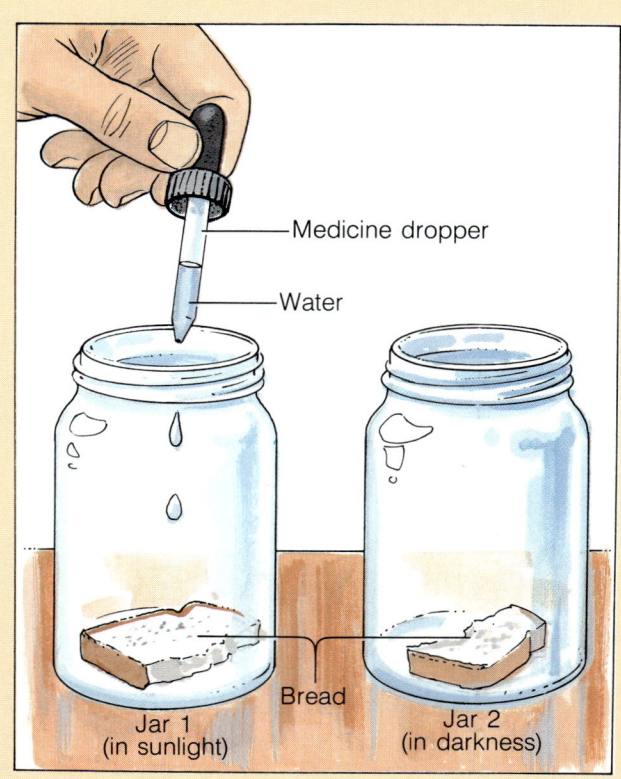

Medicine dropper

Water

Bread

Jar 1
(in sunlight)

Jar 2
(in darkness)

CHAPTER REVIEW

SUMMARY

2–1 Characteristics of Living Things

❏ The elements most abundant in living things are carbon, hydrogen, nitrogen, and oxygen.

❏ The basic characteristics of living things include movement, metabolism, growth, response, and reproduction.

❏ Metabolism is the sum of all chemical activities essential to life. Ingestion, digestion, respiration, and excretion are metabolic activities that occur in all organisms.

❏ Life span is the maximum length of time a particular organism can be expected to live.

❏ A living thing reacts to a stimulus, which is a change in the environment, by producing a response.

❏ Reproduction is the process by which organisms produce offspring.

❏ Asexual reproduction requires only one parent while sexual reproduction requires two parents.

2–2 Needs of Living Things

❏ In order to live, an organism must have energy, food, water, oxygen, living space, and a fairly constant body temperature.

❏ Living things need energy for metabolism. The primary source of energy for all living things is the sun.

❏ Oxygen in the air or dissolved in water is used by all organisms during respiration. Carbon dioxide is used by plants to make food.

❏ Competition is a struggle for survival.

❏ Homeostasis is the ability of an organism to keep conditions constant inside its body when the outside environment changes.

❏ Warmblooded animals maintain a constant body temperature. Coldblooded animals have body temperatures that change with changes in the external temperature.

2–3 Chemistry of Living Things

❏ All things are made up of matter. Matter is anything that takes up space and has mass.

❏ The basic building blocks of matter are atoms. Atoms are made up of protons, neutrons, and electrons.

❏ Elements are substances that cannot be separated into simpler substances by chemical processes. All elements are made of atoms.

❏ Two or more elements joined together form a compound. A molecule is the smallest part of a compound.

❏ Chemical formulas show the kinds and number of atoms in a molecule.

VOCABULARY

Define each term in a complete sentence.

asexual reproduction	electron	matter	response
atom	element	metabolism	sexual reproduction
coldblooded	excretion	molecule	spontaneous generation
competition	formula	neutron	stimulus
compound	homeostasis	proton	warmblooded
digestion	ingestion	reproduction	
	life span	respiration	

CONTENT REVIEW: MULTIPLE CHOICE

On a separate sheet of paper, write the letter of the answer that best completes each statement.

1. Which is *not* one of the basic characteristics of life?
   a. metabolism   b. air   c. response   d. reproduction
2. The process by which food is broken down into simpler substances is called
   a. ingestion.   b. digestion.   c. respiration.   d. excretion.
3. Organisms combine oxygen with other materials to produce energy during
   a. ingestion.   b. digestion.   c. respiration.   d. excretion.
4. The maximum length of time an animal can be expected to live is its
   a. growth.   b. development.
   c. spontaneous generation.   d. life span.
5. A signal to which an organism reacts is called a (an)
   a. stimulus.   b. response.   c. action.   d. environment.
6. Which living thing is able to reproduce asexually?
   a. human being   b. bacteria   c. crocodile   d. dinosaur
7. All organisms directly or indirectly obtain energy from
   a. excretion.   b. plants.   c. animals.   d. the sun.
8. About what percent of your body is composed of water?
   a. 30 percent   b. 65 percent   c. 80 percent   d. 99 percent
9. Competition is the struggle among living things for
   a. energy.   b. living space.   c. water.   d. all of these.
10. A neutral particle in the atom is the
    a. proton.   b. neutron.   c. electron.   d. nucleus.

CONTENT REVIEW: COMPLETION

On a separate sheet of paper, write the word or words that best complete each statement.

1. The theory that life can spring from nonlife is called _____.
2. _____ is the building up and breaking down of chemical substances necessary for life.
3. Another word for eating is _____.
4. A stimulus in the surroundings of an organism produces a _____.
5. Living things give rise to other living things of the same type in _____.
6. _____ reproduction usually involves two parents.
7. Green plants use energy from the _____ to make food.
8. The blood of animals is composed mainly of _____.
9. _____ is the struggle among living things to get their needed resources.
10. _____ is anything that takes up space and has mass.

CONTENT REVIEW: TRUE OR FALSE

Determine whether each statement is true or false. Then on a separate sheet of paper, write "true" if it is true. If it is false, change the underlined word or words to make the statement true.

1. Carbon is one of the basic elements in all living things.
2. The theory of spontaneous generation was disproved by Miller.

52

3. The process of getting rid of waste materials is called respiration.
4. Life span is the minimum length of time an organism can be expected to live.
5. A response is an action or movement of an organism.
6. Most higher forms of plants and animals reproduce asexually.
7. You could probably live for a week without water.
8. Green plants use carbon dioxide, sunlight, and water to make food.
9. A warmblooded animal is able to maintain a constant body temperature.
10. Protons are found outside the nucleus of an atom.

CONCEPT REVIEW: SKILL BUILDING

Use the skills you have developed in the chapter to complete each activity.

1. **Drawing conclusions** Is an apple seed living or nonliving? Explain your answer.
2. **Applying definitions** In hospitals, nutrients are sometimes injected directly into a patient's bloodstream. Which metabolic process is unnecessary in such a procedure?
3. **Identifying relationships** Look at the three sets of jars that describe Redi's experiments in Figure 2–2. Suggest a reason why the second set of jars did not provide enough evidence to disprove the theory of spontaneous generation.
4. **Making comparisons** Compare the growth of an icicle to the growth of a living organism.
5. **Applying concepts** Explain why the study of chemistry is important to the understanding of biology.
6. **Making generalizations** Which life function is necessary for the survival of a species but is not necessary for the survival of an individual?
7. **Relating cause and effect** In the days when people believed in spontaneous generation, one scientist developed the following recipe for producing mice: Place a dirty shirt and a few wheat grains in an open pot; wait three weeks. Suggest a reason why this recipe may have worked. How could you prove that spontaneous generations was not responsible for the appearance of mice?
8. **Applying concepts** Which form of reproduction is more likely to result in increased variety among organisms? Explain your answer.
9. **Designing an experiment** A plant salesperson tells you that plants respond to the stimulus of classical music by growing more quickly. Design an experiment to test the salesperson's claim. Be sure to include a control. What is the hypothesis in your experiment? What is the variable?

CONCEPT REVIEW: ESSAY

Discuss each of the following in a brief paragraph.

1. Describe at least five ways in which you can tell a living thing from a nonliving thing.
2. Your friend insists that plants are not alive because they do not move. Give specific examples to explain why your friend is wrong?
3. Describe four examples of metabolism.
4. Compare atoms and molecules.
5. Defend this statement: All plants and animals get their energy from the sun.

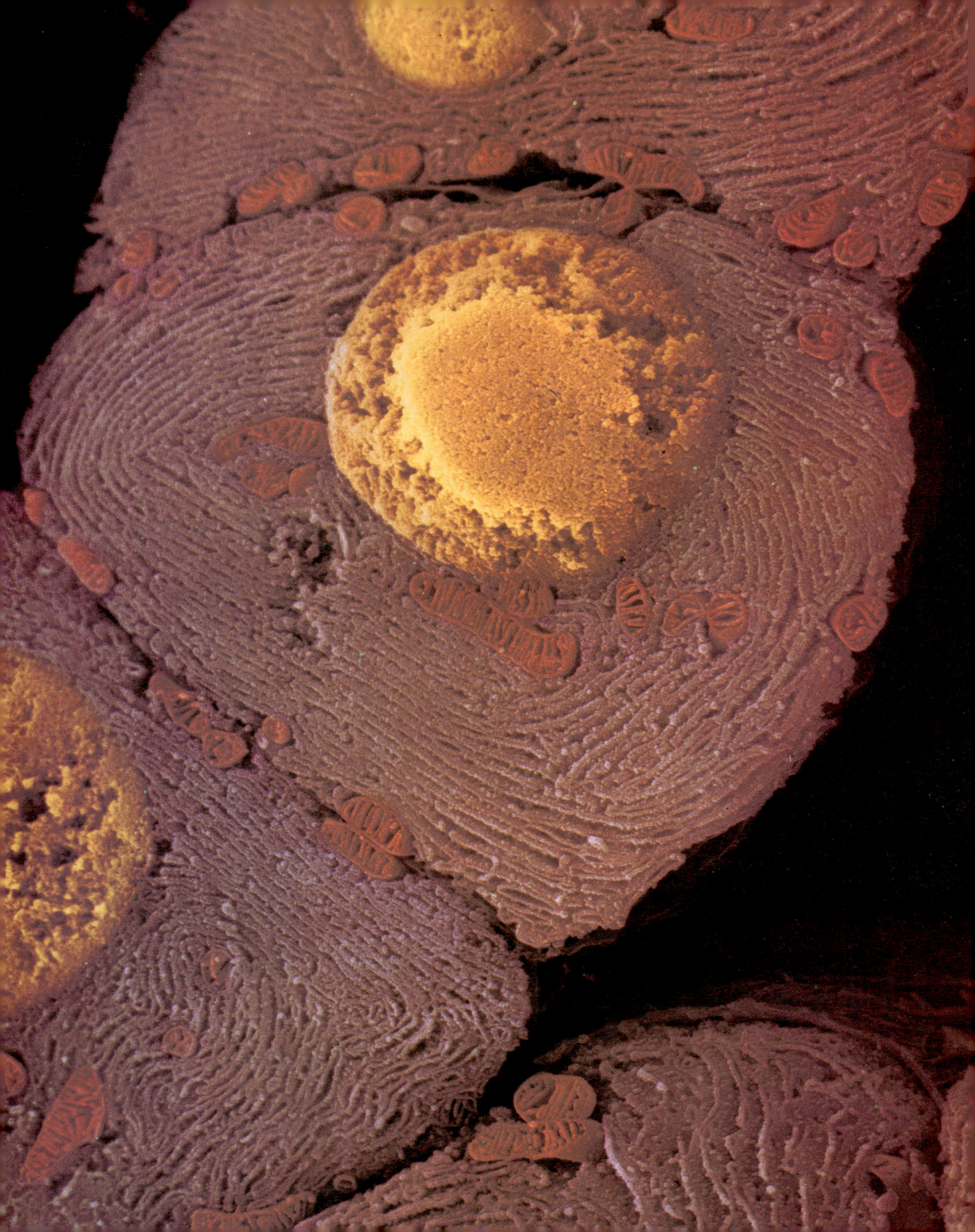

Cells, Tissues, and Organ Systems

3

CHAPTER OBJECTIVES

After completing this chapter, you will be able to

3–1 Explain the cell theory.

3–2 Describe the structures and functions of a cell.

3–2 Compare a plant cell and an animal cell.

3–3 Describe the activities of a cell.

3–3 Describe the process of mitosis.

3–4 Explain division of labor.

3–4 Describe the five levels of organization of living things.

More than one hundred trillion of these tiny structures make up your body. Some are in the shape of a rectangle, others in the shape of a sphere, and still others are spiral shaped. Some have tails. Others are star shaped. What are these structures? If you have not already guessed, they are cells.

Cells are the basic units of life. All living things—oak trees, spiders, elephants, and people—are made up of cells. Some living things are made up of only one cell. Others are made up of many cells. In some living things, cells perform all the life processes needed for the living thing's survival. In other living things, cells perform very specialized tasks. Which type of living thing are you?

Most cells are so small that they can be seen only under a microscope. The photograph on the opposite page was taken through an electron microscope, which magnified the human cells 10,000 times! To help you see some of the tinier structures inside the cells, the cells were sliced in a way that is similar to slicing bread.

As you may have already discovered, cells are fascinating and in many ways very mysterious objects. Scientists continue to probe the secrets of cells like explorers journeying through parts of an uncharted world. If you read on, you will become an explorer and travel on a fantastic journey through parts of a cell.

Human skin cells magnified 10,000 times

3–1 The Cell Theory

The basic units of structure and function of living things are **cells.** As you read in the beginning of this chapter, most cells are hard to see with the unaided eye. As a result, many of the tiny cell structures that make up the cell remained a mystery for hundreds of years. These structures are called **organelles** (or-guh-NEHLZ), or tiny organs. The invention of the microscope in the seventeenth century, however, enabled people to see some of these organelles.

In 1665, while looking at a thin slice of cork through a compound microscope Robert Hooke, an English scientist, observed tiny roomlike structures. He called these structures cells. But the cells that Hooke saw in the slice of cork were not alive. What Hooke saw were actually the outer walls of dead plant cells.

At about the same time, Anton van Leeuwenhoek (LAY-vuhn-hook), a Dutch scientist, was using a simple microscope to examine materials such as blood, rainwater, and scrapings from his teeth. In each material, van Leeuwenhoek was able to observe living cells. He even found tiny living things in a drop of rainwater. Van Leeuwenhoek called these living things "animalcules." The smallest of the ani-

Figure 3–1 *All living things, such as the red onion (left) and* Acetabularia, *a one-celled green algae (right), are made up of cells. How do these cells differ from one another?*

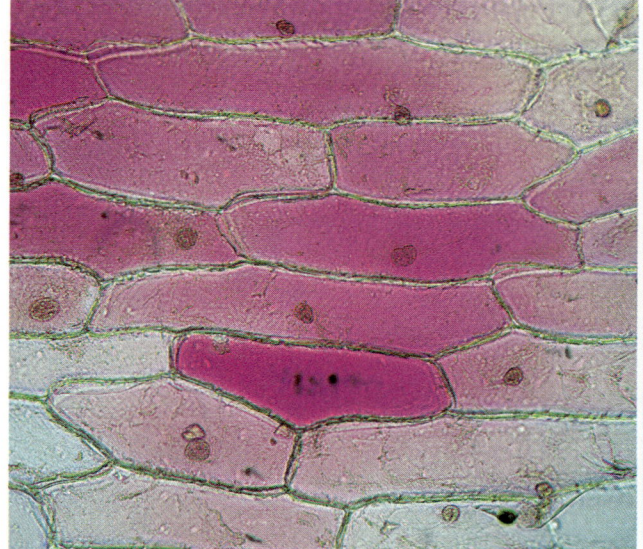

malcules are known today as bacteria. Bacteria are one-celled organisms. These discoveries made van Leeuwenhoek world-famous.

During the next two hundred years, new and better microscopes were developed. Because of these improved microscopes, Mathias Schleiden, a German botanist, was able to view different types of plant parts. Schleiden discovered that all of the plant parts he examined were made of cells. For this reason, Schleiden proposed that all plants are made of cells. One year later, Theodor Schwann, a German zoologist, made the same observations using animal parts. About twenty years later, Rudolph Virchow, a German physician, discovered that all living cells come only from other living cells.

The work of Schleiden, Schwann, Virchow, and other biologists led to the development of the cell theory. **The cell theory states that**

> **All living things are made up of cells.**
> **Cells are the basic units of structure and function in living things.**
> **Living cells come only from other living cells.**

SECTION REVIEW

1. What is the cell theory?
2. Who was the first person to use the word *cell?*
3. What three scientists contributed to the cell theory?
4. Explain the relationship between technology and the development of the cell theory.

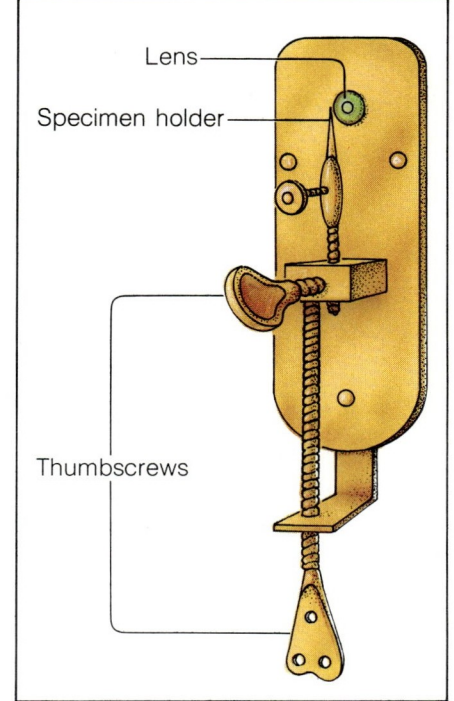

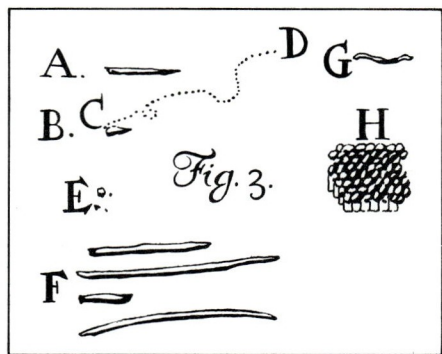

Figure 3–2 *In the mid 1600s, Anton van Leeuwenhoek made a series of simple microscopes (top) through which he observed and recorded several different types of tiny living things (bottom). How many lenses did van Leeuwenhoek's microscope have?*

3–2 Structure and Function of Cells

Section Objective

To list and describe the various functions of cell structures

You are about to take an imaginary journey. It will be quite an unusual trip because you will be traveling inside a living organism, visiting its cells.

Most cells are too small to be seen without the aid of a microscope. In fact, most cells are smaller than the period at the end of this sentence. Yet within these tiny cells are even smaller structures

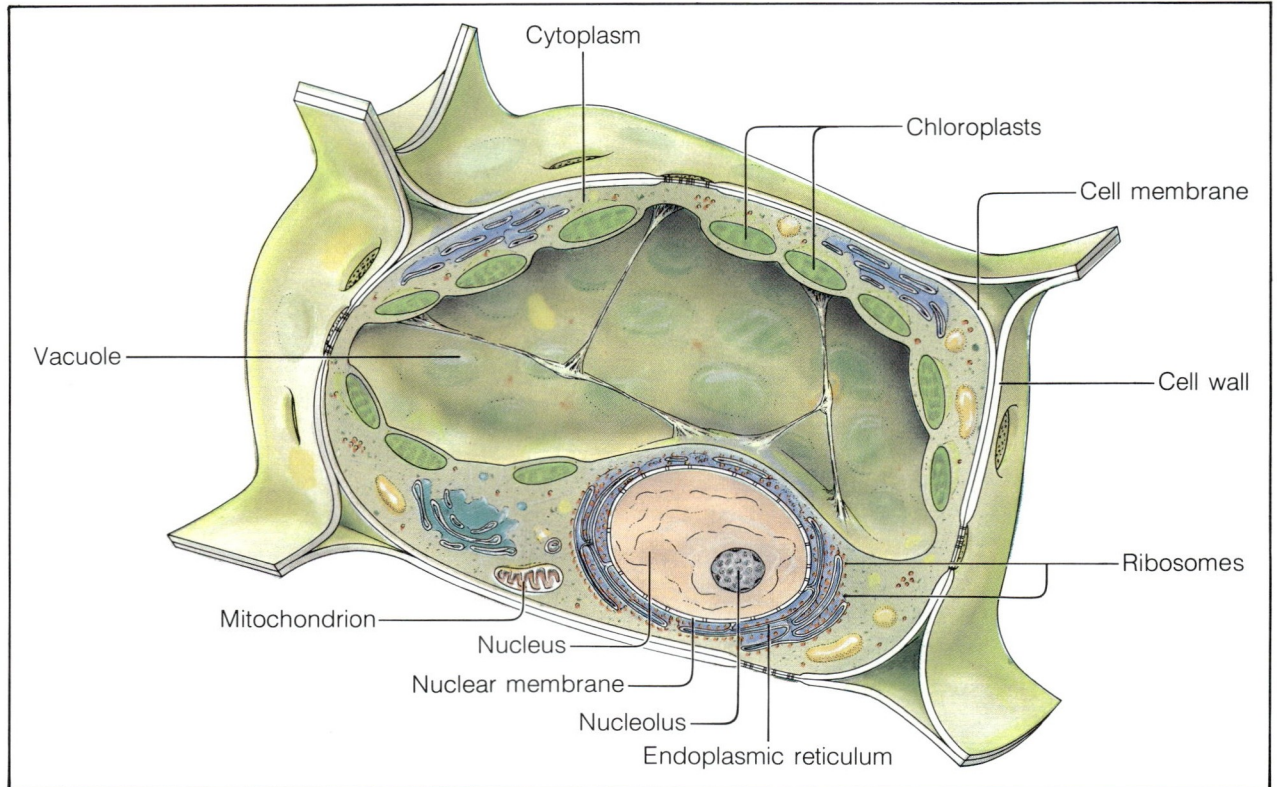

Figure 3–3 *This diagram shows the structures of a typical plant cell. What are the tiny structures within the cell called?*

that carry out the activities of the cell. **The structures within the cell function in storing and releasing energy, building and repairing cell parts, getting rid of waste materials, responding to the environment, and reproducing.**

There are many types of cells. For example, human cells include muscle cells, bone cells, and nerve cells. In plants, there are leaf cells, stem cells, and root cells. However, whether in an animal or a plant, most cells share certain similar characteristics. So hop aboard your imaginary ship and prepare to enter a typical plant cell. You will begin by sailing up through the trunk of an oak tree. Your destination is that box-shaped structure directly ahead. See Figure 3–3.

Cell Wall

Entering the cell of an oak tree is a bit difficult. First you must pass through the **cell wall.** The cell wall is made of **cellulose,** a nonliving material. Cellulose is a long chain of sugar molecules that the cell manufactures. Although the cell wall is strong and stiff, it does allow water, oxygen, carbon dioxide, and certain dissolved materials to pass through it. So sail on through.

The rigid cell wall is found only in plant cells. This structure helps give protection and support so the plant can grow tall. Think for a moment of grasses and flowers that support themselves upright. No doubt you can appreciate the important role the cell wall plays for the individual cell and for the entire plant.

Cell Membrane

Once past the cell wall, you prepare to enter the living material of the cell. Biologists call all the living material in *both* plant and animal cells **protoplasm.** Protoplasm is not a single structure or substance. It is a term used to describe all the living materials in a cell.

The first structure you encounter is the **cell membrane.** The cell membrane is a thin, flexible envelope that surrounds the cell. In a plant cell, the cell membrane is just inside the cell wall. In an animal cell, the cell membrane forms the outer covering of the cell. Look again at Figure 3–3 and at 3–4.

Figure 3–4 *This diagram shows the structures of a typical animal cell. How does an animal cell differ from a plant cell?*

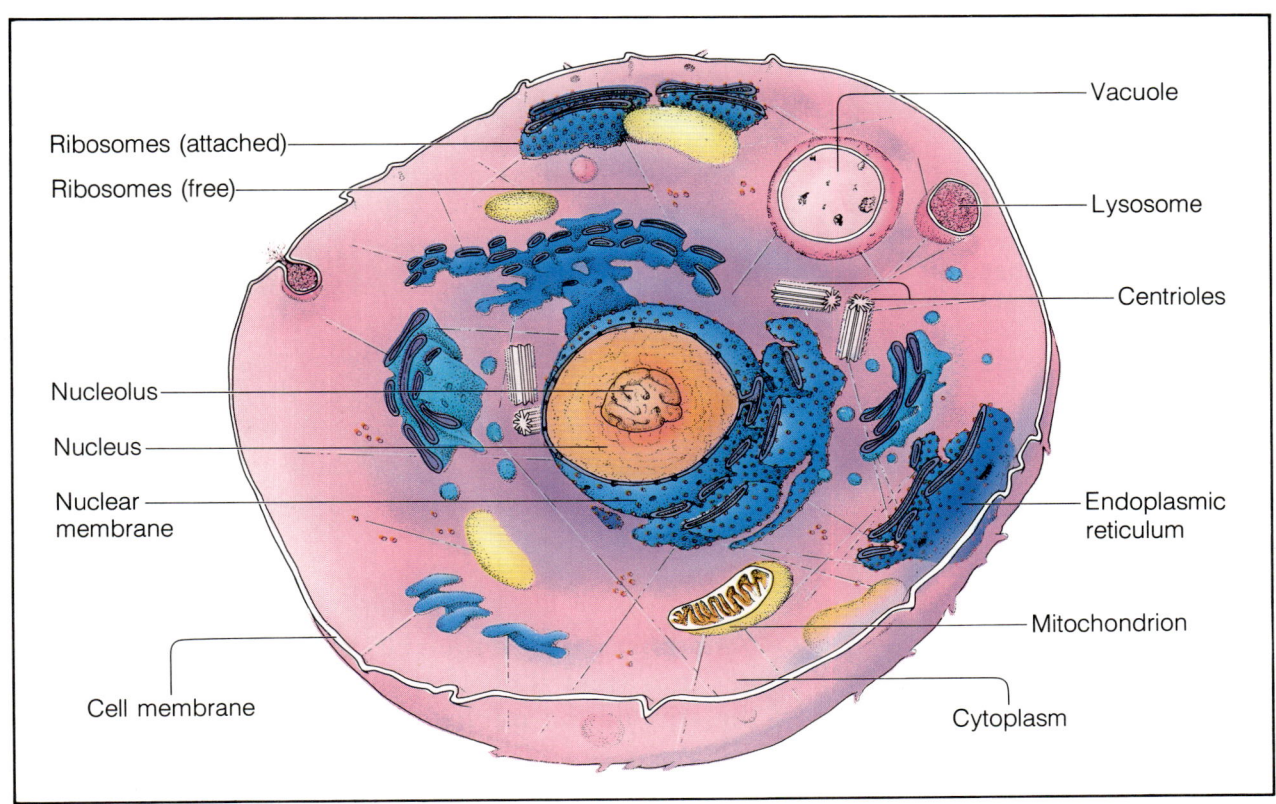

Ribosomes (attached)
Ribosomes (free)
Nucleolus
Nucleus
Nuclear membrane
Cell membrane
Vacuole
Lysosome
Centrioles
Endoplasmic reticulum
Mitochondrion
Cytoplasm

Figure 3–5 *The cell wall gives support and protection to plant cells. What is the name of the nonliving material that makes up the cell wall?*

Sharpen Your Skills

Observing Cells

1. Obtain a thin piece of cork and a few drops of rain water.

2. Place each of these on a different glass slide and cover with a cover slip.

3. Obtain a prepared slide of human blood.

4. Observe each of these slides under the low and high power of your microscope.

5. Make a labeled diagram of what you see on each slide.

Using reference books, find diagrams of these materials as seen by Hooke and van Leeuwenhoek. How do your diagrams compare with theirs?

The cell membrane has several important jobs. You can discover one of its jobs on your own. Push down on your skin with your thumb. Your skin does not break, does it? Now lift your thumb. The skin bounces back to its original position. Your skin can do this because the cell membrane around each skin cell is elastic and flexible. The cell membrane allows the cell to change shape under pressure. The flexible cell membrane also keeps the protoplasm of the cell separated from the outside environment.

As your ship nears the edge of the cell membrane, you notice that there are tiny openings in the membrane. You steer toward an opening. Suddenly, your ship narrowly misses being struck by a chunk of floating material passing out of the cell. You have discovered another job of the cell membrane. The cell membrane helps control the movement of materials into and out of the cell.

Everything the cell needs, from food to oxygen, enters the cell through this membrane. Harmful waste products exit through the cell membrane. In this way, the cell stays in smooth-running order, keeping conditions inside the cell the same even though conditions outside the cell may change. The ability of a cell to maintain a stable internal environment is called **homeostasis** (hoh-mee-oh-STAY-sihs).

Nucleus

As you sail inside the cell, a large oval structure comes into view. This structure is the control center of the cell, or the **nucleus** (NOO-klee-uhs), which acts as the "brain" of the cell. See Figure 3–6. The nucleus controls all cell activities.

The large, complex molecules that are found in the nucleus are compounds called **nucleic** (noo-KLEE-ihk) **acids.** Nucleic acids store the information that helps a cell make the substances it needs.

There are two types of nucleic acids. One of the nucleic acids is **DNA,** or deoxyribonucleic (dee-ahk-sih-right-boh-noo-KLEE-ihk) acid. DNA stores the information needed to build **proteins.** Proteins are used to build and repair cells. Some proteins, for example, are used to form structural parts of the cell, such as the cell membrane. DNA also carries "messages" about an organism that are passed from parent to offspring.

The other nucleic acid is **RNA,** or ribonucleic (right-boh-noo-KLEE-ihk) acid. RNA "reads" the DNA messages and guides the protein-making process. Unlike DNA, RNA, carrying its protein-building instructions, leaves the nucleus through a nuclear pore. Now let's look at some of the structures that make up the nucleus.

NUCLEAR MEMBRANE   The thin membrane that separates the nucleus from the rest of the cell is called the **nuclear membrane.** This membrane is similar to

Figure 3–6 *This diagram of the cell nucleus shows the nuclear membrane, chromosomes, and nucleolus. The nucleus in the photograph is magnified 20,000 times. Which structure contains RNA?*

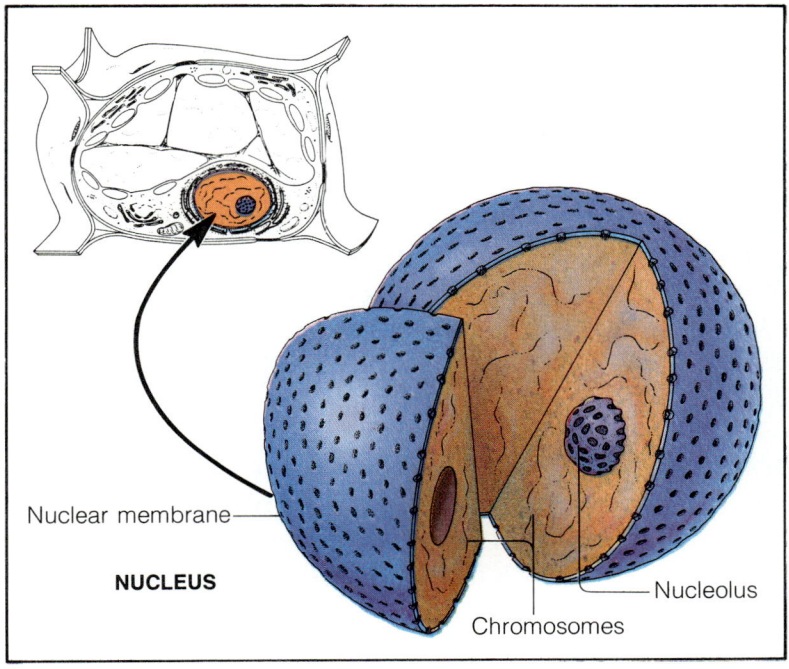

Nuclear membrane

NUCLEUS

Chromosomes

Nucleolus

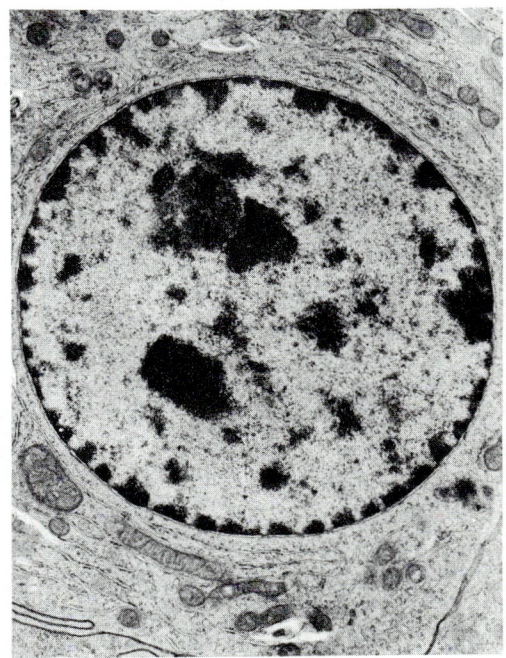

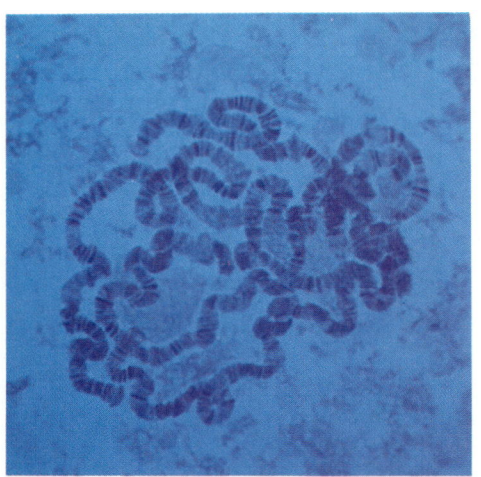

Figure 3–7 *You can clearly see the threadlike shape of a fruit fly's chromosomes in this photograph. What nucleic acid is found in the chromosomes?*

Sharpen Your Skills

Plant and Animal Cells

1. To view a plant cell, remove a very thin transparent piece of tissue from an onion.

2. Place the onion tissue on a glass slide.

3. Add a drop of iodine stain to the tissue and cover with a cover slip.

4. Observe the onion under the low and high power of your microscope. Draw a diagram of one onion cell and label its parts.

5. To view an animal cell, gently scrape the inside of your cheek with a toothpick.

6. Gently tap the toothpick on to the center of another glass slide.

7. Add a drop of methylene blue and cover with a cover slip.

8. Observe the cheek cells under the high and low power of your microscope. Draw a diagram of one cell and label its parts.

Compare the two cells.

the cell membrane in that it allows materials to pass into and out of the nucleus. Small openings, or pores, are spaced regularly around the nuclear membrane. Each pore acts as a passageway into and out of the nucleus. Set your sights for that pore ahead and carefully glide into the nucleus.

CHROMOSOMES The large, irregular mass of thin threads that take up most of the space in the nucleus is the **chromosomes.** Steer very carefully to avoid colliding with and damaging the delicate chromosomes. The chromosomes are made of DNA. Chromosomes direct the activities of the cell and pass on the traits of the cell to new cells.

NUCLEOLUS As you prepare to leave the nucleus, you see a small, dense object float past. It is the **nucleolus** (noo-KLEE-uh-luhs), or "little nucleus." The nucleolus is made up of RNA and protein. This tiny structure plays an important role in making some of the cell's RNA. Now hitch a ride on the RNA leaving the nucleus and continue your exploration of the cell.

Endoplasmic Reticulum

Outside the nucleus, floating in a clear, thick fluid, your ship needs no propulsion. For here the jellylike material called **cytoplasm** is constantly moving. Cytoplasm is the term for all the protoplasm, or living material of the cell, *outside* the nucleus.

Steering will be a bit difficult here because many particles and tubelike structures are scattered throughout the cytoplasm. The maze of tubular passageways that leads out from the nuclear membrane forms the **endoplasmic reticulum** (ehn-duh-PLAZ-mihk rih-TIHK-yuh-luhm). In some cells, these passageways connect with the cell membrane. See Figure 3–8.

The endoplasmic reticulum is involved in the transport of proteins. Proteins made in one part of the cell can pass through the endoplasmic reticulum to other parts of the cell. Other materials can be transported to the outside of the cell through this system. How is the endoplasmic reticulum suited for its job?

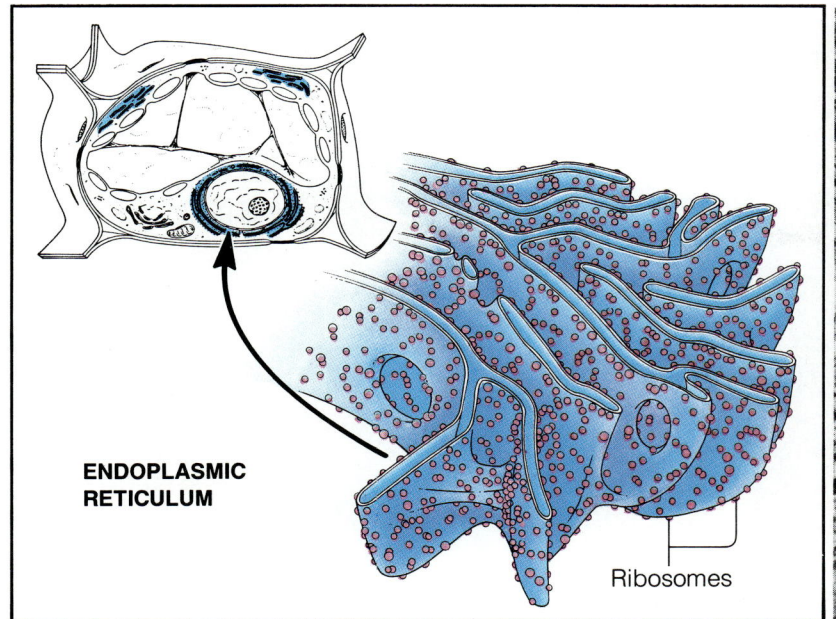

ENDOPLASMIC RETICULUM

Ribosomes

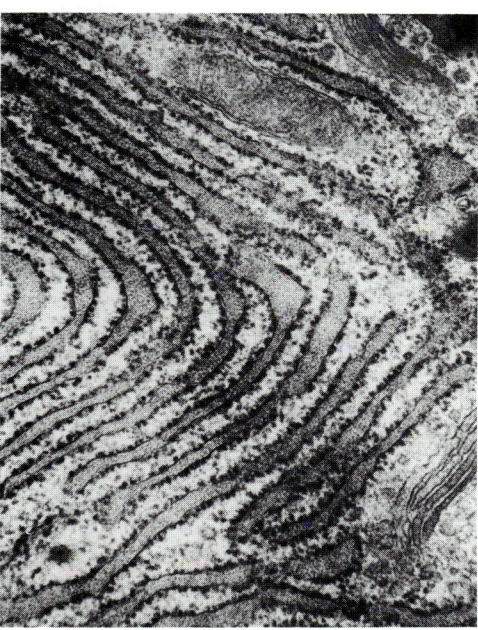

Figure 3–8 *The endoplasmic reticulum transports materials throughout the cell. The dark dots lining the tubelike passageways are ribosomes. What is the function of the ribosomes?*

Ribosomes

Look closely at the inner surface of one of the passageways in the endoplasmic reticulum. Attached to the surface are grainlike bodies called **ribosomes,** which are made up mainly of the nucleic acid RNA. Ribosomes are the protein-making sites of the cell. The RNA in the ribosomes, along with the RNA sent out from the nucleus, directs the production of proteins. Ribosomes are well positioned as they not only help make proteins, they "drop" them directly into the endoplasmic reticulum. From there the proteins go to any part of the cell that needs them.

Not all ribosomes are attached to the endoplasmic reticulum. Some float freely in the cytoplasm. Watch out! There go a few passing by. The cell you are in has many ribosomes. What might this tell you about its protein-making activity?

Mitochondria

As you pass by the ribosomes, you see other structures looming ahead. These structures are called the **mitochondria** (might-uh-KAHN-dree-uh; singular; mitochondrion). Mitochondria supply most of the energy for the cell. Somewhat larger than the ribosomes, these rod-shaped structures are often referred to as the "powerhouses" of the cell. See Figure 3–9 on page 64.

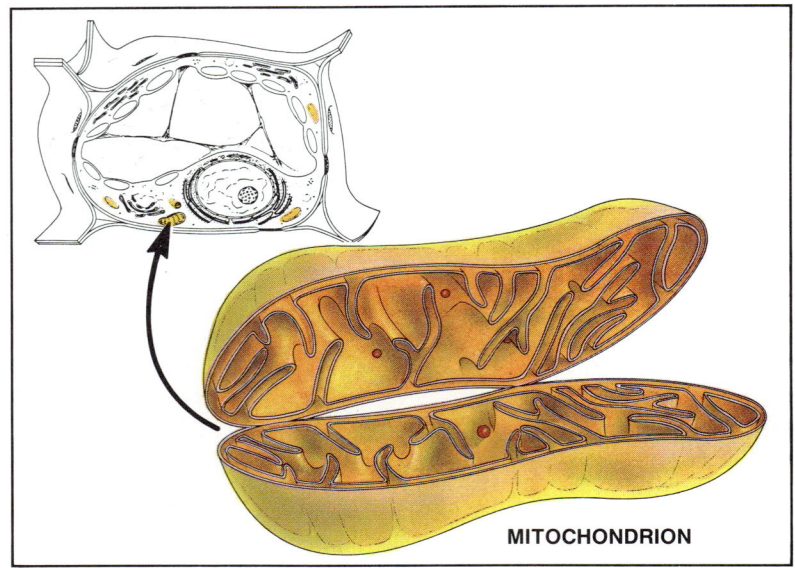

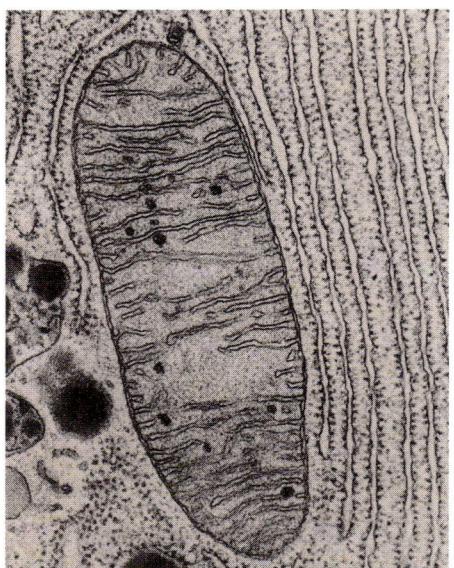

MITOCHONDRION

Figure 3–9 *The mitochondrion is the "powerhouse" of the cell. What food substances provide energy for the cell?*

Inside the mitochondria, simple food substances such as sugars are broken down to water and carbon dioxide gas. Large amounts of energy are released during the breakdown of sugars. The mitochondria gather this energy and store it in special energy-rich molecules. These molecules are convenient energy packages that the cell uses to do all its work. The more active a cell is, the more mitochondria it has. Some cells, such as a human liver cell, contain more than 1000 mitochondria. Would you expect your muscle cells or your bone cells to have more mitochondria?

Because mitochondria have a small amount of their own DNA, some scientists believe that they were once tiny organisms that invaded living cells millions of years ago. The DNA molecules in the mitochondria were passed from one generation of cells to the next as simple organisms changed into complicated ones. Now all living cells contain mitochondria. No longer invaders, mitochondria are an important part of living cells.

Vacuoles

The large, round, water-filled sac that is floating in the cytoplasm of this cell is called a **vacuole** (VA-kyoo-wohl). Most plant cells and some animal cells have vacuoles. Plant cells often have one very large vacuole and animal cells have a few small vacuoles.

Vacuoles act like storage tanks. Food and other materials needed by the cell are stored inside the vacuoles. Vacuoles can also store waste products. In plant cells, vacuoles are the main water-storage areas. When water vacuoles in plant cells are full, they swell and make the cell plump. This plumpness keeps a plant firm.

Lysosomes

If you carefully swing your ship around the lakelike vacuole, you may be lucky enough to see a **lysosome** (LIGH-suh-sohm). Lysosomes are common in animal cells but not often observed in plant cells.

Lysosomes are small, round structures involved with the digestive activities of the cell. See Figure 3–11. Lysosomes contain enzymes that break down large food molecules into smaller ones. Then these smaller food molecules are passed on to the mitochondria, where they are "burned" to provide energy for the cell.

Although lysosomes digest substances in the cytoplasm, you need not worry about your ship's safety! The membrane surrounding a lysosome keeps the enzymes from escaping and digesting the entire cell. However, lysosomes do digest whole cells when the cells are injured or dead. In an interesting process in the growth of a tadpole into a frog, lysosomes in the tadpole's tail cells digest the tail. Then this protoplasmic material is reused to make new body parts for the frog.

Chloroplasts

Your journey is just about over, and you will soon be leaving the cell. But first look around you. Floating in the cytoplasm are large, irregularly shaped green structures called **chloroplasts.** Chloroplasts contain a green pigment called **chlorophyll.** Chlorophyll captures the energy of sunlight and uses it to make food for the plant cell. In which part of the plant do you think cells with chloroplasts are located?

Chloroplasts are found *only* in plant cells. However, the sea slug, an animal, often eats plants that contain chloroplasts. After the sea slug digests the

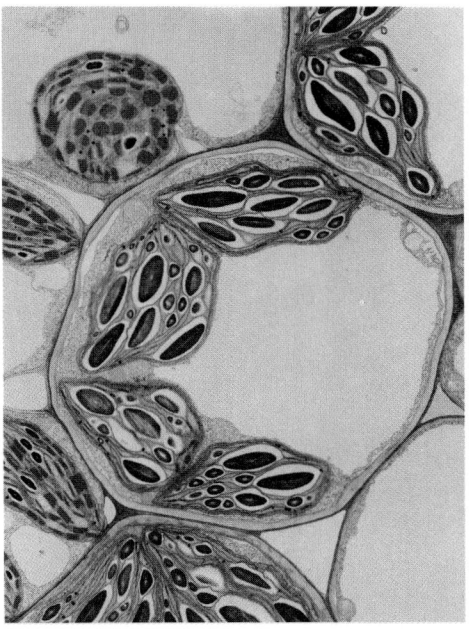

Figure 3–10 *The large, roundish empty spaces in these plant cells are vacuoles. What materials do vacuoles store?*

Figure 3–11 *The dark dots in this photograph are lysosomes. What is the function of lysosomes?*

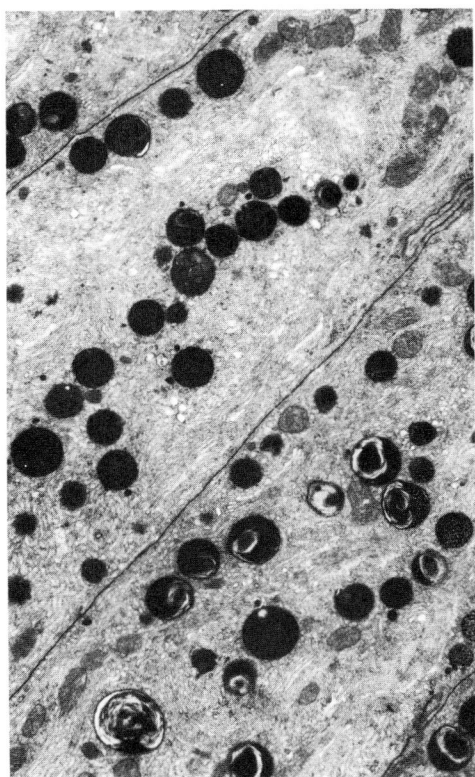

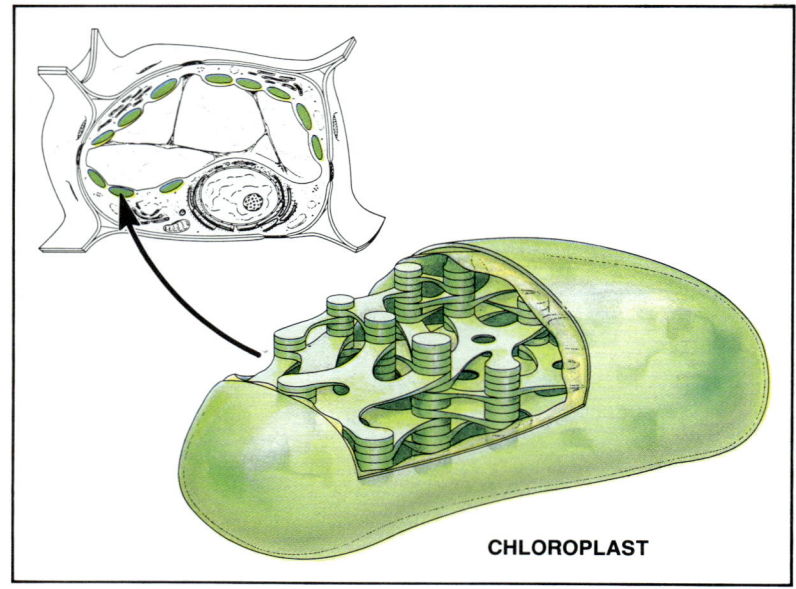

CHLOROPLAST

Figure 3–12 *The cells of green plants contain special food-making structures called chloroplasts. Why are chloroplasts green?*

plant, some of the chloroplasts get into the cells of its digestive system. There the chloroplasts continue to make food just as they do in a plant. The process goes on for a week or so and provides the sea slug's cells with food for energy.

SECTION REVIEW

1. What are five functions of the cell?
2. Why is it important that the cell membrane allow materials to enter and leave the cell?
3. Name the two structures that are found within the nucleus.
4. Explain how you would distinguish between a plant cell and an animal cell.

3–3 Cell Processes

The cell carries out a variety of chemical processes necessary to life. **Life processes performed by cells include metabolism, respiration, diffusion, osmosis, and reproduction.** And the cell as a whole carries out the chemical processes necessary to life. This tiny unit, basic to all forms of life, is like a miniature factory that produces many kinds of chemicals. Like any factory, a cell must have energy to do work. Working day and night, a cell traps, converts, stores, and uses energy. Obtaining and using energy is one of the most important activities of a living cell.

Metabolism

Even while you sleep, you need energy to keep you alive. But where does this energy come from? Cells provide it. Although cells do not *make* energy, they do *change* energy from one form to another. Cells obtain energy from their environment and change it into a usable form.

This conversion process is very complex. And it involves many chemical reactions. Some reactions break down molecules. Other reactions build new molecules. The sum of all the building-up and breaking-down activities that occur in a living cell is called **metabolism.**

Figure 3–13 *A cell, such as the plant cell, is like a miniature factory that carries out all the activities necessary to life. Which structure controls all the activities of the cell?*

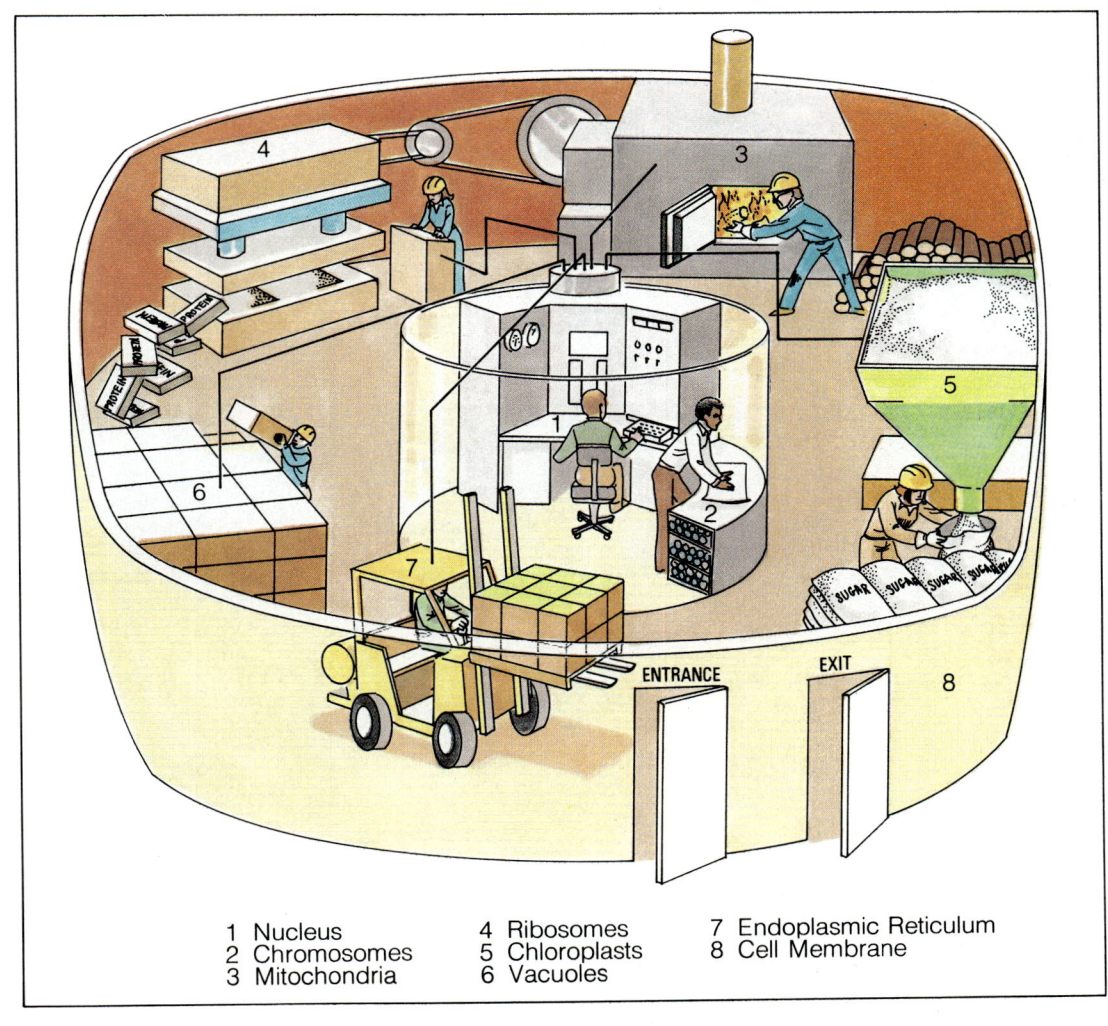

1 Nucleus
2 Chromosomes
3 Mitochondria
4 Ribosomes
5 Chloroplasts
6 Vacuoles
7 Endoplasmic Reticulum
8 Cell Membrane

Think for a moment of all the things cells do: grow, repair structures, absorb food, manufacture proteins, get rid of wastes, and reproduce. The energy for these activities is locked up in the molecules in food. As a result of metabolism, the stored energy in food is set free so it can be used to do work.

Cell Respiration

Earlier you learned that energy is released when simple food substances such as sugars are broken down inside the mitochondria. The process by which a cell releases energy from food is called **respiration.** Respiration is performed constantly by all living things.

There are two types of respiration. In aerobic (ehr-OH-bihk) respiration, food, such as the sugar glucose, is broken down with the help of oxygen. In this reaction, water and carbon dioxide are produced. The energy that is released is stored in an energy-rich molecule called ATP. The energy in ATP can then be used directly by the cell to do work when it is needed. Although aerobic respiration occurs in a series of complicated steps, the overall reaction can be shown in the following chemical equation.

$$\underset{\text{glucose}}{C_6H_{12}O_6} + \underset{\text{oxygen}}{6\,O_2} \longrightarrow \underset{\substack{\text{carbon} \\ \text{dioxide}}}{6\,CO_2} + \underset{\text{water}}{6\,H_2O} + \underset{\text{energy}}{\text{ATP}}$$

Figure 3–14 *Respiration is performed by all living things, such as yeast cells (left) and the mountain lion (right). What type of respiration occurs in each of these organisms?*

How do you get the oxygen needed for respiration? How do you get rid of the carbon dioxide that is produced during respiration?

In anaerobic (an-ehr-oh-bihk) respiration, energy is released and ATP is produced without oxygen. For example, in yeast cells, glucose is broken down to alcohol through anaerobic respiration. During strenuous exercise, your muscle cells may lack oxygen. At this time, energy is released by anaerobic respiration. Which type of respiration would produce more energy, aerobic or anaerobic?

Diffusion

Remember how you sailed through the cell membrane to enter the cell? Well, the substances that get into and out of the cell do the same thing. Food molecules, oxygen, water, and other materials enter and leave the cell through openings in the cell membrane by a process called **diffusion.**

Why does diffusion occur? Molecules of all substances are in constant motion, continuously colliding with one another. This motion causes the molecules to spread out. The molecules move from an area where there are more of them to an area where there are fewer of them. See Figure 3–15.

If there are many food molecules outside the cell, for example, some diffuse through the membrane into the cell. At the same time, waste materials built up in the cell pass out of the cell by the process of diffusion.

If substances can move into and out of the cell through the membrane, what keeps the protoplasm from oozing out? What keeps harmful materials from moving in? The cell membrane safeguards the contents of the cell because it is selectively permeable. That is, it permits only certain substances, mainly oxygen and food molecules, to diffuse into the cell. Only waste products such as carbon dioxide are allowed to diffuse out of the cell.

Osmosis

Water is the most important substance that passes through the cell membrane. In fact, about 80 percent of protoplasm is water. Water passes

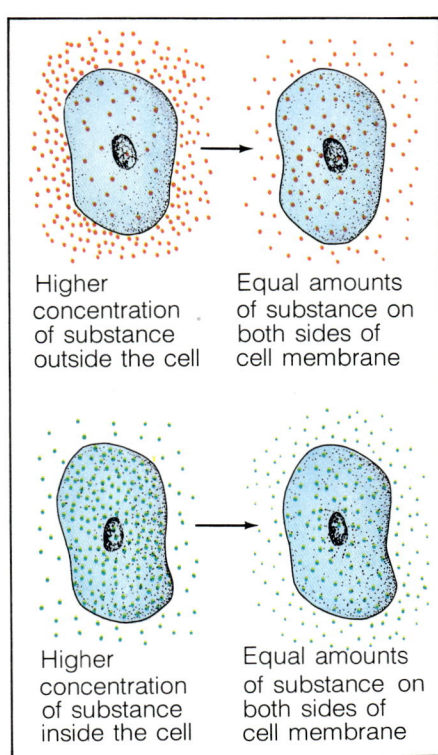

Higher concentration of substance outside the cell — Equal amounts of substance on both sides of cell membrane

Higher concentration of substance inside the cell — Equal amounts of substance on both sides of cell membrane

Figure 3–15 *Diffusion is the movement of molecules of a substance into a cell or out of a cell. Substances move from places where they are more concentrated to places where they are less concentrated. What structure controls the movement of substances into and out of the cell?*

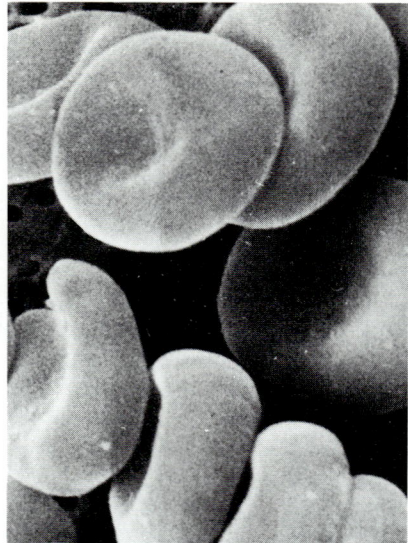
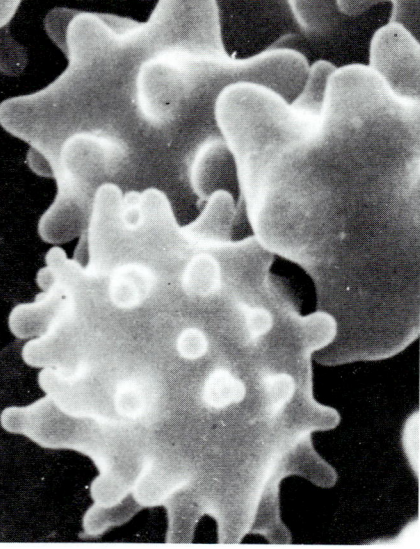
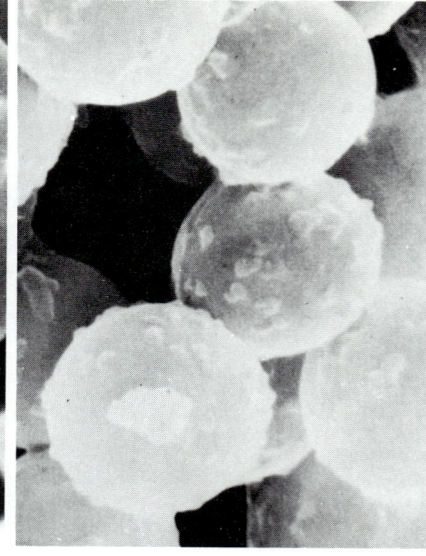
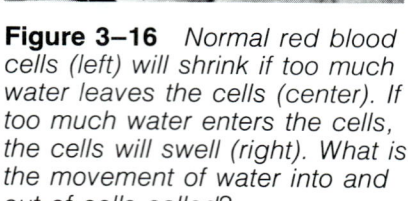

Figure 3–16 *Normal red blood cells (left) will shrink if too much water leaves the cells (center). If too much water enters the cells, the cells will swell (right). What is the movement of water into and out of cells called?*

through the cell membrane by a special type of diffusion called **osmosis.** During osmosis, water molecules move from a place of greater concentration to a place of lesser concentration. This movement keeps the cell from drying out.

Suppose you put a cell in a glass of salt water. The concentration of water outside the cell is lower than the concentration of water inside the cell. So water leaves the cell, and the cell starts to shrink. If too much water leaves the cell, it dries up and dies.

Now, if the cell was put in a glass of fresh water instead of salt water, just the opposite would occur. Water enters the cell, and the cell swells. This happens because the concentration of water is lower inside the cell than outside the cell. As you might imagine, if too much water enters the cell, the cell bursts.

Reproduction

Do you remember how many cells there are in the body of an adult human? There are approximately one hundred trillion cells. Interestingly these trillions of cells came from just one original cell. How is this possible?

In order for the total number of cells to increase and for the organism to grow, the cells must undergo **reproduction.** Reproduction is the process by which living things give rise to the same type of living thing. The cells reproduce by dividing into two new cells. Each new cell, called a daughter cell, is identical to the other and to the parent cell. How does this process occur?

If a parent cell—a skin cell, leaf cell, or bone cell, for example—is to produce two identical daughter cells, then the exact contents of its nucleus must go into the nucleus of each new cell. In other words, the "blueprints" of life in the parent cell must be passed on to each daughter cell. Now, if the parent cell simply splits in half, each daughter cell would get only half the contents of the nucleus—only half the "blueprints." It would no longer be the same kind of cell—a skin cell, leaf cell, or bone cell.

Fortunately, this is not what happens. Just before a cell divides, all the material in the nucleus is duplicated, or copied. The division of the nucleus and the cytoplasm and the duplication of the chromosomes is called **mitosis** (migh-TOH-sihs).

Mitosis is a continuous process. To form two cells, the nucleus of the dividing cell goes through four stages: **prophase, metaphase, anaphase,** and

CAREER   *Medical Artist*

HELP WANTED: MEDICAL ARTIST
Advanced degree in medical illustration required. College background in biology or zoology desirable. Needed to prepare accurate and creative drawings for medical textbooks.

Have you ever looked at a biology book or a magazine about biology that did not have diagrams? Probably not, because many concepts and objects related to biology cannot be easily understood by reading only words. For example, cell structures and functions, body parts, and organism development can best be "described" through the use of drawings. Making such drawings is the job of a special group of artists.

Artists who create visual materials dealing with health and medicine are **medical artists.** Sometimes their work involves viewing a specimen through various kinds of microscopes in order to draw it. Or a medical artist may dissect and study the parts of animals and plants. Some medical artists work closely with surgeons or

other kinds of doctors in order to prepare accurate drawings of medical conditions.

Most medical artists are employed by publishers, medical, veterinary, or dental schools, or by hospitals with programs in teaching and research. To receive more information about this field, write to the Association of Medical Illustrators, Route 5, Box 311F, Midlothian, VA 23113.

Figure 3–17 *During mitosis, each chromosome is duplicated before the nucleus divides and two new cells form. What is the period between one mitosis and the next?*

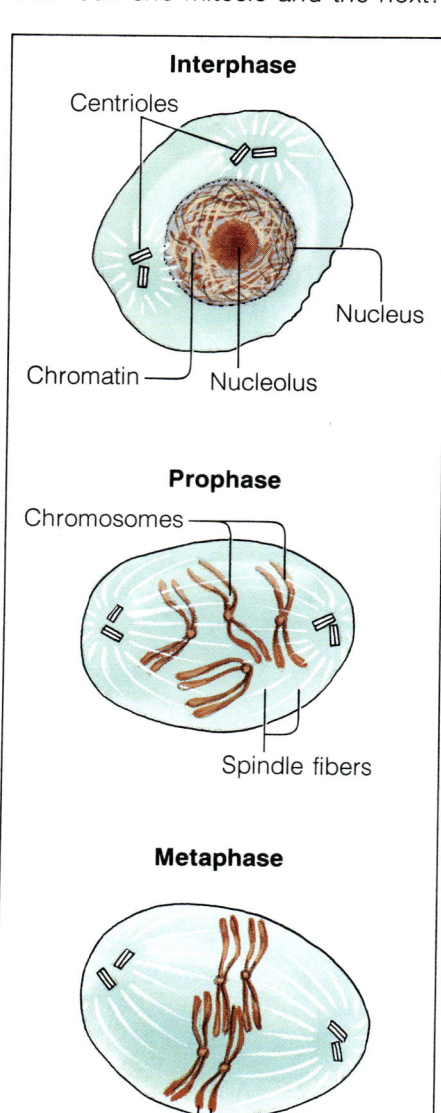

Interphase

Centrioles

Nucleus

Chromatin — Nucleolus

Prophase

Chromosomes —

Spindle fibers

Metaphase

Chromosomes

telophase. The period between one mitosis and the next is called **interphase.** During interphase, the cell is performing all of its life functions, but it is not dividing. See Figure 3–17. Notice that the chromosomes are not visible as rodlike structures. Instead they appear as threadlike coils called **chromatin.** In animal cells, two structures called **centrioles** (SEHN-tree-ohlz) can be seen outside the nucleus. They play a part in cell division. Most plant cells do not have centrioles.

PROPHASE   During prophase, the first phase of mitosis, the nuclear membrane begins to disappear. The thin threads of chromatin shorten and thicken to form the rodlike chromosomes. See Figure 3–17. Each chromosome is made up of two identical strands that remain together as a pair. The two centrioles double and then separate. Each pair of centrioles locate themselves at opposite ends of the cell. Long threads, called spindle fibers, form between the centrioles. The chromosome pairs become attached to the spindle fibers. The formation of spindle fibers and the attachment of chromosomes occur in both plant and animal cells.

METAPHASE   During metaphase, the second phase of mitosis, the chromosome pairs line up along the middle, or the equator, of the cell. The chromosome pairs are attached to the spindles.

ANAPHASE   Anaphase is the third stage of mitosis. During anaphase, the chromosomes split apart. One set of chromosomes moves toward one side of the cell. The other set of chromosomes moves to the opposite side. This stage ends with a set of chromosomes at each side of the cell.

Anaphase

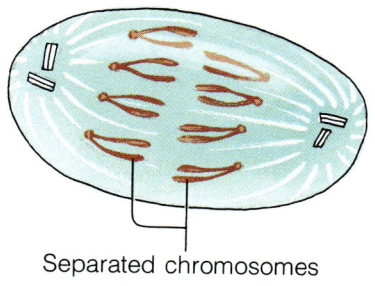

Separated chromosomes

Telophase

Nucleus

Nucleolus —

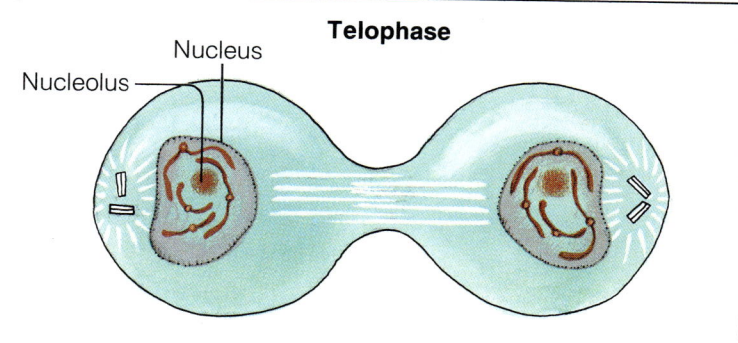

TELOPHASE   The fourth stage of mitosis, telophase, results in the formation of two cells. A new nuclear membrane forms around each set of chromosomes, forming two new nuclei. The spindle disappears, and each centriole remains near each nucleus. Mitosis ends when the cytoplasm divides in half, pinching the cell into two new cells. A cell membrane forms in the middle between the two new cells.

In plant cells, a cell plate forms and divides the cytoplasm in half. The cell plate then becomes part of the new cell wall between the plant cells.

In multicellular organisms, mitosis takes place millions of times as an organism grows and develops. Mitosis also occurs to replace dead or injured cells. Where do you think mitosis might occur in your body?

SECTION REVIEW

1. Name five processes that occur in a cell.
2. Distinguish between aerobic and anaerobic respiration.
3. How is osmosis different from diffusion?
4. Give the overall equation for respiration.
5. List the stages of mitosis.
6. During the formation of the male and female sex cells, a type of cell division called meiosis takes place and the chromosomes number is halved. Would meiosis be a suitable type of cell division to produce more skin cells? Explain.

3–4 Organization of Living Things

Section Objective

To identify the five levels of organization of living things

The whole body of a bacterium is made up of one cell. That single cell must do all the jobs that keep a bacterium alive. This accounts, in part, for the simple lives bacteria live. A panther, however, is multicellular—it is made of a great many cells. The numerous cells of the panther's body are organized into different parts, which enable the animal to

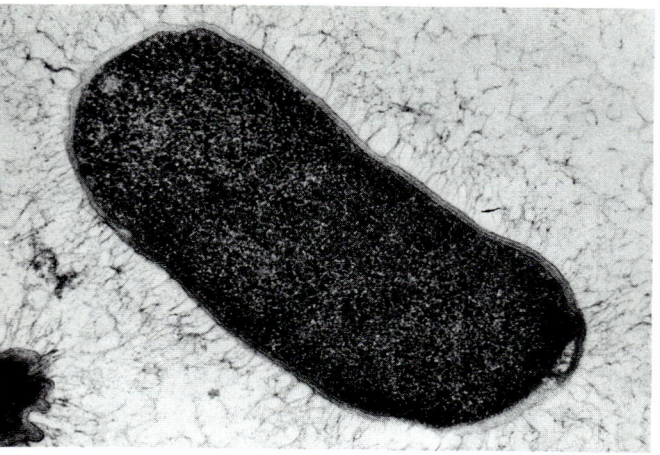

Figure 3–18 *A bacterium (right) and a honeybee (left) carry on a wide variety of complex activities. Which organism is multicellular?*

carry out more complex activities than a one-celled bacterium can.

Within the body of the panther, as within most living things, there is a **division of labor.** Division of labor means that the work of keeping the organism alive is divided among the different parts of its body. Each part has a specific job to do. And as the part does its special job, it works in harmony with all the other parts to keep the organism healthful and active.

The arrangement of specialized parts within a living thing is sometimes referred to as levels of organization. **The five basic levels of organization, arranged from the smallest, least complex structure to the largest, most complex structure are cells, tissues, organs, organ systems, and organisms.**

Figure 3–19 *Muscle cells (left), blood cells (center), and nerve cells (right) perform specialized tasks. What is the function of each of these cells?*

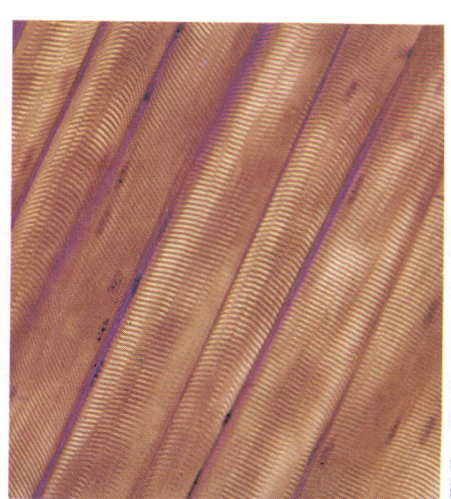
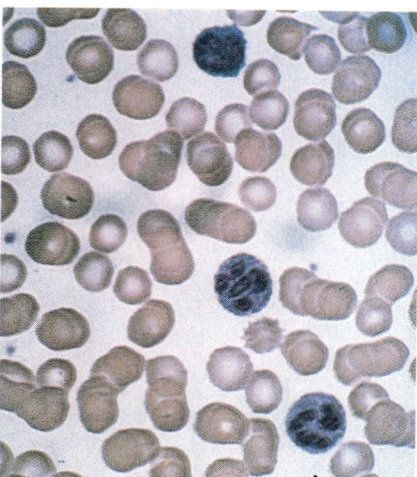
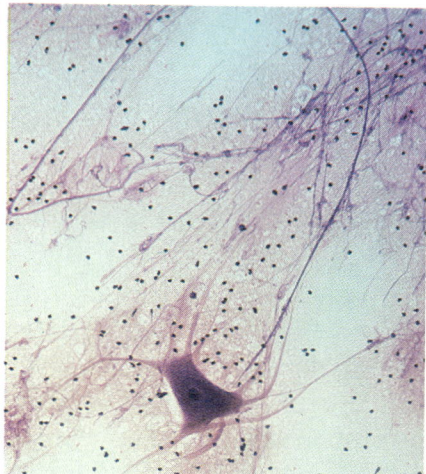

Cells—Level One

At the beginning of this chapter, you learned that all living things are made up of cells. Some living things are made of only one cell. For these organisms, the single cell exists as a free-living organism. The single cell does all the jobs that keep the organism alive.

In multicellular organisms, different cells perform specialized tasks. Muscle cells, for example, help the panther move through its environment. Nerve cells receive and send messages throughout the panther's body. Every kind of cell, however, depends on other cells for its survival and for the survival of the entire organism. Those muscle cells move only when set into action by nerve cells. And both muscle cells and nerve cells rely on red blood cells for oxygen.

Tissues—Level Two

In a multicellular organism, cells usually do not work alone. Cells that are similar in structure and function are joined together to form **tissues.** Tissues are the second level of organization.

Like the single cell of the bacterium, each tissue cell must carry on all the activities needed to keep that cell alive. But at the same time, tissues perform one or more specialized functions in an organism's body. In other words, tissues work for themselves as well as for the good of the entire living thing.

Figure 3–20 *The flower has special tissues that produce nectar, which the hummingbird uses for food (left). Various organs in the panther's body work together so that the animal can hunt for food (right). How are organs and tissues related?*

ORGAN SYSTEMS

System	Function
Skeletal	Protects and supports the body
Muscular	Supports the body and enables it to move
Skin	Protects the body
Digestive	Receives, transports, breaks down, and absorbs food throughout the body
Circulatory	Transports oxygen, wastes, and digested food throughout the body
Respiratory	Permits the exchange of gases in the body
Excretory	Removes liquid and solid wastes from the body
Endocrine	Regulates various body functions
Nervous	Conducts messages throughout the body to aid in coordination of body functions
Reproductive	Produces male and female sex cells

For example, bone cells in the panther form bone tissue, a strong, solid tissue that gives shape and support to the bodies of animals. Blood cells are part of blood tissue, a liquid tissue responsible for carrying food, oxygen, and wastes throughout the panther's body. Some muscle tissue helps move the panther's legs, neck, and other body parts.

Plants have tissues too. The leaves and stems of a plant are covered by a type of tissue called epidermis, which protects the plant and prevents it from losing water. Tissue known as xylem (ZIGH-luhm) conducts water and dissolved minerals up through the stems to the leaves. And another special tissue called phloem (FLOH-ehm) brings food made in the leaves back down to the stems and roots.

Organs—Level Three

The panther's eyes, ears, nose, and lungs are several of the **organs** that help the animal find food and stay alive. Organs are groups of different tissues working together. The panther's heart, for example, is an organ composed of nerve tissue, muscle tissue, and blood tissue. Each tissue does its special job.

Plants have organs too. Roots, stems, and leaves are common plant organs. Like animal organs, plant organs are made up of groups of tissues performing the same function. For example, various tissues in the leaf help this organ make food for the plant.

You are probably familiar with the names of many of the organs that make up your body. Your brain, stomach, kidneys, and skin are some examples. Can you name some others?

Organ Systems—Level Four

Like cells and tissues, organs seldom work alone. They "cooperate" with each other to form specific **organ systems.** An organ system, then, is a group of organs that work together to perform certain functions.

The panther is able to spot its prey because its sense organs—its eyes, ears, and nose—do their job. They receive messages in the form of sights, sounds, and smells. The panther's brain interprets this information. And its spinal cord and nerves send out messages to all parts of its body so that the animal is able to prepare to attack. Each of these organs, doing its individual job as well as working together with other organs, forms part of the panther's nervous system.

The organs that make up an organ system vary in number and complexity from one kind of living thing to another. For example, a very simple animal called the hydra has a nervous system that is a simple net of nerves spread throughout the organism. More complex animals, such as the grasshopper, have a more highly organized system of nerves and a primitive brain. The most highly developed animals, people, for example, have nervous systems consisting of a brain, a spinal cord, and nerves.

Organisms—Level Five

You, like the panther, are an **organism.** An organism is an entire living thing that carries out *all* the basic life functions. A buttercup is an organism, as are an apple tree, a deer, and a dolphin. Even a one-celled bacterium is an organism. The organism is the highest level of organization of living things.

A complex organism is a combination of organ systems. Each system performs its particular function, but all the systems work together to keep the organism alive. Without the smooth operation of any one of its systems, a living thing could not survive.

Cells, tissues, organs, organ systems, organisms. By now one thing should be clear to you: Each level of organization interacts with every other level. And the smooth functioning of a complex organism is the result of all its various parts working together.

SECTION REVIEW

1. List and define the five levels of organization of living things.
2. What is meant by the term "division of labor" in living things?
3. Explain some of the differences between a one-celled organism and one cell in a multicellular organism.

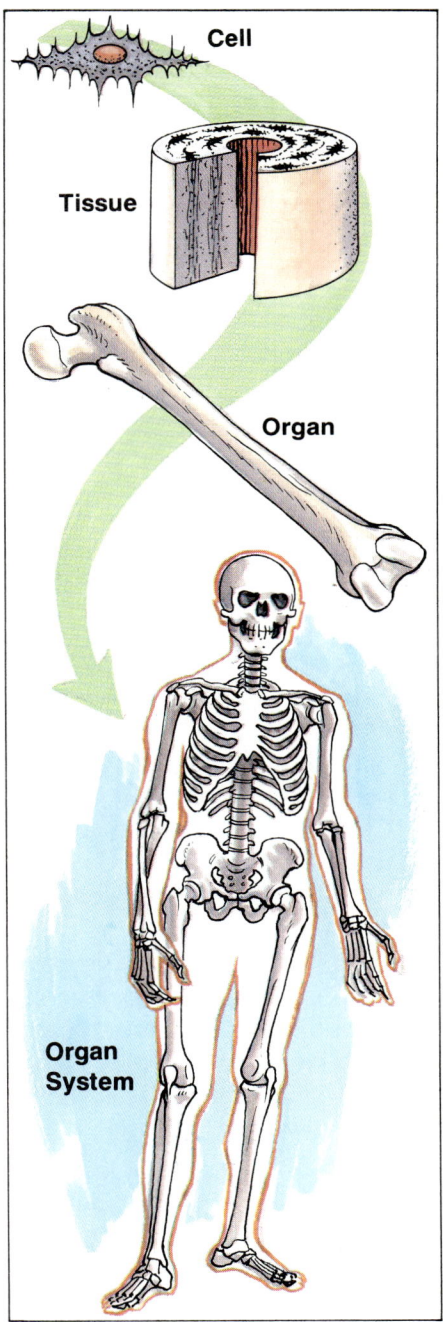

Figure 3–22 *This diagram shows how the bone cells of a human being, a complex organism, are organized to form tissues, organs, and an organ system. What is the highest level of organization?*

Using the Microscope

Problem

How do you use the microscope to observe objects?

Materials *(per group)*

microscope   water
microscope slide   cover slip
medicine dropper
small pieces of newspaper print and
    colorful magazine photographs
needle or probe

Procedure

1. To learn the parts and functions of a microscope, see Appendix B on page 609.
2. Obtain a small piece of newspaper print and place it on a clean microscope slide.
3. To make a wet-mount slide, use the medicine dropper to carefully place a drop of water over the newspaper print.
4. Carefully lower the cover slip over the newspaper print.
5. Place the slide on the stage of the microscope. The newspaper should be facing up and be in the normal reading position. **CAUTION:** *Always use a cover slip to protect the objective lens from touching the material being observed.*
6. With the low-power objective in place, follow your teacher's directions for focusing on a letter.
7. While focusing on a letter, move the slide to the left, then to the right.
8. While looking through the eyepiece, adjust the slide so that a letter is in the center of your field of view.
9. Look at the stage and objectives from the side. Turn the nosepiece until the high-power objective clicks into place.
10. Using only the fine adjustment knob, bring the letter into focus. Draw what you see.
11. Repeat steps 2 to 10 using magazine paper.

Observations

1. How do the objects appear when viewed through the microscope?
2. While looking through the microscope, in which direction do the objects appear to move when you move the slide to the left? To the right?
3. What is the total magnification of your microscope under low and high power?
4. What happens to the focus of the objects as you change from low to high power?

Conclusions

1. What conclusion can you draw about the way objects appear when viewed under the microscope?
2. How would you center an object viewed under the microscope when it is off-center to the left?
3. Explain the purpose of a cover slip.
4. Suppose an alien from another planet examined a human skin cell under the microscope. What might the alien conclude about the shape of human cells?

CHAPTER REVIEW

SUMMARY

3–1 Cell Theory

❑ Mathias Schleiden, Theodor Schwann, and Rudolf Virchow were scientists who contributed to the development of the cell theory.

❑ The cell theory states that (1) all living things are made of cells; (2) cells are the basic units of structure and function in living things; and (3) living cells come only from other living cells.

3–2 Structure and Function of Cells

❑ The cell wall gives protection and support to plant cells.

❑ The cell membrane regulates the movement of materials into and out of the cell.

❑ The nucleus is the control center of the cell.

❑ Chromosomes are found in the nucleus and are made of the nucleic acids DNA and RNA.

❑ The endoplasmic reticulum is the site of the manufacture and transport of proteins.

❑ Ribosomes are the protein-making sites.

❑ Mitochondria are the ''powerhouses'' of the cell.

❑ Vacuoles store food, water, and wastes.

❑ Lysosomes have a digestive function.

❑ Chloroplasts capture the energy from the sun and use it to make food for the plant cell.

3–3 Cell Processes

❑ The sum of all the activities that occur in a living cell is called metabolism.

❑ In aerobic respiration, energy is released from food with the help of oxygen. In anaerobic respiration, energy is released without oxygen.

❑ Food, oxygen, water, and other materials enter and leave the cell by a process called diffusion.

❑ Water passes through the cell membrane by a type of diffusion called osmosis.

❑ Mitosis is the division of the nucleus and the cytoplasm and the duplication of the chromosomes.

3–4 Organization of Living Things

❑ The least complex level of organization of living things is the cell.

❑ Cells that are similar in structure and function are joined together to form tissues.

❑ Groups of different tissues work together as organs.

❑ A group of different organs working together to perform certain functions is known as an organ system.

❑ An organism is an entire living thing that carries out all the life functions.

VOCABULARY

Define each term in a complete sentence.

anaphase	chromatin	homeostasis	nucleic acid	protein
cell	chromosome	interphase	nucleolus	protoplasm
cell membrane	cytoplasm	lysosome	nucleus	reproduction
cellulose	diffusion	metabolism	organelle	respiration
cell wall	division of labor	metaphase	organ	ribosome
centriole	DNA	mitochondrion	organism	RNA
chlorophyll	endoplasmic reticulum	mitosis	organ system	telophase
chloroplast		nuclear membrane	osmosis	tissue
			prophase	vacuole

79

CONTENT REVIEW: MULTIPLE CHOICE

On a separate sheet of paper, write the letter of the answer that best completes each statement.

1. The first scientist who used the term cell was
   a. Hooke.   b. Schleiden.   c. van Leeuwenhoek.   d. Virchow.
2. All the living material in a cell is called the
   a. mitochondria.   b. nucleus.   c. nucleolus.   d. protoplasm.
3. The chromosomes contain
   a. hormones.   b. waste products.   c. nucleic acids.   d. enzymes.
4. The "powerhouses" of the cell are the
   a. nuclei.   b. mitochondria.   c. ribosomes.   d. lysosomes.
5. The ability of a cell to maintain a stable internal environment is known as
   a. reproduction.   b. mitosis.   c. homeostasis.   d. diffusion.
6. Structures that are involved in the digestive activities of the cell are called
   a. ribosomes.   b. mitochondria.   c. vacuoles.   d. lysosomes.
7. The food-making structures in green plant cells are the
   a. vacuoles.   b. mitochondria.   c. lysosomes.   d. chloroplasts.
8. The process by which water enters and leaves the cell is known as
   a. diffusion.   b. mitosis.   c. osmosis.   d. metabolism.
9. The period between one mitosis and the next is known as
   a. interphase.   b. metaphase.   c. prophase.   d. anaphase.
10. Eyes, kidneys, and skin are examples of animal
    a. organs.   b. tissues.   c. cells.   d. organ systems.

CONTENT REVIEW: COMPLETION

On a separate sheet of paper, write the word or words that best complete each statement.

1. The cell wall is made of _____.
2. The structure that controls all of the cell's activities is the _____.
3. The structures that pass on traits to new cells are the _____.
4. DNA is a (an) _____.
5. The protein-making structures of the cell are the _____.
6. The network of passageways throughout the cell is known as the _____.
7. Food, water, and wastes are stored in structures called _____.
8. The process by which a cell releases energy from food is called _____.
9. The division of the nucleus and the cytoplasm and the duplication of the chromosomes in cells is called _____.
10. During _____, the chromosome pairs line up along the equator of the cell.

CONTENT REVIEW: TRUE OR FALSE

Determine whether each statement is true or false. Then on a separate sheet of paper, write "true" if it is true. If it is false, change the underlined word or words to make the statement true.

1. The cell <u>membrane</u> is the outer covering of an animal cell.
2. The <u>endoplasmic reticulum</u> is the "brain" of the cell.

3. Proteins are used to rebuild and repair cells.
4. The passage of food, oxygen, water, and other materials in and out of a cell is called osmosis.
5. The nucleolus is made up of RNA and protein.
6. Vacuoles are the cell's "powerhouses."
7. The sum of all the activities that occur in a living cell is called homeostasis.
8. The green pigment found in special structures of a plant cell is called chlorophyll.
9. The first stage of mitosis is called interphase.
10. The last stage of mitosis is called telophase.

CONCEPT REVIEW: SKILL BUILDING

Use the skills you have developed in the chapter to complete each activity.

1. **Classifying organelles** On a sheet of paper, list the various activities of a cell. Next to each activity, give the organelle that is involved in that activity.
2. **Sequencing events** Prepare a time line that illustrates the events that led to the cell theory.
3. **Sequencing events** On a separate sheet of paper, draw each stage of mitosis that occurs in an animal cell in order. Do the same for a plant cell. Below each drawing, list the changes that occur in that stage. How does mitosis differ in plant and animal cells?
4. **Making models** To make a model of a cell, dissolve some colorless gelatin in warm water. Pour the gelatin in a rectangular pan (for a plant cell) or a round pan (for an animal cell). Using edible materials that resemble cell structures, place these materials in the gelatin before it begins to gel. On a sheet of paper, identify each cell structure.

5. **Interpreting graphs** Enzymes are substances that control the speed at which certain reactions take place. Use the graph to answer the following questions. What two factors are described in the graph? How are these two factors related?

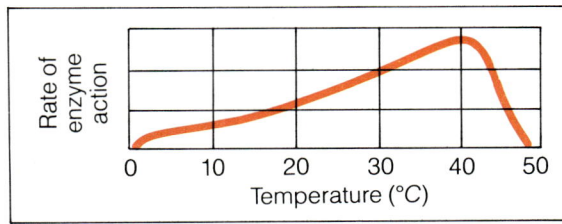

6. **Developing a hypothesis** Fill a beaker with tap water at room temperature, another with ice water, and a third with hot water. Add equal amounts of red food coloring to each beaker. In which beaker does the food coloring diffuse the fastest? The slowest? What is the variable in the experiment? Which beaker is the control? State a hypothesis for the experiment.

CONCEPT REVIEW: ESSAY

Discuss each of the following in a brief paragraph.

1. What were the contributions of Hooke, van Leeuwenhoek, Schleiden, Schwann, and Virchow to the understanding of the cell?
2. Based on your knowledge of cells, explain this statement. "One-celled organisms are often more complex than the individual cells of multicellular organisms."
3. Why is it important that the cell membrane be selective in allowing materials into and out of the cell?
4. Explain why reproduction is necessary for the survival of a species.
5. Briefly describe the changes that occur in the cell during mitosis.

Classification of Living Things

4

CHAPTER OBJECTIVES

After completing this chapter, you will be able to

4–1 Trace the history of classification.

4–1 Explain how binomial nomenclature is used to classify living things.

4–2 Identify the seven major classification groups.

4–3 List some general characteristics of the plant, animal, protist, moneran, and fungi kingdoms.

From its frightening face to the end of its fan-shaped tail, the fish stretched more than 1.5 meters. Its large body was covered with steel-blue scales. It had a powerful lower jaw and, strangest of all, its fins were attached to what seemed to be stubby legs! The unusual fish had turned up in the net of a fishing boat in the Chalumna River in South Africa. The crew of the fishing boat had never seen anything like it before!

The peculiar fish was brought to M. Courtenay-Latimer, a curator at a local museum. She searched through many books but could find no description of the fish. But M. Courtenay-Latimer was determined to identify the strange fish. So, she preserved the fish and sent it to African fish expert Professor J. L. B. Smith.

Smith was shocked when he saw the fish. Later he wrote, "I would hardly have been more surprised if I met a dinosaur on the street." The scientist knew right away that the fish was a coelacanth (SEE-luh-kanth), an animal thought to have become extinct more than 60 million years ago. The discovery was exciting because coelacanths are believed to be close relatives of fish that developed into four-legged land animals.

Smith was able to identify the fish because he knew about a special system that all scientists use to identify and name organisms. In this system of biological classification, every living thing has its own scientific name. The coelacanth was named *Latimeria chalumnae*. Can you use clues in this story to figure out why?

This is a rare photograph of a coelacanth.

4–1 The Beginning of Classification

Thousands of years ago, people began to recognize that there were different groups of living things in the world. Some animals had claws and sharp teeth and roamed the land. Others had feathers and beaks and flew in the air. Still others had scales and fins and swam in the water.

People also made observations about plants. Not only did plants vary in shape, size, and color, but some were good to eat while others were poisonous. In a similar sense, some animals, such as those with sharp teeth, were very dangerous, while others, such as those with feathers, were relatively harmless.

Without knowing exactly what they were doing, these people developed simple **classification** systems. **Classification is the grouping of living things according to similar characteristics.** Knowledge of these characteristics helped people to survive in their environment. For example, they quickly learned to fear the sharp-toothed animals and to hunt the feathered ones. Perhaps these people painted pictures on the walls of caves to communicate this knowledge. See Figure 4–1. In what other ways could a knowledge of classification have helped people survive?

Today, similar but much more complex information is found in the drawings, photographs, and

Figure 4–1 *Thousands of years ago, people drew this painting on a cave wall. What message does the painting communicate?*

Figure 4–2 *Because of its unusual characteristics and behavior, the giant panda (left) was once thought to be more closely related to the raccoon (right) than to the bear. Today giant pandas are classified as bears. What is the purpose of taxonomy?*

words in scientific books. This information is organized into the science of classification, which is called **taxonomy** (tak-SAHN-uh-mee). The scientists who work in this field are called taxonomists. The purpose of taxonomy is to group all the organisms on earth in an orderly system.

Classification helps provide a better understanding of the relationships among living things. Sometimes it even helps solve biological riddles of the past, such as the identity of the rare coelacanth. Meaningful names are given to newly discovered living things based on classification systems.

The First Classification Systems

In the fourth century B.C., the Greek philosopher Aristotle first proposed a system to classify life. He placed the animals into three groups. One group included all animals that flew, another group included those that swam, and a third group included those that walked.

Aristotle classified animals according to the way they moved. Although this system was useful, it caused problems. According to Aristotle, both a bird and a bat would fall into the same flying group. Yet in some basic respects, birds and bats are very different. Birds, for example, are covered with feathers. Bats, on the other hand, are covered with hair. In what other ways are bats different from birds?

Figure 4–3 *Grant's gazelles (left) and rhinoceroses (right) are animals. However, they are not members of the same species because they cannot interbreed. Who was the first person to scientifically use the word species?*

Although the system devised by Aristotle would not satisfy today's taxonomists, it was the first attempt to develop a scientific and orderly system of classification. Aristotle's classification system was used for almost 2000 years.

In the seventeenth century, John Ray, an English biologist, set an enormous goal for himself. He decided to collect, name, and classify all the plants and animals in England. But unlike Aristotle, Ray based his system of classification on the internal anatomy of plants and animals. He examined how they behaved and what they looked like as well. Ray was the first person to scientifically use the term **species** (SPEE-sheez). A species is a group of organisms that are able to interbreed and produce young.

John Ray achieved his goal. Moreover his work aided in the development of more complete and accurate classification systems.

Binomial Nomenclature

The eighteenth-century Swedish scientist Carolus Linnaeus spent the major part of his life developing a new system of classification. The system placed all living things into plant and animal groups according to similarities in form or structure.

Linnaeus also developed a simple system for naming organisms. Before Linnaeus developed his naming system, plants and animals had been identified by a series of Latin words. These words described the physical features of the organism. Sometimes five or six Latin words were used to describe a single organism!

Linnaeus devised the simpler naming system of **binomial nomenclature** (bigh-NOH-mee-uhl NOH-muhn-klay-cher). In this system, each organism is given two names, a **genus** (plural: genera) name and a species name. For example, you could think of the genus name as your family name. The species name could be thought of as your first name.

SECTION REVIEW

1. What is the naming system scientists use to classify organisms?
2. What is a species?
3. Why was Ray's system of classification an improvement over Aristotle's system of classification?
4. If you were shown two organisms that looked very similar, how could you tell if they were members of the same species?

4–2 Classification Today

Section Objective

To identify the seven major classification groups

Today, 200 years after Linnaeus completed his work, scientists consider many additional factors when classifying organisms. Of course, scientists still examine large internal and external structures, but they also examine microscopic structures. Scientists even analyze the chemical makeup of an organism.

Although such a classification system may at first seem complicated, it is really quite simple. The modern system of classification does two jobs. First, it groups organisms according to their *basic* characteristics. Second, it gives a *unique name* to an organism that scientists all over the world can use and understand. For example, in North and South America one large cat is called a mountain lion by some people, a cougar by others, and a puma by still others. If these people were to talk to one another about the animal, they might think they were talking about three different animals. But scientists throughout the world use only one name for this large cat, *Felis concolor*. This name easily identifies the cat to all scientists, no matter where they live or what language they speak.

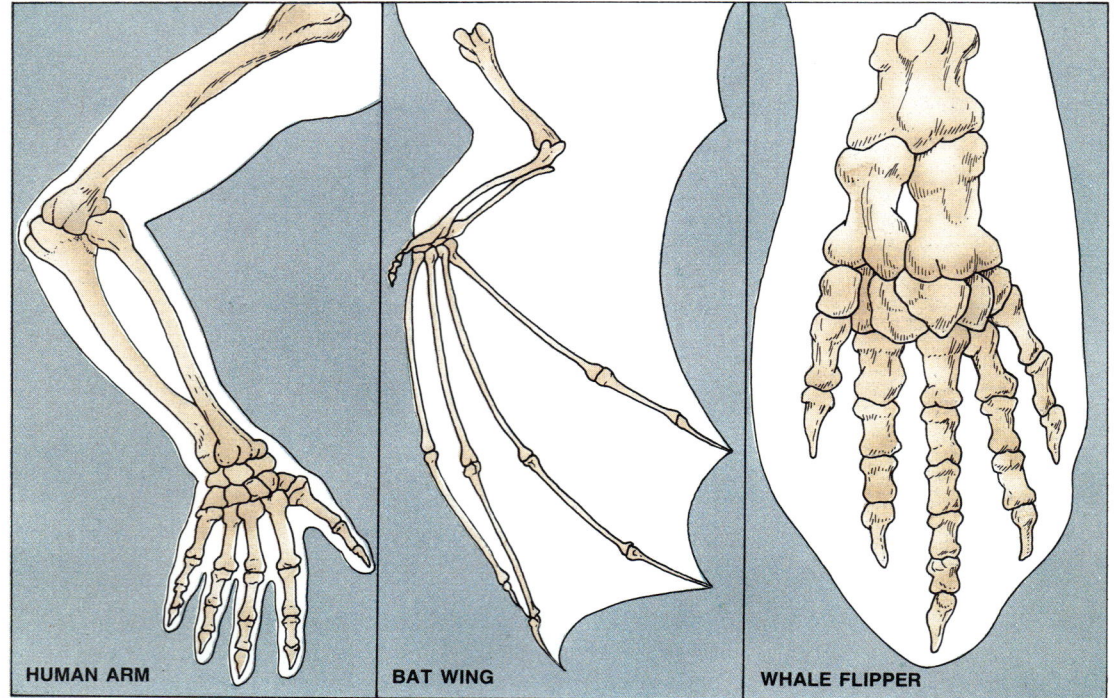

HUMAN ARM

BAT WING

WHALE FLIPPER

Figure 4–4 *Scientists often use skeletons of animals to help in classification. Notice how the similarity of forelimb structures of these animals helps to classify them as being part of the same group—in this case, mammals.*

Classification Groups

All living things are classified into seven major groups: kingdom, phylum, class, order, family, genus, and species. The largest group is the **kingdom.** All animals, for example, belong to the animal kingdom. The second largest group is called a **phylum** (FIGH-luhm; plural: phyla).

Each phylum is made up of **classes.** Within each class are **orders.** In turn, each order is divided into **families** that consist of many related genera. Each genus usually is divided into one or more species.

Figure 4–5 *The variey of living things in a rain forest, such as this one in the Amazon region, is greater than in any other place on the surface of the earth. Here a squirrel monkey shares its territory with many other plants and animals. To which kingdom does the squirrel monkey belong?*

The word for an organism's genus and species make up its scientific name. The genus name is capitalized, but the species name begins with a small letter. For example, the genus and species name for a wolf is *Canis lupus.* These two names identify the organism. Although most of these names are in Latin, some are in Greek. Scientists estimate that there are at least 3 million and perhaps as many as 10 million different species alive today. Many of them have yet to be identified and named.

Sharpen Your Skills

All in the Family

Use reference books to learn about the various families in the plant and animal kingdom. Choose the family that you find most interesting. Make a collage of the family using pictures from old magazines and newspapers. Present the collage to your class.

Classifying an Organism

Figure 4–6 shows the classification of several organisms. Each group indicates something about the organism's characteristics. For example, the lion belongs to the order Carnivora. Carnivora is the Latin word for "flesh eater." Many other familiar organisms, such as dogs, raccoons, and bears, also belong to this order. Look at Figure 4–7 on page 91 and notice that the lion is in the family Felidae. This family contains not only the lion but other cats, including ordinary house cats.

This knowledge allows you to have a very good idea of how a lion acts and looks even if you have never seen one in person. You can do this because lions and house cats are in the same family. But

Figure 4–6 *This chart shows the classification of five organisms. Which organism is a member of the order* Asterales?

CLASSIFICATION OF FIVE DIFFERENT ORGANISMS

	Lion	Dandelion	Amoeba	Yogurt-making Bacterium	Edible Mushroom
Kingdom	Animalia	Plantae	Protista	Monera	Fungi
Phylum	Chordata	Tracheophyta	Sarcodina	Eubacteriacea	Basidiomycetes
Class	Mammalia	Anthophyta	Rhizopoda	Schizomycetes	Homobasidiomycetes
Order	Carnivora	Asterales	Amoebida	Eubacteriales	Agaricales
Family	Felidae	Compositae	Amoebidae	Lactobacillaceae	Agaricaceae
Genus	*Panthera*	*Taraxacum*	*Amoeba*	*Lactobacillus*	*Agaricus*
Species	*leo*	*officinale*	*proteus*	*bulgarius*	*campestris*

lions and house cats are not in the same genus and species, which indicates they are somewhat different. For example, taxonomists include the lion in the genus *Panthera* along with other large cats. All cats in the genus *Panthera* roar; they do not purr. Other cats, including house cats, are in the genus *Felis*. They, of course, purr and do not roar. Finally, the species name for the lion is *leo*. In this case, the lion, *Panthera leo*, is the animal that you see jumping through hoops in the circus or lounging lazily in a cage at your local zoo. *Felis domesticus* is the scientific name for your playful pet cat.

SECTION REVIEW

1. Beginning with the largest, list the seven classification groups in order.
2. What two jobs does the modern system of classification do?
3. What is the scientific name of a dandelion?
4. Which have more in common, animals in the same order or animals in the same class? Explain.

CAREER   *Zoo Keeper*

HELP WANTED: ZOO KEEPER High school diploma required. Minimum one year experience as an apprentice zoo keeper needed. Must like animals and be tolerant of both animal and human behavior. Apply in person.

Today is a big day at the zoo. The arrival of a male and female giraffe is expected from Africa. For months, everyone has prepared for the new exhibit. Gardeners planted fruit trees bearing the giraffe's favorite food. In addition, zoo architects designed and built a new home.

At first, the giraffes will be kept away from other animals. Once tests show they are strong and healthy, the **zoo keeper** will help move them to their new home. The zoo keeper will also be responsible for giving food and water to the giraffes. Other responsibilities of the zoo keeper include keeping the animals' area clean and observing and recording their behavior. The zoo keeper usually is the first to notice any medical problems as well as to treat minor injuries or ailments. However, serious problems must be reported to the senior zoo keeper.

Of the many people working at the zoo, zoo keepers come into the most direct contact with the animals. In fact, zoo keepers sometimes bathe and groom the animals. Occasionally, zoo keepers answer visitors' questions about the inhabitants of the zoo. For further career information, write to the American Association of Zoo Keepers National Headquarters, 635 Gage Boulevard, Topeka, KS 66606.

CLASSIFICATION OF THE LION

Figure 4–7 *According to this chart, to which class does the lion,* Panthera leo, *belong?*

4–3 Kingdoms in the Classification System

In Figure 4–6, you may have noticed that some organisms were not in the Plantae or Animalia kingdom. The paramecium, for example, is usually placed in a kingdom called Protista. Scientists invented this kingdom to include organisms that did not seem to fit into either the plant or animal kingdom. Later, two more kingdoms called Monera (muh-NAIR-uh) and Fungi (FUHN-jigh) were added to many modern classification systems. **Today, most scientists use a system of classification that includes five kingdoms: plants, animals, protists, monerans, and fungi.**

PLANTS   Most plants are **multicellular,** or many-celled, organisms that contain specialized tissues and organs. A few plants are **unicellular,** or one-celled. Some algae are examples of unicellular plants. In general, plants are **autotrophs** (AWT-uh-trohfs). Autotrophs are organisms that can make their own food from simple substances. Most plants contain chlorophyll, a green substance needed to make food.

ANIMALS   Only multicellular organisms are in the animal kingdom. These organisms have tissues, and most have organs and organ systems. Unlike plants, animals do not contain chlorophyll and are not able to make their own food. These organisms are called **heterotrophs** (HEHT-uh-roh-trohfs).

Figure 4–8 *The ability to move from place to place is one way scientists distinguish the clouded sulfur butterfly, a member of the animal kingdom, from a thistle, a member of the plant kingdom (right). The trumpet-shaped unicellular stentors (left) are members of the Protista kingdom. Which of these organisms are autotrophs?*

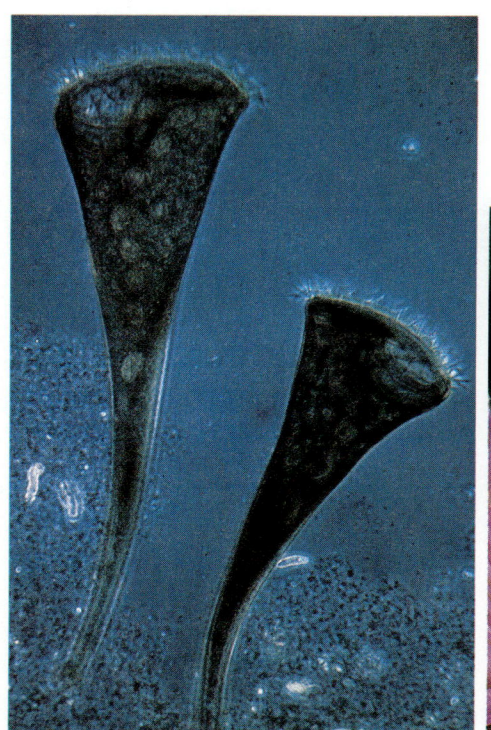

Figure 4–9 *These umbrella-shaped mushrooms are members of the Fungi kingdom. How do fungi differ from plants?*

PROTISTS   The Protista kingdom includes most unicellular organisms with nuclei. The nucleus is the control center of the cell. Microscopic protozoans such as paramecia are found in this kingdom. Why do you think protists were not included in the early classification systems?

MONERANS   All of the earth's bacteria are found in the Monera kingdom. One kind of algae, the blue-green algae, are also in the Monera kingdom. Like protists, monerans are unicellular. However, the cells of monerans do not contain a nucleus. The hereditary material is spread throughout the cell.

FUNGI   As you might expect, the world's wide variety of fungi make up the Fungi kingdom. The kinds of fungi you are probably most familiar with are mushrooms and molds. Fungi share many characteristics with plants. However, fungi do not contain the green substance chlorophyll. So, unlike plants, fungi cannot make their own food.

SECTION REVIEW

1. List the five kingdoms and give an example of an organism from each kingdom.
2. Which kingdom consists mainly of autotrophs?
3. Suppose a scientist discovers a new kind of algae. The algae is an autotroph with a nucleus. Into which kingdom should this new algae be placed?

Developing a Classification System

Problem

How can a group of objects be classified?

Materials *(per group of 6)*

students' shoes   pencil and paper

Procedure

1. At your teacher's direction, remove your right shoe and place it on a work table.
2. As a group, think of a characteristic that will divide all six shoes into two kingdoms. For example, you may first divide the shoes by the characteristic of color into the black shoe kingdom and the nonblack shoe kingdom.
3. Place the shoes into two separate piles based on the characteristic your group has selected.
4. Next, working only with those shoes in one kingdom, divide that kingdom into two groups based on a new characteristic. The shoe kingdom, for example, may be divided into shoes with rubber soles and shoes without rubber soles.
5. Further divide these groups into subgroups. For example, the shoes in the rubber-soled group may be separated into a shoelace group and a nonshoelace group.
6. Continue to divide the shoes by choosing new characteristics until you have only one shoe left in each group. Identify the person who owns this shoe.
7. Now repeat this process working with the nonblack shoes.
8. Draw a diagram similar to the one below to represent your classification system.

Observations

1. How many groups are there in your classification system?
2. Was there more than one way to divide the shoes into groups? How did you decide which classification groups to use?

Conclusions

1. Was your shoe classification system accurate? Why or why not?
2. If black and nonblack shoe groups represent kingdoms, what do each of the other groups in your diagram represent?
3. Compare your classification system to the classification system used by most scientists today.

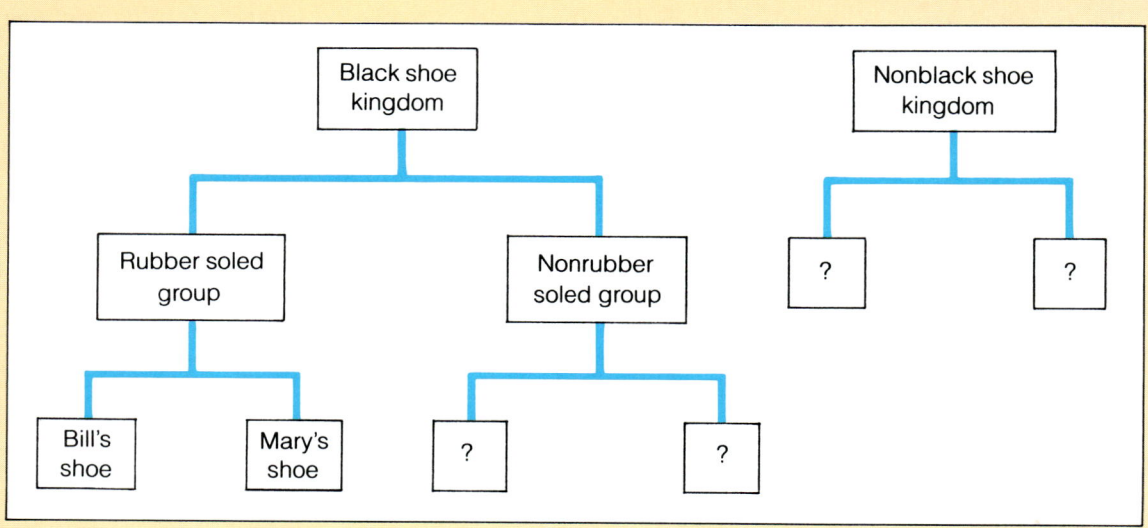

CHAPTER REVIEW

SUMMARY

4–1 The Beginning of Classification

❏ Classification is the grouping of living things according to similar characteristics.

❏ The science of classification is called taxonomy.

❏ Aristotle was the first person to develop a classification system for living things. His system was used for almost 2000 years.

❏ In the seventeenth century, John Ray, an English biologist, introduced a classification system that became the foundation of modern systems.

❏ John Ray was the first person to scientifically use the term *species*. A species is a group of organisms that are able to interbreed and produce young.

❏ In the eighteenth century, Carolus Linnaeus, a Swedish scientist, developed the system of binomial nomenclature, in which every organism is given two names. Linnaeus also developed a classification system based on the structural similarities of organisms.

4–2 Classification Today

❏ Today scientists classify organisms according to structural, chemical, and other similarities.

❏ Organisms are classified into seven groups. In order of decreasing size, these groups are kingdom, phylum, class, order, family, genus, and species.

4–3 Kingdoms in the Classification System

❏ A five-kingdom system of classification consists of plants, animals, protists, monerans, and fungi.

❏ Autotrophs can make their own food.

❏ Heterotrophs are organisms that cannot make their own food.

❏ Multicellular organisms are made up of many cells. Unicellular organisms are one-celled.

VOCABULARY

Define each term in a complete sentence.

autotroph

binomial
  nomenclature

class

classification

family

genus

heterotroph

kingdom

multicellular

order

phylum

species

taxonomy

unicellular

CONTENT REVIEW: MULTIPLE CHOICE

On a separate sheet of paper, write the letter of the answer that best completes each statement.

1. The first person to scientifically use the term species was
   a. Aristotle.   b. Ray.
   c. Linnaeus.   d. Haeckel.

2. Carolus Linnaeus classified plants and animals according to similarities in
a. color.　b. size.　c. structure.　d. habits.

3. The first word in a scientific name is the
a. genus.　b. species.
c. family.　d. phylum.

4. Which factor is *not* used by scientists when classifying organisms today?
a. internal structure　b. method of movement
c. external structure　d. chemical makeup

5. Which animal is identified by the scientific name *Felis concolor?*
a. puma　b. mountain lion　c. cougar　d. all of these

6. The second largest classification group is the
a. phylum.　b. genus.　c. order.　d. family.

7. Scientists believe that the number of different species alive today is
a. between 1 and 3 million.　b. between 3 and 10 million.
c. between 10 and 15 million.　d. less than 1 million.

8. The Latin word for "flesh eater" is
a. *Canis*.　b. *Felis*.
c. *Carnivora*.　d. *Panthera*.

9. Which organisms have cells that do *not* contain a nucleus?
a. plants　b. fungi　c. monerans　d. protists

10. In which kingdom do most of the organisms contain chlorophyll?
a. plants　b. fungi　c. protists　d. monerans

CONTENT REVIEW: COMPLETION

On a separate sheet of paper, write the word or words that best complete each statement.

1. The first person to propose a classification system was _____.

2. A group of organisms that are able to interbreed is a (an) _____.

3. The language that is most often used to name organisms in the modern classification system is _____.

4. In classification, the two-word naming system is known as _____.

5. The largest classification group is the _____.

6. In the modern classification system, each order is divided into _____.

7. Today most scientists use a classification system that includes five _____.

8. A bacterium is a member of the _____ kingdom.

9. An organism that is composed of many cells is _____.

10. An organism that can make its own food is a (an) _____.

CONTENT REVIEW: TRUE OR FALSE

Determine whether each statement is true or false. Then on a separate sheet of paper, write "true" if it is true. If it is false, change the underlined word or words to make the statement true.

1. The science of classification is called <u>taxonomy</u>.

2. The first person to propose a system to classify life was <u>Linnaeus</u>.

3. The first word in a scientific name is the species.
4. The naming system used by scientists today is binomial nomenclature.
5. The correct order of classification groups from largest to smallest is kingdom, phylum, order, class, family, genus, species.
6. In a scientific name, the species name begins with a capital letter.
7. The scientific name for a housecat is Felis domesticus.
8. Unicellular organisms are composed of only one cell.
9. A paramecium is a member of the Monera kingdom.
10. An organism that cannot make its own food is a heterotroph.

CONCEPT REVIEW: SKILL BUILDING

Use the skills you have developed in the chapter to complete each activity.

1. **Making charts** Create a chart in which you classify each of the following into its correct kingdom:
   a. spider plant
   b. honey bee
   c. bread mold
   d. *Blepharisma*
   e. straw mushroom
   f. amoeba
   g. giraffe
   h. blue-green algae
   i. maple tree
   j. bacteria
   k. toadstool
   l. paramecium
   m. hammerhead shark

2. **Making comparisons** You can use the following words to classify the place in which you live: county, city, country, house number, continent, street. Compare each word with one of the seven groups used to classify animals.

3. **Identifying relationships** Which two of the following three organisms are most closely related: *Quercus alba, Salix alba, Quercus rubra?* Explain your answer.

4. **Making generalizations** Why are the following common names of organisms confusing: starfish, seahorse, horseshoe crab, sea cucumber, jellyfish, shellfish, reindeer moss?

5. **Developing a model** Design a classification system for objects that might be found in your closet.

6. **Relating cause and effect** How do you think the invention of the compound microscope affected the classification of living things?

7. **Relating concepts** Explain why a wolf and a fox cannot interbreed and produce young, while a German shepherd and a cocker spaniel can.

8. **Applying concepts** Why do you think scientists do not classify animals by where they live or what they eat?

CONCEPT REVIEW: ESSAY

Discuss each of the following in a brief paragraph.

1. Explain why scientists classify organisms.
2. What kind of characteristics did Linnaeus use to develop his classification system?
3. Compare the classification system developed by Linnaeus with the classification system used by scientists today.
4. Describe each of the kingdoms used in the five-kingdom classification system.
5. What is the difference between an autotroph and a heterotroph? Give at least two examples of each.

Adventures in Science

James E. Lovelock

In a small town in southwestern England stands a white cabin. A half dozen pet peacocks strut proudly on the lawn. The laboratory of Dr. James E. Lovelock is attached to the cabin. Lovelock, a well-known biologist and inventor, has worked in his remote laboratory for several decades. During that time he has developed a hypothesis that has changed the way many scientists view the earth.

Lovelock's has named his hypothesis after Gaia, the Greek goddess of the earth. The Gaia hypothesis states that, in many ways, the earth is like a living system. Dr. Lovelock believes that not only does the environment of earth affect living things, but that living things also affect the earth's environment. He believes that living things help keep the earth's environment stable over time. The stability of the earth's environment, in turn, helps living things on earth survive.

For example, scientists have long known that the sun's energy output has increased more than 30 percent during the last 3.5 billion years. Yet the average temperature on earth has changed very little, even during the ice ages. "Why didn't the earth's water freeze?" asks Lovelock. "Or why hasn't it boiled away?" Lovelock's answer

is that the earth is homeostatic. Homeostasis means to remain constant or stable. Lovelock believes that the earth "follows the wisdom of the body." "The earth," states Lovelock, "functions as a living organism."

How does the earth achieve homeostasis? The answer is not quite clear and it will take more years of research before the complete picture unfolds. But that it is homeostatic seems to be a fact. Consider this: The salt content in ocean water remains fairly constant at 3.5 percent. Yet each year billions of kilograms of salt are carried into the ocean by rainwater. The rainwater falls on land and carries salts from the soil to the ocean. If the salt content of the earth's ocean was to increase to about 6 percent, almost all living things in the ocean would die. Yet, over the long course of earth's history, the salt content has remained relatively constant. This shows, says Lovelock, that the environments of the earth maintain a kind of homeostasis.

This homeostasis is an important aspect of the Gaia hypothesis. Still Dr. Lovelock's hypothesis is not accepted by all scientists. However, Lovelock remains firm. "The earth is a living organism and I'll stick by that," he says. Yet even Dr. Lovelock admits that

ℰ The Living Planet

he does not mean the earth is alive. Instead, he points out, the earth "is not living, but like a cat's fur, a bird's feathers, or the paper of a wasp's nest, it is an extension of life."

The Gaia hypothesis is certainly not the only achievement of James Lovelock. Dr. Lovelock considers himself an inventor. His most important invention is the electron capture detector. This complicated device is used to help analyze the earth's atmosphere. It was used to gather information on pesticides in the atmosphere. These data helped Rachel Carson write *Silent Spring*—the first book to call public atten-

This photograph taken by an orbiting satellite shows the earth and its atmosphere.

tion to the problems of environmental pollution due to pesticides. In the 1960s, NASA called upon Dr. Lovelock's talents to help determine if there is life on Mars.

Dr. Lovelock's work and scientific adventure continue to this day. His Gaia hypothesis has sparked a great deal of scientific debate. Yet even if it proves to be false, the Gaia hypothesis might well inspire the work of future scientists as they think about planet earth—the living planet!

Animal Experimentation

IS IT NECESSARY?

Each year, many experiments are conducted using laboratory animals. In many cases, the animals are treated with great care. However, in other cases, the animals are mistreated. Recently, a case involving cruelty to animals actually came to trial in Maryland. A well-known scientist working at a laboratory in Silver Springs was accused of mistreating laboratory animals.

For four months, a volunteer worker in the scientist's laboratory secretly photographed the laboratory animals. Then the volunteer brought the photographs to the police. Later the police raided the laboratory and charged the scientist with violating Maryland's animal cruelty law.

This was the first case of its kind in the United States. However, the issue of the treatment of laboratory animals has been debated for decades. Many animal rights activists claim that laboratory experiments involving animals are often cruel. Defenders of animal experimentation counter this argument. They point out hundreds of cases in which the use of animals to test medicines, surgical techniques, or household products has contributed to the health and well-being of millions of humans.

The controversy over animal experiments usually asks the same question: Is it moral, or even practical, to experiment on animals in order to benefit humans? Animal rights activists believe that a large proportion of animal experiments are unnecessary. They feel that similar information could be obtained by other means. They point to instances in which experiments using animals had no practical uses to humans.

But Joseph Held, a veterinarian who was a research director at the National Institutes of Health, thinks animal experimentation is necessary. "I feel we ought to be using more animals in research," he has stated. Recently Dr. Held gave a lecture in which he showed two slides. The first showed a child who was born damaged by a drug its mother had taken while she was carrying the child. The second slide showed a baby monkey with similar deformities. Dr. Held said that the deformities in human babies could have been predicted and prevented if the drug had been tested on monkeys first.

Today monkeys and apes, whose anatomy is similar to humans, are constantly being used to study such diseases as hepatitis, cancer, and malaria, as well as other serious diseases. Many important advances in medicine, psychology, and biology are partly the result of studies on animals. For example, the vaccine that protects people from polio was first tested on animals before it was given to humans.

Animals rights activists claim that many experiments using animals are not important. One of the most well-publicized experiments they consider unnecessary is the Draize test. This test has been widely used by some cosmetic companies. In this test, chemicals are dropped into the eyes of rabbits. The rabbits' eyes are checked carefully to determine the effects of the chemicals. The chemicals may be included in shampoos and soaps if they do not harm the rabbits' eyes. If the chemicals bother the rabbits' eyes, they are considered unsafe for human use. Animal rights activists claim that many rabbits suffer when they are used in the Draize test. They feel that the laboratory rabbits may be subjected to pain for an unimportant reason.

In his book *Animal Liberation*, Peter Singer argued that anyone who believes in a human rights movement also should support animal rights. Singer believes that the differences between human and animal intelligence does not mean that animals feel less pain than people.

Supporters of animal experimentation insist that it would be ridiculous to place the welfare of animals over the life of even one human. They argue that animal experimentation is justified when the experiments are meant to save human lives.

Increasing controversy over the use of laboratory animals has caused many people to rethink this issue. What do you think? Should animal experiments continue? What kinds of guidelines should be applied to animal experiments?

Today, vaccines developed through research on animals protect these young soccer players from disease.

Simple Organisms

The year: 323 B.C. Alexander the Great lies on his deathbed as soldiers from a hundred battles pass by in tribute. In his short lifetime, Alexander had conquered much of the known world, including Greece, Egypt, and Persia. And he had built the greatest city of his time—Alexandria. Yet this mighty conqueror had become victim to one of the earth's tiniest creatures.

Many historians suspect that Alexander the Great had malaria. The disease could not be treated in Alexander's time. It began with fever and chills, and usually ended in death.

Malaria is caused by a one-celled organism. Although it is not likely that you will come in contact with this organism, you are surrounded at this very moment by many simple organisms such as viruses and bacteria. Some of these simple organisms cause diseases, just as they did in the time of Alexander the Great. However, modern science has come a long way in the 2000 or more years since Alexander's death. Cures have been found for many diseases caused by simple organisms. But cures for others are still lacking. Modern science, you see, still has a way to go in controlling diseases caused by simple organisms.

In this unit, you will enter the world of simple organisms. On your journey into the world of the microscopic, you will discover many amazing facts about creatures that fill almost every corner of the earth. As you begin your journey, keep one thing in mind—even the page you are now reading has some simple organisms on it. To find out more, read on.

CHAPTERS

5 Viruses and Bacteria

6 Protozoans

Alexander the Great died from malaria, a disease caused by a one-celled microorganism.

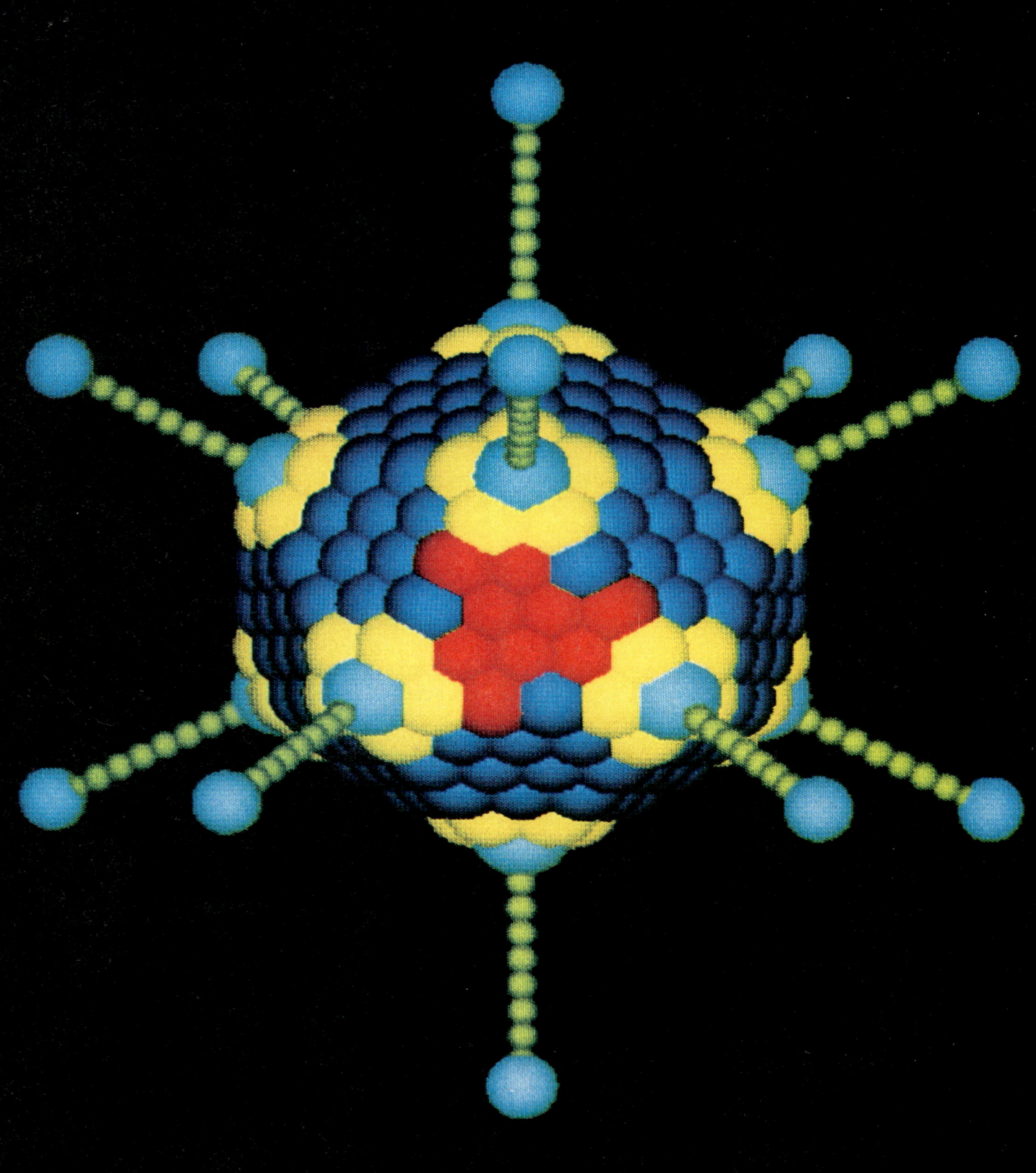

Viruses and Bacteria 5

CHAPTER OBJECTIVES

After completing this chapter, you will be able to

5–1 Identify the parts of a virus.

5–1 Describe the sequence of events in viral reproduction.

5–2 Classify bacteria according to their three basic shapes.

5–2 Identify the parts of a bacterium.

5–2 Compare autotrophic and heterotrophic bacteria.

5–2 Describe the helpful and harmful effects of bacteria.

Achoo! And so it begins—the misery of the most common illness—the cold. You are probably quite familiar with the symptoms of a cold—runny nose, teary eyes, sore throat, and aching bones. Perhaps you can add a few symptoms to this list. The point is, the common cold is rarely a serious health problem. But it is a major annoyance to millions of people each year.

Colds can be caused by any of more than 100 kinds of viruses, one of which is shown on the opposite page. Because you can be infected by one cold virus one month and another the next, you can get several colds every year. The viruses that cause colds are remarkably small, often measuring little more than 10 billionths of a meter across.

Tired of catching colds? Do not despair, for help is on the way. In 1986, scientists working at the University of Virginia School of Medicine were successful in stopping the attacks of some common cold viruses. The scientists are seeking a way to stop the cold virus before it can cause an infection. To do so, they have developed antibodies, or disease-fighting substances, in the laboratory. These antibodies keep the cold virus from invading unsuspecting cells. By acting like a protective shield, they guard the body from invaders.

Will this technique stop all colds? Probably not. But researchers do believe preventing people from catching colds—rather than trying to cure a cold—will be an important step in the battle against the common cold virus.

As you read through the pages of this chapter, you will learn a great deal about viruses and tiny living cells called bacteria. Now turn the page and begin your journey into a strange, microscopic world.

This image of a typical cold virus was produced using a computer.

To describe the structure and reproduction of a virus

5–1 Viruses

Living things come in many shapes and sizes. But all living things seem to have one thing in common. They all contain cells. Some organisms consist of only a single cell. Other organisms, such as you, are made up of many cells. Unlike these organisms, **viruses** are not cells and do not contain cells. Viruses are actually tiny particles that contain hereditary material.

Because viruses are not cells, they cannot perform the life functions of living cells. They cannot, for example, take in food or get rid of wastes. In fact, about the only similarity that viruses share with cells is that viruses are able to reproduce. However, viruses cannot reproduce on their own. They need the help of living cells. These living cells are **hosts,** or living things upon which the virus lives. For this reason, many scientists consider viruses an unusual form of life. Other scientists strongly disagree and do not classify viruses as living things. Thus, it might help if you think of viruses as being on the threshold of life.

Sharpen Your Skills

Discovery of Viruses

Using reference materials in the library, prepare a report on the discovery of viruses. In your report include the work of Dimitri Iwanowski, Martinus Beijerinck, and Wendell Stanley. Describe their work and how it benefited the science of microbiology.

Figure 5–1 *The red spherical structures (left) are viruses that cause influenza. The rod-shaped objects (right) are tobacco mosaic viruses. What type of microscope is used to study viruses?*

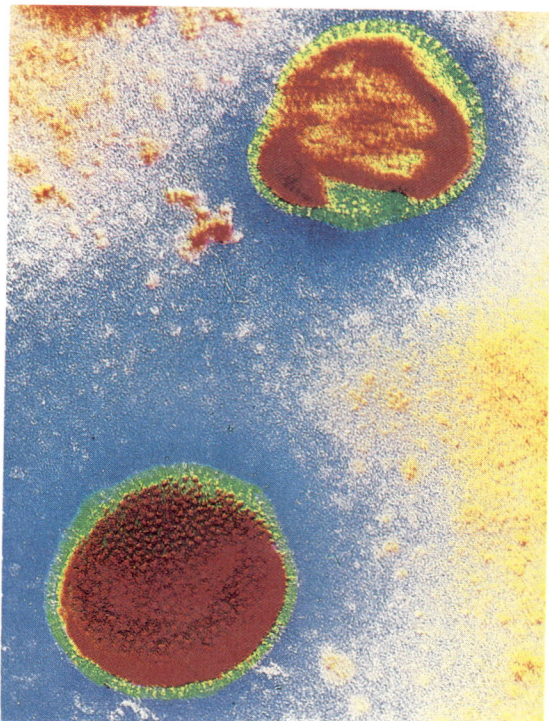

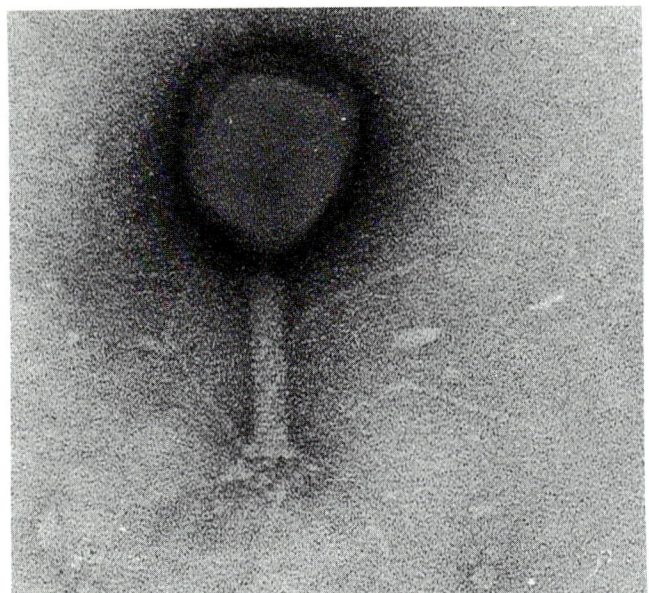
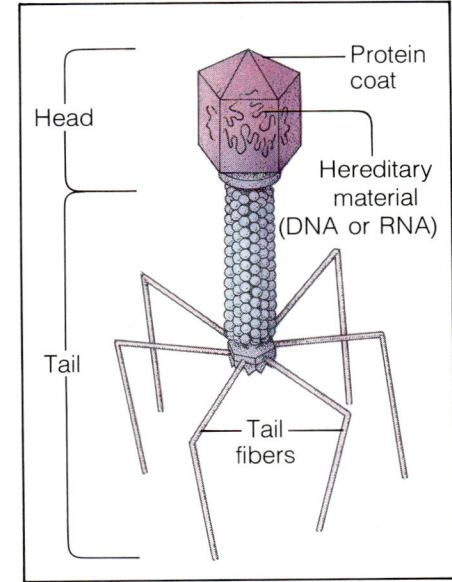

Figure 5–2 *This drawing shows the structure of a bacteriophage, which attacks bacterial cells. The photograph is of the same virus magnified 190,000 times. What structures are contained in the head portion of the bacteriophage?*

Structures of Viruses

A virus has two basic parts: a core of hereditary material and an outer coat of protein. The hereditary material in the virus's core may be either **DNA,** deoxyribonucleic acid, or **RNA,** ribonucleic acid. DNA and RNA are **nucleic acids.** Nucleic acids control the production of new viruses.

Surrounding the DNA or RNA in the virus is a protein coat. The coat is much like the shell of a turtle. Like the turtle's shell, the protein coat encloses and offers protection to the virus. In fact, the protein coat is so protective that some viruses survive after being dried and frozen for years.

With the invention of the electron microscope in the 1930s, scientists were able to see and to study the shapes and sizes of certain viruses. Some viruses, such as those that cause the common cold, have 20 surfaces. Each of these surfaces is in the shape of a triangle that has equal sides. Other viruses look like small threads, and still others resemble small spheres. Some even resemble small spaceships. See Figure 5–3. How many times larger is the bacterium than the smallpox virus?

Figure 5–3 *Viruses vary in size from about 10 to 250 nanometers (nm). A nanometer is equal to one millionth of a millimeter. How many times larger is the smallpox virus than the conjunctivitis virus?*

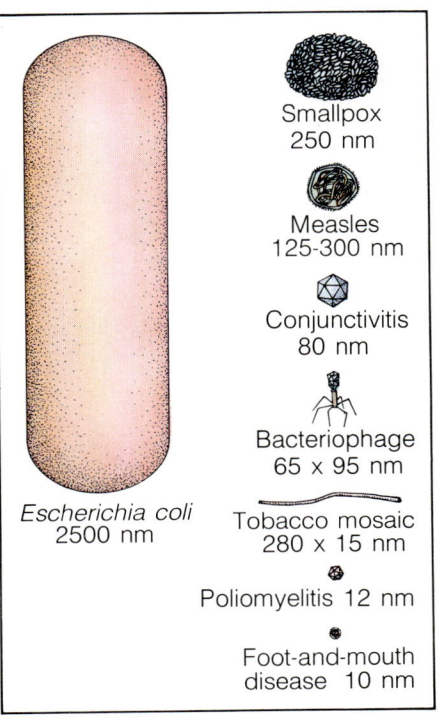

Escherichia coli 2500 nm

Smallpox 250 nm

Measles 125-300 nm

Conjunctivitis 80 nm

Bacteriophage 65 x 95 nm

Tobacco mosaic 280 x 15 nm

Poliomyelitis 12 nm

Foot-and-mouth disease 10 nm

Reproduction of Viruses

In order to understand how a virus reproduces, causing disease, you can examine the activities of one kind of virus called a **bacteriophage** (bak-TEER-ee-uh-fayj). A bacteriophage is a virus that infects **bacteria** (singular: bacterium). Bacteria are unicellular microorganisms. The word *bacteriophage* means "bacteria eater."

In Figure 5–4, you can see how a bacteriophage attaches its tail to the outside of a bacterium. The virus quickly injects its hereditary material directly into the living cell. The protein coat is left behind and discarded by the virus. Once inside the cell, the virus's hereditary material takes control of all of the bacterium's activities. As a result, the bacterium is no longer in control. The bacterium begins to produce new viruses rather than its own chemicals.

Soon the bacterium fills up with new viruses, perhaps as many as several hundred. Eventually, the bacterium bursts open. The new viruses are released and infect nearby bacteria. This continues until all living bacteria have been infected by the virus.

The viruses that attack plants, animals, and people may vary in size and shape. But all viruses act in much the same way as bacteriophages. In Chapter 18, you will learn more about viruses that cause diseases in people.

Figure 5–4 *Viruses reproduce in cells. A virus attaches to the cell and injects its hereditary material. The cell then makes more viruses, which burst from the cell. What part of the virus is injected into the cell?*

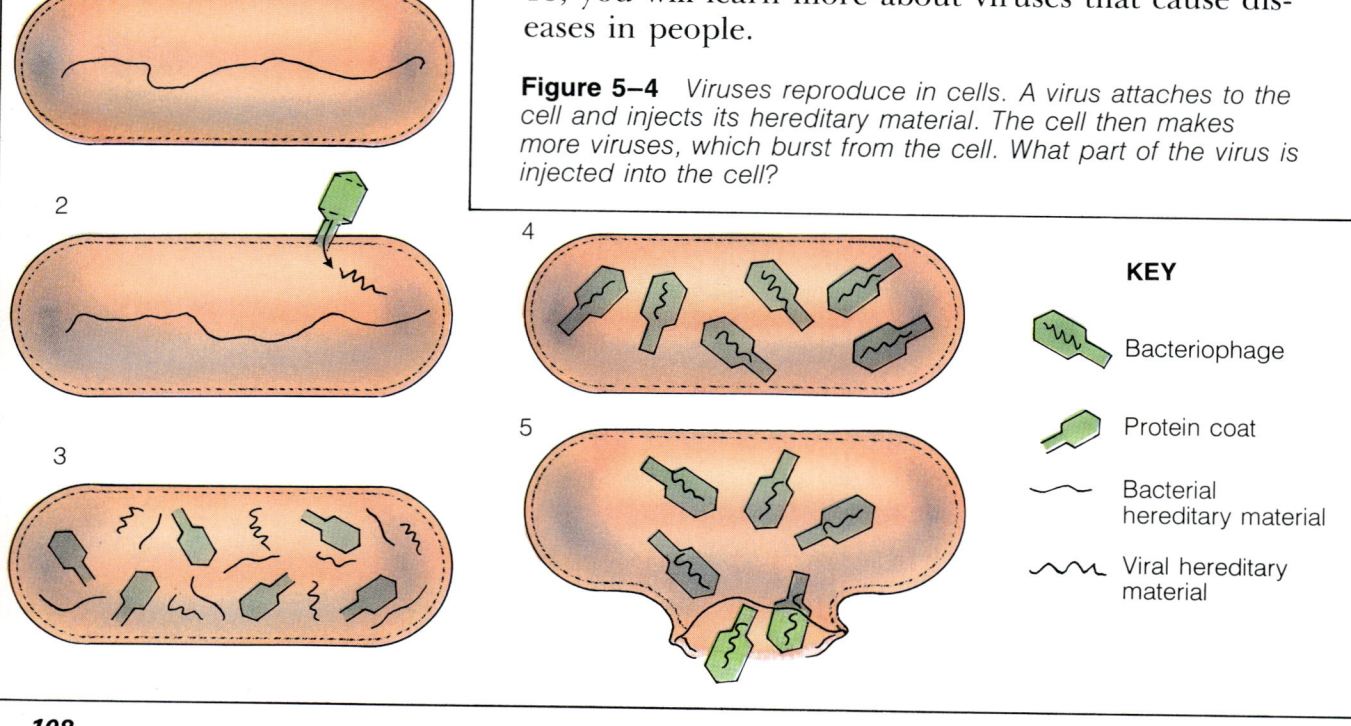

KEY

Bacteriophage

Protein coat

Bacterial hereditary material

Viral hereditary material

SECTION REVIEW

1. Describe the structure of a virus?
2. Would you classify viruses as living or nonliving? Explain.
3. What is a bacteriophage?
4. How does a virus reproduce?
5. Why were scientists not able to study the structure of viruses until after the electron microscope was invented?

5–2 Bacteria

Section Objective

To describe the structures and life functions of bacteria

Unlike viruses, bacteria clearly are living organisms. At one time, scientists placed bacteria in the plant kingdom because, like plants, bacteria have **cell walls.** The cell wall is the outermost boundary of plant and bacterial cells. The cell wall helps the cells of these organisms keep their shape. **All bacteria have a cell wall, cell membrane, cytoplasm, and hereditary material.** Today most classification systems consider bacteria as **monerans.** You may recall from Chapter 4 that monerans are unicellular organisms that do not have a nucleus. Monerans are members of the Monera kingdom.

Bacteria are among the most numerous organisms on earth. Scientists estimate there are about 2.5 billion bacteria in a gram of garden soil. And the total number of bacteria living in your mouth is greater than the number of people who have ever lived.

Bacteria are considered the simplest organisms. However, bacteria are more complex than they may appear. Each bacterial cell performs the same basic functions that more complex organisms, including you, perform.

Scientists classify bacteria into three basic groups according to shape. Bacteria that resemble spheres are called **cocci** (KAHK-sigh; singular: coccus). Those bacteria that look like rods are called **bacilli** (buh-SIHL-igh; singular: bacillus). Still others resemble spirals. These bacteria are called **spirilla** (spigh-RIHL-uh; singular: spirillum). See Figure 5–5 on page 110 for the shapes of these bacteria.

Sharpen Your Skills

Observing Bacteria

1. Obtain some plain yogurt. Add water to the yogurt to make a very thin mixture.
2. With a medicine dropper, place a drop of the yogurt mixture on a glass slide.
3. With another dropper, add one drop of methylene blue to the slide.
4. Carefully place a cover slip over the slide.
5. Observe the slide under the low and high power of the microscope.

Describe what you see under the microscope. What are the shapes of the bacteria? Why do you think you had to use methylene blue?

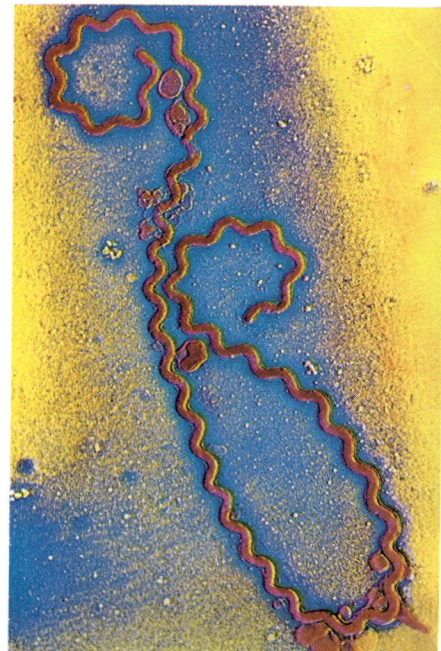
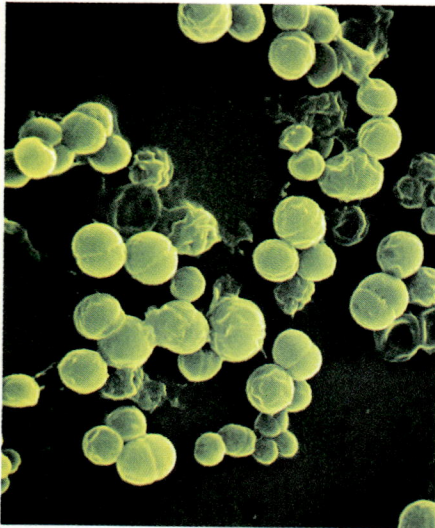
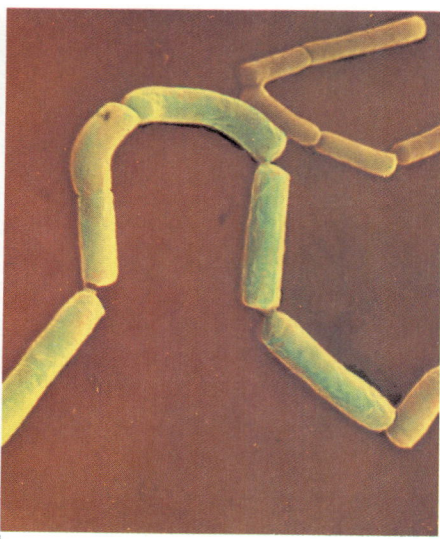

Figure 5–5 *Bacteria are grouped and named according to basic shape. Spiral-shaped bacteria (left) are called spirilla, spherical bacteria (center) are called cocci, and rodlike bacteria (right) are called bacilli. To which kingdom do bacteria belong?*

Structure of Bacteria

All bacteria have a **cell membrane** on the inside of the cell wall. The cell membrane controls which substances enter and leave the bacterium. In some bacteria, there is a coating on the outside of the cell wall. This coating is called the capsule. How might the capsule provide protection for a bacterium?

Unlike most other cells, the hereditary material of bacteria is not confined in a **nucleus** (NOO-klee-uhs; plural: nuclei). The nucleus is a spherical struc-

Figure 5–6 *The drawing shows the typical structure of a bacterium. The bacterium* Proteus mirabilis, *magnified 17,650 times, is shown in the photograph. What are the long, thin, whiplike structures called?*

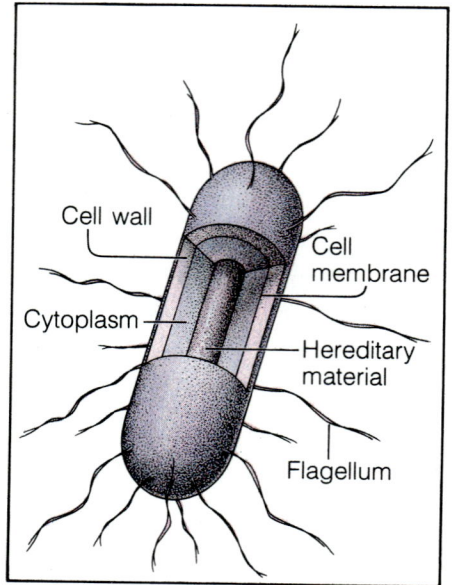

Cell wall
Cell membrane
Cytoplasm
Hereditary material
Flagellum

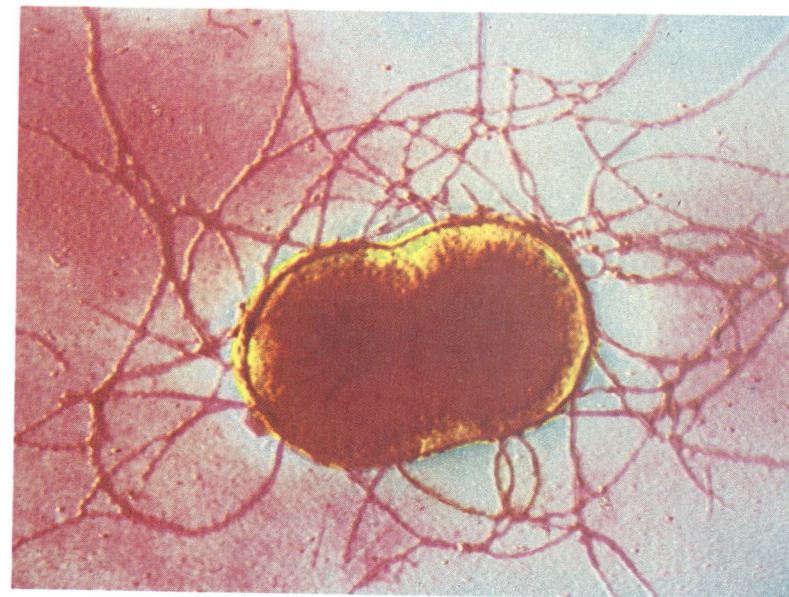

HELP WANTED: BACTERIOLOGIST
College degree required; Master's or
Ph.D. desirable. Strong background in
science necessary. Applicant should be
creative and innovative. Position involves
testing and research. Past experience as
lab helper/technician an asset.

The specially equipped plane landed at the
Atlanta airport. Within minutes, white-coated
technicians rushed the test tube the plane was
carrying to the Centers for Disease Control.
There, under strict laboratory conditions, the
contents of the test tube were analyzed.
Through the powerful lens of the microscope, a
tiny rod-shaped object became visible. Was it a
living organism? The answer soon became obvious. Twenty minutes later a second rod-
shaped object became visible under the lens of
the microscope. The organism had reproduced.

Now a more important question had to be
answered. Could this rod-shaped bacteria cause
disease in human beings? Was it related to a
type of bacteria that had caused diseases such
as diphtheria, typhoid fever, and plague in the
past? A **bacteriologist** would determine the

type of bacteria it was and whether it was dangerous. In San Francisco, a sick patient and her
doctor waited eagerly for the results.

Many bacteriologists specialize in identifying
unknown microorganisms from the 2000 or so
known types of bacteria. Others try to devise
methods to combat harmful bacteria. Still others study how disease-causing bacteria may be
spread in our environment. For further information on this career, write to the American
Society of Microbiology, 1913 I Street N.W.,
Washington, DC 20006.

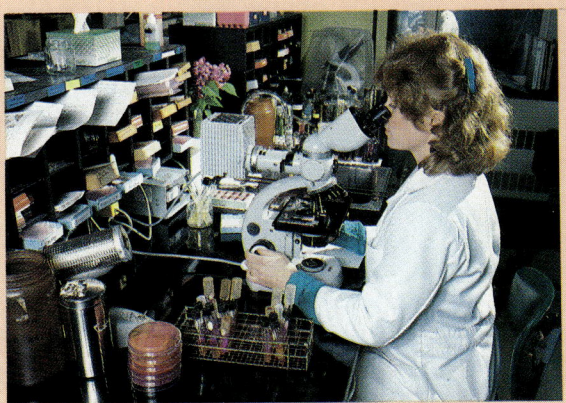

ture within a cell that directs all the activities of the
cell. Instead, the hereditary material is spread
throughout the **cytoplasm** in the cell. The cytoplasm
is the jellylike material inside the cell membrane.

Many bacteria are not able to move on their
own. They can be carried from one place to another
by air and water currents, clothing, and other objects. Other bacteria have special structures such as
flagella (fluh-JEHL-uh; singular: flagellum) that help
them move in watery surroundings. A flagellum is a
long, thin, whiplike structure that propels the bacterium. Some bacteria may have many flagella.

Life Functions of Bacteria

Bacteria are in water, air, and the upper layers
of soil. In fact, bacteria live almost everywhere, even
in places where other living things cannot survive.
For example, some bacteria were discovered living

Sharpen Your Skills

Shapes of Bacteria

1. Obtain three prepared
slides containing bacteria.

2. Observe each slide under
low and high power of the
microscope.

Draw and label a few cells
from each slide. Identify each
slide as containing the following bacteria shapes: cocci,
bacilli, or spirilla.

Figure 5–7 *Bacteria of decay break down material in a dead tree and return important substances to the soil. Are such bacteria parasites or saprophytes?*

in volcanic vents at the bottom of the Pacific Ocean south of Baja, California. Temperatures here are as high as 250°C. Unlike most living things, some bacteria can thrive without oxygen. However, all bacteria need water or they eventually die.

Most bacteria are **heterotrophs** (HEHT-uhr-oh-trafs). A heterotroph is an organism that feeds on other living things or on dead things. Bacteria that feed on living organisms are called **parasites** (PAR-uh-sights). These bacteria cause infections in people, animals, and plants. Bacteria that feed on dead things are called **saprophytes** (SAP-ruh-fights). Why are saprophytes among the most important organisms on earth?

Some bacteria are **autotrophs** (AWT-uh-trahfs). Autotrophs are organisms that can make their own food. Like green plants, some of these food-making bacteria use the energy of sunlight to produce food. Other bacteria use the energy in substances such as sulfur and iron to make food.

From time to time, conditions may become unfavorable for bacteria. If the food, water, and air supply of bacteria is used up, some bacteria form **endospores.** An endospore is a spherical structure that has a thick outer coat. The endospore protects the DNA inside it. The thick outer coat enables the bacteria to survive long periods during which the environment is not suitable for bacterial growth. Endospores, for example, can protect some bacteria from boiling water and disinfectant chemicals. When the environment improves and food, air, and water

Figure 5–8 *These bacteria are reproducing by splitting in two, a process called binary fission. To which basic group do these bacteria belong?*

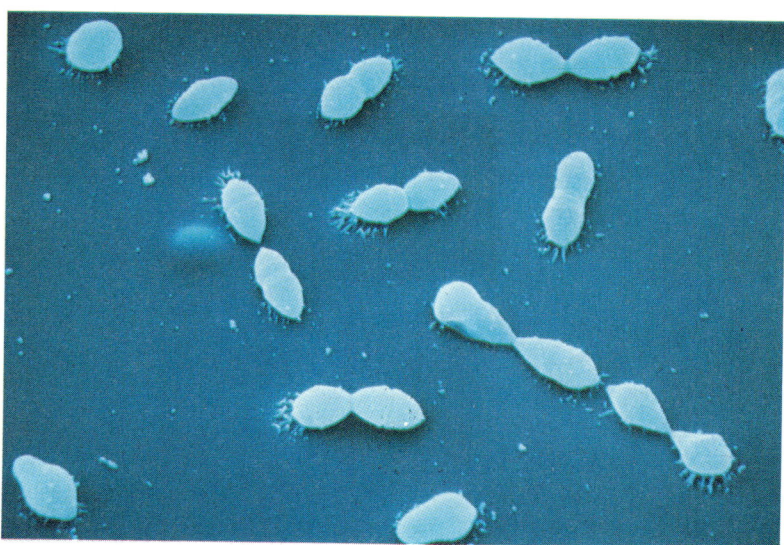

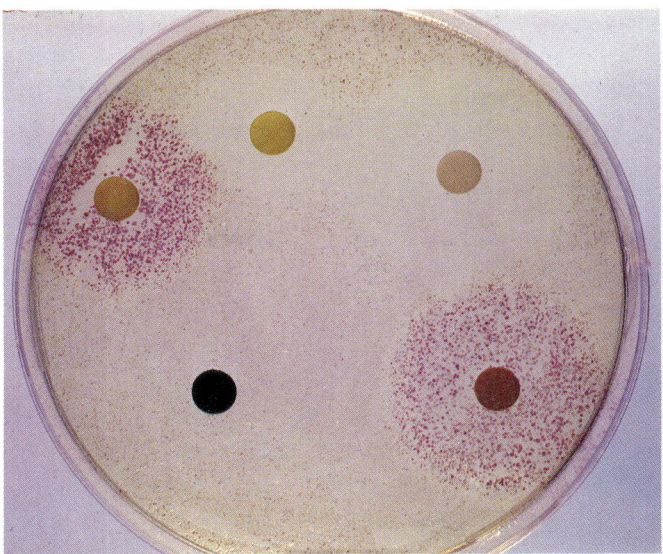

again become available, the endospores develop into active bacteria.

Most bacteria reproduce by **binary fission.** In binary fission, a cell divides into two cells. Under the best conditions, most bacteria reproduce quickly. Some types can double in number every 20 minutes. At this rate, after about 24 hours the offspring of a single bacterium would have a mass greater than 2 million kilograms, or as much as 2000 small cars. In a few more days, their mass would be greater than that of the earth. Obviously, this does not happen. Why do you think this does not happen?

Helpful Bacteria

Most types of bacteria are not harmful and do not cause disease. For example, many food products, especially dairy products, are produced with the help of bacteria. Bacteria are needed to make cheeses, butter, yogurt, and sour cream. Some bacteria are used in the process of tanning leather.

Some helpful bacteria are used to fight other, harmful bacteria. These helpful bacteria can produce **antibiotics.** Antibiotics are chemicals that destroy or weaken disease-causing bacteria. Scientists also have found a way to turn certain types of bacteria into "chemical factories." This process is called **genetic engineering.** Through genetic engineering, the DNA within these bacteria is changed. The bacteria then produce large amounts of substances such

Figure 5–9 *One of the products of helpful bacteria is antibiotics. Each colored disk in the petri dish (left) contains a particular antibiotic. As you can see, some of the antibiotics are effective in destroying harmful bacteria. Other products of helpful bacteria are produced through genetic engineering. The vats (right) contain hundreds of liters of genetically engineered bacteria.*

113

Figure 5-10 *The tiny round structures on the roots of this bean plant contain nitrogen-fixing bacteria. What is the function of these bacteria?*

as insulin. Insulin helps to control the rate at which the human body breaks down sugar. People who do not produce enough of their own insulin can use the insulin produced by the bacteria. Can you think of any other uses of bacteria?

Certain bacteria, called nitrogen-fixing bacteria, can make nitrogen compounds from nitrogen gas in the atmosphere. Nitrogen-fixing bacteria turn nitrogen gas, which plants cannot use for food, into nitrogen compounds that plants can use for food. These nitrogen-fixing bacteria often appear as lumps on the roots of plants such as alfalfa. See Figure 5-10. Nitrogen-fixing bacteria also help replace the nitrogen compounds in the soil. Without such nitrogen-fixing bacteria, most nitrogen compounds in the soil would be quickly used up and plants could no longer grow.

Harmful Bacteria

Although most bacteria are harmless, and some are helpful, a few can cause trouble. The trouble comes in a number of forms—food spoilage, diseases of people, diseases of farm animals and pets, and diseases of food crops. Fortunately, there are a number of defenses against attacks by harmful bacteria.

FOOD SPOILAGE Food spoilage can be prevented or slowed down by heating, drying, salting, cooling, or smoking foods. Each one of these processes prevents or slows down the growth of bacteria. Milk, for example, is heated to 71°C for 15 seconds before it is placed in containers and shipped to the grocery or supermarket. This process, called **pasteurization,** destroys most of the bacteria that would cause the milk to spoil quickly. Heating and then canning foods such as vegetables, fruits, meat, and fish are also used to prevent bacterial growth. But if the foods are not sufficiently heated before canning, bacteria can grow inside the can and produce poisons called **toxins.**

HUMAN DISEASES Some human diseases caused by bacteria include strep throat, certain kinds of pneumonia, diphtheria, tuberculosis, and whooping

Figure 5–11 *Leaves on this soybean plant have been attacked by a bacterial disease called blight.*

cough. Some of these diseases can be prevented by an immunization shot. Others can be treated with antibiotics. You will learn more about human bacterial diseases in Chapter 18.

ANIMAL DISEASES Bacteria cause diseases in animals. These diseases include anthrax, which attacks sheep, horses, cattle, and other large farm animals, and fowl cholera, which attacks chickens. Vaccinations can prevent both diseases.

PLANT DISEASES Some plant diseases caused by bacteria include fire blight of apples and pears and soft rot of vegetables. There are many ways of treating plant diseases, including the use of antibiotics and various chemicals.

SECTION REVIEW

1. Describe the structure of a bacterium.
2. Name and describe the three basic shapes of bacteria.
3. What is an endospore? How does it help protect bacteria?
4. Describe two ways in which bacteria are helpful and two ways in which they are harmful.
5. Certain bacteria produce antibiotics that slow the growth of other bacteria. Explain why this action is an advantage to the antibiotic-producing bacteria.

LABORATORY INVESTIGATION

Examining Bacteria

Problem

Where are bacteria found?

Materials *(per group)*

5 petri dishes with sterile nutrient agar
glass-marking pencil
pencil with eraser
soap and water

Procedure

1. Turn each petri dish bottom side up on the table. **Note:** *Be careful not to open the petri dish.*
2. With the glass-marking pencil, label the bottom of the petri dishes containing the sterile nutrient agar A to E. Turn the petri dishes right side up.
3. Remove the lid of dish A and lightly rub a pencil eraser across the petri dish. Close the dish immediately.
4. Remove the lid of dish B and leave it open to the air until the class period ends. Then close the lid.
5. Remove the lid of dish C and lightly rub your index finger over the surface of the agar. Then close the lid.
6. Wash your hands thoroughly. Remove the lid of dish D and lightly rub the same index finger over the surface of the agar. Then close the lid.
7. Do not open dish E.
8. Place all five dishes upside-down in a warm, dark place for three or four days.
9. After three or four days, examine each dish. **CAUTION:** *Do not open dishes.*
10. On a sheet of paper, construct a table similar to the one on this page. Fill in the table.
11. Return the petri dishes to your teacher. Your teacher will properly dispose of the petri dishes.

Observations

1. How many colonies, or similar types of bacteria, appear to be growing on each petri dish? How can you distinguish between different bacterial colonies?
2. Which petri dish has the most bacterial growth? Which has the least?

Petri Dish	Source	Description of Bacterial Colonies
A		
B		
C		
D		
E		

Conclusions

1. Which petri dish was the control?
2. Did the dish that you touched with your unwashed finger contain more or less bacteria than the one that you touched with your washed finger? Explain.
3. Explain why the agar was sterilized before the investigation.
4. Design an experiment to show if a particular antibiotic will inhibit bacteria grown in a petri dish.
5. Suggest some methods that might stop the growth of bacteria.
6. What kinds of environmental conditions seem to influence where bacteria are found?

CHAPTER REVIEW

SUMMARY

5–1 Viruses

❑ Viruses cannot carry out any life processes unless within a living cell called a host.

❑ Viruses have two basic parts: a core of hereditary material and an outer coat of protein.

❑ The hereditary material in the core of viruses may be either DNA, deoxyribonucleic acid, or RNA, ribonucleic acid.

❑ Surrounding the DNA or RNA in the virus is a protein coat.

❑ With the invention of the electron microscope in the 1930s, scientists were able to see and to study the shapes and sizes of certain viruses.

❑ Reproduction of viruses occurs only when the virus invades a living cell.

❑ Bacteriophages are viruses that infect bacteria.

❑ When a virus comes into contact with a living cell, reproduction of the virus begins. The virus injects its hereditary material directly into the living cell. The protein coat is left behind and is discarded by the virus.

❑ Once inside the cell, the virus's hereditary material takes control of all the cell's activities. The cell begins to produce new viruses, which are released from the cell when the cell bursts. These viruses infect nearby cells.

5–2 Bacteria

❑ Bacteria are simple unicellular monerans.

❑ Bacteria are classified into three groups according to shape: sphere-shaped cocci, rod-shaped bacilli, and spiral-shaped spirilla.

❑ Bacteria have a cell wall and a cell membrane surrounding the cytoplasm. The hereditary material of bacteria is spread throughout the cytoplasm.

❑ Some bacteria have flagella, or whiplike structures, that enable them to move.

❑ Bacteria that must obtain their food from other organisms are called heterotrophs. Those that feed on living things are parasites. Those that feed on dead things are saprophytes.

❑ Some bacteria called autotrophs use energy from the sun or chemicals to make food.

❑ When conditions become unfavorable for some bacteria, they form thick protective membranes called endospores.

❑ Bacteria usually reproduce by binary fission, or splitting in two.

❑ Many bacteria are helpful and have uses in the food industry, in medicine, in tanning leather, and in agriculture.

❑ Some bacteria are harmful and cause food spoilage and diseases in people, animals, and plants.

VOCABULARY

Define each term in a complete sentence.

antibiotic	cell wall	genetic engineering	parasite
autotroph	coccus	heterotroph	pasteurization
bacillus	cytoplasm	host	RNA
bacteriophage	DNA	moneran	saprophyte
bacterium	endospore	nucleic acid	spirillum
binary fission	flagellum	nucleus	toxin
cell membrane			virus

CONTENT REVIEW: MULTIPLE CHOICE

On a separate sheet of paper, write the letter of the answer that best completes each statement.

1. Viruses contain
   a. DNA.   b. RNA.   c. either DNA or RNA.   d. neither DNA nor RNA.
2. Viruses are
   a. cylinder shaped.   b. tadpole shaped.   c. sphere shaped.   d. any of these shapes.
3. A virus that infects a bacterium is called a (an)
   a. parasite.   b. bacillus.   c. endospore.   d. bacteriophage.
4. The outer coat of a virus is made up of
   a. hereditary material.   b. protein.   c. DNA.   d. RNA.
5. Bacteria belong to the kingdom
   a. Monera.   b. Plantae.   c. Protista.   d. Fungi.
6. Rod-shaped bacteria are called
   a. spirilla.   b. endospores.   c. bacilli.   d. cocci.
7. Sphere-shaped bacteria are called
   a. spirilla.   b. endospores.   c. bacilli.   d. cocci.
8. The jellylike material inside the cell membrane is called the
   a. endospore.   b. flagellum.   c. cytoplasm.   d. antibiotic.
9. The whiplike structures that propel bacteria are called
   a. flagella.   b. bacilli.   c. endospores.   d. spirilla.
10. Bacteria that feed on living things are called
   a. autotrophs.   b. saprophytes.   c. parasites.   d. cocci.

CONTENT REVIEW: COMPLETION

On a separate sheet of paper, write the word or words that best complete each statement.

1. Viruses are composed of an outer protein coat and a (an) _____.
2. DNA and RNA are _____.
3. Viruses can be observed by using a (an) _____ microscope.
4. _____ feed on dead things.
5. To survive unfavorable conditions, a bacterium may form a (an) _____.
6. Most bacteria reproduce by _____.
7. _____ are chemicals that destroy or weaken disease-causing bacteria.
8. The process by which scientists turn certain bacteria into chemical factories is known as _____.
9. _____ is the process by which milk is heated to kill bacteria.
10. The poisons produced by some bacteria are called _____.

CONTENT REVIEW: TRUE OR FALSE

Determine whether each statement is true or false. Then on a separate sheet of paper, write ''true'' if it is true. If it is false, change the underlined word or words to make the statement true.

1. Viruses <u>are</u> cells.

2. A <u>virus</u> has a protein coat.

118

3. A bacteriophage injects <u>nucleic acids</u> into a bacterium.
4. A <u>host</u> is a living thing upon which a virus lives.
5. Rod-shaped bacteria are called <u>cocci</u>.
6. <u>All</u> bacteria have flagella.
7. <u>Heterotrophs</u> are organisms that make their own food.
8. <u>Parasites</u> feed on dead things.
9. <u>Antibiotics</u> are poisons produced by some bacteria.
10. Bacteria are <u>monerans</u>.

CONCEPT REVIEW: SKILL BUILDING

Use the skills you have developed in the chapter to complete each activity.

1. **Making diagrams** Draw a diagram showing a bacteriophage infecting a bacterium. Label each structure. What structure aids in the attachment of the bacteriophage to the bacterium? What structures enter the bacterium? What structures remain outside the bacterium?
2. **Relating concepts** Explain why it is more difficult to grow poliomyelitis viruses than it is to grow the bacilli that cause diphtheria.
3. **Designing an experiment** Penicillin is an antibiotic that kills certain types of bacteria but not others. If you had five different bacteria cultures, how could you find out which cultures could be destroyed by penicillin? What controls would you use?
4. **Designing an experiment** Design an experiment to show that the growth of bacteria is slowed down by cooler temperatures. Be sure you have a control and a variable in your experiment. How can you apply this information in order to control food spoilage?
5. **Interpreting a graph** The graph shows the growth of bacteria. Describe the growth of the bacteria using the numbers given for each growth stage. Hypothesize why the growth leveled off in stage 3 and then fell in stage 4.

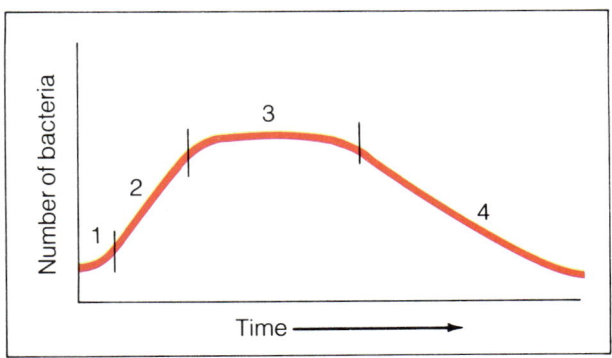

CONCEPT REVIEW: ESSAY

Discuss each of the following in a brief paragraph.

1. Explain the differences between viruses and cells.
2. Describe the structure of a bacteriophage.
3. Briefly explain how viruses reproduce.
4. Can viruses be grown in the laboratory on synthetic material? Explain.
5. Explain two ways in which bacterial growth could be slowed down.
6. Explain how bacteria are able to get from one place to another.
7. What is the difference between bacteria that are heterotrophs and those that are autotrophs?
8. Explain how low temperatures can prevent food from spoiling.
9. Explain why some bacteria are harmful.
10. Explain why a farmer may choose to plant a field of alfalfa every few years.
11. What diseases can bacteria cause?

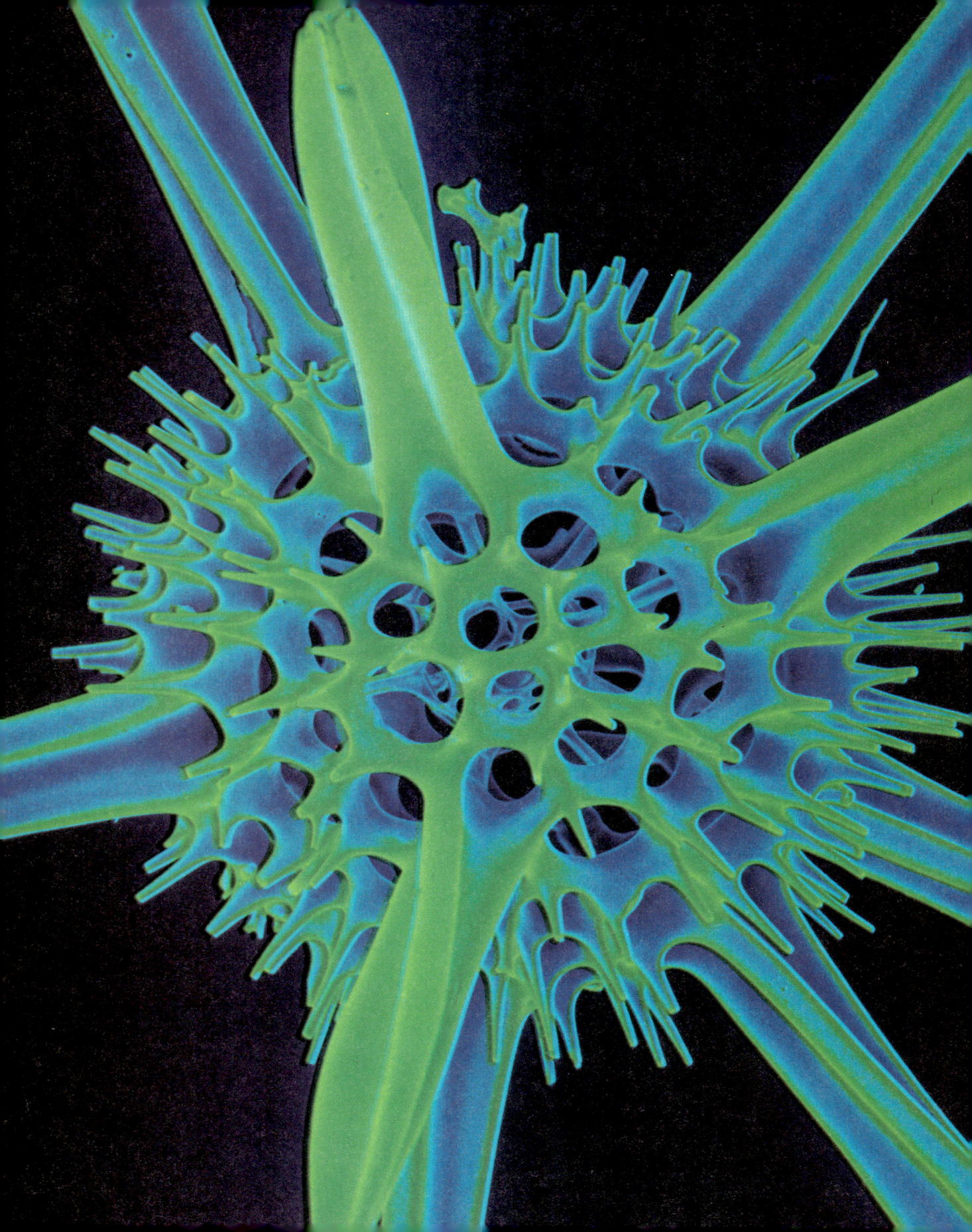

Protozoans

6

CHAPTER OBJECTIVES

After completing this chapter, you will be able to

6–1 Identify the characteristics of protozoans.

6–1 Relate the method of movement to the classification of protozoans.

6–2 Compare the movement of sarcodines and ciliates

6–2 Describe the structures and functions of the amoeba and the paramecium.

6–3 Distinguish between flagellates and sporozoans.

6–3 Describe the *Euglena.*

Two billion years ago the earth was a strange and barren place. No animals roamed the land, swam in the ocean, or flew in the air. No trees, shrubs, or grasses grew from the soil. From the air, the earth looked gray, brown, and blue. But even though the earth looked lifeless, an unseen new life was developing in the blue waters. This form of life, the protozoans, represented a giant step in the parade of living things that would follow.

Today two billion years later, many forms of life have appeared and disappeared from the face of the earth. Yet the protozoans still inhabit the world's oceans, lakes, rivers, and ponds.

One such inhabitant, a radiolarian, is pictured on the opposite page. Actually what you are seeing is the lacy, glasslike shell that houses this one-celled creature. The complicated and beautifully sculptured shell is made of the mineral silica, a component of sand. The long, needlelike spines are used to catch food. When a tiny organism accidentally touches one of the spines, other spines quickly join in the attack and help trap the unknowing victim.

Interestingly, when radiolarians die, their glassy skeletons sink to the ocean floor. In many parts of the ocean, millions of these skeletons litter the floor. In fact, one area in the eastern part of the Pacific Ocean north of the equator contains about 750 thousand square kilometers of radiolarian skeletons!

Radiolarians are only one group of protozoans. So read on and find out more about the other groups of fascinating one-celled life forms.

This photograph of a radiolarian's skeleton has been magnified thousands of times.

6–1 Characteristics of Protozoans

The **protozoans** are members of the Protista kingdom. For this reason, protozoans are often called protists. **Protozoans are unicellular, animal-like organisms.** Like animals, protozoans cannot make their own food and must get it from their surroundings. Interestingly, the term *protozoan* comes from the Greek words for "first animal."

Protozoans differ in many ways from another group of unicellular organisms—the bacteria. Bacteria, as you may recall, are members of the Monera kingdom. For one thing, protozoans do not have cell walls. But they do have a nucleus, which contains hereditary material. Protozoans are also larger than bacteria. Yet most protozoans cannot be seen without the aid of a microscope.

Most protozoans live in a watery environment. They can be found in salty oceans and in bodies of fresh water. Some protozoans live in moist soil. Others live inside other organisms and feed off them. These types of protozoans can cause serious diseases in animals and human beings.

Like many bacteria, most protozoans reproduce by **binary fission.** Binary fission is the splitting of a cell into two identical cells. In other cases, two protozoans may share their hereditary material in a process called **conjugation** (kahn-juh-GAY-shuhn).

Figure 6–1 *Protozoans are unicellular, animallike organisms. Like animals, protozoans cannot make their own food and must get it from their surroundings. To which kingdom do protozoans belong?*

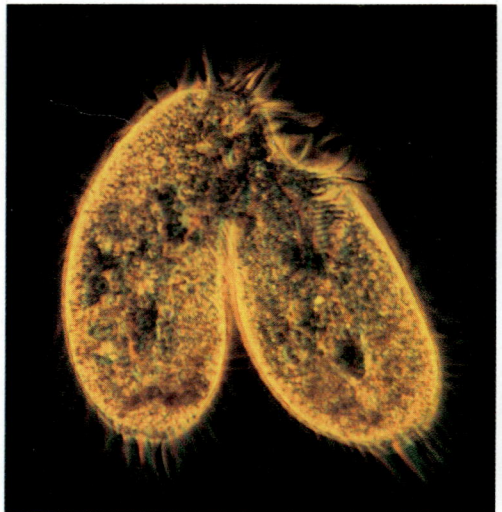
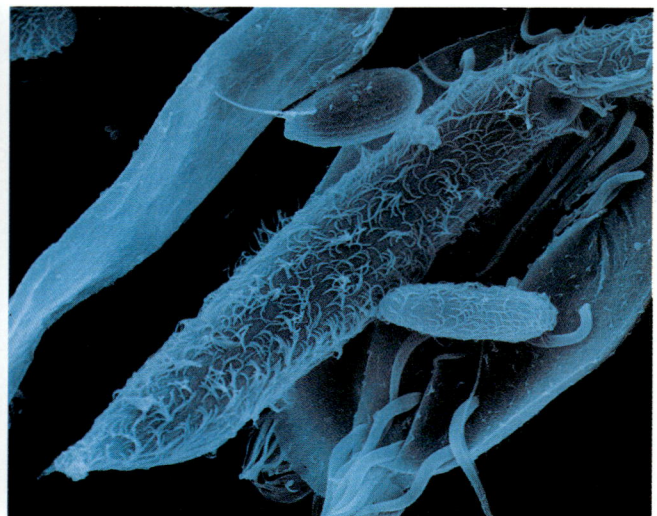

During conjugation, the two protozoans join together and exchange hereditary material. A third type of reproduction involves the production of **spores.** Spores are tiny reproductive cells that develop into protozoans.

Most protozoans have the ability to move from place to place. In fact, protozoans are divided into four main groups based upon the way they move about. These four groups are the **sarcodines** (SAHR-kuh-deenz), the **ciliates** (SIHL-ee-ayts), the **flagellates** (FLAJ-uh-layts), and the **sporozoans** (SPOR-uh-zoh-uhnz). The following sections will discuss the characteristics of these four protozoan groups.

SECTION REVIEW

1. What are the characteristics of protozoans?
2. Compare bacteria and protozoans.
3. List and describe three methods of reproduction in protozoans.
4. Explain why protozoans are not classified in the animal kingdom.

6–2 Sarcodines and Ciliates

Section Objective

To classify sarcodines and ciliates by their method of movement

Sarcodines and ciliates are protozoans that have different ways of moving. **The sarcodines move by means of pseudopods, and the ciliates move by means of cilia.** As sarcodines move, they extend their cytoplasm and cell membrane. These extensions are called **pseudopods** (soo-duh-pahdz), or "false feet." When the pseudopod is fully extended, it pulls the sarcodine along.

Many sarcodines have shell-like outer coverings. For example, the foraminiferans (fuh-ram-uh-NIHF-uhr-uhnz) have a chambered outer shell made of calcium carbonate, or limestone. The shell of the foraminiferan has many tiny pores, or openings. The pseudopods move in and out of these pores as the organism feeds.

The ciliates have small hairlike projections called **cilia** (SIHL-ee-uh; singular: cilium) on the outside of their cells. The cilia act like tiny oars and help these organisms move. The beating of the cilia also helps sweep food toward the ciliates. In addition, the cilia function as sensors. When the cilia are touched, the ciliate receives information about its environment. How does this help the organism?

Cilia may cover the entire surface of a ciliate or may be concentrated in certain areas. For example, one type of ciliate has rows of cilia on the outside of its cell. These cilia beat forward or backward causing the organism to move in a spiraling motion.

Amoeba: A Typical Sarcodine

The most familiar sarcodine to most people is the bloblike **amoeba.** Amoebas are found in fresh water and move by means of pseudopods. They also use the pseudopods to obtain food. As the amoeba nears a small piece of food, such as a smaller protozoan, the amoeba extends its pseudopod around the food. Soon the food particle is engulfed, or surrounded, by the pseudopod.

A spherical **food vacuole** (VAK-yoo-wohl) in the pseudopod forms around the engulfed food particle. Like a human stomach, the food vacuole releases special digestive chemicals, called enzymes, that break down the captured food. The digested food can then be used by the amoeba. The waste products left behind in the food vacuole are eliminated through the cell membrane.

To supply itself with energy, the amoeba requires oxygen as well as food. Oxygen passes from the watery environment into the amoeba through the cell membrane. At the same time, waste prod-

Figure 6–2 *The foraminiferan (left) is a sarcodine that has an outer shell made of calcium carbonate, or limestone. Ciliates (right) have small hairlike projections on the outside of their cells. What are these hairlike projections called?*

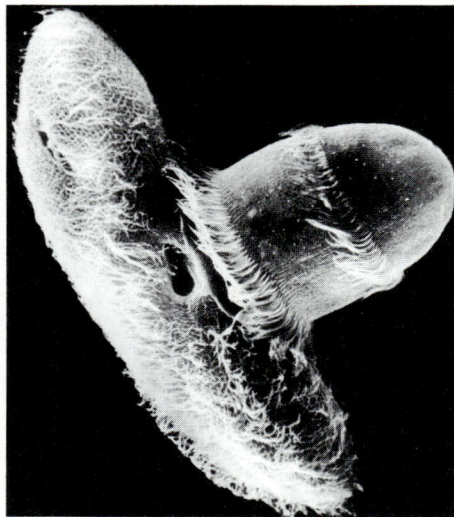

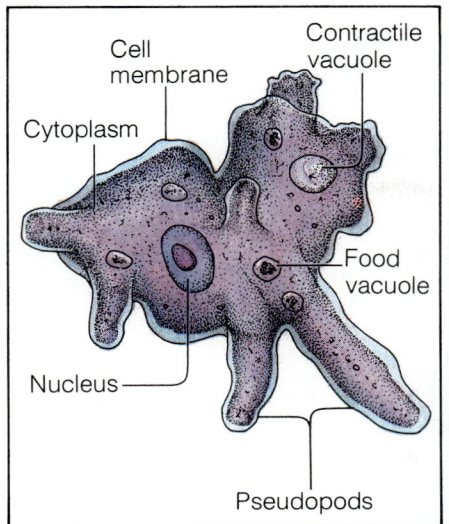

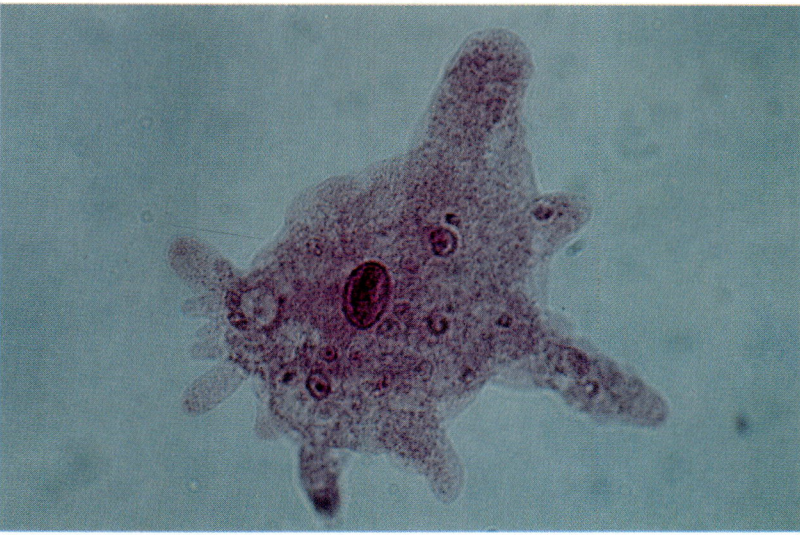

Figure 6–3 *This drawing shows the structure of a typical amoeba. The photograph of the amoeba was magnified 100 times. How do amoebas move?*

ucts, such as carbon dioxide, pass out of the amoeba through the cell membrane. Excess water in the amoeba is pumped out through the cell membrane by a **contractile vacuole.** As you might expect, a contractile vacuole contracts to force excess water out of the amoeba. What do you think would happen to the amoeba if the excess water was not pumped out?

Like bacteria, amoebas reproduce by binary fission. A parent cell divides into two new identical cells. Each of these new cells has the same amount and kind of hereditary material as the parent cell.

Amoebas respond simply to changes in their environments. They are sensitive to bright light and move to areas of dim light. Amoebas are also sensitive to certain chemicals, moving away from some and toward others.

Paramecium: A Typical Ciliate

One of the most interesting ciliates is the **paramecium** (par-uh-MEE-see-uhm; plural: paramecia). A paramecium has a hard membrane called the **pellicle** (PEHL-ih-kuhl) that covers its outer surface. This membrane gives the paramecium its slipper shape. See Figure 6–5 on page 126. The cilia of the paramecium move food particles in the water into its **oral groove.** The oral groove is an indentation in the paramecium. Food goes from the oral

Figure 6–4 *This drawing shows the wavelike motion of a cilium.*

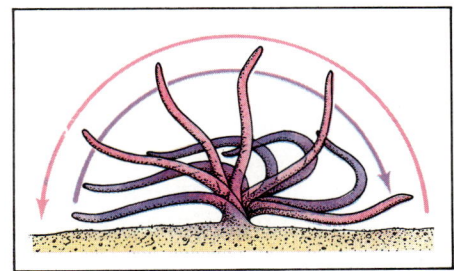

125

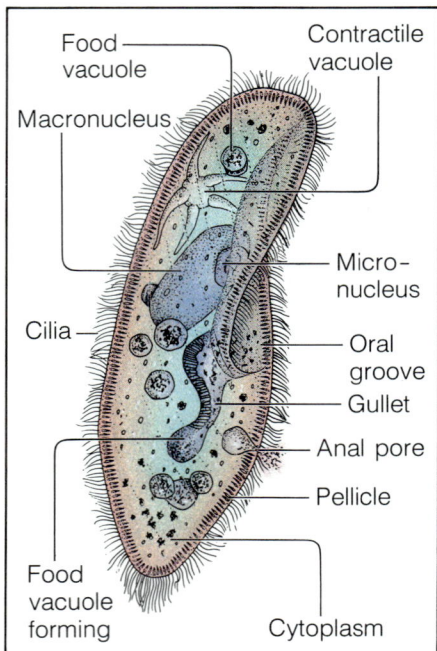

Figure 6–5 *This drawing shows the structure of a typical paramecium. The photograph of the paramecium was magnified 140 times. Which structure controls reproduction?*

Labels in the drawing:
Food vacuole
Contractile vacuole
Macronucleus
Micro-nucleus
Cilia
Oral groove
Gullet
Anal pore
Pellicle
Food vacuole forming
Cytoplasm

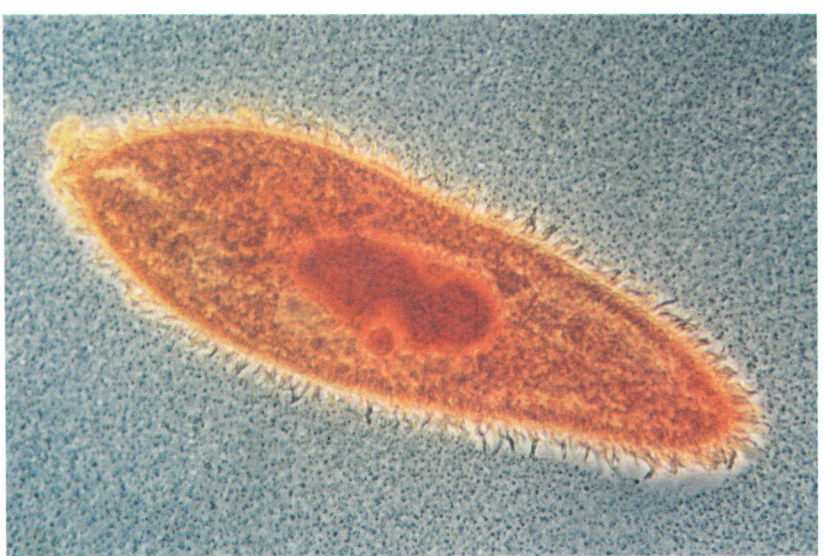

groove into the **gullet.** The gullet is a funnel-shaped structure that ends at the food vacuole. The food vacuole distributes food particles as it moves through the organism. Undigested material is eliminated through the **anal pore.**

Located in the cytoplasm of paramecia are two nuclei. The **micronucleus,** or small nucleus, controls reproduction. The **macronucleus,** or large nucleus, controls all other life functions.

Reproduction in paramecia occurs in two ways. Like an amoeba, a paramecium reproduces by binary fission. A paramecium may also share its hereditary material with another paramecium through conjugation. During conjugation, two paramecia join together. Soon the macronuclei disappear and the micronuclei divide in two. One of the micronuclei from each pair passes into a special tube joining the paramecia. After exchanging the hereditary material, the two paramecia separate and move away from each other. A new micronucleus and a new macronucleus form in each of the two paramecia. Conjugation is beneficial because it allows paramecia to share hereditary characteristics that may help the paramecia become better able to survive a changing environment.

SECTION REVIEW

1. Compare the methods of movement in sarcodines and ciliates.

2. Compare the way in which food is obtained in the amoeba and paramecium.
3. Explain how amoebas and paramecia reproduce. How do these methods of reproduction differ?
4. Design an experiment to show an amoeba's response to salt.

6–3 Flagellates and Sporozoans

Section Objective

To classify flagellates and sporozoans by their method of movement

The two remaining groups of protozoans are the flagellates and the sporozoans. **The flagellates move by means of flagella. Because the sporozoans have no means of movement, they are carried from place to place by their hosts.**

Flagellates include protozoans that have **flagella** (fluh-JEHL-uh; singular: flagellum), or whiplike structures. These structures propel the organism through its watery environment. Most of the flagellates are unicellular. But some flagellates form colonies, or clusters of cells.

Unlike the other protozoans, sporozoans have no pseudopods, cilia, or flagella by which to move. All sporozoans are **parasites.** A parasite is an organism that lives off of living organisms called **hosts.** Parasites absorb all their food from their hosts.

Flagellates

There are two groups of flagellates. One group contains food-making structures called **chloroplasts.** Chloroplasts contain the green substance chlorophyll. These flagellates are autotrophs because they can make their own food. *Volvox* is an example of this type of flagellate. See Figure 6–6.

The second group of flagellates does not contain chloroplasts. They are heterotrophs because they cannot make their own food. Most of these heterotrophic flagellates are parasites. Some of them cause diseases, such as African sleeping sickness. The flagellates that cause African sleeping sickness are transmitted to people and animals by the tiny tsetse fly.

A flagellate that is both an autotroph and a heterotroph is *Euglena. Euglena* is a tiny oval-shaped

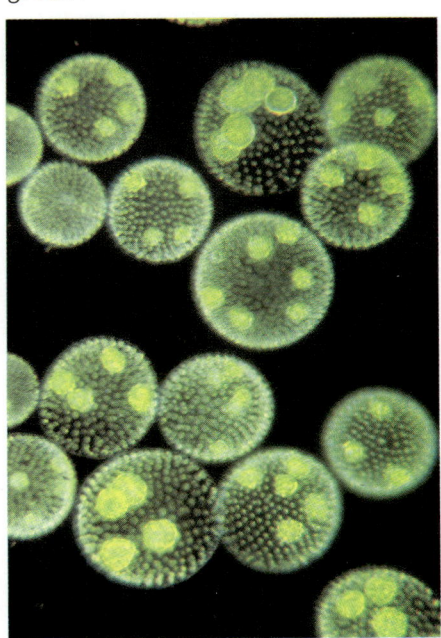

Figure 6–6 Volvox *are flagellates that form colonies. Volvox are autotrophs because they can make their own food. Why are Volvox green?*

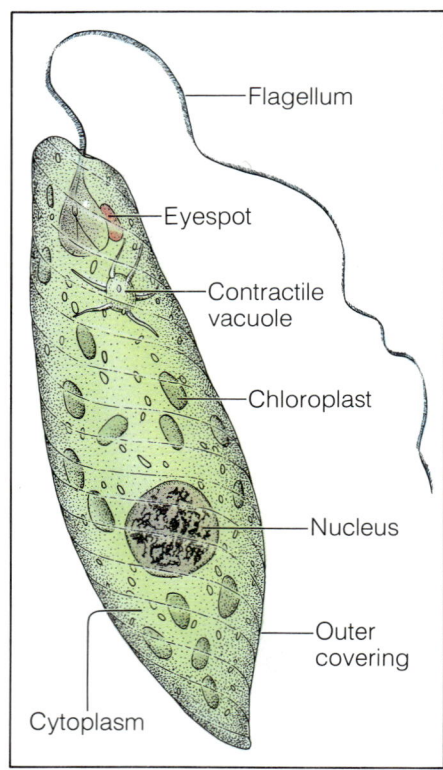

Flagellum

Eyespot

Contractile vacuole

Chloroplast

Nucleus

Outer covering

Cytoplasm

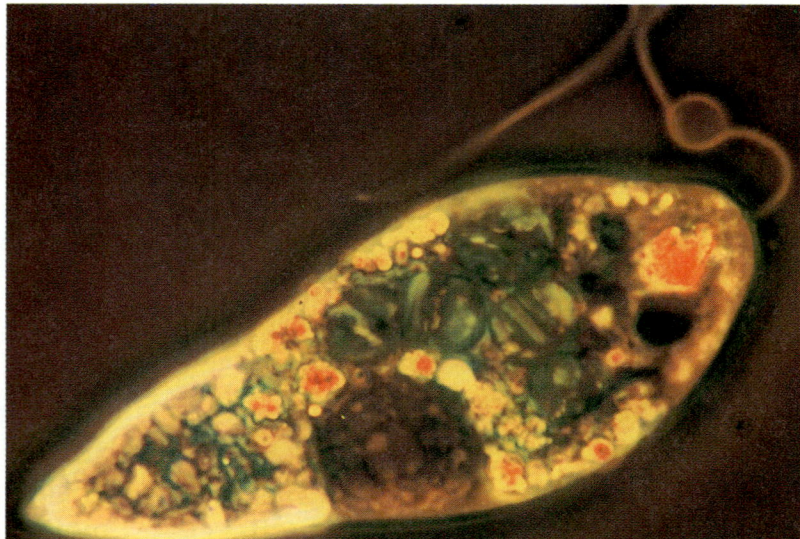

Figure 6–7 *This drawing shows the structure of the autotrophic flagellate called* Euglena. *The photograph is of a typical* Euglena *magnified 500 times. How does the* Euglena *move?*

Figure 6–8 *The sporozoans that cause malaria are carried by the* Anopheles *mosquito. How do sporozoans differ from other protozoans?*

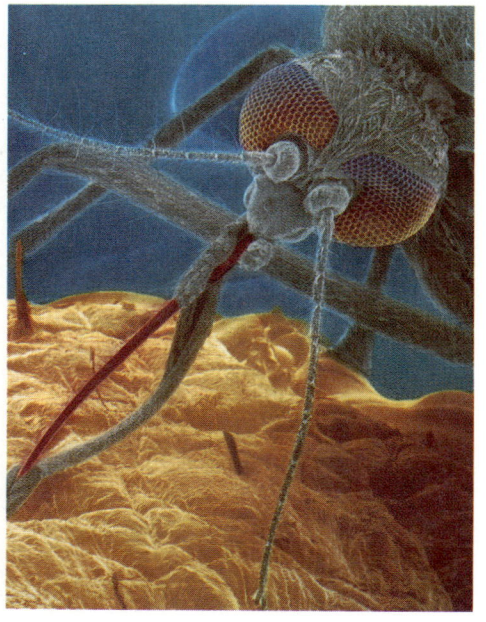

organism with one pointed end and one rounded end. The rounded end contains the flagellum. Also located in this area in the cytoplasm is the **eyespot.** The eyespot is a reddish structure that is sensitive to light. *Euglena* tends to move toward areas where there is enough light. *Euglena* use light energy to make their own food.

Like the amoeba and paramecium, *Euglena* reproduces by binary fission. But, unlike most protozoans, *Euglena* splits in two lengthwise. This type of division produces two cells that are mirror images of each other.

Sporozoans

Many sporozoans have complicated life cycles. During these life cycles, sporozoans form spores. Each spore contains hereditary material and a small amount of cytoplasm. Eventually, the sporozoans release the spores. Each spore can develop into a mature sporozoan.

Perhaps the most familiar type of sporozoan is the organism that causes the disease malaria. This organism is known as *Plasmodium* (plaz-мoн-dee-uhm). *Plasmodium* is carried by the *Anopheles* mosquito. See Figure 6–8.

When an infected mosquito bites a human being, the mosquito injects its saliva, and the *Plasmodium*

HELP WANTED: LABORATORY TECH-NICIAN Position involves preparing water and food samples for microscope study and identifying microorganisms that cause disease. Completion of a formal training program or an associate degree program required.

People who perform complicated micro-scopic tests in laboratories are called **laboratory technicians.** They prepare blood and other body fluid samples by adding stains or dyes. After this is done, they examine the slide for abnormal cell growth or the presence of a parasite. Then the laboratory technician pre-pares a report on his or her findings.

Laboratory technicians also work in veteri-nary hospitals and help in the detection of infec-tious diseases in pets and farm animals. Some may work in agriculture to help study the ef-fects of microorganisms on farm crops. Other laboratory technicians work for water and waste-treatment plants that use microorgan-isms to break down wastes into harmless sub-stances. They help to make sure that the water that leaves the treatment plants is safe.

If you are interested in this career, write to the National Association of Trades and Techni-cal Schools, Accrediting Commission, 2021 L Street, N.W., Washington, D.C. 20036.

sporozoans enter the person's bloodstream. The sporozoans pass into the person's liver, where they form spores. They then invade red blood cells, and grow and multiply. Eventually they burst through the red blood cells, destroying them. See Figure 6–9. This bursting releases more spores into the blood. If the infected person is then bitten by an-other mosquito, the spores along with the blood enter that mosquito. In the mosquito, the spores develop into sporozoans and the cycle begins again.

Several medicines have been found effective in treating people with malaria. One such medicine is quinine (KWIGH-nighn). Can you suggest a way to stop the spread of malaria?

SECTION REVIEW

1. Compare the methods of movement in flagellates and sporozoans.
2. What is the function of the eyespot in *Euglena?*
3. If all the chloroplasts were removed from a *Euglena,* do you think it would survive? Explain.

Figure 6–9 *Once inside the body, the malaria-carrying sporozoans invade the red blood cells, grow, and multiply. Soon the sporozoans burst through the red blood cells and destroy them. What do spores contain?*

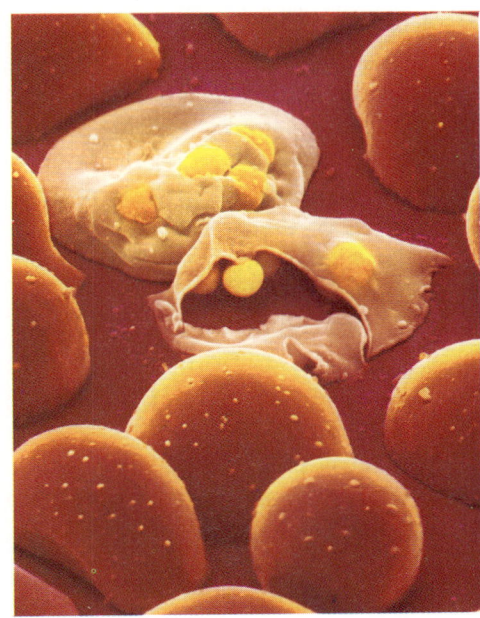

6 LABORATORY INVESTIGATION

Collecting Protozoans

Problem

What are the characteristics of protozoans?

> **Materials** *(per group)*
>
> large glass jar
> tap water
> medicine dropper
> 3 microscope slides and 3 cover slips
> microscope
> cotton fibers
> glass-marking pencil
> graph paper

Procedure

1. Collect leaves and small stones from around your neighborhood.
2. Loosely pack the leaves and stones into the large glass jar. Then add tap water so that the leaves and stones are barely covered.
3. Set the jar near a window, but not in direct sunlight.
4. With the glass-marking pencil, number the slides from 1 to 3. Place each number in the lower left-hand corner of the slide. Place a few cotton fibers on each slide.
5. Using the medicine dropper, take a water sample from the top of the jar. Place a drop of water on top of the cotton fibers on slide 1. Cover with a cover slip.
6. Then take a water sample from the middle of the jar and place a drop on slide 2. From the bottom of the jar, take a water sample and place it on slide 3. Cover the slides with cover slips.
7. Examine the slides under the low and high powers of the microscope.
8. Repeat steps 5 to 7 once a day for a total of 10 days.

Observations

1. How many different kinds of protozoans were you able to observe? Draw and label each type of protozoan.
2. How did each of the ten protozoans move?
3. Draw a graph that shows the number of protozoans seen each day.

Conclusions

1. Using reference material, identify the ten protozoans that you drew.
2. What characteristics do all protozoans have in common?
3. Devise a way to count the number of protozoans that you see with the microscope.
4. Where would you expect to find autotrophic protozoans? Give a reason for your answer.
5. What is the function of the cotton fibers?

Large jar

Tap water

Medicine dropper

Leaves and stones

Cotton fibers and water

Cotton fibers

Glass slide

Cover slip

CHAPTER REVIEW

SUMMARY

6-1 Characteristics of Protozoans

❑ Protozoans are members of the Protista kingdom. Protozoans are unicellular animallike organisms.

❑ Unlike bacteria, protozoans have a nucleus and no cell wall.

❑ Protozoans reproduce by binary fission, conjugation, or by producing spores.

❑ Protozoans are divided into four groups: the sarcodines, the ciliates, the flagellates, and the sporozoans.

6-2 Sarcodines and Ciliates

❑ Sarcodines, such as amoebas, move by means of pseudopods, or extensions of cytoplasm and cell membrane.

❑ Amoebas obtain food by extending their pseudopods around the food. A food vacuole forms and releases enzymes that break down the captured food.

❑ Oxygen passes from the watery environment into the amoeba through the cell membrane. Waste products pass out through the cell membrane. Excess water is pumped out through the cell membrane by the contractile vacuole.

❑ Amoebas reproduce by binary fission.

❑ Ciliates, such as paramecia, have hairlike cilia used for obtaining food and for movement.

❑ The cilia are used to sweep food particles into the paramecium's oral groove. Food then moves into the gullet, which ends in a food vacuole. Undigested material is eliminated through the anal pore.

❑ Paramecia reproduce by binary fission and conjugation.

6-3 Flagellates and Sporozoans

❑ Flagellates use whiplike flagella to move.

❑ The flagellates are divided into two groups. One group has chloroplasts and makes its own food. The other group does not have chloroplasts and obtains food from its environment.

❑ The *Euglena* is an example of a flagellate that is both a heterotroph and an autotroph.

❑ The *Euglena* has a light-sensitive eyespot and moves toward areas of bright light.

❑ *Euglena* reproduce by binary fission.

❑ The sporozoans have no means of movement. They are carried from place to place by their hosts.

❑ All sporozoans are parasites. Sporozoans have complex life cycles in which they reproduce by means of spores. Each spore can develop into a mature sporozoan.

❑ *Plasmodium* is a common sporozoan that causes malaria in human beings.

VOCABULARY

Define each term in a complete sentence.

amoeba	contractile vacuole	host	protozoan
anal pore		macronucleus	pseudopod
binary fission	eyespot	micronucleus	sarcodine
ciliate	flagellate	oral groove	spore
cilium	flagellum	paramecium	sporozoan
chloroplast	food vacuole	parasite	
conjugation	gullet	pellicle	

CONTENT REVIEW: MULTIPLE CHOICE

On a separate sheet of paper, write the letter of the answer that best completes each statement.

1. Protozoans are classified according to their method of
   a. reproduction.   b. getting food.   c. movement.   d. eliminating wastes.
2. Which sarcodine has a glassy shell?
   a. amoeba   b. sporozoan   c. radiolarian   d. foraminiferan
3. Paramecia move by means of
   a. flagella.   b. pseudopods.   c. cilia.   d. sporozoans.
4. The cellular structure that releases enzymes to digest food is the
   a. food vacuole.   b. gullet.   c. oral groove.   d. pseudopod.
5. The hard membrane that covers the outer surface of the paramecium is the
   a. cell membrane.   b. pellicle.   c. gullet.   d. contractile vacuole.
6. In the paramecium, food is swept into an indented structure called a (an)
   a. gullet.   b. vacuole.   c. oral groove.   d. anal pore.
7. In the paramecium, undigested material is eliminated through the
   a. oral groove.   b. anal pore.   c. eyespot.   d. gullet.
8. A flagellate that is both an autotroph and a heterotroph is the
   a. amoeba.   b. sporozoan.   c. paramecium.   d. *Euglena*.
9. The structure in *Euglena* that responds to changes in light is the
   a. eyespot.   b. flagellum.   c. chloroplast.   d. nucleus.
10. *Plasmodium* is transmitted by
    a. the air.   b. a mosquito.   c. water.   d. people.

CONTENT REVIEW: COMPLETION

On a separate sheet of paper, write the word or words that best complete each statement.

1. Amoebas move by _____.
2. Pseudopod means _____.
3. A (An) _____ is a sarcodine with a calcium carbonate shell.
4. A paramecium contains two _____.
5. In the paramecium, the _____ controls reproduction.
6. _____ reproduce by binary fission and conjugation.
7. In *Euglena,* chlorophyll is contained in structures called _____.
8. *Euglena* is an example of a (an) _____.
9. _____ are a group of protozoans that cannot move by themselves.
10. The most familiar type of sporozoan causes _____.

CONTENT REVIEW: TRUE OR FALSE

Determine whether each statement is true or false. Then on a separate sheet of paper, write ''true'' if it is true. If it is false, change the underlined word or words to make the statement true.

1. Unlike bacteria, protozoans have a distinct <u>cell wall</u>.
2. Protozoans belong to the <u>Monera</u> kingdom.
3. <u>Pseudopod</u> means ''first animal.''
4. The <u>oral groove</u> pumps excess water out of an amoeba.

5. Amoebas reproduce by conjugation.
6. In the paramecium, the gullet is the funnel-shaped structure that ends in a food vacuole.
7. In the paramecium, the macronucleus controls reproduction.
8. Cilia are the whiplike structures that enable *Euglena* to move.
9. Sporozoans are all parasites.
10. During their life cycles, sporozoans form spores.

CONCEPT REVIEW: SKILL BUILDING

Use the skills you have developed in the chapter to complete each activity.

1. **Applying concepts** Explain why it would not be wise to drink water from a pond.
2. **Making charts** *Entamoeba histolytica, Stentor coeruleus,* and *Trypanosoma gambiense* are examples of protozoans. Find out the following information about each organism: phylum to which it belongs, presence or absence of cell membrane, nucleus, vacuoles, and chloroplasts, color, method of movement, and shape. Arrange your information in the form of a chart. Include a labeled diagram of each protozoan.
3. **Making comparisons** Using prepared slides, compare the following protozoans: *Stentor, Chilomonas, Blepharisma,* and *Pelomyxa (Chaos chaos).* Record your observations in a chart that includes the shape, internal structure, method of movement, method of obtaining food, and method of reproduction for each protozoan. Which of these characteristics can you observe on the prepared slides? Which requires further research?
4. **Designing an experiment** A student is given the materials listed below and told to show the response of paramecia to certain substances such as cotton fibers and vinegar.

> medicine droppers
> paramecium culture
> microscope
> cover slips

Help the student out by describing an experiment to determine the paramecia's response to each substance. Be sure that your experimental design allows you to answer the following questions: What should the response of the paramecia to the cotton fibers and vinegar be? What should the variables be? What should the control be?

CONCEPT REVIEW: ESSAY

Discuss each of the following in a brief paragraph.

1. How does an amoeba obtain its food?
2. Compare conjugation and binary fission in paramecia.
3. The sporozoans have no means of movement. How have they managed to live for thousands of years?
4. Describe the two functions of cilia in the paramecium.
5. Describe how reproduction occurs in the amoeba, paramecium, and *Euglena.*
6. Suppose a biologist reported the discovery of a sarcodine that moves by means of pseudopods and has a thick cell wall made of cellulose. Could this report be true? Explain.

Adventures in Science

Claire Veronica Broome:

DISEASE DETECTIVE

Ask Claire Veronica Broome about her work at the Centers for Disease Control in Atlanta, Georgia, and she is apt to break into a satisfied grin. "It's very exciting," she says, "because the answers you get are practical." Along with other scientists in her team, Dr. Broome travels to different locations to study outbreaks of disease.

One of the most confusing cases this "disease detective" has ever solved was an outbreak of listeriosis in Halifax, Nova Scotia. Listeriosis is a disease that affects membranes around the brain. This disease is caused by a type of bacterium called *Listeria monocytogenes*. Often the disease affects the elderly. However, it can infect an infant before it is born. In fact, it was an epidemic of listeriosis in newborn infants that brought this case to Dr. Broome's attention.

Before going to Halifax, Dr. Broome gathered as much data on listeriosis as possible. She discovered that the bacteria that cause listeriosis were identified as a cause of human disease in 1929. At that time, it was learned that the bacteria live in soil and can infect animals. But as late as 1981, no one was certain how the bacteria are transmitted to unborn humans.

When Claire Broome and her team of researchers arrived in Halifax, they set to work reviewing hospital records and talking to physicians about past occurrences of listeriosis. Soon, they discovered the hospital in Halifax was experiencing an epidemic.

Dr. Broome compared two groups of people. The mothers of infants born with listeriosis made up one group. Mothers who gave birth to healthy babies made up the second group. People in both groups were interviewed. Dr. Broome collected data that covered several months of the new mothers' lives before they gave birth. Dr. Broome soon noticed a definite trend. The women with sick babies had eaten more cheeses than the women with healthy babies. Could the cheese have been contaminated by the bacteria that made the babies ill?

During her research, Dr. Broome became aware of another case of listeriosis. This case occurred in a Halifax man who had not spent time in a hospital. After examining the contents of the man's refrigerator, Dr. Broome added coleslaw to her list of possible transmitters of listeriosis. She reasoned that coleslaw is made from uncooked cabbage, and cabbage grows in soil—a place in which listeria bacteria had already been discovered. Careful chemical analysis revealed that listeria bacteria were indeed present in the coleslaw found in the sick man's refrigerator.

Dr. Broome and her team traced some of the contaminated coleslaw to a cabbage farm. They found that the farmer used sheep manure to fertilize the cabbage. When the sheep were tested, they were also found to contain the listeria bacteria.

Now it was time for Dr. Broome to interview the mothers of the sick babies again. Not surprisingly, she found that many of them had eaten coleslaw during the time they were carrying their child. Later Dr. Broome discovered that the bacteria that causes listeriosis can also live in milk and milk products, including cheese.

What made the packaged coleslaw and the milk products a good environment for the bacteria to live in? Actually it was the refrigerator these foods were stored in that contributed to this disease outbreak. Refrigerators are used to keep foods from spoiling quickly. However, *Listeria monocytogenes* can survive, and even multiply, in cold temperatures. Even under proper storage conditions, the coleslaw and the cheese contained enough bacteria to make people ill.

The mystery of the Halifax listeriosis outbreak was solved. This case was closed. Dr. Broome returned to her office in Altanta, but she barely had time to unpack before she was off to discover the cause of another "mystery" disease.

In her laboratory, Dr. Broome uses a computer to analyze data about epidemics.

Issues in Science

ANTIBIOTICS IN ANIMAL FEED:

Are They A Danger To Human Health?

When they were first discovered in 1928, antibiotics were hailed as "wonder drugs." Indeed they were a wonderful discovery. The powerful antibiotic drugs were capable of killing many kinds of bacteria, including bacteria that cause disease. Finally, a method had been discovered to effectively treat many life-threatening diseases caused by bacteria. Antibiotics have saved countless lives since their discovery in 1928. However, some scientists fear that antibiotics may now pose a serious danger to human health.

Today, many scientists are investigating the effects of antibiotic misuse. They have discovered that the widespread use of antibiotics in animal feed can create serious health problems for humans. How? The overuse of antibiotics in animal feed has resulted in the development of strains of bacteria that are resistant to antibiotics. A resistant strain develops when one bacterium is not affected by an antibiotic. Because bacteria reproduce quickly, this single bacterium produces other bacteria that are also not affected by the antibiotic.

You may be surprised to learn that all turkeys, 60 percent of all cattle, and 30 percent of the chickens raised in the United States are raised on feed that contains antibiotics. These antibiotics are added to animal feed to increase the rate of animal growth. Over time, some bacteria that are naturally present in the animals become resistant to antibiotics. And, as more antibiotics are used, more resistant kinds of bacteria develop. Today, scientists have found that more than 40 percent of the bacteria they have studied are resistant to at least one kind of antibiotic.

When meat reaches the supermarket, antibiotics and most bacteria are usually no longer present. However, resistant strains of bacteria may remain. These are the bacteria that concern scientists. Many of these bacteria can cause infection and disease. And because they are resistant to antibiotics, an infection or disease caused by these bacteria is very difficult to treat.

One might be quick to demand an end to the adding of antibiotics to animal feed. But critics say that there are enough benefits to justify the continued use of antibiotics in animal feed. They say that antibiotics are useful in preventing certain animal diseases. Left unchecked, such diseases could spread rapidly in a herd of animals. Farmers and ranchers would suffer economic losses if their herds died.

It is also argued that antibiotics help animals gain weight faster. A quick weight gain makes it less expensive to produce

Antibiotics added to animal feed keep the animals healthy and help them gain weight faster.

animals of market size. Farmers and ranchers are quick to point out that such reduced costs are eventually passed on to consumers.

However, the Food and Drug Administration, individual scientists, and an international organization called Alliance for the Prudent Use of Antibiotics urge that a ban be placed on some, if not all, antibiotic use in animal feeds. They stress that research in Great Britain has shown that the reduced use of antibiotics has decreased the number of resistant strains of bacteria. These scientists are convinced that the continued use of antibiotics in animal feed will cause new kinds of bacteria to develop. These new kinds of bacteria could cause untreatable diseases in people, diseases that might cause serious epidemics.

Clearly something must be done. Should we immediately ban the use of antibiotics in animal feed? Would you be willing to pay more for meat that would be more expensive to produce?

Plants

Early on a spring morning, a large pampered pig is pushed in its own wheelbarrow into an oak forest near Perigord, France. There its master gently puts the pig on a leash. The pig is now ready for the hunt!

Soon the pig catches a whiff of a wonderful odor—one that people cannot even smell. The pig begins to dig, but its master quickly stops it. He does not want the animal to destroy the buried treasure it has found. This "treasure" is not gold or silver. It is an ugly round black fungus known as a truffle. Because of its flavor, this thick-skinned, warty cousin of the mushroom is considered a luxury. In fact, truffles can sell for more than $1400.00 a kilogram!

Truffles are one of many plantlike organisms. Plants and plantlike organisms are used by people for food, clothing, medicine, and shelter. In Unit Three, you will learn more about some of the unusual—and not so unusual—plants that contribute to our daily lives. You will also discover why all animals, including people, could not exist without plants and plantlike organisms.

CHAPTERS

7 Nonvascular Plants and Plantlike Organisms

8 Vascular Plants

In France, people use pigs to aid in the hunt for truffles. Truffles (inset) are considered by many to be a rare and delicious treat.

Nonvascular Plants and Plantlike Organisms

7

CHAPTER OBJECTIVES

After completing this chapter, you will be able to

7–1 Define nonvascular plants.

7–1 Identify the characteristics of algae.

7–1 Classify algae into six groups.

7–1 Describe the process of photosynthesis.

7–2 Identify the characteristics of fungi.

7–2 Describe mushrooms, yeasts, and molds.

7–3 Describe lichens and slime molds.

7–4 Compare mosses and liverworts.

The time: 2.5 billion years ago. The place: the earth. An erupting volcano spews a tall column of smoke and ash into a cloudless sky. Fierce air currents blow in from the ocean bringing huge waves with them. As the waves move toward the shore, they crash against an enormous sand bar. The sand bar transforms the powerful waves into calm ripples, which gently flow into a large pool of water. There is no evidence of any life here, except for hundreds of rocklike columns scattered throughout the pool. Over 2.5 billion years later, scientists would discover that these columns were built by the earth's first living things.

Today scientists call the rocklike columns stromatolites. The word *stromatolite* comes from Greek and French words meaning "bed of stone." But what living thing could have formed these stone beds 2.5 billion years ago? The answer is a very simple form of life— the blue-green algae. Within the sheltered pools, the blue-green algae grew rapidly, and they produced lime, a rocky mineral. Over the years, the lime piled up and formed the stone beds. As you can see from the photograph on the opposite page, this process is still going on in certain places in the world such as in Shark Bay on the northwest coast of Australia. Now turn the page to read more about other types of algae and plantlike organisms.

These stromatolites in Shark Bay, Australia were built by blue-green algae about 2.5 billion years ago.

7–1 Algae

In a northern California forest, a giant redwood tree more than 200 years old stands through a violent rainstorm. Deep in the soil the tree's roots soak up water. Soon thin tubes in the tree's trunk will carry the water nearly 60 meters to the top of the tree. Without these tubes, the redwood could not bring water to its millions of living cells.

If you were walking through this forest, you would probably marvel at the height of the redwood tree. But unless you look closely, you might not notice a smaller plant growing on the bark near the bottom of the tree. Unlike the redwood, this plant cannot grow to majestic heights. It is one of the earth's **nonvascular** (nahn-VA-skyuh-luhr) **plants.** Nonvascular plants do not have transportation tubes to carry water and food throughout the plant.

The plant living on the bark of the redwood tree is a type of **alga** (AL-guh; plural: algae—AL-jee). **Algae are nonvascular plants that contain chlorophyll.** Because algae have no way of transporting water and food over long distances, they must live near a source of water and food.

Nonvascular plants such as the algae are often called simple plants, or plants that lack true stems, leaves, and roots. However, do not let the term *simple* fool you. Nonvascular plants have managed to

Figure 7–1 *The green patches floating on the surface of this lake (right) are colonies of algae. Other algae grow in the soil, on the sides of houses, and on the bark of trees (left). Why are algae classified as nonvascular plants?*

survive for hundreds of millions of years while other forms of life have come into being and then died off. In fact, through their long history, nonvascular plants have become well adapted to many different environments on the earth.

Algae live in many different places. Most algae live in watery environments such as oceans, ponds, and lakes. See Figure 7–1. Other algae grow in the soil, on the sides of houses, and at the base of trees. Certain species of algae grow in the near-boiling water of hot springs such as those in Yellowstone National Park. Other algae grow in snow as well as in the icy waters of such places as the Antarctic continent, home of the South Pole.

Structure of Algae

Some species of algae are **unicellular,** or one-celled, and can only be seen with a microscope. Quite often unicellular algae group together to form colonies. These colonies may even attach themselves to one another and form chains of algae.

Beaches often become littered with seaweed after a heavy storm. Seaweed is one type of **multicellular,** or many celled, algae. Multicellular algae can grow quite large. The giant kelp, for example, may stretch over 30 meters.

Whether unicellular or multicellular, all algae are autotrophs, or organisms that can make their own food. All algae contain **chlorophyll,** a green substance found in green plant cells. Chlorophyll is used in the process of **photosynthesis** (foh-tuh-SIHN-thuh-sihs). In photosynthesis, plants use the energy in sunlight to make their own food. Plants make food by combining carbon dioxide from the air with water and minerals from the soil. During the process of photosynthesis, oxygen is released from the plant.

Early in the earth's history there was very little oxygen in the atmosphere. In time, algae and other simple green plants, along with certain bacteria, released vast amounts of oxygen into the air. Eventually there was enough oxygen in the air to allow other forms of life to develop. What would happen if green plants were to suddenly vanish from the surface of the earth?

Figure 7–2 This photograph shows several kinds of kelp, a type of algae, washed up on a California beach.

Figure 7–3 Most types of blue-green algae form threadlike colonies. This photograph of Anabaena is magnified 400 times. Why is this alga green?

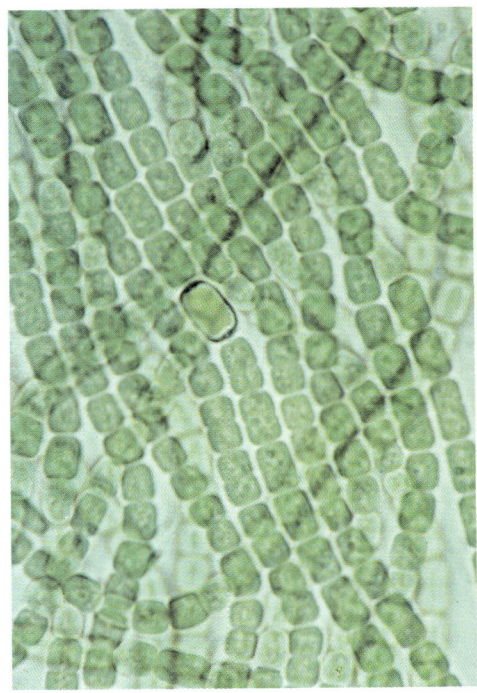

Most algae reproduce by binary fission. Others have very complicated life cycles. Regardless of these differences, scientists have placed the algae into six groups. The algae are classified according to their color.

Blue-Green Algae

In 1883, on the island of Krakatoa in Indonesia, a great volcano exploded. The sound of the explosion was heard by people thousands of kilometers away. In an instant, all living things on the island were destroyed, leaving behind only bare rock. Yet a few years later, one form of life had begun to grow on the barren island. This living thing was a blue-green algae.

Soon a layer of blue-green algae carpeted much of the island. The layer of algae eventually became so thick that plants were able to grow on top of it. Today, as they did before the volcanic eruption, many kinds of plants live and flourish on the island of Krakatoa.

Unlike other groups of algae, blue-green algae are not considered plants. They are placed in the Monera kingdom, along with bacteria. In fact, many scientists now call them the blue-green bacteria and do not consider them algae at all. All blue-green algae need to live is sunlight, nitrogen and carbon dioxide gas from the air, and a few minerals in their water supply. It is likely that they were the first organisms to grow on land.

Figure 7–4 *Algae are found in many environments on the earth. These blue-green algae (left) are in a hot springs in Yellowstone National Park, Wyoming. Some blue-green algae (right) grow below the frozen surface of Lake Hoare in Antarctica. To which kingdom do blue-green algae belong?*

Some blue-green algae can remove nitrogen gas from the air and combine the nitrogen with other substances to make nitrogen compounds. In this way, blue-green algae help provide the nitrogen compounds that plants need to live. Normally, farmers must use fertilizers to replace nitrogen compounds used by green plants. However, in places like Asia where blue-green algae live in the water, farmers can plant rice year after year without the need of fertilizers. The blue-green algae fertilize the rice paddies naturally.

Not all blue-green algae are blue-green in color. In some algae, the blue-green substance is masked by another substance. For example, one species of blue-green algae is actually red. It lives in the Red Sea, which accounts for the sea's name.

Green Algae

The year is 2001. Deep in space, a silvery ship is on its second year in a four-year journey to study the planets. On board, the two astronauts are about to finish dinner. Although they brought along enough food to last the entire journey, they could not carry enough oxygen to last several years. Are the astronauts doomed to suffocate in space? Of course not.

In a tiny well-lit room near the back of the ship lies a tub of water filled with green algae. The green algae use the carbon dioxide wastes from their environment to make food during photosynthesis. In return, the algae release their own waste, oxygen, into the environment. If you looked closely at the tub, you could see bubbles containing oxygen floating toward the surface. In this way, the astronauts and the green algae support each other's lives.

This scene is one that will not be played out for some time. Scientists today are hard at work developing methods of growing green algae in closed environments. These methods will allow the algae to produce the oxygen future travelers will use in their explorations of space.

Right now you can find green algae only at home on the earth. Most green algae live in the water and make up a part of **plankton**. Plankton consists of all the small organisms that float or swim

Figure 7–5 Spirogyra *(left) is a green alga that forms filaments and is found in pools of fresh water. The green alga Ulva (right) anchors itself to rocks at the bottom of a pond. What substance do algae release during photosynthesis?*

near the surface of water. Larger organisms, such as fish, feed on plankton. Other green algae grow on the branches, stems, and leaves of trees. A few types of green algae anchor themselves to rocks and pieces of wood. Look at Figure 7–5. What differences do you notice between these two types of green algae?

Golden Algae

Every morning and evening you probably brush your teeth. Chances are that the last thing you would consider brushing with is algae. Yet a part of many toothpastes is made of golden algae. Do not let their name fool you. Golden algae all contain the green substance chlorophyll. Their golden color comes from a mixture of orange, yellow, and brown substances in their cells.

The most common, and certainly the most attractive, golden algae are the **diatoms** (DIGH-uh-tahmz). If you looked at diatoms through a microscope, they would resemble tiny glass boxes. See Figure 7–6. The appearance of diatoms is due to a glassy material in their cell walls. When diatoms die, their tough glasslike walls remain. In time, the walls collect in layers and form deposits of diatomaceous (digh-uh-tuh-MAY-shuhs) earth. Diatomaceous earth is a coarse, powdery material. For this reason, it is often added to toothpastes to help polish teeth. Moreover, because diatomaceous earth reflects light, it is also added to the paint used to indicate separate traffic lanes on highways. Why might diatomaceous earth be added to scouring cleansers?

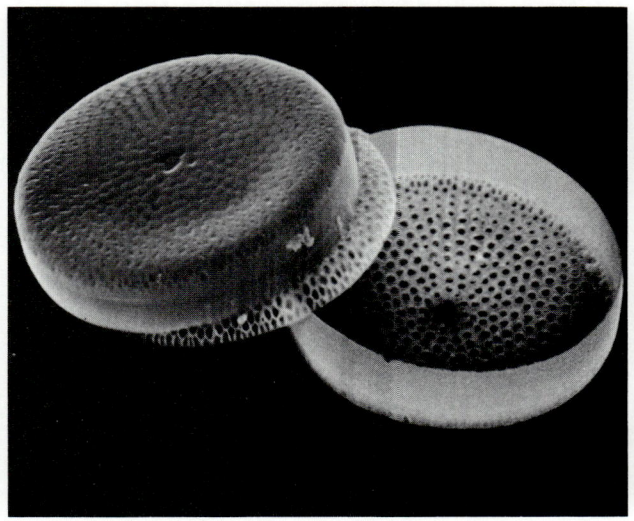

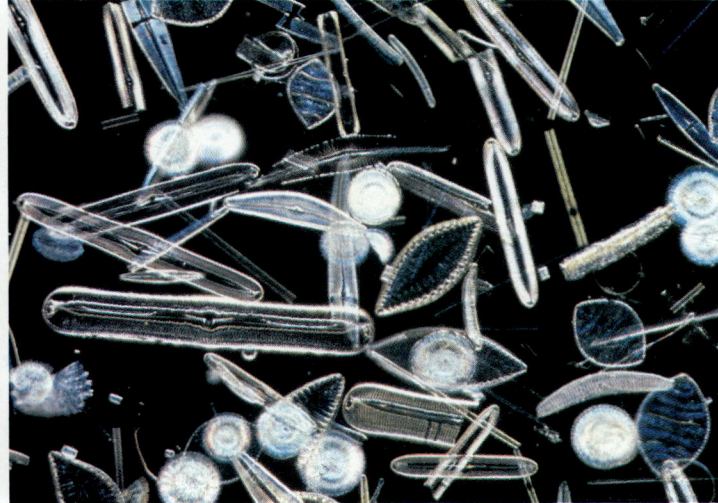

Brown Algae

For centuries, sailors whispered tales of a huge sea within a sea in the Atlantic Ocean southeast of Bermuda. This sea, the Sargasso Sea, the sailors said was filled with endless tangles of brown seaweed. The name of the sea comes from the brown *Sargassum* algae that grow in great amounts in this part of the Atlantic Ocean. Legend had it that ships trapped in this thick growth of seaweed would remain there forever, forming a fleet of dead ships guarded by skeletons. Today, some people believe that the sailors traveling with Christopher Columbus mutinied not out of fear of falling off the edge of the earth, but out of fear of being trapped in the Sargasso Sea.

The legends about the Sargasso Sea are merely a product of people's strong imagination. However, the brown algae that float there are very real. *Sargassum* lie on or near the ocean's surface. There they are able to receive enough sunlight to perform photosynthesis. They stay near the surface because they have **air bladders.** Air bladders are tiny grape-shaped structures that act like inflatable life preservers and keep the algae afloat.

Brown algae, such as *Sargassum,* are the largest and most complex of the algae. Most brown algae live in salt water and are called seaweeds or kelps. Some brown algae float freely. Others are attached to a rock or other object on the ocean floor by **holdfasts.** A holdfast is a rootlike structure.

Figure 7–6 *Diatoms (right) which have many different shapes, have cell walls made of silica, a glasslike substance. The single diatom (right) is magnified 2400 times. How does diatomaceous earth form?*

Figure 7–7 *Some brown algae, such as rockweed, are attached to rocks or other objects on the ocean floor by rootlike structures. What are these structures called?*

Figure 7–8 *Like most red algae, this coralline red alga grows attached to rocks on the ocean floor. What is the jellylike substance produced by red algae?*

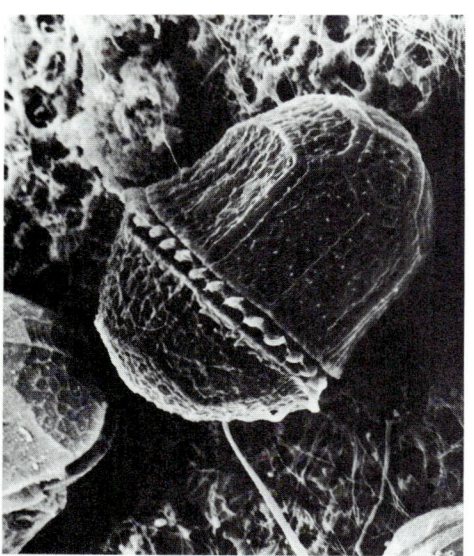

Figure 7–9 *Notice that the cell wall of this fire alga looks like armor plates. Why are fire algae also called dinoflagellates?*

Red Algae

Red algae, named for a red substance in their cells, make up another large group of algae. Some red algae produce a jellylike material called **agar** (AG-uhr). Agar is a substance on which scientists grow bacteria.

Red algae can grow to be several meters long, but they never reach the size of brown algae. Red algae usually grow attached to rocks on the ocean floor. Their red color allows them to absorb a part of sunlight that can penetrate deep into the ocean. For this reason, some species of red algae grow as far as 200 meters beneath the ocean's surface.

Fire Algae

In the winter and spring of 1974, the waters near the west coast of Florida suddenly turned brown and red. Soon thousands of dead fish washed up on shore. One young boy became very ill after eating clams taken from the area. This was not the first time a "red tide" had struck a Florida beach, or for that matter, other beaches in the United States. Red tides have swept onto beaches in such places as New England and Los Angeles.

The red tide is not a tide at all. It is a large group of unicellular fire algae. Fire algae range in color from yellow-green to orange-brown. Some of these fire algae produce poisons. These poisons can injure or kill living things.

If you observed fire algae through a microscope, you would notice something interesting. Most fire algae have cell walls that look like plates of armor. In addition, these fire algae have two **flagella** (fluh-JEHL-uh; singular: flagellum), or whiplike structures, that help propel them forward. For this reason, they are called **dinoflagellates** (digh-nuh-FLAJ-uh-lihts). Because of these structures, dinoflagellates look like hundreds of tiny submarines spinning through the water like tops.

Some fire algae have a characteristic that has amazed sailors since they first set sail on the world's oceans. These sailors often saw a strange glow at night in the water behind their ships. The glow, or **bioluminescence** (bigh-oh-loo-muh-NEHS-uhns), is

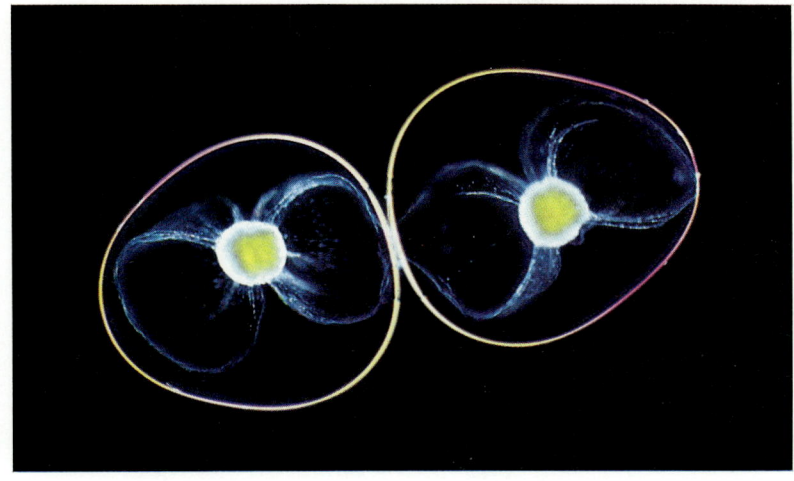

Figure 7–10 *These two fire algae produce a cold light, or a glow. What is this glow called?*

produced by fire algae. Bioluminescence means "having living light." The glow of the fire algae is similar to the glow produced by fireflies, or "lightning bugs."

SECTION REVIEW

1. What are nonvascular plants?
2. Describe the characteristics of algae.
3. Describe the process of photosynthesis.
4. Name the six groups of algae.
5. What is diatomaceous earth?
6. Explain why some algae do not appear green even though they contain chlorophyll.

7–2 Fungi

Section Objective

To describe the characteristics of fungi

In 1927, scientists discovered a deadly elm tree disease in England. The disease, called Dutch elm disease, caused the leaves of the trees to wither. Eventually, the trees died. You would think that elm trees in the United States, separated from England by the Atlantic Ocean, would be safe. However, despite repeated warnings, wood of the English elm trees was shipped to this country to make furniture. By 1930, Dutch elm disease had struck trees in Ohio and a few other states.

Today, Dutch elm disease has spread to almost every state in the country. The disease is carried from tree to tree by insects called bark beetles. Actually Dutch elm disease is caused by a type of

Figure 7–11 *The elm tree has been attacked by a fungus that causes Dutch elm disease.*

Figure 7–12 *The umbrella-shaped structure growing on the dead leaves is the gemmed amanita fungus. Is this fungus a parasite or a saprophyte?*

fungus (FUHNG-guhs; plural: fungi) that is carried by the beetles.

Fungi are nonvascular plantlike organisms that have no chlorophyll. Therefore, fungi cannot perform photosynthesis. Because fungi must obtain their food from living or once-living organisms, they are heterotrophs. As mentioned in Chapter 4, fungi are classified into their own kingdom.

Fungi live in moist, shady areas. Some species of fungi live and grow on other living things. They are called **parasites.** Parasites are organisms that obtain their food from living organisms. The fungus that kills elm trees is an example of a parasite. Other parasitic fungi grow on and destroy crop plants, which people depend upon for food. Fungi can also attack people directly, as anyone who has ever suffered from the itching of athlete's foot can tell you. Animals, too, can be victims of fungi. Some fungi, for example, infect fishes, including those that are raised in home aquariums.

Other species of fungi get their food from dead matter. These fungi are called **saprophytes.** They, along with many types of bacteria, are the earth's "clean-up crew." These fungi decompose, or break down, dead plant and animal matter. These broken-down products become the foods of other living things. Without saprophytes, dead plants and animals would soon litter and pollute the earth.

Structure of Fungi

Fungi have several types of structures. Most members of the Fungi kingdom have threadlike structures called **hyphae** (HIGH-fee; singular: hypha). These hyphae produce special chemicals called enzymes. The enzymes digest, or break down, the cells of living or dead organisms. The broken-down cells and their chemicals are used by fungi as food. Fungi absorb food through the walls of the hyphae.

Most fungi reproduce by means of **spores.** Spores are small reproductive cells. These spores are contained in special **fruiting bodies.** Fruiting bodies are structures that form from the closely packed hyphae. When the spores are released and land in moist areas, such as on the forest floor, they grow into new fungi.

Figure 7–13 *Fungi have many shapes and are very colorful. The chicken-of-the-woods (top left), red cup (top right), sulfur polypore (bottom left), and orange cup (bottom right) are different kinds of fungi. Are fungi autotrophs or heterotrophs?*

Mushrooms

Have you ever ordered a pizza with all the trimmings? If so, you probably ate one type of fungus called the mushroom. All you may know about the mushrooms on pizza is that they taste good. But the pizza maker has to know a lot more because certain mushrooms are poisonous, and poisonous mushrooms on a pizza certainly would not improve sales.

Fortunately, the mushrooms on pizzas and stocked on grocery shelves come from special farms and are safe to eat. However, mushrooms that grow in the wild often are not safe to eat. One of the most poisonous wild mushrooms is the death cup. This well-named mushroom produces a poison that can kill you as surely as the bite of a rattlesnake. Because most people cannot tell the difference between poisonous and nonpoisonous mushrooms, wild mushrooms should never be picked or eaten.

Mushrooms belong to a special group of fungi called the **club fungi.** The name comes from the club-shaped structure that produces the spores.

The umbrella-shaped part of the mushroom you probably are most familiar with is the mushroom's

Figure 7–14 *This photograph of a parasol mushroom shows the gills on the underside of the mushroom, the stalk, and the ring. What is the function of the gills?*

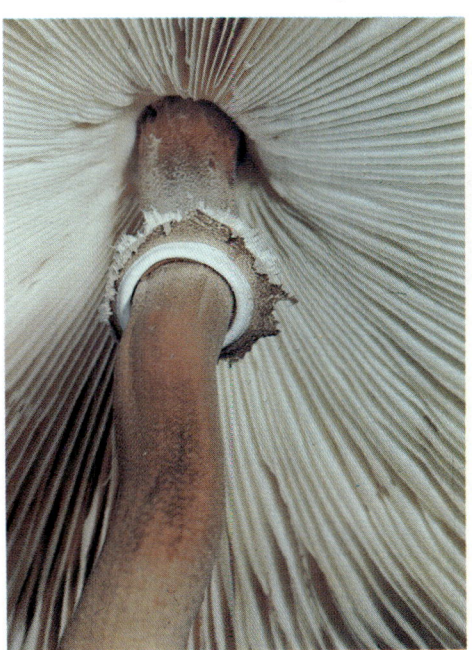

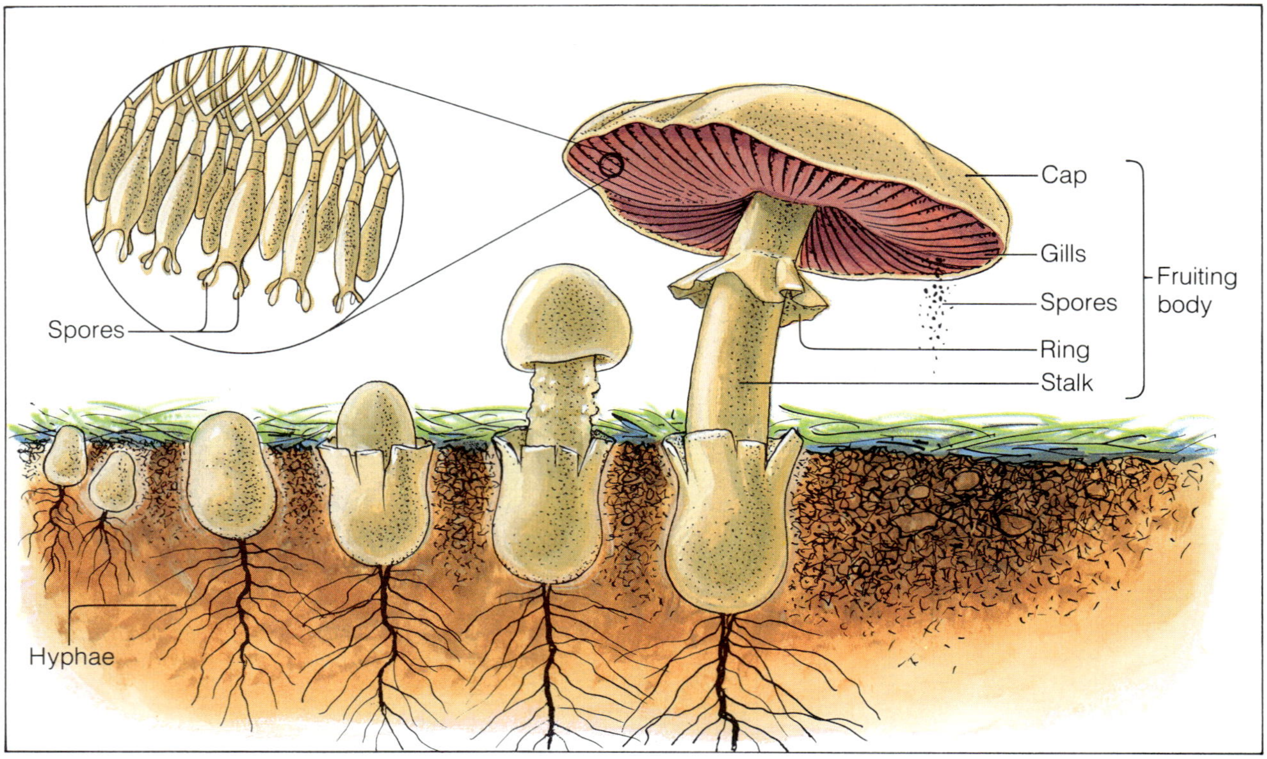

Figure 7–15 *As a mushroom develops, the fruiting body rises from under the ground and releases spores. These spores then develop into new mushrooms. What is the stemlike structure on the mushroom called?*

Figure 7–16 *When a water droplet hits this puffball fungus, the fungus releases a cloud of spores. To what special group of fungi do puffballs belong?*

cap. The cap is part of the mushroom's fruiting body. It contains spores. See Figure 7–15. The cap can be any color, depending on the species of mushroom. The cap is at the top of a **stalk,** a stem-like structure that has a ring near its top. If you turned the mushroom upside down, you would see its **gills.** The gills are the mushroom's spore factories.

Not all mushrooms resemble the types that you see in the grocery store. The puffball, for example, looks like a giant softball and can grow up to 60 centimeters in diameter. When a raindrop hits the puffball, a tiny puff of "smoke" is given off. The puff of smoke contains thousands of spores.

Yeasts

Yeasts belong to a group of fungi called **sac fungi.** Sac fungi produce spores in a saclike structure. Powdery mildews, morels, truffles, and most molds belong to this group of fungi.

Yeasts, like other fungi, have no chlorophyll. They obtain their energy through a process called **fermentation.** During fermentation, sugars and

152

starches are changed into alcohol and carbon dioxide gas. At the same time, energy is released.

Bakers add yeast to bread dough. As the dough bakes, the yeast produces carbon dioxide. This causes the bread to rise. The carbon dioxide also produces hundreds of tiny bubbles in the bread, which you can see as holes in a slice of bread.

Although yeasts are unicellular, they can clump together to form chains of cells. Yeasts reproduce by forming spores or by **budding.** See Figure 7–17. During budding, a portion of the yeast cell pushes out of the cell wall and forms a tiny bud. In time, the bud forms a new yeast.

Molds

Centuries ago people treated infections in a rather curious way. They often placed decaying breads, cheeses, and fruits such as oranges on the infection. Although the people did not have a scientific reason to do this, every once in a while the infection was cured. What these people did not and

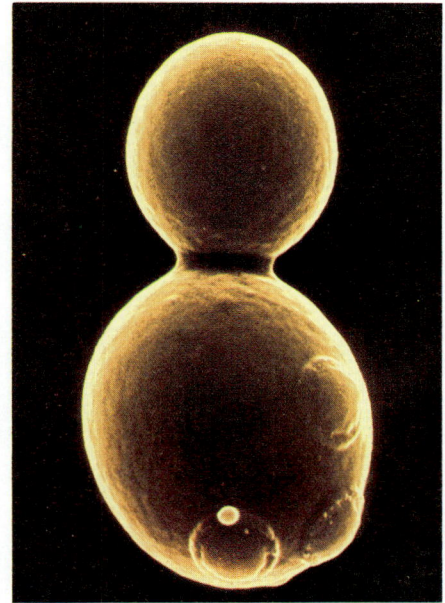

Figure 7–17 *Yeast cells reproduce by budding. The three tiny bumps at the bottom of the yeast cell will develop into three new yeast cells. How else do yeast cells reproduce?*

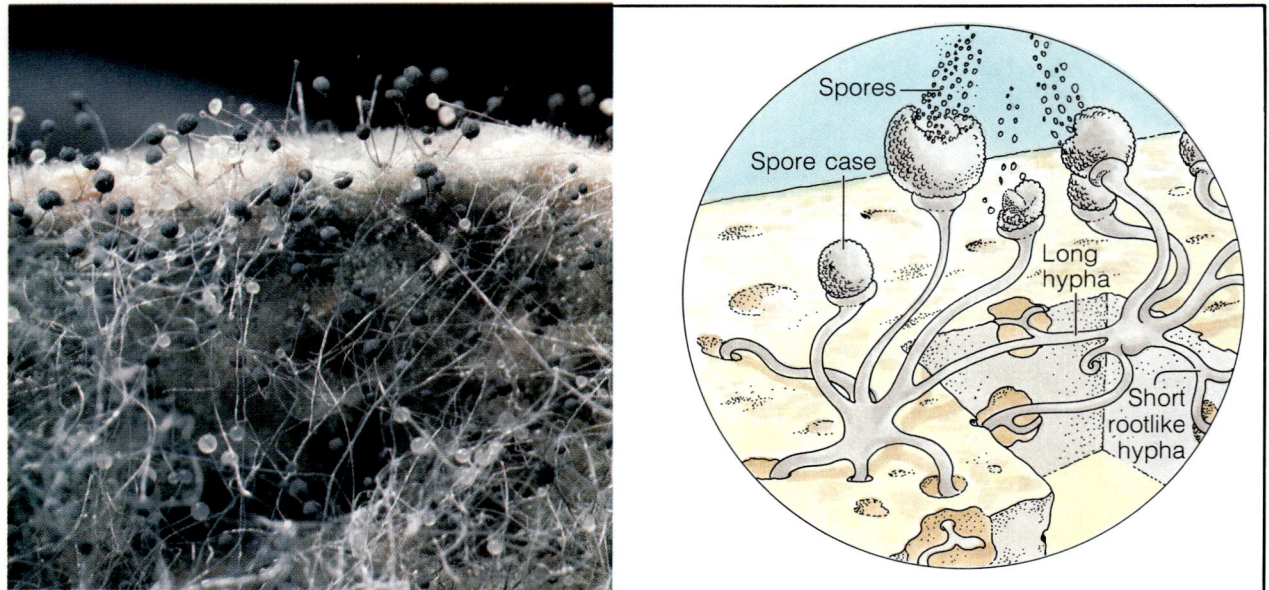

Figure 7–18 *Bread mold reproduces by producing spores, which are released from structures called spore cases. Long hyphae anchor the mold to the bread. Short hyphae absorb food from the bread. How are spores carried from one place to another?*

could not know was that the cure was due to a type of fungus, called mold, which grows on some foods.

In 1928, the Scottish scientist, Sir Alexander Fleming found out why this treatment worked. Fleming discovered that a substance produced by the blue-green mold *Penicillium* could kill certain bacteria that caused infections. Fleming named the substance penicillin. Since that time penicillin, an antibiotic, has saved millions of lives.

Have you ever seen mold growing on bread? Bread mold looks like tiny fluffs of cotton. The fluffs are groups of long hyphae that grow over the surface of bread. Shorter hyphae grow down into the bread and resemble tiny roots. The shorter hyphae release the enzymes that break down chemicals in the bread. The broken-down chemicals are foods for the mold and are absorbed by the hyphae. See Figure 7–18.

The tiny black spheres on bread mold are spore cases, which produce spores. The spores are carried from one place to another through the air. When the spores land on food, they begin to develop into a new mold. On what other types of surfaces have you seen mold growing?

SECTION REVIEW

1. Describe the characteristics of fungi.
2. Compare fungi and algae.
3. List the functions of the hypha, fruiting body, cap, stalk, and gill in a mushroom.
4. What is fermentation?
5. Explain why there are more fungi in a shady forest than in a sunny field.

7–3 Lichens and Slime Molds

Section Objective

To describe the characteristics of lichens and slime molds

Suppose someone asked you what kind of organism can live in the hot, dry desert as well as the cold, wet Arctic. What if the person added that this organism can also survive on bare rocks, wooden poles, the sides of trees, and even the tops of mountains? You might reply that no one organism can survive in so many different environments. In a way, your response would be right. For although **lichens** (LIGH-kuhnz) can actually live in all of these environments, they are not one organism but two. **A lichen is made up of a fungus and an alga that live together.** Combined, these two organisms can live in many places that neither could survive in alone.

Figure 7–19 *The British soldiers (left) and the reindeer moss (right) are examples of lichens. What two organisms make up lichens?*

155

Figure 7–20 *During its life cycle, a slime mold goes through two stages. In one stage, the slime mold resembles a fungus (top). In the other stage, the slime mold looks like a protozoan (bottom). Where are slime molds found?*

The fungus part of the lichen provides the alga with water and minerals that the fungus absorbs from whatever the lichen is growing on. The alga part of the lichen uses the minerals and water to make food for the fungus and itself. This type of relationship, in which two different organisms live together, is called **symbiosis** (sihm-bigh-OH-sihs).

Lichens are called pioneer plants because they are often the first plants to appear in rocky, barren areas. Lichens release acids that break down rock and cause it to crack. Dust and dead lichens fill the cracks, which eventually become fertile places for other organisms to grow. In time, the rocky area may become a lush, green forest.

While walking through a forest, you may have seen a bloblike organism growing in moist areas, on dead leaves, rotting logs, and other types of decaying material. This organism is called a **slime mold** because of its appearances during its life cycle. **The slime mold resembles a protozoan and a fungus during the two stages of its life cycle.**

In one of the slime mold's stages, it resembles a jellylike mass of protoplasm, engulfing and digesting food like a giant amoeba. In the other stage, the slime mold develops fruiting bodies like its cousin the mushroom. The fruiting bodies produce spores. The spores develop into new slime molds.

SECTION REVIEW

1. Describe the characteristics of a lichen and a slime mold.
2. Define symbiosis.
3. Explain why slime molds could not be classified as fungi or protozoans.

Section Objective

To compare the characteristics of mosses and liverworts

7–4 Mosses and Liverworts

In a barren cold part of northern Europe, above the Arctic Circle, is an area called Lapland. Its rocky landscape has a few small trees and beds of lichens and **mosses.** The people who live in this icy place are called Lapps. For them, everything in their environment is important, including the

Figure 7–21 *The rounded structures at the top of these mosses (left) are capsules, which release spores (right). Where are mosses found?*

mosses. Because the mosses are soft and keep in warmth, the Lapps use them to line the inside of their baby cradles.

Mosses are small green nonvascular plants that have stemlike and leaflike parts. They are found near or on the ground, close to sources of water. Mosses absorb water through rootlike structures called **rhizoids** (RIGH-zoids). Because mosses contain chlorophyll, they make their own food through photosynthesis. They live in damp and cool places such as the shaded surfaces of trees and rocks.

Liverworts are small green nonvascular plants that have flat leaflike parts. Each **liverwort** looks like a tiny green leaf, although it is not a real leaf. Some liverworts look like miniature livers, which explains how these simple plants got their name. Like mosses, liverworts grow in moist places and have rhizoids. Certain liverworts have a special way of reproducing. New plants grow from pieces broken off older plants. See Figure 7–22.

Figure 7–22 *The small, flat, green plants in this photograph are great scented liverworts. What structure anchors liverworts to rocks and other objects?*

SECTION REVIEW

1. Compare the characteristics of mosses and liverworts.
2. What are rhizoids?
3. Would you expect to find mosses that were two meters tall? Explain.

Examining a Slime Mold

Problem

What are the characteristics of a slime mold?

Materials *(per group)*

petri dish containing agar
filter paper containing slime mold
crushed oatmeal flakes
water
glass slide
masking tape
cover slip
forceps
microscope
medicine dropper
dissecting needle

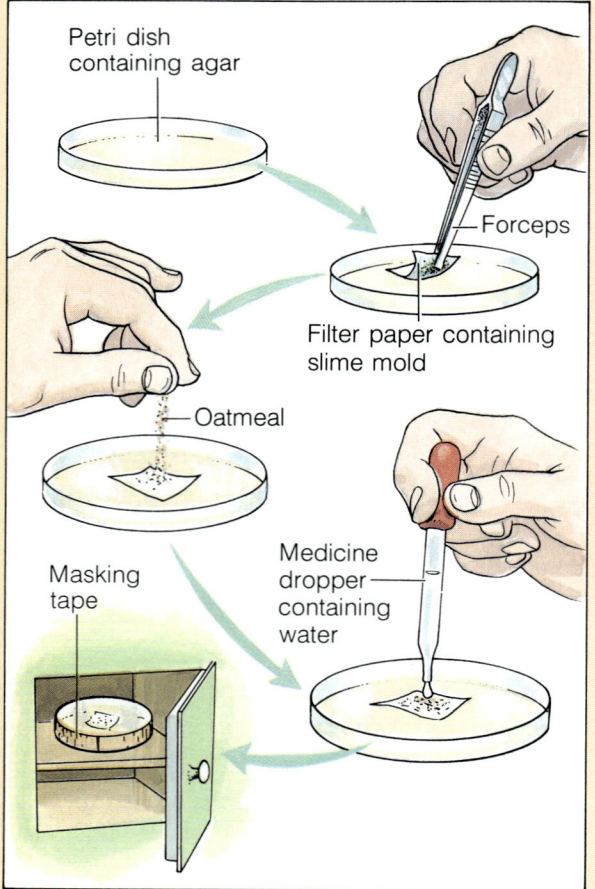

Petri dish containing agar

Forceps

Filter paper containing slime mold

Oatmeal

Masking tape

Medicine dropper containing water

Procedure

1. With a forceps, place the small piece of filter paper containing the slime mold in the center of the petri dish.
2. Sprinkle some crushed oatmeal flakes next to the piece of filter paper.
3. Add two to three drops of water to the slime mold and oatmeal flakes.
4. Cover the petri dish. Seal the dish with masking tape.
5. Place the sealed petri dish in a cool, dark place.
6. Examine the petri dish each day for three days. Record your observations.
7. After three days, remove a small amount of the slime mold from the petri dish and place it on a glass slide.
8. Examine the slime mold under the low power of a microscope.
9. With a dissecting needle, puncture a branch of the slime mold. **CAUTION:** *Be careful when using a dissecting needle.* Observe the slime mold for a few minutes.

Observations

1. Describe the changes that took place in the slime mold during the three-day observation period.
2. What activity did you observe in the slime mold after placing it on the glass slide?
3. Describe what happened to the puncture that you made in the slime mold.

Conclusions

1. Explain why oatmeal was added to the petri dish.
2. Is the slime mold a heterotroph or an autotroph? Explain.
3. Based on your observations, describe the characteristics of a slime mold.
4. Describe an experiment to determine the response of the slime mold to certain substances, such as salt or sugar.

CHAPTER REVIEW

SUMMARY

7–1 Algae

❑ Algae are nonvascular plants that contain chlorophyll, a green substance needed for photosynthesis.

❑ Algae are classified into six groups according to their color. They are grouped as blue-green, green, golden, brown, red, and fire algae. All but the blue-green are classified as plants.

7–2 Fungi

❑ Fungi are nonvascular plantlike organisms that have no chlorophyll.

❑ Fungi are heterotrophs. Some fungi are parasites, others are saprophytes.

❑ Mushrooms belong to a special group of fungi called club fungi.

❑ Yeasts, powdery mildews, morels, truffles, and most molds belong to the group of fungi called sac fungi.

7–3 Lichens and Slime Molds

❑ Lichens are made up of two organisms, a fungus and an alga, that live together. This type of relationship is called symbiosis.

❑ Slime molds are found in moist areas on once-living leaves, rotting logs, and other types of decaying material.

❑ During their life cycle, slime molds alternate between a funguslike stage and a protozoanlike stage.

7–4 Mosses and Liverworts

❑ Mosses are small, green nonvascular plants that have stemlike and leaflike parts. Mosses also have rootlike structures called rhizoids, which help absorb water from the soil.

❑ Liverworts are small, green nonvascular plants that have flat leaflike parts. Some liverworts reproduce when pieces of the plant break off from older plants.

VOCABULARY

Define each term in a complete sentence.

air bladder	dinoflagellate	liverwort	sac fungus
agar	fermentation	moss	saprophyte
alga	flagellum	multicellular	slime mold
bioluminescence	fruiting body	nonvascular plant	spore
budding	fungus	parasite	stalk
cap	gill	photosynthesis	symbiosis
chlorophyll	holdfast	plankton	unicellular
club fungus	hypha	rhizoid	
diatom	lichen		

CONTENT REVIEW: MULTIPLE CHOICE

On a separate sheet of paper, write the letter of the answer that best completes each statement.

1. All plants that lack true roots, stems, and leaves are said to be
   a. photosynthetic.   b. nonvascular.   c. parasitic.   d. saprophytic.
2. The algae that produce red tides are called
   a. diatoms.   b. *Sargassum*.   c. dinoflagellates.   d. brown algae.

3. Algae are classified by their
   a. color.   b. shape.   c. size.   d. environment.
4. Which are capable of bioluminescence?
   a. lichens   b. diatoms   c. blue-green algae   d. dinoflagellates
5. Structures that enable brown algae to remain attached to objects on the ocean floor are
   a. rhizoids.   b. holdfasts.   c. air bladders.   d. hyphae.
6. In most fungi, spores are contained in
   a. fruiting bodies.   b. stalks.   c. hyphae.   d. rhizoids.
7. The umbrella-shaped part of a mushroom is called the
   a. rhizoid.   b. cap.   c. hypha.   d. stalk.
8. Which is an example of a club fungus?
   a. yeast   b. mushrooms   c. mold   d. truffle
9. A moss absorbs water from its environment through its rootlike
   a. hyphae.   b. stalks.   c. rhizoids.   d. fruiting bodies.
10. A plant that is very similar to a liverwort is a (an)
    a. moss.   b. alga.   c. mushroom.   d. slime mold.

CONTENT REVIEW: COMPLETION

On a separate sheet of paper, write the word or words that best complete each statement.

1. The green substance in plants needed for photosynthesis is _____.
2. The tough, glasslike skeletons of _____ algae form a substance used in many toothpastes.
3. Structures that enable brown algae to float near the surface of the water are _____.
4. The structures that fire algae use for movement are _____.
5. Yeasts reproduce by _____.

6. _____ is an example of a club fungus.
7. _____ are the threadlike structures that produce enzymes in most fungi.
8. During _____, sugars and starches are changed into alcohol and carbon dioxide and energy is released.
9. Penicillin is a (an) _____.
10. _____ is the relationship in which two different organisms live together.

CONTENT REVIEW: TRUE OR FALSE

Determine whether each statement is true or false. Then on a separate sheet of paper, write "true" if it is true. If it is false, change the underlined word or words to make the statement true.

1. Algae are heterotrophs.
2. In bioluminescence, plants use energy in sunlight to make their own food.
3. Blue-green algae are members of the plant kingdom.
4. Diatoms are examples of golden algae.
5. Diatomaceous earth is a substance on which scientists grow bacteria.
6. Fire algae are also called diatoms.
7. Fungi are autotrophs.
8. A mushroom is a sac fungus.
9. Yeasts are multicellular fungi.
10. A lichen is made up of a fungus and an alga that live together.

CONCEPT REVIEW: SKILL BUILDING

Use the skills you have developed in the chapter to complete each activity.

1. **Applying concepts** Explain why most algae live in shallow water or float on the surface of the water.

2. **Interpreting a graph** An algae population was used in a photosynthesis experiment. The results of the experiment are shown in the following graph.

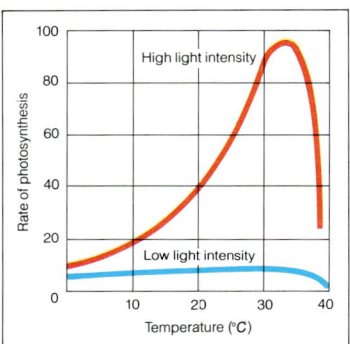

What was the purpose of the experiment? What conclusions can be drawn about the algae that received high light intensity? Low light intensity?

3. **Designing an experiment** Design an experiment to show the effects of temperature, light, and dryness on the growth of bread mold. Be sure to include a control. Where do the mold spores that started the mold growth come from? What conclusions can you draw from the experiment?

4. **Developing a hypothesis** In many forests, mushrooms suddenly spring up out of the soil after a rainstorm. Using this information and your knowledge of fungi, develop a hypothesis to explain this.

5. **Making diagrams** Using reference materials, draw diagrams that show the methods of reproduction in the green alga, *Ulothrix,* and the bread mold. What are some of the similarities between each reproductive method? What are some of the differences? Label all structures.

6. **Making graphs** Using the data in the table, construct a bar graph. What does the bar graph show?

Species	Common Name	Height of Fruiting Body
A. aurantia	Orange-peel fungus	12
C. argillacca	Moor-club fungus	6
L. lubrica	Jellybaby	6
C. vermicularis	Coral fungus	10
P. vesiculosa	Early-cup fungus	7
X. hypoxylon	Candle-snuff fungus	8
L. molle	——	5
M. caninus	Smaller-dog stinkhorn	2
C. visocosa	Stag's horn fungus	8
R. flava	Yellow-coral fungus	3

7. **Applying concepts** Imagine a forest from which all of the fungi have somehow been removed. What would be the effects of this action?

CONCEPT REVIEW: ESSAY

Discuss each of the following in a brief paragraph.

1. List and describe the six groups of algae.
2. List and describe three types of fungi.
3. Compare a diatom and a dinoflagellate.
4. Explain why fungi are heterotrophs.
5. How do fungi take in food?
6. Describe the structures of a mushroom. What is the function of each structure?
7. Explain why yeast cells are used in bread making.
8. Describe how lichens show symbiosis.

Vascular Plants

8

CHAPTER OBJECTIVES

After completing this chapter, you will be able to

8–1 Compare vascular and nonvascular plants.

8–1 Describe the characteristics of ferns.

8–2 Compare the functions of roots, stems, and leaves in seed plants.

8–3 Describe the characteristics of gymnosperms.

8–4 Describe the characteristics of angiosperms.

8–4 Identify the parts of the flower.

8–5 Describe photosynthesis.

8–6 Define tropism and describe the actions of plant hormones.

Look closely at the photograph on the opposite page. Does the insect have eyes? Does it have wings? What type of insect is it?

Actually it is not an insect at all. It is a flower called the spider orchid. As you can see, the spider orchid looks very much like an insect. In fact, it resembles a female wasp, complete with eyes, antennae, and wings. The spider orchid even smells like a female wasp.

You are not the only one to be tricked by the spider orchid. It fools male wasps too. The flower does this by giving off a scent that attracts a male wasp. As the male wasp touches the spider orchid, it may leave behind some pollen that it has collected from another spider orchid. At the same time, the male wasp may pick up a fresh supply of pollen from this flower and carry it to another spider orchid plant. This begins the process of plant reproduction.

Can you imagine what would happen to the spider orchid plant, if the male wasps could not smell the orchid? Eventually, these plants might disappear off the face of the earth. What if butterflies that find flowers by their brilliant colors were colorblind? These flowers would probably disappear too because butterflies carry pollen from one plant to another.

You can see that insects play a role in the life cycle of flowering plants. As you read this chapter, you will learn more about the pollination of flowering plants, as well as other interesting plant groups.

The spider orchid resembles a female wasp, complete with eyes, antennae, and wings.

8–1 Ferns

Imagine walking through a rain forest, such as Washington State's Olympic National Park. As you enter, you immediately feel the cool dampness of the air. Tall trees, such as firs, spruces, and cedars, are everywhere. Mosses drape the branches and trunks of trees and hold moisture like a sponge.

As you continue your walk, you notice feathery green leaves above some moss plants. Many of these leaves belong to ferns. Unlike mosses, which are nonvascular plants, ferns are **vascular plants.** Vascular plants contain transporting tubes that carry material throughout the plant. Remember that mosses must live close to the ground. Mosses have no transporting tubes to carry water and nutrients to their stemlike and leaflike parts.

Ferns are among the oldest plants on the earth. They appeared more than 300 million years ago and soon became the most abundant type of plant. These ferns were as large as trees. Today some ferns the size of trees are still found in areas that have tropical rain forests.

Most ferns disappeared as the earth's climate changed. All that remained of some ferns was dead

Figure 8–1 *The photograph on the left shows tree ferns as they appear today in a rain forest in Hawaii. The photograph on the right shows a model of a fern forest during the Carboniferous Period. Why is this period of time so named?*

plant material. This material later formed great coal deposits. Most living or once-living material contains carbon. Because coal is made up mostly of carbon, the geological period in which ferns were abundant is called the Carboniferous Period.

Fern Structure

Like all vascular plants, ferns have true leaves, stems, and roots. In fact, a fern's **fronds,** or leaves, are the part of the plant you usually notice. Most developing fern leaves are curled at the top and resemble the top of a violin. Because of their appearance, these developing leaves are called fiddleheads. See Figure 8–2. As they mature, the fiddleheads uncurl until they reach their full size.

Following the leaves down toward the ground, you find the **rhizome** (RIGH-zohm), or underground stem of the fern. In some ferns, the rhizome looks like a large, fuzzy, brown caterpillar with leaves. The rhizomes of many ferns grow along the surface of the soil. In other types of ferns, the rhizomes grow beneath the ground. Roots grow from the rhizomes and anchor the ferns to the ground. The roots also absorb water and minerals for the plant.

Transporting tubes travel throughout the fern's leaves, stems, and roots. These tubes carry materials throughout the plant.

Fern Reproduction

As the fern grows, it goes through two stages in its life cycle. If you were to look at both stages of the same plant, you would think that each was a different plant.

The fern uses a different type of reproduction in each stage. In the first stage of a fern's life cycle, small brown structures appear on the underside of the fronds. These structures are called **sori** (SOH-ree; singular: sorus), or spore cases. Sori contain **spores.** A spore is a special reproductive cell. It is exactly like the parent cell. That is, each spore has the same type of hereditary material as the parent. During this stage, the entire plant is called a sporophyte (SPOR-uh-fight).

Figure 8–2 *The curled structures at the tops of these ferns are called fiddleheads. Into what structures do fiddleheads develop?*

Figure 8–3 *The tiny brown spots on the underside of this fern frond are spore cases. How are spores carried from place to place?*

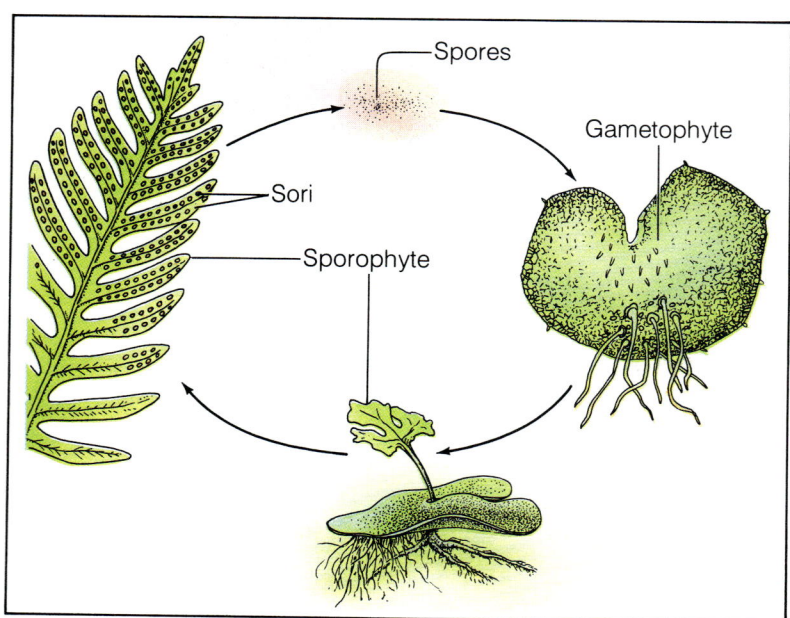

Figure 8–4 *The fern's life cycle has two stages. One stage is called the sporophyte stage and the other is called the gametophyte stage. Which stage produces the male and female sex cells?*

When the spore cases are ripe, they open and release spores. Spores are swept away by gusts of wind. Eventually, the spores fall to the ground. If the ground is fertile and moist, the spores will begin to grow. Reproduction in which spores are formed is called **asexual reproduction.** Asexual reproduction is the formation of an organism from a single parent cell.

As the spores grow into tiny, green heart-shaped structures, the second stage of a fern's life cycle begins. In each of these heart-shaped structures, special tissues develop that produce male and female sex cells called gametes (GAM-eets). The male gamete is called the sperm, and the female gamete is called the egg. Both sperm and eggs are produced by the heart-shaped structures. These structures are called the gametophytes (guh-MEET-uh-fights). When dew and rainwater are present, the sperm swim toward the eggs. Each sperm joins with an egg.

After a sperm joins with an egg, a new fern plant begins to grow. The new fern plant was produced by **sexual reproduction.** Sexual reproduction is the formation of an organism from the uniting of two different sex cells. The new plant will grow and sprout fiddleheads that will uncurl into fronds. After a while, sori will appear on the underside of the fronds. And the cycle will go on.

1. What are the characteristics of ferns?
2. Describe the life cycle of ferns.
3. If you were to come across a tall plant in a forest, would you guess it was a vascular or nonvascular plant? Explain why.

8–2 Seed Plants

If you wanted to plant a vegetable garden, you would begin by loosening the soil. Also you would have to make sure that your garden had a source of water. Only then would you plant your **seeds.** A seed contains a young plant, stored food, and a seed coat. With proper care, the seeds in your garden would begin to grow and, after a few months, produce adult plants loaded with vegetables.

Seed plants are vascular plants that produce seeds. Like ferns, seed plants have true roots, stems, and leaves. Seed plants go through the same two reproductive stages as ferns. Seed plants are divided into two groups based on the structure of their seeds. In **gymnosperms** (JIHM-nuh-spermz), the seeds *are not* covered by a protective wall. In **angiosperms** (AN-jee-uh-spermz), the seeds *are* covered by a protective wall.

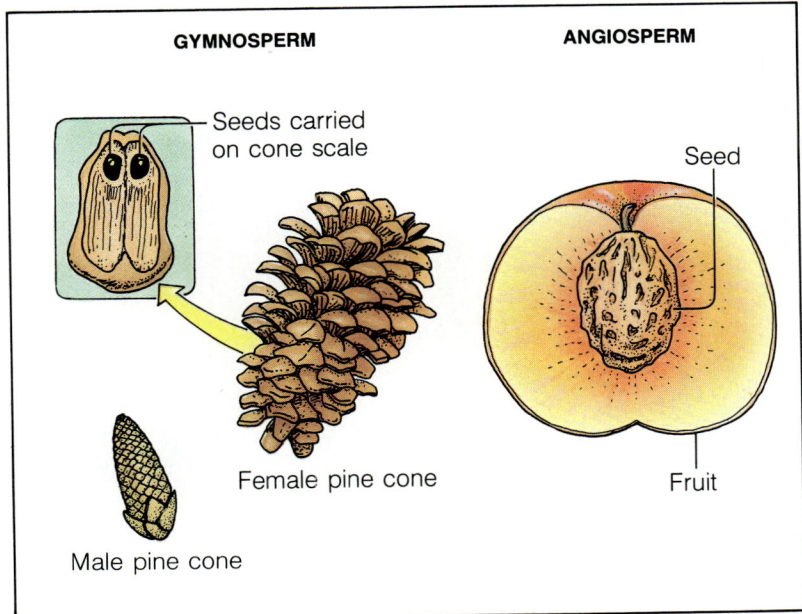

Figure 8–5 *Seed plants are divided into two groups based on the structure of their seeds. In gymnosperms (left), the seeds are not covered by a protective wall. In angiosperms (right), the seeds are covered by a protective wall. What is a seed?*

Seed plants are among the most numerous plants on the earth. Also, they are the plants with which most people are familiar. The trees in the forest, the vegetables in a garden, and the cotton plant are all examples of seed plants. What other seed plants can you name?

Roots

Roots anchor a plant in the ground and absorb water and minerals from the soil. In some plants, such as sweet potatoes, roots also store food for the plant. Plants such as grasses have slender roots that spread out in many directions. Others, such as radishes, have roots that grow straight down into the ground. See Figure 8–6.

Many roots have tiny **root hairs** that cover the surface of the roots. Root hairs are microscopic extensions of individual cells. The root hairs greatly increase the root's surface area, or amount of surface exposed to the surrounding soil. They help the plant absorb water and minerals from the soil. These materials then pass into tissue called **xylem** (ZIGH-luhm). The xylem tissue carries water and minerals from the roots up through the plant. Another kind of tissue contained in the roots is called **phloem** (FLOH-uhm). Phloem tissue carries food substances down from the plant's leaves. The xylem and phloem tissue form the two main transporting tubes in vascular plants.

Figure 8–6 *Although all roots absorb materials from the soil and anchor a plant to the soil, they do not all look the same. Notice the difference in the roots of the stilt palm (left), corn plant (center), and mangrove (right).*

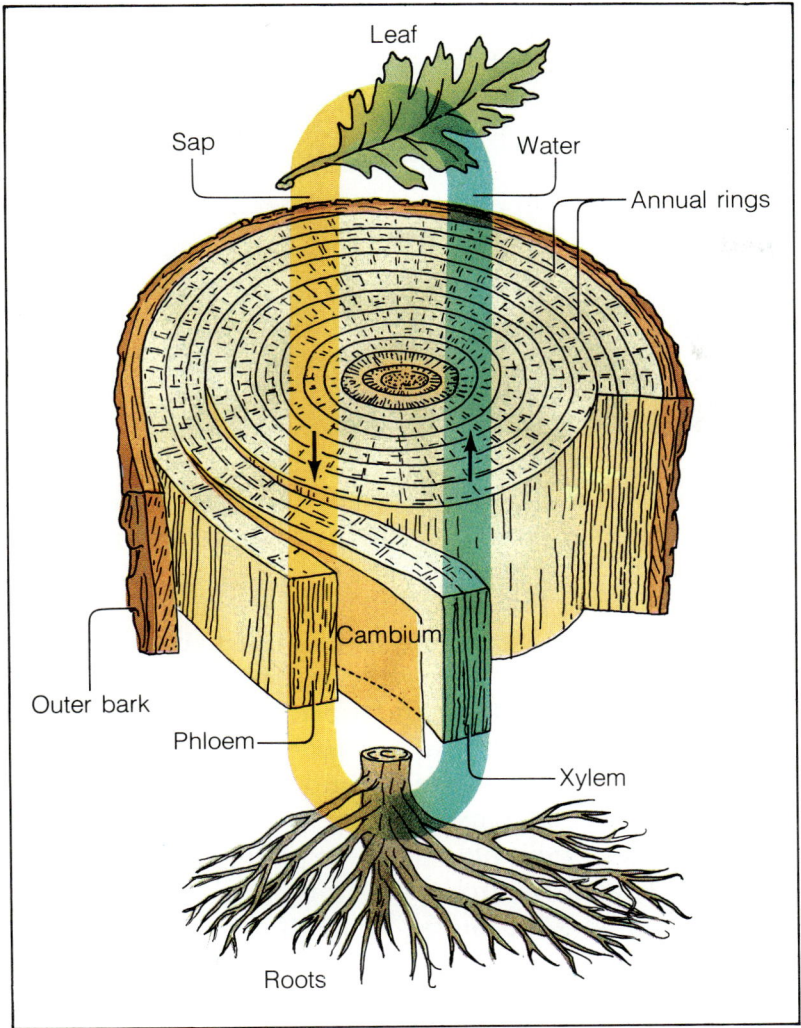

Figure 8–7 *In a stem, the xylem transports water and other materials from the roots up to the leaves. The phloem conducts sap from the leaves down to the roots. What is the cambium?*

Stems

In addition to xylem and phloem tissue, stems also contain a tissue called **cambium** (KAM-bee-uhm). Cambium is the growth tissue of the stem and produces new xylem and phloem cells. The addition of new cells causes the wood in the stem to become thicker. Each year, as a new layer of xylem cells grows, it wraps around the layer before it. Because each layer is one year's growth of xylem cells, these layers are called **annual rings.** By counting these annual rings, scientists can estimate a tree's age.

The xylem and the phloem tissue in the stems transport water, minerals, and food throughout the

Figure 8–8 *The age of a tree, such as this western hemlock, can be determined by counting the number of rings in its stem. What does each ring represent?*

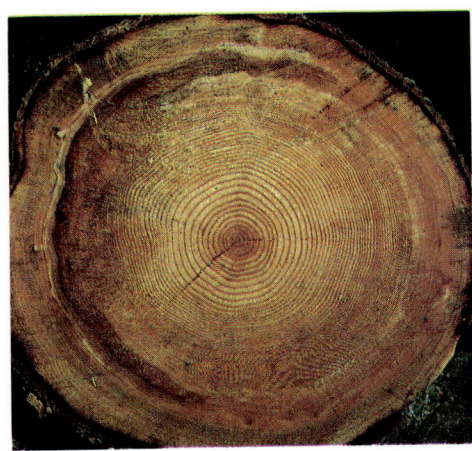

Figure 8–9 *Cactus stems (left) store water. A potato stem (right) stores food. The structures growing out of each side of the potato are actually new potato plants.*

Sharpen Your Skills

Plant Transport

1. Fill a medium-sized jar one-fourth full of water. Then add a few drops of vegetable coloring and stir.

2. Place a freshly cut twig from a tree and a stalk of celery in the jar so that only the cut part of each is underwater.

3. After 24 hours, remove the twig and the stalk.

4. With a knife cut off the portion of each plant that was underwater. **CAUTION:** *Be very careful when using a knife.* Discard these portions.

5. Cut another small section across the bottom of each remaining portion. Examine.

How do the two sections differ? How are they alike? What structure transports the colored water up the plants?

plant. Stems also support the plant's leaves and flowers. Some stems, such as the stem in the cactus, store water. See Figure 8–9.

Most stems grow vertically and above the ground, but some grow underground like roots. In some plants, such as the white potato, the stems grow under the ground and store food. These stems are called **tubers.** The "eyes," or tiny growths, on the white potato are buds. A bud produces plant structures such as leaves and stems. Each bud can develop into a new plant. Other plants, such as onions and hyacinths, produce short, thick, underground stems called bulbs. Bulbs have thick leaves that contain food for the plant. Ginger, grasses, and lilies of the valley have yet another kind of underground stem. This stem is called a rhizome. Rhizomes grow horizontally.

Stems are divided into two groups based on their hardness. The soft green stems of plants, such as corn, tomato, and geraniums, are **herbaceous** (huhr-BAY-shuhs) **stems.** Most plants with herbaceous stems complete their life cycle within one growing season and are called annuals. A plant's life cycle begins when the plant starts to grow from a seed and ends when the plant produces its own seeds and then dies. Dandelions and grasses are examples of annuals.

The stems in the other group are called **woody stems.** Unlike herbaceous stems, woody stems are rigid and have large amounts of xylem tissue. Most plants with woody stems, such as trees and shrubs, are called perennials (puh-REHN-ee-uhlz). A perennial plant lives for *more* than two growing seasons. Plants that live for *only* two growing seasons are

known as biennials (bigh-EHN-ee-uhlz). Biennials may be either herbaceous or woody. Turnips, beets, and cabbages are examples of biennials.

Leaves

Most leaves have a stalk and a blade. The stalk connects the leaf to the stem. The blade is the thin, flat part of the leaf. Most of the cells involved in food making are located in the blade.

Leaves are divided into two groups. Some leaves, such as oak and beech leaves, have only one blade. These are simple leaves. Other leaves have two or more blades and are known as compound leaves. The Virginia creeper and ailanthus (ay-LAN-thuhs) have compound leaves.

Leaves are very important structures. In the leaves, the largest and most important manufacturing process in the world takes place. This process is called **photosynthesis** (foh-tuh-SIHN-thuh-sihs). During photosynthesis, the sun's light energy is trapped and used to produce food. The sun's light energy is captured by **chlorophyll,** the green substance in plants. The chlorophyll is contained in oval-shaped structures called **chloroplasts.** Then the energy is used to combine water from the soil and carbon dioxide from the air to produce food and oxygen. You will learn more about photosynthesis later on in this chapter.

If you closely examine a leaf section under a microscope, you will see that it contains four layers

Figure 8–10 *Stems are either herbaceous or woody. The bluebells (left) have herbaceous stems. The stem of the General Sherman tree (right) is a woody stem. Which type of stem is rigid?*

Figure 8–11 *Leaves come in a variety of types and shapes. The big tooth aspen leaf (top) is a simple leaf. The staghorn sumac (bottom) is a compound leaf. What is the function of leaves?*

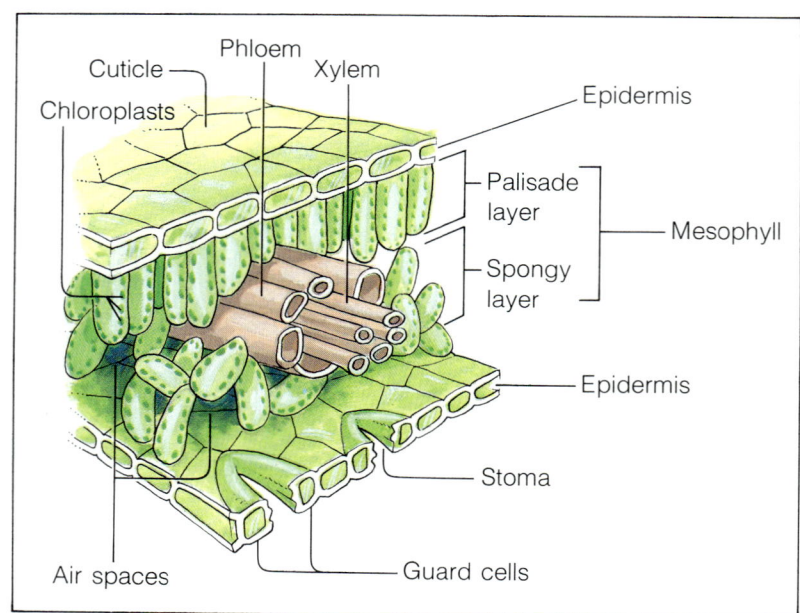

Figure 8–12 *This drawing of a leaf shows the tissues and structures found in various parts of the leaf. Which layers contain the food-making cells?*

Figure 8–13 *Plants release excess moisture through the leaves, as can be seen by the tiny droplets on the inside of the container. What is this process called?*

of cells. The outer, protective layer of the leaf is called the **epidermis** (ehp-uh-DER-mihs). The epidermis is located on both the upper and lower surfaces of the leaf. The cells in this layer do not contain chlorophyll. As a result, they cannot make food. This layer of cells is only one cell thick and allows light to pass through to the food-making cells below.

Some leaves have a waxy **cuticle** (KYOOT-ih-kuhl) that covers the epidermis. The cuticle prevents the loss of too much water from the leaf. Between the upper and lower epidermis is the mesophyll (MEHS-uh-fihl). The mesophyll is made up of cells that contain chloroplasts. In most plants, the mesophyll is made of two types of cells. The upper layer of cells forms the **palisade** (pal-uh-SAYD) **layer.** The cells in this layer are long and cylindrical. The cells of the lower mesophyll form the **spongy layer.** The spongy layer consists of irregularly shaped cells and many air spaces. Phloem and xylem tissue are also found in the mesophyll.

In most plants, the lower surface of the epidermis contains many openings called **stomata** (STOH-muh-tuh; singular stoma). Some plants have stomata on the top surface of the leaves. Stomata permit carbon dioxide to enter the leaf and allow oxygen to pass out of the leaf into the air. Stomata also play a role in **transpiration.** Transpiration is the process by which excess water is released through the leaves.

Each stoma is surrounded by two sausage-shaped structures called **guard cells.** The guard cells regulate the opening and closing of the stoma. When the guard cells swell, the stoma opens and carbon dioxide enters the leaf. When the guard cells relax, the stoma closes. Usually stomata are open during the day and closed at night.

SECTION REVIEW

1. What are the characteristics of seed plants?
2. Which plant tissues transport food? Which transport water and minerals?
3. Name three types of plants based on their growing seasons.
4. Name and describe two processes that occur in leaves.
5. Most leaves are broad and flat. How are these characteristics useful to plants?

CAREER   *Nursery Manager*

HELP WANTED: NURSERY MANAGER
Requires two- or four-year degree in horticulture. Business background in marketing and management desirable. Nursery or greenhouse experience helpful. Apply in person.

Nearly every community has a garden center that provides its customers with the plants and supplies they need for their homes and gardens. If you pass by a garden center in April, you see pots of tulips and daffodils on display. In July, petunias and other colorful flowers catch your eye. Piles of pumpkins appear in October, and rows of pine trees emerge in December. However, all these plants are not grown at the garden center. They are ordered and delivered by a nursery.

Usually nurseries are run under the direction of **nursery managers.** The managers supervise the growing of plants and flowers from seeds, seedlings, or cuttings until they are large enough to sell. Nursery managers make sure nursery workers properly fertilize, weed, transplant, and apply pesticides to the plants. In addition, nursery managers make sure that proper inventory records are kept and that the orders placed by garden centers are filled quickly.

Nursery managers also make many business decisions, such as choosing new equipment, deciding how much money to spend on the growing of various plants, and determining the kinds of plants to sell. Nursery managers also are involved in planning chemical use, in deciding which new plant varieties should be planted, and in directing the building of storage areas. If you think you would enjoy working with plants, write to the American Society for Horticultural Science, 701 North Saint Asaph Street, Alexandria, VA 22314.

8–3 Gymnosperms

The word *gymnosperm* comes from two Greek words meaning "uncovered" and "seed." **Gymnosperms are seed plants that produce uncovered seeds.** The three main groups of gymnosperms are conifers (KAHN-uh-fuhrz), cycads (SIGH-kadz), and gingkoes (GIHNG-kohz).

The largest group of gymnosperms is the conifers. The word *conifer* means "cone-bearing." Cones are structures that contain the seeds of the plant. Conifers are woody plants that live in areas of the world that are neither too hot nor too cold. Most conifers, such as pines, cedars, and hemlocks, are called evergreen because they appear to keep their leaves all year. Actually, evergreens lose leaves while adding new ones. The leaves or needles may remain on an evergreen tree for 2 to 12 years. What are some other examples of evergreen trees?

Conifers produce male and female cones on the same tree. The male cones contain **pollen** and the female cones contain **ovules** (OH-vyoolz). Pollen contain male sex cells called sperm. Ovules contain female sex cells called eggs. The pollen is carried to the female cones by wind. After the sperm and the egg join, a seed develops. As the seed matures, part of the cone dries up and releases the seeds. When a seed falls to the ground, the seed may develop into a new conifer.

Figure 8–14 *Conifers, such as the pines (left), produce male cones (top center) and female cones (bottom center). The cycad (right), another gymnosperm, grows in tropical areas. What is the structure in the middle of the cycad called?*

The second group of gymnosperms are the cycads. These tropical trees look like small palm trees or ferns, though some grow as tall as 18 meters. Cycads have large leaves that look like a feathery crown on top of their trunks. Like conifers, cycads have cones. But in cycads, the trees are either male or female and the cones are located in the center of the leaves.

The third group of gymnosperms are gingkoes. There is only one species of gingko. This species is commonly called the maidenhair tree. Although gingkoes originally came from China, they are often found growing along many streets in the United States. Gingkoes have fan-shaped leaves and can grow as tall as 30 meters. Like cycads, male and female cones are found on separate trees. After male and female sex cells join, yellowish seeds form. The seeds are about the size of cherries and have an awful odor. The odor is so unpleasant that only male gingkoes are planted on city streets.

Figure 8–15 *This gingko, or maidenhair tree, has fan-shaped leaves and can grow as tall as 30 meters.*

SECTION REVIEW

1. What are gymnosperms?
2. List the three groups of gymnosperms.
3. What is contained inside male cones? Female cones?
4. What advantage is it to have male and female cones on the same tree?

8–4 Angiosperms

Section Objective

To describe the characteristics of angiosperms

To many people, especially those with lawns and gardens, the bright yellow dandelion is a weed and is difficult to get rid of. Actually, the dandelion is a **flower.** A flower is a structure that contains the reproductive organs of an angiosperm. **Angiosperms are seed plants that produce covered seeds.** Angiosperms are also called flowering plants.

Angiosperms differ in size and live in many environments. The saguaro cactus lives in a desert, the water lily lives in lakes and ponds, and the horse chestnut may live in your local park. The plants range in size from the duckweed plant, a tiny water

Figure 8–16 *The most noticeable difference between gymnosperms (left) and angiosperms (right) is that angiosperms produce flowers. What is a flower?*

plant about 1 millimeter long, to a giant redwood tree 100 meters tall. Can you name some other angiosperms?

Angiosperms are similar to gymnosperms in that both groups of plants form seeds and have leaves, stems, and roots. But angiosperms differ from gymnosperms in several ways. The most noticeable difference is that only angiosperms produce flowers.

Flowers

Have you ever leaned over to sniff a brightly colored flower and been startled to find a bee inside the flower? The beautifully shaped and colored leaflike structures of a flower are the **petals.** The petals help to attract insects and other animals. These animals play a vital role in the reproduction of flowering plants.

If you examine a flowering plant, you may see that some of the flowers have not yet blossomed. Instead, some may be enclosed in green, leaflike structures called **sepals** (SEE-puhlz). The sepals protect the flower when it is a bud. When they open, they reveal the flower's brightly colored petals.

STAMEN   Within the petals are the male and female reproductive organs of the flower. The male organs are called the **stamens** (STAY-muhnz). In most flowers, the stamens have two parts, the filament and the anther. The filament is a stalklike structure that supports the anther. The anther produces the pollen, which contains the sperm.

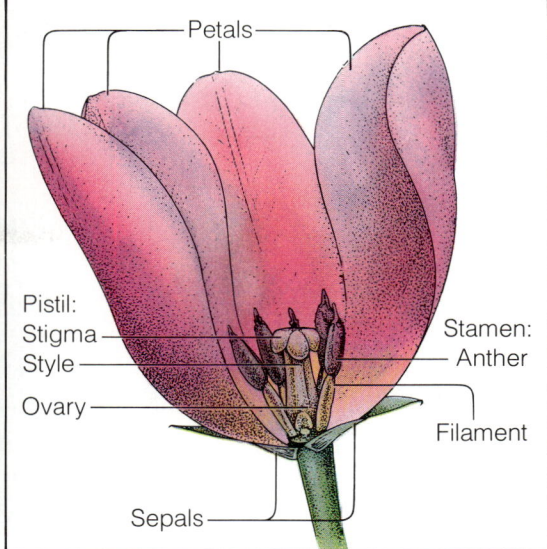

Petals

Pistil:
Stigma
Style
Ovary

Stamen:
Anther

Filament

Sepals

Figure 8–17 *In the photograph, some of the flower's petals have been removed to show the structures inside a flower. The drawing of the flower indicates the names of these structures. What is the function of the petals?*

PISTIL The female organs of the flower are the **pistils.** They are located in the center of the flower. Most flowers have at least two pistils. And in most flowers, the pistil has three parts. At the base of the pistil is the hollow **ovary** (OH-vuhr-ee). It contains the egg cells. A slender tube called the **style** connects the ovary to a sticky structure at the top of the pistil. This sticky structure is called the **stigma** (STIHG-muh).

POLLINATION In flowers, there are two stages of reproduction. The first stage is called **pollination.** In this stage, the pollen is transferred from an anther to a stigma. In *self*-pollination, pollen is transferred from the anther to a stigma within the same flower or to another flower on the same plant.

Figure 8–18 *Animals, such as this butterfly, help transfer pollen from one flower to another. What is this process called?*

Figure 8–19 *Ripened ovaries develop into fruits, which may contain one or more seeds. What is the function of a fruit?*

Figure 8–20 *This bean seed is an example of a dicot, which contains two cotyledons. Within the seed is a young plant that is made up of the epicotyl, radicle, and hypocotyl. What plant structures do these parts of the seed become?*

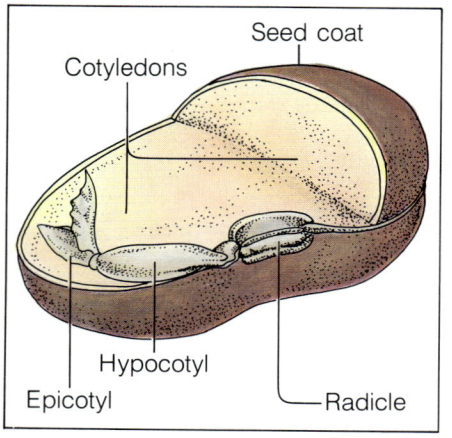

In *cross*-pollination, pollen is transferred from the anther of one plant to the stigma of another plant. In this kind of pollination, pollen grains are carried from flower to flower by wind and animals.

FERTILIZATION Because the surfaces of the stigmas are sticky, pollen grains cling to them. Once on a stigma, the pollen forms a tube, which pushes its way down the style to an ovule in the ovary. Then, sperm from the pollen grain travel down the tube to the ovule. When the sperm unites with the egg, the second stage in flower reproduction occurs. This stage is called **fertilization.**

After fertilization has occurred, the fertilized egg becomes a seed and the ovary develops into a **fruit.** A fruit is the ripened ovary that encloses and protects the seed. Raspberries, peaches, and pears are examples of fruits.

Seeds

The seeds of different types of flowering plants vary greatly in size. For example, one kind of coconut tree can produce a seed that has a mass of almost 23 kilograms! On the other hand, orchid seeds are so tiny that 200,000 of them have a mass of about 7 grams. But no matter its size, every seed contains all that is necessary for the development of a new plant.

The seed is surrounded by a protective seed coat that develops from the ovule's wall. Inside the seed is a tiny young plant and an endosperm. The endosperm is a tissue that supplies food to the embryo plant until it is able to make its own food.

If you were to remove the seed coat of a bean seed carefully, the seed would divide into two halves. These halves are called **cotyledons** (kaht-uhl-EED-uhnz). A cotyledon is the young plants' leaflike structure that stores food. Angiosperms have one or two cotyledons. Some angiosperms, such as corn and grasses, have only one cotyledon. These plants are called monocotyledons, or monocots. Angiosperms that have two cotyledons are called dicotyledons, or dicots. Lima beans and roses are examples of dicots. Figure 8–21 shows some other characteristics of monocots and dicots.

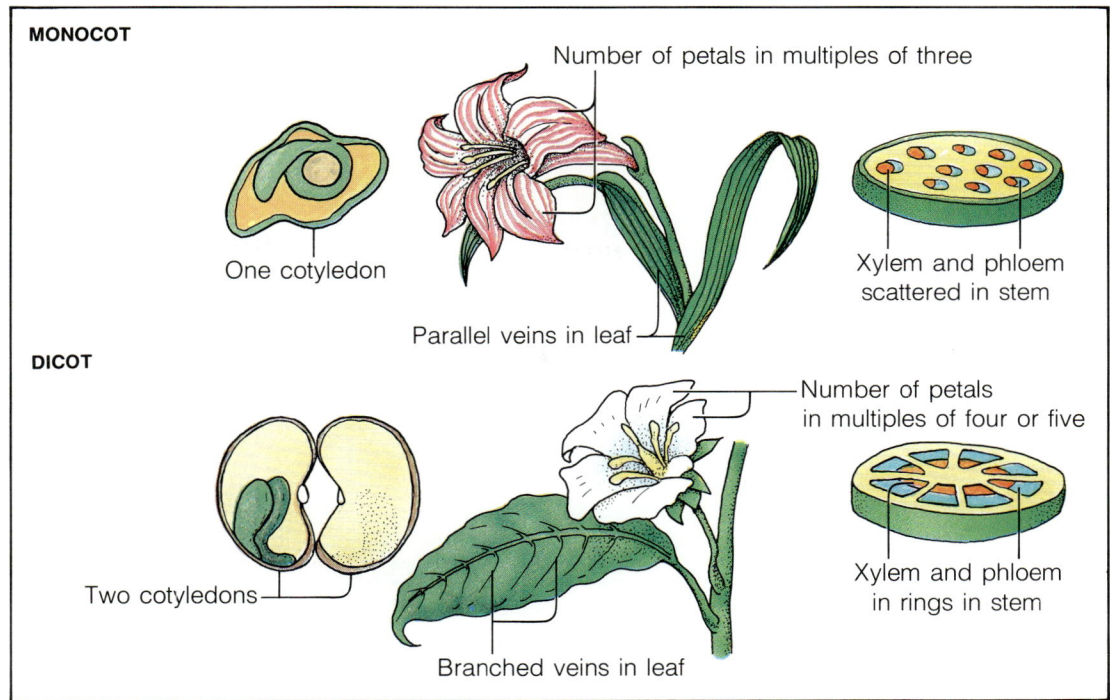

	MONOCOT	
One cotyledon	Number of petals in multiples of three Parallel veins in leaf	Xylem and phloem scattered in stem

DICOT

Two cotyledons — Branched veins in leaf — Number of petals in multiples of four or five — Xylem and phloem in rings in stem

After seeds have been scattered they usually remain dormant, or inactive, for some time. Then, if the temperature is just right and if the seeds receive the proper amount of moisture and oxygen, **germination** occurs. Germination is the early growth stage, or the "sprouting," of a young plant.

When a seed germinates, the lower part of the young plant grows down into the soil and becomes the stem. This part is called the hypocotyl (high-puh-KAHT-uhl). The upper part of the young plant grows up into the air to become the leaves. The

Figure 8–21 *This drawing shows the characteristics of monocots and dicots. Are the leaves in Figure 8–11 on page 171 those of dicots or monocots?*

Figure 8–22 *Seeds are dispersed, or scattered, in many ways. Seeds of the milkweed (left) are dispersed by the wind. Seeds of the burdock (right) are carried by people's clothes or animal fur. What happens to the seeds if they land in an area that has a proper temperature and a proper amount of moisture and oxygen?*

upper part is called the epicotyl (ehp-uh-KAHT-uhl). Another part, called the radicle (RAD-ih-kuhl), becomes the roots.

SECTION REVIEW

1. Describe the characteristics of angiosperms.
2. What are the male reproductive organs of a flower? The female reproductive organs?
3. What is contained in a seed?
4. Why is it an advantage to a maple tree to produce many seeds?

Section Objective

To describe the process of photosynthesis

Figure 8–23 *In green plants, photosynthesis takes place in the leaves and the stems (top), which contain cell structures called chloroplasts (bottom). What substance do chloroplasts contain?*

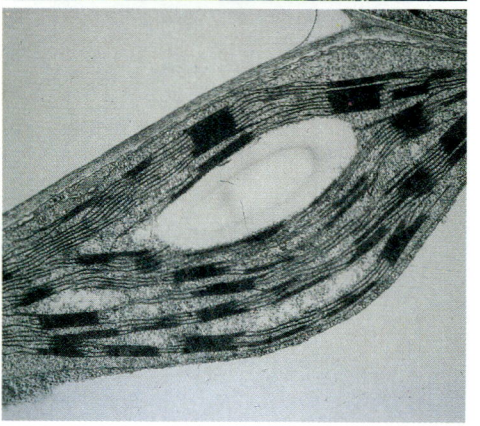

8–5 Photosynthesis

Photosynthesis is the process by which green plants use carbon dioxide, water, and light energy to produce glucose and oxygen. Glucose is a simple sugar that provides organisms with energy. To the plants, oxygen is a waste product. What value does this waste product have to other living things?

At the beginning of this chapter, you learned that most photosynthesis takes place in the leaves of green plants. Some photosynthesis may also occur in the stems of green plants. Within the leaves and the stems are cell structures called chloroplasts. Chloroplasts contain a green substance called chlorophyll. During the process of photosynthesis, chlorophyll traps the light energy.

Because photosynthesis is a complicated process, scientists use an equation to describe what occurs. An equation is scientists' way to describe a chemical reaction. The equation for photosynthesis is:

$$\text{carbon dioxide} + \text{water} \xrightarrow[\text{chlorophyll}]{\text{sunlight}} \text{glucose} + \text{oxygen}$$

$$6CO_2 + 6H_2O \xrightarrow[\text{chlorophyll}]{\text{sunlight}} C_6H_{12}O_6 + 6O_2$$

Photosynthesis takes place in two major stages. The first stage takes place in the chloroplasts, where the chlorophyll absorbs light energy. This energy is then used to split each water molecule, H_2O, into hydrogen and oxygen. At this point, oxygen is given off. In the second stage, hydrogen quickly combines

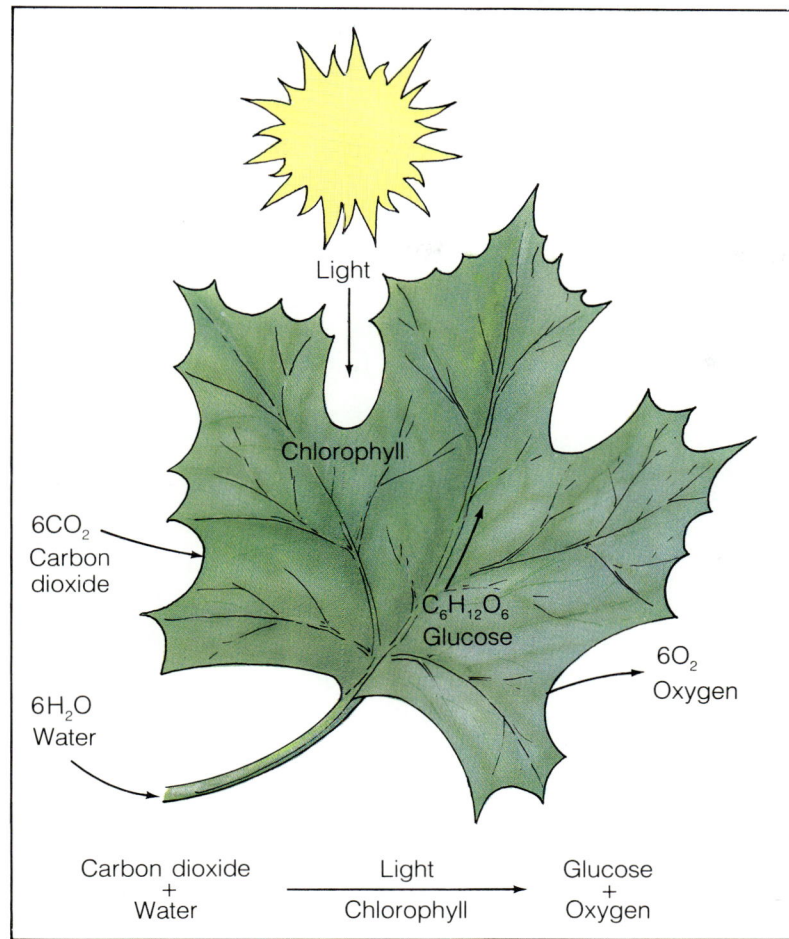

Figure 8–24 *During the process of photosynthesis, a green plant uses the raw materials carbon dioxide and water in the presence of light energy and chlorophyll to produce glucose and oxygen. How is oxygen produced?*

Light

Chlorophyll

$6CO_2$
Carbon
dioxide

$C_6H_{12}O_6$
Glucose

$6O_2$
Oxygen

$6H_2O$
Water

Carbon dioxide + Water	Light ———→ Chlorophyll	Glucose + Oxygen

with the carbon dioxide, CO_2, to form the simple sugar glucose, $C_6H_{12}O_6$. As a result of photosynthesis, light energy is converted into the chemical energy in the sugar molecules.

In addition to the simple sugars that are formed during the process of photosynthesis, plants can also make other compounds. These include complex sugars, starches, fats, and proteins. Complex sugars and starches are made by joining simple sugars together. These materials provide energy for plants and animals. Fats, on the other hand, are made by breaking down simple sugars into smaller carbon-containing molecules. These molecules are then rearranged to form more complex molecules called fatty acids and glycerol (GLIHS-uhr-awl). These molecules combine to form fats. Like sugars and starches, fats provide organisms with energy. Proteins are formed when fatty acids join with the element nitrogen to form amino acids. Amino acids are the building blocks of proteins. Proteins are used in the building and repair of cells.

SECTION REVIEW

1. Define photosynthesis.
2. Discuss the role of chlorophyll in the process of photosynthesis.
3. Write the word equation and the chemical equation for photosynthesis.
4. Explain how photosynthesis results in the production of complex sugars, fats, and proteins.
5. What are two factors that may affect the rate of photosynthesis?

Section Objective

To list and describe the factors that affect plant growth

Sharpen Your Skills

Observing a Tree

1. Choose a tree that you see every day in your neighborhood or on your way to school.
2. Observe the tree carefully. Record all the characteristics that distinguish your tree from other trees.
3. Draw a diagram of your tree.
4. Observe your tree at least three times a week for the next six months.

In a journal, write down the changes that occur in your tree. Make sure you record the date that you noticed the changes. How do changes in the environment affect your tree?

8–6 Plant Growth

Just as a seed needs a proper environment to germinate, a plant needs a proper environment to grow and develop during its life. **Plants need water, light, proper temperatures, air, and special chemicals called hormones to grow.** Because these factors cause a response in plants, they are called stimuli (singular: stimulus). In this section, you will discover how plants respond to these various stimuli.

Tropisms

Some of the responses in plants involve movement of a plant part toward or away from a stimulus. This is known as a **tropism** (TROH-pihz-uhm). A plant may have a positive or a negative tropism. A positive tropism is the movement of a plant toward a stimulus. A negative tropism is the movement of a plant away from a stimulus.

Tropisms are named according to the stimuli that cause them. Phototropism is bending caused by light, geotropism is caused by gravity, and hydrotropism is caused by water. For example, the stems and leaves of a plant placed in a window will grow toward the light. This is a positive phototropism. Roots, on the other hand, show a negative phototropism and grow away from the light. But roots show positive geotropism. That is, the roots grow downward in the direction of the pull of gravity. Which plant parts show negative geotropism?

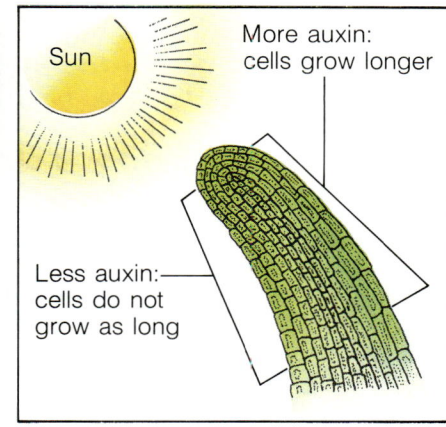

Figure 8–25 *When a plant is placed near a sunny window, the plant's stem grows toward the light. The drawing shows how auxins stimulate the growth of stems toward the light. Is this an example of a positive or negative tropism? Explain.*

Plant Hormones

Tropisms are controlled by substances called auxins (AWK-sihnz). Auxins are types of plant hormones, or growth-regulating substances. Some auxins may stimulate cell growth in certain parts of a plant, and other auxins may prevent cell growth. For example, if there is light on one side of a plant, the auxin will move away from the lighted stem. This causes an increase in auxin on the shaded side. As a result, the cells on the shaded side will grow faster than those on the lighted side. The unequal rates of growth in the stem cause it to bend toward the light. See Figure 8–25.

Some auxins have been made in laboratories. Known as 2, 4-D and 2, 4, 5-T, these auxins greatly speed up the growth of some plants. As a result, these auxins kill the plant. For this reason, these auxins are used as weed killers.

Another type of plant hormone is the gibberellins (jihb-uh-REHL-ihnz). The gibberellins cause stems to grow longer. For example, if gibberellins are given to dwarf corn plants, the plants will grow unusually tall.

Figure 8–26 *Gibberellins have been given to the plants on the right, while the plants on the left have been left untreated. Which part of the plant was affected by gibberellins?*

SECTION REVIEW

1. What are five factors that influence plant growth?
2. Describe a tropism.
3. Name two plant hormones. How do they affect plant growth?
4. Design an experiment to show which plant parts show a positive response to a chemical fertilizer in soil. What would you call this response?

Observing a Tropism

Problem

How do germinating seeds respond to gravity?

Materials *(per group)*

4 corn seeds soaked in water for
  24 hours
paper towels   clay
petri dish   scissors
masking tape   water
glass-marking pencil   medicine dropper

Procedure

1. Arrange four soaked corn seeds in a petri dish. The pointed ends of the seeds should all point toward the center of the dish. One of the seeds should be at the 12 o'clock position of the circle, and the other seeds at 3, 6, and 9 o'clock.
2. Place a circle of paper towel over the seeds. Then pack the dish with enough pieces of paper towel so that the seeds will be held firmly in place when the other half of the petri dish is put on.
3. Moisten the paper towels with water. Cover the petri dish and seal the two halves together with a strip of masking tape.
4. With the glass-marking pencil, draw an arrow on the lid pointing toward 12 o'clock. Label the lid with your name and the date.
5. With pieces of clay, prop the dish up so that the arrow is pointing up. Place the dish in a completely dark place.
6. Predict what will happen to the seeds. Then observe the seeds each day for about one week. Make a sketch of them each day. Be sure to return the dish and seeds to their original position when you have finished.

Observations

1. What happened to the corn seeds? In which direction did the roots and the stems grow?
2. Which part of the germinating seeds showed positive geotropism? Negative geotropism?

Conclusions

1. What would happen to the corn seeds if the dish was turned so that the arrow was pointing toward the bottom of the dish? To the right or left? What is the variable in this investigation?
2. Why is it important that the petri dish remain in a stable position throughout the investigation?
3. Suppose you planted all your corn seeds in the soil upside down. In which direction would the stems grow? The roots?
4. Explain why the seeds were placed in the dark rather than near a sunny window.

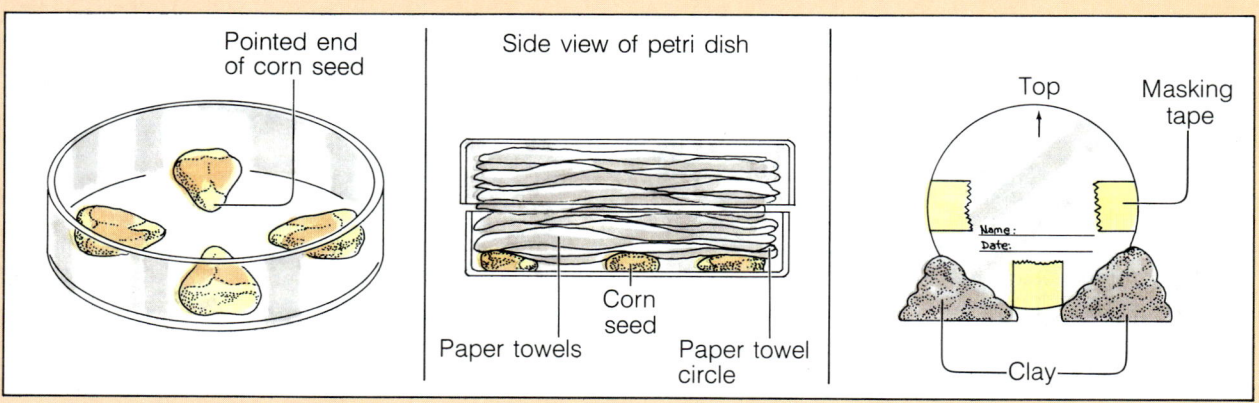

CHAPTER REVIEW

SUMMARY

8–1 Ferns

❏ Vascular plants contain transporting tubes that carry material throughout the plant.

❏ Ferns have true leaves, stems, and roots.

❏ In asexual reproduction, new fern plants develop from spores.

❏ In sexual reproduction, fern plants develop from the uniting of a sperm and an egg cell.

8–2 Seed Plants

❏ A seed contains a young plant, stored food, and a seed coat.

❏ Roots are structures that anchor plants and absorb water and minerals.

❏ Xylem tissue carries minerals and water from the roots up the plant. Phloem tissue carries food down from the leaves.

❏ Stems support plant parts.

❏ The epidermis forms a protective covering around the leaf. The lower surface of the epidermis contains stomata, or openings. Guard cells regulate the opening and closing of the stomata. They regulate the amount of water lost through the plant leaves.

8–3 Gymnosperms

❏ Gymnosperms produce uncovered seeds.

❏ Conifers, cycads, and gingkoes are gymnosperms.

8–4 Angiosperms

❏ Angiosperms produce covered seeds.

❏ Flowers contain stamens, or the male reproductive organs, and pistils, or the female reproductive organs.

❏ After fertilization, the ovary develops into a fruit, which encloses the seed.

8–5 Photosynthesis

❏ During photosynthesis, green plants use carbon dioxide, water, and light energy to produce glucose and oxygen.

❏ Photosynthesis occurs in chloroplasts, which contain chlorophyll.

8–6 Plant Growth

❏ A tropism is a growth response to stimuli.

❏ Plant hormones, such as auxins and gibberellins, are growth regulators.

VOCABULARY

Define each term in a complete sentence.

angiosperm	frond	photosynthesis	spore
annual ring	fruit	pistil	stamen
asexual reproduction	germination	pollen	stigma
cambium	guard cell	pollination	stoma
chlorophyll	gymnosperm	rhizome	style
chloroplast	herbaceous stem	root hair	transpiration
cotyledon	ovary	seed	tropism
cuticle	ovule	sepal	tuber
epidermis	palisade layer	sexual reproduction	vascular plant
fertilization	petal	sorus	woody stem
flower	phloem	spongy layer	xylem

CONTENT REVIEW: MULTIPLE CHOICE

On a separate sheet of paper, write the letter of the answer that best completes each statement.

1. The leaf of a fern is called a
   a. frond.   b. rhizome.
   c. fiddlehead.   d. biennial.
2. Ferns produce spores during
   a. germination.   b. fertilization.
   c. sexual reproduction.   d. asexual reproduction.
3. Plant tissue that carries food substances down from the plant's leaves is called
   a. stoma.   b. cambium.   c. phloem.   d. xylem.
4. Which of these structures is *not* a stem?
   a. bulb   b. rhizome   c. frond   d. tuber
5. Angiosperms produce
   a. cones.   b. uncovered seeds.
   c. covered seeds.   d. spores.
6. Which structure produces eggs?
   a. stamen   b. anther
   c. ovary   d. style
7. The union of the sperm and the egg is called
   a. germination.   b. transpiration.
   c. pollination.   d. fertilization.
8. How many cotyledons does a dicot contain?
   a. one   b. two   c. three   d. four
9. During photosynthesis, carbon dioxide enters the leaf through the
   a. stomata.   b. rhizoids.   c. guard cells.   d. rhizomes.
10. Auxins are
    a. enzymes.   b. sugars.   c. hormones.   d. fruits.

CONTENT REVIEW: COMPLETION

On a separate sheet of paper, write the word or words that best complete each statement.

1. The small brown structures on the underside of fern fronds are called _____.
2. The fern structures that absorb water and minerals from the soil are called _____.
3. In the stem, new xylem and phloem cells are produced by the _____.
4. Leaves that have one blade are called _____ leaves.
5. The _____ is the waxy covering of a leaf.
6. Cycads belong to a group of seed plants called _____.
7. During pollination, pollen grains cling to the stigma, which is the sticky structure at the top of the _____.
8. When conditions are just right, a seed may go through a process of growth called _____.
9. The gas that is released during photosynthesis is _____.
10. _____ are the plant hormones that cause stems to grow longer.

CONTENT REVIEW: TRUE OR FALSE

Determine whether each statement is true or false. Then on a separate sheet of paper, write "true" if it is true. If it is false, change the underlined word or words to make the statement true.

1. In <u>asexual</u> reproduction, an organism develops from the union of an egg and a sperm.
2. Seed plants absorb water and minerals through microscopic extensions of single cells called <u>root hairs</u>.
3. One year's growth of <u>phloem</u> cells produces a layer called an annual ring.
4. An <u>annual</u> is a plant that lives for more than two growing seasons.
5. The food made by a plant during <u>transpiration</u> is called glucose.
6. The <u>cuticle</u> prevents the loss of too much water from the leaf.
7. Conifers produce <u>cones</u>.
8. Angiosperms produce <u>flowers</u>.
9. When a sperm unites with an egg in a flower's ovary, <u>pollination</u> occurs.
10. <u>Auxins</u> stimulate or prevent growth in plants.

CONCEPT REVIEW: SKILL BUILDING

Use the skills you have developed in the chapter to complete each activity.

1. **Classifying leaves** Collect some different types of leaves from your neighborhood. Classify each as a fern, a gymnosperm, or an angiosperm. If the leaf is from an angiosperm, determine whether it is a compound or a simple leaf. Make a chart of your findings.
2. **Making predictions** What would happen to the petals of a freshly cut carnation, if you were to place the stem in a container of colored water? Explain.
3. **Relating concepts** Pesticides, or chemicals that are designed to kill harmful insects, sometimes kill beneficial insects as well. What effect could this have on angiosperms?
4. **Making predictions** What would happen to a plant if its leaves were covered with petroleum jelly?
5. **Applying concepts** Girdling is the complete removal of bark from a section of a tree. Explain why girdling can kill a tree.
6. **Designing an experiment** Imagine that you are a scientist and must determine if light is needed for seed germination. Describe your experimental setup. Be sure to include a control.

CONCEPT REVIEW: ESSAY

Discuss each of the following in a brief paragraph.

1. Describe the two stages in the reproductive cycle of a fern.
2. Compare the functions of roots and stems.
3. What is the difference between herbaceous and woody stems?
4. Photosynthesis takes place in the leaves of a plant. How does the carbon dioxide and water needed for photosynthesis get into the leaves? Explain how the oxygen formed during photosynthesis is released.
5. Describe the similarities and differences between gymnosperms and angiosperms.
6. What is the difference between self-pollination and cross-pollination?
7. What is a tropism? Name and describe three types of tropisms.
8. How does an auxin cause a plant to grow toward the light?

Virginia Walbot:

Crop Designer

Imagine walking through a desert in Arizona or New Mexico and finding tall corn plants growing among the cactus. Or suppose you were riding in a camel caravan making its way through the hot Sahara desert and you spotted a field of wheat moving gently in the wind. Your first reaction might be to pinch yourself to see if you were dreaming. For surely corn and wheat cannot grow in a desert with little or no water.

Of course these scenes are a dream. But they may become reality in the future because of the work of Virginia Walbot, a plant biologist at Stanford University in California. Today, Dr. Walbot is working to develop food crops that can be grown in desert areas. In the future, her work may make food production in the desert a reality.

In many countries, crops are planted in deserts because that is the only land available. But crops planted on desert land must be irrigated. Water must be supplied for these plants to grow well and produce a good harvest. Dr. Walbot is working to develop crop plants that can grow in dry areas without irrigation.

In 1986, Dr. Walbot received a fellowship from the Guggenheim Foundation. At that time, she was working on a project that would change the characteristics of corn plants. Her project involved removing genes from one kind of plant and inserting them into another corn plant. Genes are the hereditary materials that control the characteristics of organisms. Genes are found in the nuclei of most cells. Dr. Walbot thought that the corn plant might grow and develop the traits of the inserted genes.

Suppose that genes from a cactus could be introduced into corn or wheat as Dr. Walbot hopes. The corn or wheat might be able to inherit the cactus's ability to grow in a dry, desert environment. If this could be done, corn and wheat could be grown in a desert without the need for expensive irrigation projects.

Dr. Walbot has developed a special technique for inserting foreign genes into corn cells. In 1985, Dr. Walbot found that she could increase the ability of corn cells to absorb new genes by first soaking the corn cells in salt solutions. Next, she passed a strong burst of electricity through the corn cells. The electricity created tiny openings in the corn cell walls. The foreign genes were then able to pass through the openings. Only about 20 percent of the corn cells took in the foreign genes. And only about 1 percent of these altered cells were able to pass on characteristics inherited from the new genes to the next generation. But this is a good beginning.

Dr. Walbot's next goal is to find a way to grow mature plants from the altered cells. In order to accomplish this, Dr. Walbot and other members of her team are working with chemicals that stimulate plant growth. They expect to be able to produce crops of genetically altered corn in about two years.

Many people applaud Dr. Walbot's achievements in altering corn cells. They look forward to a time when these techniques will expand world food production. But other people are afraid that altering the genetic makeup of plants may prove dangerous. They fear that the changed plant may eventually be a danger to the environment. Still others are afraid that this system might make the breeding of new crop plants too expensive. They feel that only a few large companies could afford the very expensive techniques necessary to produce new kinds of plants with this method.

Although Dr. Walbot feels that most of these fears are unfounded, she has cautiously kept her experiments confined to her laboratory. But she sees the day when she can move her experiments out of the laboratory and into the fields. She hopes to alter dramatically the way crops are produced. Her dream of "farming" the desert with plants that can grow without additional water may result in more food for hungry people.

Dr. Walbot says, "Custom designing plants will have profound consequences. Things we thought would take years to develop are taking months. Everything is moving so quickly."

Issues in Science

Rain Forests: Will They Be Saved?

A visitor stands for a moment listening to the clatter that surrounds her. Strange chirps, clicks, and howls blend in an eerie song. A constant rustle of leaves fills the air. The noise level rises and falls, but it is never silent in a rain forest.

Scientists believe that 3 million of the approximately 4 million known species of plants and animals in the world live in tropical rain forests. And scientists estimate that they have not seen or named another 2.5 million species of plants and animals that live there. Biologists consider the world's rain forests to be wondrous zoos, greenhouses, and living museums.

When undisturbed, the natural sounds in a tropical rain forest continue as they have for millions of years. But, in the last 25 years, new sounds have been added to the natural sounds. These new sounds are made by the chain saws and bulldozers used to cut down the world's tropical rain forests at an astounding rate.

Tropical rain forests are cut down for many reasons. People use the wood for fuel and for building materials. People use the cleared land to plant crops for themselves or to feed the animals they raise. People also build towns on the cleared lands. And, of course, they use additional land to construct roads that connect the towns they build.

Today, two-thirds of the more than 4 billion people on earth live in countries with tropical rain forests. Much of the cleared land is used to produce food. At first, crops grow quickly on the newly cleared land. But, after a few years, the land becomes much less productive. The nutrients in the soil are used up. In about three years, farmers are forced to abandon the land they cleared and move to another piece of newly cleared land.

Some ecologists have expressed alarm at the rate at which tropical rain forests are being destroyed. They point out that the plants in a tropical rain forest, besides providing homes and food for many of the animals that live there, also produce a great deal of the earth's oxygen. As the plants grow, they give off oxygen that may be used by animals in places on the earth far away from a tropical rain forest. Some of the oxygen in the air you breathe might have been produced by a banana plant growing in Brazil.

Destroying tropical rain forests might also affect rainfall. Rain forests normally grow in areas that receive 200 to 500 centimeters of rain a year. The rain forest canopy, or upper layer of treetops, is so dense that only 20 percent of the rain that strikes the canopy reaches the forest floor. Most of the water remains trapped in the leaves of the canopy. Eventually the water evaporates and forms new rain clouds. Throughout the year, the tropical rain forest remains warm and moist. The water cycle would be interrupted if vast areas of forest lands were cleared of trees. In time, the climate of areas near tropical forests would also change.

Since the early 1960s, almost half of the rain forests in Central America have been destroyed. Each year, nearly 12,000 square kilometers of tropical rain forests—an area about the size of Pennsylvania—is leveled. Most countries will lose all of their tropical rain forests within three decades if the current rate of destruction continues.

Some countries limit the amount of rain forest that may be cleared each year. Other countries have set up national parks to protect their forest areas. But human needs are great and an ever-increasing human population continues to put a great deal of pressure on the remaining forest lands.

How can the world's tropical rain forests be saved while providing food and space for a growing human population? What do you think should be done?

Today as large machines clear tropical rain forests, many plants and the home they provide for animals are destroyed.

Animals

On a small pond, tiny water beetles glide along the slate-blue surface of the water. The water beetles pay little attention to a deer standing near the pond. But, beneath the water, a hungry popeyed minnow is following every move of the water beetles. The minnow is about to strike an unsuspecting beetle and catch its daily meal.

As a soft breeze blows over the pond, the pleasant odors of pond grasses drift toward the twitching nostrils of a cottontail rabbit feeding near the water's edge. Just then a group of migrating geese land feet first in the pond water. As their feet hit the water, a speckled salamander quickly swims out of their way. Nearby, a croaking frog can be heard calling to its mate. In the bushes near the frog, a tiny black snake slithers along the ground in search of food.

You can see that a wide variety of animals may live in the area surrounding a small pond. Yet even these animals are only a tiny fraction of the different kinds of animals that live on the earth. As you read this unit, you will discover a great deal about the animals that inhabit earth's lands, seas, and air. Read on and you will discover why scientists call earth the living planet.

CHAPTERS

Canada geese are among the many organisms that visit this pond in New Jersey.

Invertebrates 9

CHAPTER OBJECTIVES

After completing this chapter, you will be able to

9–1 Compare vertebrates and invertebrates.

9–2 Describe the characteristics of sponges.

9–3 Describe the characteristics of coelenterates.

9–4 Compare flatworms, roundworms, and segmented worms.

9–5 Describe the characteristics of three groups of mollusks.

9–6 Describe echinoderms.

9–7 Identify the four major groups of arthropods.

9–8 Describe the characteristics of insects.

Long before the first dinosaur appeared on earth, bird-sized dragonflies soared through the skies. These prehistoric insects had wings that spread 76 cm across. Of course, the dragonflies you see today are much smaller. But they are still some of the most amazing fliers in the animal kingdom!

Like a helicopter, a dragonfly can fly straight up or straight down. It can move to the right or to the left. It can glide forward or backward, or just hover in the air. And it can land on a lily pad in a pond without causing a ripple.

The average speed of a dragonfly in flight is 40 kilometers per hour. But some have been clocked at speeds of nearly 100 kilometers per hour. And a dragonfly does not even have to stop flying to catch a meal. The insect simply forms a basket with its legs and scoops its prey out of the air. Such an accomplished flier is the dragonfly, that it actually mates in the air!

The one thing the dragonfly cannot do is fold its gauzelike wings. In this way, the insect is similar to an airplane. And believe it or not, engineers have been studying dragonfly flight in the hope that they can learn to build safer, quieter, and more efficient airplanes.

Insects such as the dragonfly belong to a fascinating group of animals called invertebrates. You will read more about the world of invertebrates in the pages that follow. Read carefully, for it is unlikely a day passes in your life that you do not encounter one of the earth's invertebrates.

A hawker dragonfly, newly emerged from its old skin, sits on an iris.

9–1 Characteristics of Invertebrates

Animals! Just hearing the word brings a different image to almost everyone. Some people think of the fierce great cats in Africa. To others, the word brings to mind the friendly porpoise, certainly among the most intelligent of all animals. Of course, for a great many people the word *animal* reminds them of that cuddly puppy they had, or have, or have always wanted to get. All of these animals, no matter how different they may seem, have one thing in common. They are **vertebrates** (VER-tuh-brihts). **Vertebrates are animals with a backbone.**

On the other hand, almost nobody thinks of the earthworm crawling through the soil when the word *animals* is mentioned. Or the mosquito about to raise a bump on another victim. Or the jellyfish with its stinging cells always ready to poison a passing fish. And yet, these organisms are among a group of animals called **invertebrates** that makes up more than 90 percent of all animal species. **Invertebrates are**

Figure 9–1 *The red sun star (top left), pink scallop (top right), white booted shrimp (bottom left), and katydid (bottom right) are all invertebrates. How do invertebrates differ from vertebrates?*

animals without a backbone. They are found in just about every corner of the world. Invertebrates can be classified into different groups, or phyla, according to their body structure.

SECTION REVIEW

1. What characteristic is used to classify an organism as a vertebrate or an invertebrate?
2. How are invertebrates classified into phyla?

9–2 Poriferans: Sponges

Section Objective

To identify the main characteristic of poriferans

Sponges are the simplest group of invertebrates, and they lead very simple lives. They grow attached to one spot on the ocean floor, usually in shallow water. A sponge stays in that one spot its entire life unless a strong wave or current washes it somewhere else.

Because the body of a sponge has many pores, it is classified in the phylum of poriferans. The word *poriferan* (po-RIHF-uhr-uhn), means "pore-bearing." Moving ocean water carries food and oxygen through the **pores,** or holes, into the sponge. The sponge's cells remove food and oxygen from the water and, at the same time, release waste products into the water. Because the pores of a sponge are small, most sponges live in clear water that is free of floating matter that could clog the pores.

Figure 9–2 *The body of the vase sponge, like that of all sponges, contains pores. What is the function of the pores?*

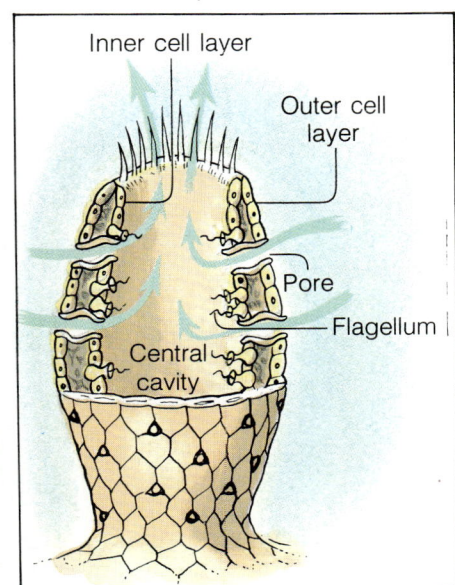

Inner cell layer

Outer cell layer

Pore

Flagellum

Central cavity

197

1. Obtain a natural sponge from your teacher.

2. Use a hand lens to examine the surface and the pores. Make a drawing of what you see.

3. Tear off a small piece of the sponge and place it on a glass microscope slide. Draw what you observe under the microscope. Spicules are hard, pointed structures that support the body of some sponges. Can you see any spicules? Draw what you observe.

4. Now repeat this activity with a synthetic kitchen sponge. How does it compare with a natural sponge?

Sponge cells are unusual in that they function on their own without any coordination with one another. In fact, some people think of sponges as a colony of cells living together. However, despite their independent functioning, sponge cells have a mysterious attraction to one another. This attraction can be demonstrated easily by passing a sponge through a fine filter so that it is broken into clumps of cells. Within hours, these cells reform into the shape of the original sponge. No other animal species shares this amazing ability of sponge cells to reorganize themselves.

SECTION REVIEW

1. What is the main characteristic of sponges?
2. How do food and oxygen enter a sponge's body?
3. Predict what would happen to a sponge if it lived in water that contained a lot of floating matter.

Figure 9–3 *The jellyfish is a coelenterate. What is the main characteristic of coelenterates?*

9–3 Coelenterates

The **coelenterates,** (sih-LEHN-tuh-rayts), are a phylum of invertebrates that includes corals, jellyfish, hydra, and sea anemones. The word *coelenterate* means "hollow body cavity." **All coelenterates contain a central cavity with only one opening.**

You can think of coelenterates as being cup-shaped animals with an open mouth. Tentacles with stinging cells called **nematocysts** (NEHM-uh-toh-sihsts) surround the mouth of most coelenterates. Coelenterates use nematocysts to stun or kill other animals. It is no surprise, then, that the stinging cells are near the mouth. For after capturing an animal with their tentacles, coelenterates pull it into their mouth and then into their central body cavity. Once the food is digested, coelenterates release waste products through their mouth.

Unlike the sponges, coelenterates contain groups of cells that perform special functions. That is, coelenterates have specialized tissues. For example, coelenterates such as jellyfish use muscle tissue to move about in the water. Others, like the corals, remain in one place.

Figure 9–4 *The hard outer covering of corals (left) forms coral reefs, such as this one in the Red Sea (right).*

Corals

Corals, like all coelenterates, are soft-bodied organisms. But corals use minerals in the water to build a hard protective covering of limestone. When the coral dies, the hard outer covering is left behind. Year after year, for many millions of years, generations of corals live and die, each adding a layer of limestone. In time, a coral reef such as the Great Barrier Reef forms. The outer layer of the reef, then, contains living corals. But underneath this "living stone" are the remains of corals that may have lived when dinosaurs walked the earth.

Corals live together in colonies that can grow into a wide variety of shapes and colors. Some corals look like antlers, others like fans swaying in the water, while still others look like underwater brains.

At first glance, a coral appears to be little more than a mouth surrounded by stinging tentacles. See Figure 9–5. But there is more to a coral than meets the eye. Algae live inside the coral's body. The algae help make food for the coral. And, since algae need sunlight to make food, corals must live in shallow water where sunlight can reach them. This relationship between a coral animal and an alga plant is among the most unusual in nature.

Figure 9–5 *The mouth of this coral is surrounded by tentacles. What is the function of tentacles?*

Figure 9–6 *Notice the clownfish swimming among the tentacles of the sea anemone. How do these organisms help each other?*

Sea Anemones

Can you spot the clownfish swimming through the plant in Figure 9–6? You might be surprised to discover that the "plant" is actually an animal—a type of coelenterate called a sea anemone. Sea anemones look like underwater flowers. However, the "petals" are really tentacles, and their brilliant coloring helps attract passing fish. When a fish passes over the anemone's stinging cells, the cells poison the fish. The tentacles then pull the fish into the anemone's mouth, and the stunned prey soon is digested.

The clownfish, however, is not harmed by the anemone. It is not affected by the anemone's poison and swims safely through the anemone's tentacles. In this way, the clownfish is protected from other fish that might try to attack it. At the same time, the clownfish serves as a kind of living bait for the anemone. Other fish see the clownfish, come closer, and are trapped by the anemone. You can see how the sea anemone and the clownfish help each other.

Jellyfish

If you were ever swimming in the ocean and saw a jellylike cup floating in the water, you probably knew enough to swim away. Most people quickly recognize this coelenterate, the jellyfish. Although the jellyfish may look harmless, it can deliver a painful poison through its stinging cells. In fact, even when they are broken up into small pieces, the stinging cells remain active and can sting a passing swimmer who accidentally bumps into them. Of course, jellyfish do not have stinging cells merely to bother passing swimmers. Like the other coelenterates, jellyfish use the stinging cells to capture prey.

SECTION REVIEW

1. What is the main characteristic of coelenterates?

2. What is the function of nematocysts?

3. How do waste products leave a coelenterate's body?

4. Which group of animals is more complex, the poriferans or the coelenterates? Explain.

9–4 Worms

Most people think of a worm as a slimy, squiggly creature. And, in fact, many are. However, there are other kinds of worms that look nothing like the worms used to bait fishing hooks. See Figure 9–7. The shape of a worm helps you classify it. **Worms can be classified into three main groups, or phyla: the flatworms, the roundworms, and the segmented worms.**

Figure 9–7 *Not all worms are slimy and squiggly. How do you think the feather duster worm (left) and the plume worm (right) got their names?*

Platyhelminths: Flatworms

You have probably never seen worms like the ones in Figure 9–8 on page 202. This worm, a planarian, is a kind of flatworm. Flatworms, as you might expect, have flat bodies. They are grouped in the phylum of platyhelminths (plat-ee-HEHL-mihnths). The word *platyhelminthes* comes from two Greek words, *platy* and *helminthes*. Can you guess what these two words mean?

Most planarians live in ponds and streams, often on the bottom of plants or on underwater rocks. Planarians feed on any dead plant or animal matter.

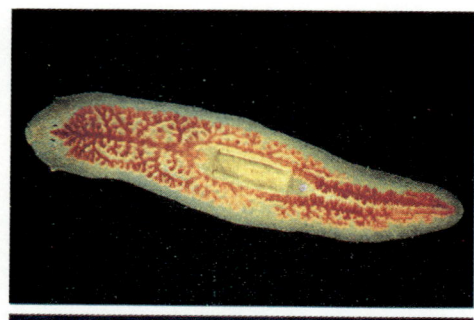

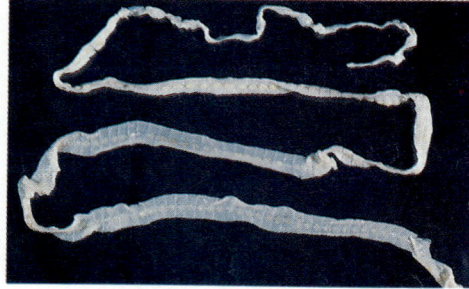

Figure 9–8 *The planarian (top) and the tapeworm (bottom) are examples of flatworms. In which phylum do flatworms belong?*

Figure 9–9 *Trichina worms can live in the muscle tissue of a human being. Here you can see how individual worms have imbedded themselves in the muscle. How do the worms enter the human body?*

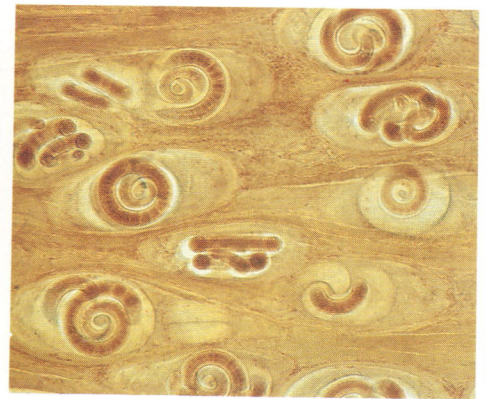

However, when there is little food available, some planarians do a rather unusual thing. They break down their own organs and body parts for food. Later, the missing parts regrow when food is available again. In fact, if a planarian is cut into pieces, each piece eventually grows into a new planarian! The ability of an organism to regrow lost parts is called **regeneration.**

Not all flatworms are found on dead or nonliving matter. Some are parasites and grow on or in living things. Tapeworms, for example, look like long flat ribbons and live in the bodies of many animals, including human beings.

The head of a tapeworm has special hooks that it uses to attach itself to the tissues of the **host,** or animal in which it lives. The tapeworm then takes food and water from its host. In return, the tapeworm gives back nothing.

Tapeworms can grow quite large inside the host. One of the largest tapeworms can live in human beings and grow to be 18 meters long. However, size is not always a good indicator of danger. The most dangerous tapeworm known to humans is only about 8 millimeters long and enters the body through microscopic eggs in food.

Nematodes: Roundworms

You probably have been told time and time again never to eat pork unless it is well cooked. Did you know that the reason had to do with the *Trichina* (trih-KEE-nuh) worm? This worm lives in the muscle tissue of pigs. If a piece of pork is cooked improperly, the worm may survive and enter a human's body. In the body, the worm lives in the muscle tissue. The disease it causes, called trichinosis (trihk-uh-NOH-sihs), can be very painful and is difficult to cure.

Trichina is a kind of roundworm. All roundworms resemble strands of spaghetti with pointed ends. Roundworms belong to the phylum of nematodes. The word *nematode* means "threadlike."

Roundworms live on land or in water. Many roundworms are animal parasites, although some live on plants. One roundworm, the hookworm, infects more than 600 million people in the world each year. The worms enter the body by burrowing through the skin. Eventually, they end up in the intestines where they live on the blood of the host.

Roundworms, like other worms, have both a head end and a tail end. In fact, worms were the first organisms to develop with distinct head and tail ends. Connecting the two ends is a tube called the digestive tube. Food enters the tube in the head end and waste products leave through the tail end. Although this system may not seem very unusual, it is far more advanced than that of the coelenterates, which use a mouth for both taking in food and releasing wastes.

Annelids: Segmented Worms

The worm you are probably most familiar with is the common earthworm. Earthworms belong to the phylum of annelids. The word *annelid* means "ringed." As you might expect, the earthworm's body is divided into numerous rings, or segments, usually at least 100.

Earthworms, of course, live in the soil. But other segmented worms live in salty oceans or fresh water lakes and streams. Segmented worms also include sandworms and leeches.

If you have ever felt an earthworm, you know it has a slimy outer layer. This layer, made up of slippery mucus, helps the earthworm glide through the soil. Small **setae** (singular: seta), or bristles, on the segments of the earthworm also help it to pull itself along the ground.

Earthworms are good to have in the garden. As they pass through the soil, the earthworms feed on

Figure 9–10 *This drawing shows the digestive system of the earthworm, a segmented worm. The photograph is of an earthworm moving through the soil. What structures help the earthworm move?*

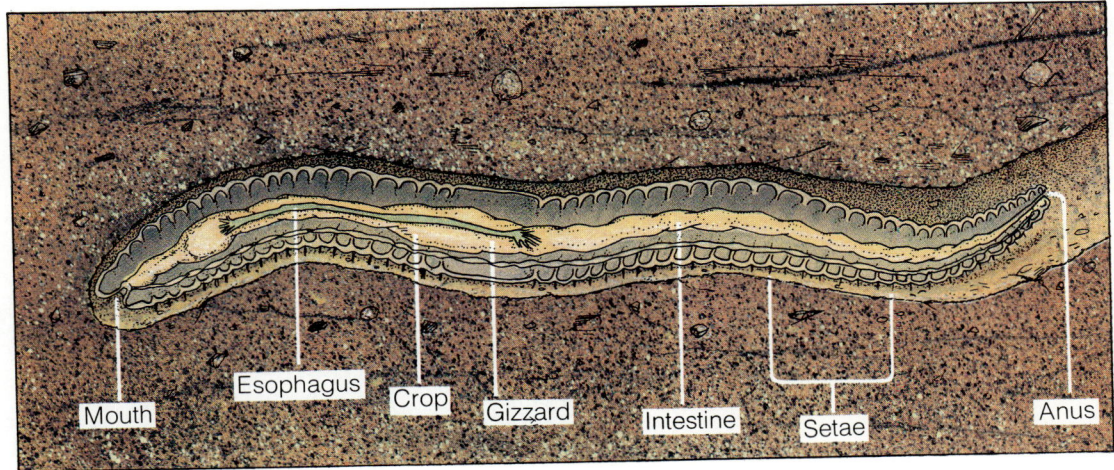

Mouth   Esophagus   Crop   Gizzard   Intestine   Setae   Anus

A Worm Farm

Here is a way to observe earthworms in soil.

1. Fill a clear plastic box with about 2 cm of sand.

2. Place about 7 cm of loosely packed topsoil over the sand.

3. Use pond water or tap water that has been standing for a day to slightly moisten the soil. Add more water whenever the soil appears dry.

4. Place between six and twelve earthworms on top of the soil.

5. Cover the box with clear plastic wrap. Put a few holes in it so that the worms can get air.

6. Observe the worms for a week. Record your observations.

How do the earthworms move through the soil?

dead plant and animal matter. Earthworms create tiny passages as they move. Air enters these passages and improves the quality of the soil. In addition, waste products released by earthworms help fertilize the soil.

Earthworms have a well-developed digestive system. Food enters the earthworm's body through its mouth. Then the food passes through a muscular pharynx (FAR-ihngks), and a tube called the esophagus (ih-SAHF-uh-guhs), before reaching the worm's **crop.** The crop is a saclike organ that stores food. From the crop, the food travels into the worm's muscular **gizzard.** The gizzard grinds up food and then passes it into the worm's intestine. Nutrients are removed from the intestine and enter the worm's circulatory system.

The earthworm has a closed circulatory system. All of its body fluids are contained within small tubes. The fluids are pumped throughout the earthworm's body by a series of special vessels that act like a tiny "heart."

In the respiratory system of the earthworm, oxygen enters through the skin. The earthworm produces carbon dioxide gas, which leaves through the animal's skin. However, the skin must remain moist in order for gases to pass through. If an earthworm remains out in the heat of the sun, the skin dries out and the animal suffocates.

The earthworm's reproductive system has both male and female structures. In one part of the earthworm, sperm cells are produced. In another part, egg cells are made. But although earthworms

Figure 9–11 *Individual earthworms contain both male and female structures. When two earthworms mate, they exchange sperm cells. The sperm of each worm unites with the eggs of the other worm. What is this process called?*

have both types of reproductive cells, reproduction occurs through sexual reproduction. Earthworms mate when two worms join together and exchange sperm cells. The sperm of each worm unite with the eggs of the other worm. This process is called **fertilization.** After the worms move apart, each worm lays a number of fertilized eggs in a slimy shell or cocoon. Eventually, the young worms break out of the shell and crawl away.

Although earthworms have a simple nervous system, they are very sensitive to their environment. Earthworms seem to be able to both sense danger and warn other earthworms. They do so by releasing a sweatlike material that helps them glide away much faster and warns other worms in the area at the same time.

Sharpen Your Skills

Flukes

Flukes are parasitic flatworms that have interesting life cycles. Visit the library and look up the life cycle of the liver fluke. On a sheet of white paper, make a diagram of its life cycle. Use arrows to show its life path through the various hosts. How many hosts are there in one life cycle?

SECTION REVIEW

1. List the three phyla of worms and give an example of each.
2. Name two worms that cause disease in humans.
3. List in order the structures through which food passes in an earthworm's body.
4. In what way are worms more complex than coelenterates?
5. Describe a method that would prevent people from being infected by hookworms.

9–5 Mollusks

Section Objective

To identify the main characteristic of mollusks

If you enjoy seafood, you probably are familiar with such **mollusks** (MAHL-uhsks) as clams, oysters, mussels, octopuses, and squids. The word *mollusk* comes from the Latin word meaning "soft." **Mollusks have a soft, fleshy body, which, in many species, is covered with a hard shell.** See Figure 9–12 on page 206. Which of the mollusks that you eat have shells?

The bodies of most kinds of mollusks that have outer shells are similar. Most mollusks have a thick muscular foot. The foot of some mollusks opens and closes their shell. In other species, the foot is used for movement. Some mollusks even use their foot to bury themselves in the sand or mud.

Figure 9–12 *Mollusks such as the scallop (right) have a soft body. And, like the scallop, many are also covered by a hard shell. The shells of mollusks, such as whelks and mussels, are often found scattered along the beach. (left). The outer coverings of another kind of animal called a sea urchin can also be seen in this photograph.*

The head region of a mollusk generally contains the mouth and sense organs such as the eyes. The rest of the body contains various organs involved in reproduction, circulation, digestion, and other important processes. Covering much of the body is a soft **mantle.** The mantle is the part of a mollusk that produces the material that makes up the hard shell. As the animal grows, the mantle enlarges the shell, providing more room for its occupant.

Mollusks are grouped according to certain characteristics. These characteristics include the presence of a shell, the type of shell, and the type of foot.

One-Shelled Mollusks

Many kinds of mollusks have only one shell and are called univalves. *Uni-* is the prefix that means "one" and *valve* is the word scientists use to mean "shell." There are many kinds of univalves, which live in the ocean, in fresh water, or on land. However, univalves that live on land still must have a moist environment to survive.

Univalves have an interesting feature called a **radula** in their mouth. The radula resembles a file used by carpenters to file wood. The radula files off bits of plant matter into small pieces that can be swallowed by the univalve. Some species of univalves

Figure 9–13 *Univalves such as the snail have two eyes located on stalks sticking out of their heads. How many shells do most univalves have?*

can inject a poison through the radula that can even be a danger to people.

The land univalve you are probably most familiar with is the common garden snail. This type of snail moves slowly along a trail of mucus that it produces. Because it releases this slippery mucus, it is able to travel easily over rough surfaces.

It may seem strange to you, but there is one kind of univalve that does not have any visible shell. It is the slug. Slugs have reduced shells, internal shells, or no shells at all. See Figure 9–15. Many slugs, called sea slugs, live in ocean water. However, some slugs live on land. Most people are familiar with the simple slugs they often find in moist areas.

Two-Shelled Mollusks

If you have been to the beach, you may have seen clam and mussel shells along the shore. Clams and mussels belong to a group of mollusks called bivalves, or two-shelled mollusks. Bivalves move through the water by clapping their two shells together, which forces water out between the shells. The force of the moving water propels the bivalve.

Bivalves do not have radula. Instead, they feed on small organisms in the water. Bivalves are often called filter feeders because they spend most of their time straining the water for food.

Figure 9–14 *This bay scallop has two shells. The blue dots are its eyes. What is the name for a two-shelled mollusk?*

Figure 9–15 *The sea slug is an example of a univalve. How does it differ from other univalves?*

Figure 9–16 *The squid (top left), nautilus (top right), and octopus (bottom) are head-footed mollusks. How do these animals move about?*

Head-Footed Mollusks

The most highly developed mollusks are the head-footed mollusks. These mollusks include the octopus, the squid, and the nautilus. Most head-footed mollusks do not have an outer shell, but do have some part of a shell within their body. An exception is the chambered nautilus. The nautilus shell consists of many chambers, or rooms. These chambers are small when the animal is young but increase in size as the animal grows. The nautilus constructs a new chamber as it grows. It lives in the outer chamber.

All of these mollusks have tentacles that are used to capture food and to move. But these mollusks differ in the number and type of tentacles they possess. Octopuses, of course, have eight tentacles. Squids have ten tentacles, although two of them differ in shape from the other eight. These two tentacles are shaped like paddles.

Octopuses, squids, and nautiluses use a water propulsion system for movement. They force water out of a tube in one direction, which pushes them along in the opposite direction. Using this "jet" system, these animals can move rapidly away from danger. Squids and octopuses also produce a purple dye. When they squirt this dye into the water, it hides them and confuses predators. The squid or octopus can then make good its escape.

SECTION REVIEW

1. What is the main characteristic of mollusks?
2. What characteristics are used to group mollusks?
3. List the three groups of mollusks and give an example of each.
4. Suppose you found a mollusk that had one shell and eyes located on two stalks sticking out of its head. Into which group of mollusks would you place it?

9–6 Echinoderms: Spiny-Skinned Animals

Section Objective

To identify the main characteristics of echinoderms

Earlier you read about the Great Barrier Reef. Today the living corals on the surface of the reef are in danger of being destroyed. If they are destroyed, the reef will grow no larger. What is damaging the corals? A kind of starfish called the crown-of-thorns sea star is responsible for the damage. See Figure 9–17. The crown-of-thorns eats the soft body parts of the living coral and leaves an empty shell behind.

The crown-of-thorns belongs to a phylum of animals called **echinoderms** (ih-KEE-nuh-dermz). The word *echinoderm* means "spiny-skin." **Echinoderms are invertebrates with rough, spiny skins.**

Starfish

Starfish are *not* fish. But most *are* shaped like stars. They have five or more arms that extend from a central body. On the bottom of these arms are thousands of tube feet that resemble tiny suction cups. These tube feet not only help the animal to move about, but do much more. For example, when a starfish passes over a clam, one of its favorite foods, the tiny tube feet grasp the clam's shell. The tube feet exert a tremendous force on the clam's shell, and eventually the shell opens. Then the starfish can eat a leisurely meal.

People who harvest clams from the ocean bottom have long been at war with starfish that destroy their clam beds. In the past, when starfish were captured near clam beds, they were cut into pieces and

Figure 9–17 *The crown-of-thorns starfish is an echinoderm that eats the soft body parts of living coral. What is the main characteristic of echinoderms?*

Figure 9–18 *A starfish can use its suctionlike tube feet to open a mussel's shell (right) or to attach itself to surfaces. This one-meter starfish (left) has its arms wrapped around a small submarine.*

Figure 9–19 *Like the starfish, the sea cucumber (top) and the purple sea urchin (bottom) are echinoderms. What do these animals have in common?*

thrown back into the water. But starfish have the ability to regenerate. So by cutting them up, all people did was make sure there would always be more and more starfish than before—exactly the opposite of what was wanted.

Other Echinoderms

Other echinoderms vary widely in appearance. Some, such as the sea cucumber, resemble vegetables. These animals usually are found lying on the bottom of the ocean. Sea cucumbers lack arms. They slowly move along the ocean bottom with tube feet or by wiggling back and forth.

Sea lilies have many arms that look like flower petals. These animals grow on stalks attached to the ocean bottom. Most sea lilies remain in one spot for their entire lives.

Sea urchins and sand dollars are round and lack arms. They may be flat like the sand dollar, or dome-shaped like the sea urchin. Some species have long poisonous spines used for protection.

SECTION REVIEW

1. What is the main characteristic of echinoderms?
2. What are two functions of the starfish's tube feet?

3. Why can clam harvesters not kill starfish by cutting them up?
4. What are some similarities and differences between echinoderms and mollusks? Which group do you think is more complex? Explain.

9–7 Arthropods

The most numerous and successful phylum of invertebrates is the **arthropods** (AHR-thruh-pahdz). The word *arthropod* means "jointed legs." **Jointed legs and an exoskeleton are the main characteristics of arthropods.** An **exoskeleton** is a rigid outer covering. In some ways it is similar to the armor worn by knights as protection in battle. One drawback of an exoskeleton is that it does not grow as the animal grows. So the arthropod's protective suit must be shed and replaced from time to time. While the exoskeleton is being replaced, the arthropod is more vulnerable to attack from other animals.

There are more species of arthropods than of all the other animal species combined. Arthropods live in air, on land, and in water. Wherever you happen to live, you can be sure arthropods live there too. In fact, arthropods are our main competitors for food

Figure 9–20 *These spider crabs have jointed legs and exoskeletons. To which phylum do they belong?*

Figure 9–21 *A Caribbean hermit crab, a crustacean, peers out from its borrowed shell. How do crustaceans obtain oxygen?*

and, if left alone and unchecked, they would eventually take over the world.

Arthropods include a variety of animal groups. Among these groups are crustaceans, centipedes and millipedes, arachnids, and insects.

Crustaceans

Do you see the two eyes peering out of the shell in Figure 9–21? These eyes belong to a hermit crab, a crab that uses discarded shells for its home. Crabs, along with lobsters, crayfish, and shrimp, are arthropods that are included among the crustaceans (kruhs-TAY-shuhnz). All of these animals live in a watery environment. Crustaceans obtain their oxygen from the water through special structures called **gills.** Even the few land-dwelling crustaceans have gills and must live in damp areas to get oxygen.

The bodies of crustaceans are divided into segments. A pair of limbs or other body parts, such as claws, are attached to each segment. These limbs have many different functions, depending on the organism. The claws of some crabs, for example, are strong enough to enable them to open their favorite food, coconuts. Crabs use their claws to cut through the tough outer husk of a coconut to reach the tender flesh inside. Other limbs are adapted for walking. The female lobster even attaches clusters of eggs to the limbs beneath her tail. In this way, she carries the eggs.

Crustaceans are able to regenerate certain parts of their body. A crab for example, can grow new claws. The stone crab lives in the waters off the coast of Florida. Its claws are considered particularly good eating. When a stone crab is caught, one of its claws is broken off and the crab is returned to the water. In about a year's time, the broken claw has regenerated. If the crab is unlucky enough to be caught again, that claw may once again be removed.

Centipedes and Millipedes

Both centipedes and millipedes have been described as worms with legs. In fact, you probably think the main difference between centipedes and millipedes is the number of legs. Actually, both types of arthropods have many legs, and you cannot easily tell them apart by counting. Centipedes have one pair of legs in a segment, while millipedes have two pairs of legs in a segment. However, if you were a tiny earthworm crawling through the soil you would certainly know the difference between the two. Millipedes live on plants and simply would pass you by. Centipedes, however, are carnivorous, or meat eating, and are active hunters. The centipede would use its claws to inject a poison into your body. So, for the earthworm, the difference between the two can mean the difference between life and death.

Arachnids

In Greek mythology, there is a legend of a young woman named Arachne who challenged the goddess Athena to a weaving contest. When the mortal Arachne won, Athena tore up her tapestry. Arachne hanged herself in sorrow. Athena, the legend goes, then changed Arachne into a spider. Her tapestry became a spider's web. Today spiders, as well as scorpions, ticks, and mites, are all included in a group of arthropods called arachnids (uh-RACK-nihdz). As you can guess, arachnids are named for Arachne.

Figure 9–22 *Centipedes are carnivorous arthropods. Here a Thai centipede uses its claws to inject its poison into a frog. How many pairs of legs does a centipede have in each segment?*

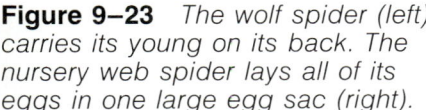

Figure 9–23 *The wolf spider (left) carries its young on its back. The nursery web spider lays all of its eggs in one large egg sac (right).*

The bodies of arachnids are divided into two main sections: a head–chest section and an abdominal section. Although arachnids vary in size and shape, they all have eight legs. So, if you ever find a small "bug," one way you can tell if it is an arachnid is to count its legs.

Arachnids live in many environments. Most spiders, for example, live on land. However, one kind of spider lives underwater inside a bubble of air it brings from the surface. Scorpions, for the most part, are found in dry desert areas. Ticks and mites live on other organisms. They may live on a plant and stay in one place, or they may live on an animal and go wherever the animal goes.

Spiders catch their prey in different ways. Many spiders make webs of fine, yet very strong, strands of silk. Special glands in the abdomens of spiders secrete this silk. Spiders' webs are often constructed in complicated and beautiful patterns. You may be surprised to learn that the spider actually spins two different kinds of silk. Some strands of the silk are sticky. These strands catch and hold prey until the spider is able to kill it. Other strands are not sticky. These are the strands the spider uses when it walks around its web. Many spiders weave a new web every day. At night the spiders eat the strands of that day's web, recycling this material the following day when they produce a new web.

Some spiders hide and spring out to surprise their prey. The trapdoor spider lives in a hole in the ground covered by a door made of silk and hid-

Figure 9–24 *Spiders have different methods of catching their prey. The black and yellow spider (top) traps a grasshopper in its web. The goldenrod crab spider (bottom left) blends in with a flower to surprise a butterfly. And the burrowing wolf spider (bottom right) hides in the sand waiting for unsuspecting prey.*

den by dirt and pieces of plants. When an unsuspecting insect passes close to the trapdoor, the spider rushes out to catch it.

Once a spider catches its dinner, it uses a pair of fangs to inject venom, or poison, into its prey. Sometimes the venom kills the prey immediately. Other times it paralyzes the prey so that the spider can save living creatures trapped in its web for another day when it needs more food.

Scorpions are active mostly at night. They capture and hold their prey with their large front claws while they inject venom through the stinger in their tail. During the day, scorpions hide under logs, stones, or in holes in the ground. Campers have to be careful when they put their boots on in the morning because a scorpion may have mistaken a boot lying on the ground for a suitable place to hide to escape the heat of the day.

Ticks and mites live off the body fluids of animals and plants. Many ticks suck blood from larger animals. When they do this they may spread disease. Rocky Mountain spotted fever, for example, is spread to people through the bites of ticks. Some ticks and mites live on insects. Other mites live by sucking juices from plant stems and leaves. They may also spread disease.

Figure 9–25 *The scorpion (top) and mite (bottom) are examples of arachnids. How many legs do arachnids have?*

SECTION REVIEW

1. What are the two main characteristics of arthropods?
2. What is the drawback of an exoskeleton?
3. List four groups of arthropods.
4. What is the difference between a centipede and a millipede?
5. What are three advantages of an exoskeleton?

9–8 Insects

Section Objective

To describe the characteristics of insects

By now you may have noticed that there is one group of arthropods, or animals with jointed legs, that has not been discussed. This group, of course, is the insects. It would be hard to overlook insects for too long because there are more kinds of insects than all other animal species combined. In fact, it has been estimated that there may be as many as 300 million insects for every single person alive on the earth!

Along with birds and bats, certain insects are the only animals that can fly. Insects vary greatly in appearance. If you examine different insect species, for example, you will find that the mouth parts vary greatly. A mosquito's mouth resembles a tiny needle. The mosquito can push its needlelike mouth through an animal's skin and remove some of the animal's body fluids. Other insects have mouths that are adapted to eating parts of plants. If you have ever watched a caterpillar eating a leaf, you know how efficient this insect's mouth is.

Anatomy of Insects

Although both are arthropods, insects are different from arachnids. An arachnid's body is divided into two sections. **An insect's body is divided into three main sections: a head, a thorax, and an abdomen. And, an insect has three pairs of legs.**

In the grasshopper, a typical insect, the three pairs of legs are not identical. In order for the grasshopper to jump, one pair of legs is much larger than the other two pairs. Which pair of legs do you think would be larger?

If you took a look at the head of a grasshopper, you would find five eyes peering at you. The grasshopper has three simple eyes on the front of its

Figure 9–26 *Weevil beetles (right) and the praying mantis (left) are insects. How many pairs of legs do insects have?*

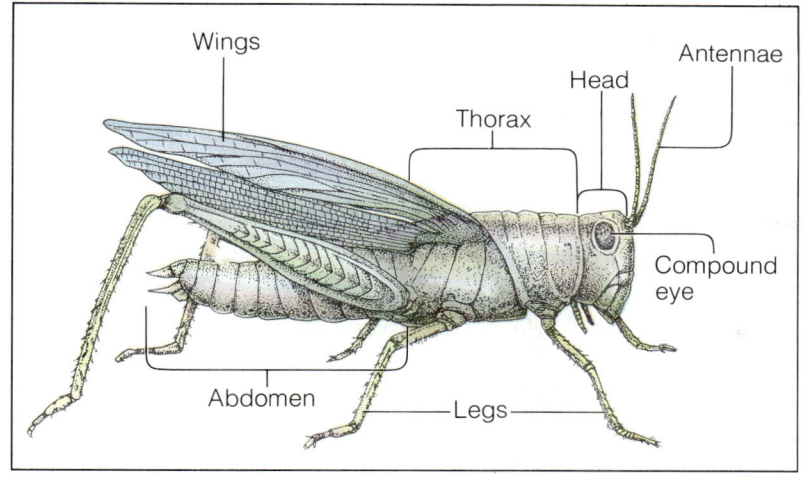

Figure 9–27 *Compare this rainbow grasshopper with the diagram of the structure of a grasshopper. What are the three body parts of an insect?*

head. These eyes can detect only light and dark. However, on the sides of its head are two compound eyes. Compound eyes contain many lenses. See Figure 9–28. These compound eyes can detect some colors, but they are best at detecting movement. This ability is important to an animal that is hunted daily by other animals for food.

Most insects have wings. The grasshopper has two pairs of wings, and it can fly quite well for short distances. Some kinds of grasshoppers can fly great distances in search of food. Insect flight varies from the gentle fluttering flight of a butterfly to the speedy flight of the hawkmoth, an insect that can fly as fast as 50 kilometers per hour.

Insects have an open circulatory system. Their blood usually is not contained within a system of blood vessels, as it is in humans. The blood moves around the inside of the insect's body, bathing the internal organs. An insect's blood carries food. But the blood does not carry oxygen. Insects do not have a well-developed system for moving oxygen into the body and waste gases out. Instead, they have a system of tubes that pass through the exoskeleton and into the insect's body. Gases move into and out of the insect's body through these tubes.

Figure 9–28 *Insects, such as this fly, have compound eyes. Within the eyes are many lenses that enable the insect to detect the slightest movement of an object.*

Growth and Development of Insects

Insects spend a great deal of time eating, and they grow rapidly. Like other arthropods, insects must shed their exoskeleton as they grow. As insects develop, they pass through several stages. Some species of insects change their appearance completely as they pass through the different stages. This change in appearance due to development is known as **metamorphosis** (meht-uh-MOR-fuh-sihs).

During metamorphosis, an insect passes through four distinct stages. The first stage is the egg. When the egg hatches, a **larva** emerges. A caterpillar, for example, is the larva of an insect that will one day become a butterfly or a moth. The larva spends almost all of its time eating and can eat all the leaves of a plant in a short time.

Eventually, the caterpillar begins the next phase, the **pupa** (PYOO-puh) phase. In this phase, the cater-

Figure 9–29 *The monarch butterfly begins life as an egg (top left) and then becomes a larva, also known as a caterpillar (top right). During the pupa stage (bottom left), it wraps itself into a cocoon. Finally the butterfly emerges (bottom right). What is the name for this series of changes due to development?*

HELP WANTED: BEEKEEPER No special education required. Courses in beekeeping, agriculture, and business helpful. Will train to harvest honey from honeycombs and clean, repair, and inspect beehives.

People who maintain hives of bees from which they collect honey and beeswax are called **beekeepers.** They sell the honey to bakeries for use in cookies and crackers or package it to sell in markets by the jar. The beeswax is sold for use in making candles, lipsticks, and polishes.

Beekeepers construct wooden boxes with removable frames as hives for the bees. The bees build honeycombs of wax on the frames. The insects bring nectar from flowers back to these honeycombs. The nectar has mixed with chemicals from the bees' stomachs. In the hive, water evaporates from the nectar mixture and it becomes honey. Bees used stored honey for food during the winter.

Anyone interested in beekeeping might begin by keeping a hive or two in a backyard or on a roof. Beginner colonies can usually be bought or rented from local beekeepers. For more information on a career working with these fascinating insects, write to the New Jersey Beekeepers Association, 157 Five Point Road, Colts Neck, NJ 07722.

pillar secretes a covering made of silk or another material. It wraps itself in this cocoon and appears to be sleeping. But inside the cocoon remarkable changes take place. The pupa changes into an adult insect with a completely different appearance. It is often difficult to believe that a beautiful butterfly was once a creeping caterpillar.

Insect Behavior

Most insects live alone. In this way, they do not compete directly with other members of their species for available food. They come together only to mate and to produce fertilized eggs. Insects attract mates in different ways. One way involves the giving off of a special scent, which you might think of as a kind of perfume. These scents are called **pheromones** (FAIR-uh-mohnz), extremely powerful chemicals that cannot be smelled by a human. However, even a small amount of a pheromone can be noticed by a potential mate over great distances. Some pheromones produced by female insects can attract a male located more than 11 kilometers away.

Other insects, known as social insects, cannot survive alone. These insects form colonies or hives. Ants, termites, some wasp species, and bees are social insects. They survive as a society of individual insects that perform different jobs. Many of these colonies are highly organized.

A beehive is a marvel of organization. Worker bees, actually infertile females, perform all of the work needed to ensure the survival of the hive. For example, worker bees supply the hive with food. They make the honey and the combs to store it. They feed the queen bee, whose only function is to lay huge numbers of eggs. Worker bees keep the queen bee clean as well as do the housekeeping for the hive. They also protect the hive. Male bees, whose only function is to fertilize the queen, are unable to feed themselves and so are also dependent upon the workers.

Defense Mechanisms of Insects

Insects have many defense mechanisms to ensure their survival. Wasps and bees have stingers that

Figure 9–30 *Can you believe this mound (top) was built by termites? Termites, ants (bottom left), and honey bees (right) are all social insects. How are they different from most other insects?*

Figure 9–31 *Insects defend themselves in many ways. The bombardier beetle (bottom) sprays a foul-smelling chemical. The peacock moth (top left) has eyespots that startle predators. How does the tropical walking stick (top right) defend itself?*

they use to defend themselves against enemies. Other insects are masters of **camouflage** (KAM-uh-flahzh), that is, they can hide from their enemies by blending into their surroundings. These insects survive because their bodies are not easily seen. Some insects, for example, resemble sticks and twigs. Other insects resemble leaves or the thorns of plants. Some insects have the ability to spray foul-smelling chemicals at an enemy. Other insects have markings that frighten birds and other animals that might eat them. In Figure 9–31, you can see two large "eyespots" on the wings of the moth. These spots startle other animals and may confuse the moth's enemies long enough for the insect to escape.

SECTION REVIEW

1. How are insects different from other arthropods?
2. What is metamorphosis?
3. List the four stages of metamorphosis.
4. Predict how scientists might use pheromones to control insect pests.

Investigating Isopod Environments

Problem

What type of environment do isopods (pill bugs) prefer?

> **Materials** *(per group)*
>
> collecting jar
> 10 isopods
> shoe box with a lid
> aluminum foil
> 2 paper towels
> masking tape
> water

Procedure

1. With your collecting jar, gather some isopods. These are usually found under loose bricks or logs. Observe the characteristics of the isopods.
2. Line a shoe box with aluminum foil.
3. Tape down two paper towels side by side in the bottom of the shoe box. Separate them with a strip of masking tape.
4. Moisten the paper towel on the left side of the box only.
5. Place the ten isopods on the masking tape. Put the lid back on the shoe box.
6. Predict what will happen when you open the lid. Wait five minutes. Meanwhile, make a data table.
7. After five minutes, open the lid and quickly count the number of isopods on the dry paper, on the masking tape, and on the moist paper. Record the results in your data table.
8. Repeat the procedure two more times. Before each trial, be sure to place the isopods on the masking tape. Record your results in your data table.
9. After you have completed the three trials, find the average result for each column (dry, tape, moist). To do this, add up each column and divide by 3. Record your average results on a class chart.

Observations

1. How did the isopods react when you opened the lid of the box? How did this reaction compare with your prediction?
2. What was the variable in this experiment?
3. Were there other variables in the experiment that could have affected the outcome? If so, what were they?
4. What was the control in this experiment?
5. How did your results compare to the class results?

Conclusions

1. From the class results, what conclusions can you draw about the habitats isopods prefer?
2. Into which phylum of invertebrates would you classify isopods? Into which group within that phylum would you place isopods? What characteristics led you to your conclusions?
3. What was the purpose of the masking tape in the experiment?
4. Why did you go through the procedure three times?
5. Design another experiment in which you test the following hypothesis: Isopods prefer dark environments over light environments. Be sure to include a variable and a control in your design.

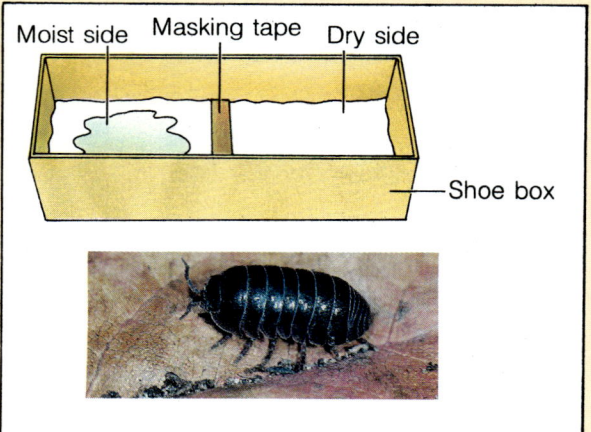

Moist side   Masking tape   Dry side

Shoe box

CHAPTER REVIEW

SUMMARY

9–1 Invertebrates

❏ Vertebrates are animals with a backbone, and invertebrates are animals without a backbone.

❏ Invertebrates are grouped into phyla according to their body structure.

9–2 Poriferans: Sponges

❏ Sponges are classified as poriferans because their bodies have many pores.

❏ The cells of sponges remove food and oxygen from ocean water as the water flows through pores. The water flowing out carries away waste products.

9–3 Coelenterates

❏ Coelenterates have a central body cavity with one opening.

❏ Most coelenterates have stinging cells called nematocysts on their tentacles. Coelenterates sting and capture their prey, which they digest in the central body cavity. Waste products are released through the mouth.

9–4 Worms

❏ Worms are classified into three phyla.

❏ Flatworms, or platyhelminths, have flat bodies and live in ponds and streams. Flatworms can regenerate missing or cut-off parts.

❏ Roundworms, or nematodes, resemble strands of spaghetti. Food passes from the head end to the tail end through a digestive tube. Trichinosis is a disease caused by eating pork that contains roundworms.

❏ Segmented worms, or annelids, such as earthworms, have segmented bodies and live in soil, salt water, or fresh water.

9–5 Mollusks

❏ Mollusks have soft bodies covered by a shell-producing mantle. Their muscular foot opens and closes the shell and permits movement. Snails, clams, and squids are examples of mollusks.

9–6 Echinoderms: Spiny-Skinned Animals

❏ Invertebrates with rough, spiny skin are classified as echinoderms. The group includes starfish, sea cucumbers, sea lilies, sea urchins, and sand dollars.

9–7 Arthropods

❏ Arthropods have jointed legs and an exoskeleton. The group includes crustaceans, centipedes and millipedes, arachnids, and insects.

9–8 Insects

❏ Insects have three body parts, three pairs of legs, and an open circulatory system.

VOCABULARY

Define each term in a complete sentence.

arthropod	gizzard	pheromone
camouflage	host	poriferan
coelenterate	invertebrate	pore
crop	larva	pupa
echinoderm	mantle	radula
exoskeleton	metamorphosis	regeneration
fertilization	mollusk	seta
gill	nematocyst	vertebrate

On a separate sheet of paper, write the letter of the answer that best completes each statement.

1. The word coelenterate means
   a. pore-bearing.   b. hollow body cavity.
   c. soft body.   d. spiny skin.

2. Which animal is *not* a coelenterate?
   a. coral   b. sea anemone   c. sponge   d. jellyfish

3. The animal that causes trichinosis is a
   a. roundworm.   b. flatworm.
   c. segmented worm.   d. mosquito.

4. All mollusks have a (an)
   a. outer shell.   b. soft body.   c. radula.   d. exoskeleton.

5. A mollusk's thick muscular foot is used for
   a. movement.   b. digging.
   c. opening and closing the shell.   d. all of these.

6. Starfish belong to a group of invertebrates called
   a. fish.   b. arthropods.
   c. coelenterates.   d. echinoderms.

7. Which is *not* a characteristic of all arthropods?
   a. jointed legs   b. exoskeleton
   c. gills   d. no backbone

8. Which group of invertebrates includes animals that can fly?
   a. coelenterates   b. arthropods   c. crustaceans   d. insects

9. In which stage of metamorphosis does the insect wrap itself in a cocoon?
   a. egg   b. larva   c. pupa   d. adult

10. Which is *not* an example of a defense mechanism in insects?
   a. camouflage   b. eyespots   c. pheromones   d. stinger

CONTENT REVIEW: COMPLETION

On a separate sheet of paper, write the word or words that best complete each statement.

1. Animals that have backbones are called _____.

2. Food and oxygen enter a sponge through its _____.

3. _____ built the Great Barrier Reef.

4. The organism from which a parasite obtains its food is called a (an) _____.

5. The _____ worm is a roundworm that lives in the muscle tissue of pigs.

6. The process in which sperm unite with an egg is called _____.

7. The _____ of a mollusk produces the material that makes up the hard shell.

8. _____ are two-shelled mollusks.

9. Crustaceans obtain their oxygen from the water through special structures called _____.

10. In the metamorphosis of an insect, the egg stage is followed by the _____ stage.

CONTENT REVIEW: TRUE OR FALSE

Determine whether each statement is true or false. Then on a separate sheet of paper, write "true" if it is true. If it is false, change the underlined word or words to make the statement true.

1. Nematocysts are found around the mouths of most <u>coelenterates</u>.
2. Flatworms are classified as <u>nematodes</u>.
3. In earthworms, the <u>gizzard</u> is a saclike organ that stores food.
4. All <u>poriferans</u> have soft, fleshy bodies.
5. <u>Univalves</u> have a radula in their mouth that files off bits of plant matter.
6. The octopus is a <u>head-footed</u> mollusk.
7. *Crustacean* means "jointed legs."
8. Arthropods have a rigid outer covering called a <u>mantle</u>.
9. Spiders are <u>insects</u>.
10. The special scents produced by insects to attract mates are called <u>pheromones</u>.

CONCEPT REVIEW: SKILL BUILDING

Use the skills you have developed in the chapter to complete each activity.

1. **Making charts** Construct a chart in which you list each phylum of invertebrates, the major characteristics of the phylum, and three animals from each group.
2. **Classifying objects** Your friend said he found a dead insect with two body parts and four legs. Is this possible? Explain.
3. **Making generalizations** In what ways are insects beneficial to human beings?
4. **Relating cause and effect** People with tapeworm infections eat a lot but still feel hungry and tired. Why?
5. **Making predictions** Would it be safe to eat clams from polluted water? Give a logical reason for your answer.
6. **Applying concepts** The makers of horror movies invent giant insects that terrorize human beings. Why is it impossible for such insects to exist?
7. **Relating concepts** Insects are often described as the most successful group of animals. What characteristics of insects could account for this description?
8. **Applying technology** Pesticides are chemicals used to kill harmful insects. Describe some advantages and disadvantages of pesticide use.
9. **Designing an experiment** The eyes of a squid are similar in structure to the eyes of a vertebrate. Design an experiment to determine whether or not squids are able to see color. Include a hypothesis, control, and variable in your design.

CONCEPT REVIEW: ESSAY

Discuss each of the following in a brief paragraph.

1. Describe the defense mechanism each of the following uses against predators: sea anemone, earthworm, squid, insect.
2. Explain the process of metamorphosis in an insect.
3. Describe the similarities and differences among the four groups of arthropods.
4. List the different methods invertebrates use to get food. For each method, give examples of those invertebrates that use it.

Coldblooded Vertebrates

10

CHAPTER OBJECTIVES

After completing this chapter, you will be able to

10–1 Describe the characteristics of vertebrates.

10–1 Compare warmblooded and coldblooded vertebrates.

10–2 List the three classes of fish and give an example of each.

10–3 Describe the main characteristics of amphibians.

10–3 Explain how metamorphosis occurs in frogs.

10–4 Describe the adaptations that allow reptiles to live their entire lives on land.

Imagine finding yourself face to face with a dragon! Not the fire-breathing, winged dragon that you read about in fairy tales, but a dragon nonetheless.

In 1912, while exploring Komodo Island in the South Pacific, scientists discovered a very unusual and fierce animal. It had dry, scaly skin, a forked tongue, and large teeth. It also had clawed feet and a powerful tail. Its legs were short but strong. The scientists were reminded of the long extinct dinosaurs. Usually the "dragon" would eat dead and decaying animals that it found. But it also could catch and eat goats and wild boars that lived on the island. Even a water buffalo with a mass of hundreds of kilograms could become food for this strange animal.

The Komodo dragon, as the animal came to be called, is the largest known lizard. It can grow to a length of more than three meters with a mass of more than 135 kilograms. Since 1912, Komodo dragons have been found on three other islands. Biologists think that there are only about 600 Komodo dragons living today. As you read the following pages, you may want to compare the characteristics of Komodo dragons with those of other coldblooded vertebrates.

The Komodo dragon is the largest known lizard alive today.

10–1 Characteristics of Vertebrates

Bats, snakes, and turtles—what do they have in common? Each of these animals is a vertebrate. Can you name some other vertebrates? **A vertebrate is an animal that has a backbone.** Because vertebrates belong to the **chordate** (KOR-dayt) phylum, they are also called chordates.

The bones that make up a vertebrate's backbone are called **vertebrae** (singular: vertebra). See Figure 10–1. The backbone is part of the vertebrate's **endoskeleton,** or internal skeleton. The skeleton provides support and helps give the body of a vertebrate its shape. One important advantage of an internal skeleton is that it increases in size as the animal grows. It does not have to be shed, as does the exoskeleton of an insect.

The backbone of a vertebrate is also important because it protects the nerves of the spinal cord. The spinal cord runs down through the center of the backbone. The nerves in the spinal cord connect the vertebrate's well-developed brain to other nerves that carry information to and from every part of its body.

There are seven classes, or groups, of vertebrates within the chordate phylum. Two of the classes of vertebrates are **warmblooded** and five are **coldblooded.** A warmblooded animal's body temperature remains the same despite the temperature of its surroundings. Its temperature is said to be constant. You will read more about these animals in

Sharpen Your Skills

Tunicates

There is a group of primitive chordates commonly called tunicates. Use reference materials in the library to find out about these animals. Present your findings to the class in an oral report.

Figure 10–1 *Like all vertebrates, fish have a series of bones called vertebrae that make up their backbone. What is the name for a vertebrate's internal skeleton?*

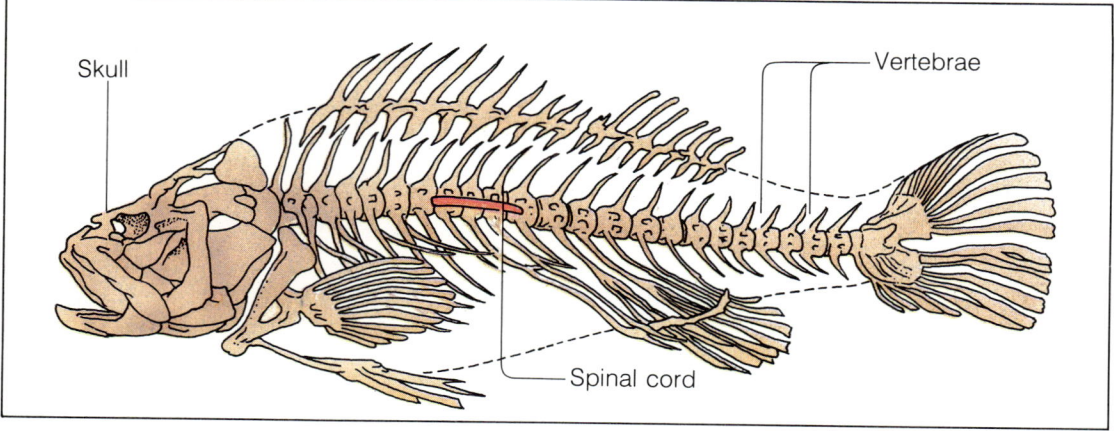

Skull · Vertebrae · Spinal cord

Chapter 11. A coldblooded animal's body temperature, on the other hand, changes somewhat with the temperature of its surroundings. But the body temperature must stay within a certain range in order for the animal to survive.

A lizard is an example of a coldblooded animal. When the temperature of its surroundings changes, the lizard must use certain behaviors to keep its body temperature at a level at which it can survive. During a day when the temperature is high, the lizard might stay under rocks where it is shady and cool. If the temperature is too cold, the lizard might sit on top of a rock with its body facing the sun. How would this behavior be helpful to the lizard?

Figure 10–2 *These marine iguanas are sunning themselves on the shore of a Galapagos island located off the west coast of South America. Iguanas are a type of lizard. Are these animals warmblooded or coldblooded?*

SECTION REVIEW

1. What are the characteristics of vertebrates?
2. Compare warmblooded and coldblooded vertebrates.
3. Are you a warmblooded or a coldblooded vertebrate? Explain.

10–2 Fish

About 500 million years ago, a tiny animal first appeared in the earth's oceans. This strange animal had no jaws. It did have fins, but they were not like the fins of modern fish. There was something very special about this animal—something that would group it with the many kinds of fish that were to come in later years. This early fish was the first animal with a backbone. In other words, it was the first vertebrate.

Fish are the vertebrates that are best adapted to a life underwater. Their smooth bodies, usually covered with scales, are streamlined, allowing them to glide through the water. Most fish have fins. Some fins keep a fish upright so that it does not roll over on its side. Other fins help the fish steer and stop. The side-to-side movement of the large tail fin helps the fish move through the water.

Like some invertebrates, fish take oxygen from the water through gills. As a fish swims, water

Figure 10–3 *Off the coast of Florida, angel fish, red snappers, and grunts swim amid the rocks. How do these animals take in oxygen?*

Figure 10–4 *Although sockeye salmon live in salt water for part of their lives, they swim upstream to lay their eggs in fresh water. Here you see a male and female salmon at a nest site (left) and a young salmon hatching (right).*

Figure 10–5 *The regal lion fish (top) lives in the Sulu Sea near the Philippines. The goat fish (bottom) lives in the Red Sea. What would happen to these tropical fish if you were to place them in the waters off the coast of Maine? Why?*

passes through its gills. Oxygen passes from the water into blood vessels in the gills. Carbon dioxide wastes also pass out of the fish through the gills. Fish have well-developed digestive, circulatory, and nervous systems, as do the other vertebrates you will read about.

Fish reproduce in an interesting manner. In most species, female fish release jelly-coated eggs directly into the water. Then, male fish release sperm to fertilize the eggs. Because this type of fertilization occurs outside the body of the female it is called **external fertilization.**

Fish live in all of the earth's waters, in both fresh and salt water. As you might expect, the same fish usually cannot live in both fresh and salt water. Also, while some fish such as the cod, live in the cold waters of the Arctic, others, such as the fierce piranha, live in warm, tropical waters. If you took a tropical piranha and placed it in arctic waters, it would soon die. Why? Because fish are coldblooded, changes in their environment can affect their body temperature. If the change is severe, as from tropical to arctic waters, most fish will not survive. But because the water temperature in an area remains relatively constant compared to the temperature on land, fish have little trouble maintaining a constant body temperature as long as they remain in their native waters. This results in a saving of energy, for fish do not have to use food energy to keep warm.

Fish are divided into three classes: the jawless fish, the cartilaginous (kahrt-uhl-AJ-uh-nuhs) fish, and the bony fish. In the following pages, you will read about these fish, beginning with the most primitive class of fish.

Jawless Fish

Jawless fish are the most primitive of all fish. They are so primitive that they lack scales and fins as well as jaws. In Figure 10–6, you can see the most common jawless fish—the lamprey. The lamprey looks like a snake with a suction-cup mouth at one end. Even though the fish has no jaws, this suction-cup mouth is very efficient. Using its mouth, the lamprey attaches itself to the soft belly of some other fish such as a trout. Then, with its teeth and rough tongue, the lamprey drills a hole into the fish and sucks out its blood and other body fluids.

There is something else that makes jawless fish unusual. To find out what it is, take your ear between your fingers and move it back and forth a few times. Nothing breaks, does it? The reason is that your ear contains a flexible material called **cartilage.** In fact, before you were born your entire skeleton was made of cartilage. In time, the cartilage was replaced by bone, except in some places like your ear and the tip of your nose. The entire skeleton of a jawless fish is made of cartilage. The skeleton is never replaced by bone tissue. As you might expect, such a fish is very flexible and can be bent so that its head touches its tail without causing any damage to the fish.

Figure 10–6 *The lamprey attaches its suction cup mouth to the sides of other fish. After drilling a hole into its prey, the lamprey sucks out the blood and other body fluids from the fish. To which class of fish do lampreys belong?*

Cartilaginous Fish

When you think of sharks, you probably think of the great white shark, which has a reputation for eating people. Great white sharks, along with a few other types, have been known to attack people occasionally. But for the most part, sharks leave people alone and prefer to be left alone as well.

Sharks are included in a class of fish called the cartilaginous fish. Like those of jawless fish, sharks' skeletons are made of cartilage. But unlike the

Figure 10–7 *In spite of their enormous size, whale sharks are harmless to people. Whale sharks eat only microscopic organisms. In what way is a shark's skeleton similar to that of a jawless fish?*

Figure 10–8 *The toothsome great white shark (top) and the stingray (bottom) are examples of cartilaginous fish. What is the purpose of the poisonous spine at the end of the stingray's tail?*

jawless fish, sharks definitely have jaws. See Figure 10–8. In fact, sharks are probably the most successful predators on earth. And they have been so for hundreds of millions of years.

Also included among the cartilaginous fish are skates and rays. These fish have two large, broad fins that stick out from their sides. They beat these fins to move through the water, much as a bird beats its wings to fly through air. Rays and skates often lie on the ocean bottom, where they hide by using their "wings" to cover their bodies with sand. When an unsuspecting fish or invertebrate comes near, the hidden skate or ray is ready to attack. Some rays have a poisonous spine at the end of their long, thin tail. They use this for defense, not to catch prey. Other rays are able to produce small charges of electricity to stun and capture prey.

Bony Fish

Anyone who has eaten a flounder or a trout knows why such fish are called bony fish. Their skeleton is made of hard bones, many of which are quite small and sharp. Some bony fish, such as the tuna, travel in groups called schools. Because of this schooling behavior, these fish can be caught in large numbers at one time by people in fishing boats.

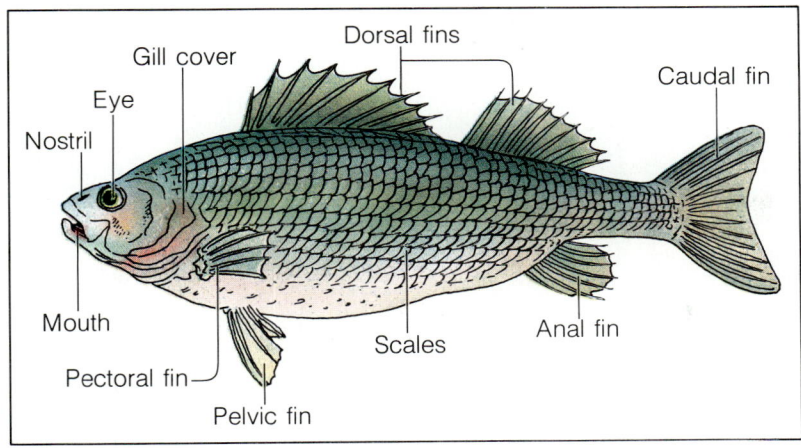

Figure 10-9 *This diagram shows the external structure of a bony fish. Which structures allow the fish to adapt to life underwater?*

Labels on figure: Gill cover, Dorsal fins, Caudal fin, Eye, Nostril, Mouth, Pectoral fin, Pelvic fin, Scales, Anal fin

One important characteristic of bony fish is their **swim bladder.** The swim bladder is a sac filled with air that the fish can deflate or inflate. The swim bladder acts in much the same way as a life preserver that keeps people afloat. By letting air in and out of the bladder, a fish can float at any level in the water. That is why a fish can sleep underwater and not sink, even though it is not moving its fins.

There are many kinds of bony fish. Some have made remarkable adaptations to life in the water. The electric eel, for example, can generate large

---

CAREER

Fish Farmer

HELP WANTED: FISH FARMER High school diploma desired. Courses in fish behavior, fish physiology, fish diseases, and business helpful. Job involves outdoor work.

In early America, when you sat down to a tasty fish dinner, it meant that somebody had had good luck fishing that day. But now, chances are that the fish you eat come from a fish farm. **Fish farmers** throughout America raise and breed catfish, salmon, trout, bass, and other types of fish to be sold at a profit to restaurants and markets.

At a fish farm, farmers place eggs from female fish into moist pans. Then they fertilize them with sperm cells from male fish. The fertilized eggs are then kept warm to ensure growth. When the fish hatch and grow to be as long as a person's finger, they are called finger-

lings. The fingerlings are moved to rearing ponds until fully grown.

You can learn more about fish farming by writing to the Marine Resources Research Institute, South Carolina Wildlife and Marine Resources Department, P.O. Box 12559; Charleston, SC 29412.

Figure 10–10 *The flounder (left) and the deep-sea fish (right) are examples of bony fish. How has each adapted to its surroundings?*

Sharpen Your Skills

Observing a Fish

Here is a way for you to study the structure of a fish.

1. Obtain a preserved fish from your teacher and place it in a tray.

2. Hold the fish in your hands and observe it. Note the size, shape, and color of the fish. Also note the number and location of the fins. To which class does your fish belong?

3. Draw a diagram of the fish and label as many structures as you can.

4. Locate the fish's gill cover. Lift it up and examine the gills with a hand lens. If necessary, use a scissor to cut away the gill cover. **CAUTION:** *Be careful when using sharp instruments.* How many gills do you see?

5. Remove one of the scales from the fish.

6. Examine the scale under a microscope. Each year a new dark ring is added to the scale. How old is your fish?

amounts of electricity, which it uses to stun its prey. The remora, a saltwater fish, uses a sucker to attach itself to a shark and feeds on bits of food the shark leaves behind.

Another fish that has adapted to its surroundings is the flounder. It is able to change color to match the ocean bottom. In this way, the flounder hides from its predators. What is the name for this blending into the surroundings?

Fish that live deep in the ocean have adaptations that allow them to live where there is little light. Some have light-producing organs that flash on and off to attract prey. Others have huge eyes to help them see in the dark.

Still other fish, such as the mudskipper and the walking catfish, have adaptations that allow them to come out of the water and spend some time on land. And the lungfish can bury itself in mud to survive a dry season during which the water of the stream or pond in which it lives evaporates.

SECTION REVIEW

1. List the three classes of fish and give an example of each.

2. What is the advantage of an endoskeleton?

3. Describe an important function of the backbone.

4. What is the function of the swim bladder?

5. The sturgeon, a bony fish, can lay up to 6 million eggs. Why do you think it is necessary for most fish to produce so many eggs?

10–3 Amphibians

Amphibians are another class of coldblooded animals in the phylum of chordates. The word *amphibian* means "double life." And most amphibians do live a double life. **Amphibians spend part of their lives in water and part on land.** All amphibians, for example, are born from eggs laid in water. And even those amphibians that spend their entire lives on land must return to the water to reproduce. Why? The eggs of amphibians lack a hard outer shell and would dry out if they were not deposited in the water.

There is another reason amphibians cannot stray too far from water—or at least from a moist area. Amphibians breathe through their skin. However, the skin must remain damp in order for them to take in oxygen.

Figure 10–11 *This Chinese newt is an example of an amphibian. What does the word* amphibian *mean?*

Frogs and Toads

Have you ever wondered where frogs and toads go in the winter when the temperature drops? Frogs and toads, like all amphibians, are unable to move to warmer climates. But they can survive. Frogs often bury themselves beneath the muddy floor of a lake during the winter. Toads dig through dry ground below the frost line. Then these amphibians go into a winter sleep called **hibernation.** During hibernation, all body activities slow down so that the

Figure 10–12 *Brazilian horned frogs (left) are similar in shape to toads, such as the Southern toad (right). Is there a difference between a frog and a toad?*

Figure 10–13 *Unlike your tongue, the sticky tongue of this frog is attached to the front of its mouth, so the frog can quickly flick out its tongue and catch a tasty meal.*

animal can live on food stored in its body. The small amount of oxygen needed during hibernation passes through the amphibian's skin as it sleeps. Once warmer weather comes, the frog or toad awakes. If you live in the country, you can usually tell when this happens. The night is suddenly filled with the familiar peeps, squeaks, chirps, and grunts that male frogs and toads use to attract their mates.

Frogs and toads appear similar in shape. But if you touch them, you can tell one difference immediately. Frogs have a smooth, moist skin. Toads are drier and are usually covered with small wartlike bumps. In many toads, the bumps behind the eyes contain a poisonous liquid, which the toad releases when attacked. The attacking animal quickly becomes sick and may even die.

Neither frogs nor toads have tails as adults. But they do have tails when they hatch from eggs in the water. In this stage of their lives, they are called tadpoles or polliwogs. A tadpole has gills to breathe underwater and feeds on plants. Eventually the tadpole begins to undergo remarkable changes. Its tail begins to disappear. Two pairs of legs take shape. And its gills begin to close. Inside the tadpole's body, lungs are forming. Soon the tadpole will be an adult toad or frog, ready for its life on land.

Figure 10–14 *This diagram shows the life cycle of a frog. What is a young frog called?*

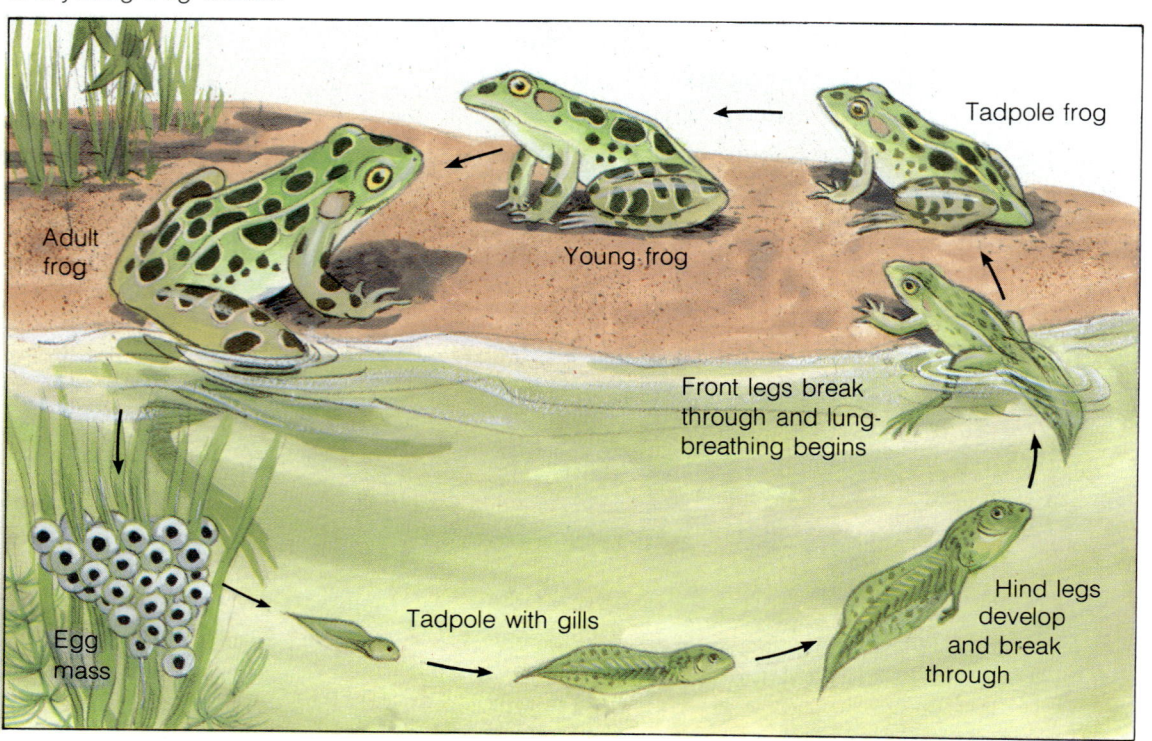

Adult frog

Young frog

Tadpole frog

Front legs break through and lung-breathing begins

Hind legs develop and break through

Tadpole with gills

Egg mass

If there is one thing most people know about adult toads and frogs, it is that they are excellent jumpers. The main reason for this is that the hind legs of a frog or toad are much larger than the front legs. It is these powerful hind legs that allow the animals to jump so well and help them escape from enemies.

Salamanders and Newts

Newts and salamanders are amphibians with tails. Like frogs and toads, these animals have two pairs of legs. But their hind legs are not developed like those of a frog. Newts and salamanders cannot jump.

Because they are amphibians, salamanders and newts must live in moist areas. Some live in the water all their lives. Others may spend most of their life under a single tree stump. Like frogs and toads, salamanders and newts must lay their eggs in water.

Figure 10–15 *This spotted salamander has just laid its eggs. Like all amphibians, it must lay its eggs in water. Why?*

SECTION REVIEW

1. What is the main characteristic of amphibians?
2. Define hibernation.
3. Compare a tadpole with an adult frog. List at least three differences between them.
4. Scientists think that amphibians may have developed from a fishlike ancestor. What characteristics of amphibians provide evidence for this belief?

10–4 Reptiles

Section Objective

To identify the main characteristics of reptiles

Reptiles are a class of coldblooded chordates that have dry, scaly skin and lay eggs on land. This group of animals includes snakes, lizards, turtles, and alligators.

Although many of these reptiles live in or near water, as do amphibians, they do not have to go through a water-dwelling stage in their lives. And, because their scaly skin is resistant to drying out, reptiles do not have to live in a moist environment. In fact, many reptiles are found in hot, dry deserts all over the world. But some of these reptiles pay a

Figure 10–16 *Many reptiles, such as the albino corn snake (left) and the green anole lizard (right), periodically shed their skin. What is this process called?*

Sharpen Your Skills

Snakes

Visit a pet store that sells snakes. Observe the various types of snakes that are for sale. Interview the pet shop owner or one of the workers. Find out the feeding habits of each snake and how to care for one at home. Present your findings to the class in an oral report.

Figure 10–17 *Beneath this green snake are its newly laid eggs. How do reptile eggs differ from amphibian eggs?*

price for their scaly covering. In order to grow larger, snakes and lizards must **molt,** or periodically shed their skin. To do so, they actually crawl out of their old covering, leaving behind an empty, thin layer of skin.

Reptiles do not have to return to the water to lay their eggs because the eggs have a leathery shell that prevents the contents from drying out. And, because the eggs are enclosed within a shell, reptiles have developed a system of **internal fertilization.** The eggs are fertilized within the female's body before the shell forms around the egg. What is the advantage of this type of fertilization?

Snakes and Lizards

Most people are naturally fearful of snakes. They mistakenly think that all snakes are dangerous. But most snakes are not poisonous and will not harm people. In fact, most snakes are helpful. They eat small animals like mice and rats.

Those snakes that *are* poisonous have developed special glands that produce their **venom,** or poison. Snakes inject their venom into their prey through special teeth called **fangs.** Four kinds of poisonous snakes are found in the United States: rattlesnakes, copperheads, water moccasins, and coral snakes. Other poisonous snakes, such as the king cobra, the largest poisonous snake in the world, are found on other continents.

Snakes have developed several remarkable ways of finding prey. Many snakes are able to detect the body heat produced by their prey. They have special pits on the sides of their head that are extremely sensitive to heat. Their tongue is also used as a sense organ. When a snake flicks its tongue in and out of its mouth, it is actually tasting the air, trying to detect those molecules in the air that tell the snake that food is nearby. Snakes are deaf and have poor eyesight. But their other senses make up for these limitations.

Lizards differ from snakes in several ways. The most obvious difference is that lizards have legs. Lizards also have ears and can detect sounds. Most lizards are small and eat insects. What does the largest known lizard eat?

Lizards have developed various ways to trap prey. For example, the Gila (HEE-luh) monster, a lizard that lives in the American Southwest, poisons its prey. The Gila monster bites the prey, holds on, and rolls over on its side. Then, poison released from the lizard's lower jaw flows into the wound, aided only by the force of gravity. How does this differ from the way a snake poisons its prey?

Some lizards have developed special ways to protect themselves from becoming another animal's dinner. The chameleon is one of several kinds of lizards that can change color to match its surroundings. In this way, the chameleon hides from predators. Another lizard has an even stranger way of escaping. If caught by its tail, it will quickly shed the tail. The tail remains behind, wriggling on

Figure 10-18 *The northern copperhead (top) uses its fangs to inject venom into its prey. The boa constrictor (bottom) squeezes its victim until it suffocates.*

Figure 10-19 *In Arizona, a hungry Gila monster (left) eats a mouse it has poisoned. In Kenya, Africa, a chameleon (right) rests on a bush. How is the chameleon protected from its enemies?*

the ground. This action confuses the predator, and the lizard scampers to safety. Later, the lizard's body regenerates the missing tail.

Turtles and Tortoises

Turtles and tortoises are two reptiles that look alike but have adapted to different environments. Turtles spend most of their time in the water. Their legs are shaped like paddles and are used for swimming. Turtles can swim quite well. However, on land they move slowly. Tortoises spend most of their time on land. They have solid, stumpy legs used for walking. Tortoises also have claws on their feet that are used for digging.

Turtles and tortoises carry around with them shells made of plates. Although the shells offer the animals some protection, they are heavy and slow the animal down. The backbones of these animals are fused to their top shell. Turtles and tortoises have no teeth. They have a beak similar in structure to the beak of a bird. Many of these animals eat plants as well as animals.

Turtles and tortoises can live for a very long time. The Galapagos tortoise, for example, may live for as long as 200 years. When these animals were first discovered, they were so numerous that it was said that a person could walk for long distances on their shells and never touch the ground! But sailors used the giant tortoises as a source of fresh meat. Then people settled on the Galapagos Islands and wiped out many of the tortoises. Today the Galapagos tortoise is an endangered species.

Alligators and Crocodiles

Although alligators and crocodiles are very similar in appearance, they have some differences. Alligators have broad snouts, and crocodiles have narrower, more pointed snouts. When a crocodile's mouth is closed, some of its bottom teeth are visible. When an alligator closes its mouth, none of its teeth are visible. Most people, of course, do not feel the need to examine the mouth of either of these animals closely enough to tell them apart!

Figure 10-20 *The loggerhead turtle (top) and the Galapagos tortoise (bottom) are reptiles with shells. What is the major difference between turtles and tortoises?*

Alligators and crocodiles spend most of the time submerged in water with only their eyes and nostrils above the surface. Both kinds of animals eat meat. Their diet consists mainly of fish and other animals that venture too close to their large mouths.

Alligators and crocodiles have unusual reproductive behavior. For example, the female alligator builds a nest of rotting plants in which to lay her eggs. The rotting plants give off heat as they decay. This heat keeps the eggs warm and helps them to develop. The alligators inside the eggs make low chirping sounds when they are about to hatch. The sounds are heard by the mother, who has remained nearby guarding her eggs. When she hears the hatchlings, she uncovers the eggs. The tiny alligators, looking like small copies of their parents, come out into the light of day. The female continues to care for the young for some time after they hatch. The male alligator also helps care for the young. This behavior is unlike the behavior of most reptiles, which usually leave their eggs after they are laid. Their young must care for themselves as soon as they hatch.

Figure 10-21 *This crocodile (right) is sunning itself on a rock. The alligator (left) has chosen a grassier spot. How can you tell these two reptiles apart?*

SECTION REVIEW

1. What are the two main characteristics of reptiles?
2. List two reasons why reptiles do not have to live near water.
3. Compare turtles and tortoises. List some similarities and some differences.
4. What is the difference between an alligator and a crocodile?

Setting Up an Aquarium

Problem

What type of environment is best for guppies?

Materials *(per group)*

rectangular aquarium (15 to 20 liters)
aquarium light (optional)
aquarium filter
gravel   dip net
metric ruler   thermometer
water plants   aquarium cover
snails   guppy food
guppies

Procedure

1. Wash the aquarium with lukewarm water and place it on a flat surface in indirect sunlight. Do not use soap.
2. Rinse the gravel and use it to cover the bottom of the aquarium to a height of about 3.5 cm.
3. Fill the aquarium about two-thirds full with tap water.
4. Gently place water plants into the aquarium by pushing their roots into the gravel.
5. If you have a filter, place it in the aquarium and turn it on.
6. Add more water until the water level is about 5 cm from the top of the aquarium. Let it stand for two days.
7. Add the snails and guppies to the aquarium. Use one guppy and one snail for every four liters of water.
8. Place the cover on top of the aquarium.
9. Keep the temperature of the aquarium between 23°C and 27°C. Feed the fish a small amount of food each day. Add tap water that you let stand for 24 hours to the aquarium when needed. Remove any dead plants or animals.
10. Observe the aquarium every day for two weeks. Record your observations.

Observations

1. Do the guppies swim alone or in a school?
2. What do you see when you observe the gill covers of the guppies?
3. Describe the reaction of the guppies when food is placed in the aquarium.
4. Describe the method the snails use to obtain their food.
5. Was there any growth in the water plants? How do you know?

Conclusions

1. To what phylum of animals do snails belong? To what phylum of animals do guppies belong? How do you know?
2. How do fish obtain oxygen?
3. What is the function of the water plants in the aquarium?
4. What is the function of the snails in the aquarium?
5. What do you think would happen if you placed the aquarium in direct sunlight? In darkness?
6. Why is it important that you do not overfeed the guppies?
7. Predict how the aquarium would be affected if 10 more guppies were added.
8. Predict what would happen to a saltwater fish if you added it to the aquarium.
9. Why did you allow the tap water to stand for 24 hours?

CHAPTER REVIEW

SUMMARY

10–1 Characteristics of Vertebrates

❏ All vertebrates have a backbone made up of bones called vertebrae. Vertebrates belong to the phylum of chordates.

❏ The backbone is part of a vertebrate's endoskeleton, or internal skeleton.

❏ The skeleton provides support and gives shape to the animal.

❏ Two of the seven classes of vertebrates within the chordate phylum are warmblooded and five are coldblooded.

❏ A warmblooded animal's body temperature remains the same despite the temperature of its surroundings.

❏ A coldblooded animal's body temperature changes with the surrounding temperature.

10–2 Fish

❏ External fertilization occurs outside the body of the female.

❏ Jawless fish, such as the lamprey, are the most primitive fish.

❏ The skeletons of cartilaginous fish and jawless fish are made of flexible cartilage. Sharks, skates, and rays are cartilaginous fish.

❏ Bony fish have a swim bladder that helps them float at different levels in the water.

10–3 Amphibians

❏ The term *amphibian* means "double life." Amphibians spend part of their life in water and part on land.

❏ The skin of amphibians must remain damp in order for them to take in oxygen through their skin.

❏ The eggs of amphibians do not have hard shells. They must be laid in water or they will dry out.

❏ Frog and toad eggs hatch into tadpoles. Tadpoles have gills and a tail.

❏ As adults, frogs and toads develop lungs and legs. The powerful hind legs of these animals allow them to leap great distances.

❏ Salamanders and newts are amphibians with tails.

10–4 Reptiles

❏ Reptiles have dry, scaly skin that does not dry out, so they do not have to live in a moist environment.

❏ The eggs of reptiles have leathery shells and will not dry out when laid on land.

❏ Internal fertilization occurs within the body of the female.

❏ Most snakes are not poisonous. Those snakes which are poisonous inject venom into their prey through special teeth called fangs.

❏ Snakes and lizards must shed their skin when they grow larger.

❏ Tortoises are adapted for life on land and turtles are adapted for life in water.

❏ Crocodiles have narrower and more pointed snouts than alligators.

VOCABULARY

Define each term in a complete sentence.

cartilage	external	internal	venom
chordate	fertilization	fertilization	vertebra
coldblooded	fangs	molt	warmblooded
endoskeleton	hibernation	swim bladder	

CONTENT REVIEW: MULTIPLE CHOICE

On a separate sheet of paper, write the letter of the answer that best completes each statement.

1. All vertebrates have
   a. a bony skeleton.   b. scales.
   c. a backbone.   d. an exoskeleton.
2. Which is *not* a vertebrate?
   a. snake   b. earthworm
   c. shark   d. lizard
3. The vertebrates best suited for life in the water are
   a. crocodiles.   b. amphibians.   c. fish.   d. lizards.
4. The fins of fish help them
   a. move through the water.   b. steer.
   c. swim upright.   d. all of these.
5. Bony fish can float at different levels in the water because of
   a. cartilage.   b. swim bladders.   c. fins.   d. wings.
6. Amphibians must lay their eggs
   a. on land.   b. in water.
   c. in nests.   d. in shells.
7. During a frog's life, it breathes through its
   a. lungs.   b. skin.
   c. gills.   d. all of these.
8. Which is *not* an amphibian?
   a. frog   b. toad   c. newt   d. lizard
9. The egg of a reptile is enclosed in
   a. jelly.   b. liquid.   c. a shell.   d. none of these.
10. A reptile that can change color is the
    a. Gila monster.   b. crocodile.
    c. chameleon.   d. turtle.

CONTENT REVIEW: COMPLETION

On a separate sheet of paper, write the word or words that best complete each statement.

1. The bones that make up a chordate's backbone are called _____.
2. The backbone helps protect the nerves of the _____.
3. The first vertebrates on the earth were the _____.
4. The lamprey is classified as a _____ fish.
5. The skeleton of skates and rays is made of a flexible material known as _____.
6. In fish, oxygen passes into blood vessels in the _____.
7. In their early stages, frogs and toads are called _____.
8. A winter sleep during which an animal's body activities are slowed down is called _____.
9. To grow, some reptiles must periodically shed their skin, or _____.
10. Some snakes produce a poison known as _____.

CONTENT REVIEW: TRUE OR FALSE

Determine whether each statement is true or false. Then, on a separate sheet of paper, write ''true'' if it is true. If it is false, change the underlined word or words to make the statement true.

1. Vertebrates are members of the chordate kingdom.
2. All vertebrates have an endoskeleton.
3. The cartilaginous fish are the most primitive group of fish.
4. Sharks are bony fish.
5. The skin of a toad is drier than that of a frog.
6. In the tadpole stage, frogs and toads feed on plants.
7. Adult amphibians obtain most of their oxygen through their lungs.
8. Newts and salamanders are reptiles with tails.
9. In reptiles, fertilization is internal.
10. The shell of a turtle is made of cartilage.

CONCEPT REVIEW: SKILL BUILDING

Use the skills you have developed in the chapter to complete each activity.

1. **Applying definitions** Your friend shows you a small four-legged coldblooded vertebrate that she found. How can you tell whether it is an amphibian or a reptile?
2. **Relating facts** Why do you never find frogs living in Antarctica?
3. **Making inferences** People who fish often use a variety of artificial lures. Explain how these lures could attract fish.
4. **Developing a hypothesis** Some fish have light coloring on their bottom surfaces and dark coloring on their top surfaces. Develop a hypothesis to explain how this coloration could be an advantage.
5. **Relating facts** When a raccoon catches a toad, it usually wipes the amphibian along the ground before eating it. Suggest a reason for this strange behavior.
6. **Applying concepts** A female bullfrog can produce as many as 25,000 eggs in a year. Explain why the world is not overrun with bullfrogs.
7. **Identifying relationships** Why do you think that some people are more seriously injured than others by poisonous snake bites?
8. **Making inferences** The poisonous coral snake has alternating bands of black, bright yellow, and bright red. The harmless scarlet king snake has a very similar color pattern. Explain why this distinctive pattern may be an advantage to the king snake.
9. **Designing an experiment** Design an experiment in which you determine whether salamanders are able to detect sound. Be sure to include a variable and a control in your experiment.

CONCEPT REVIEW: ESSAY

Discuss each of the following in a brief paragraph.

1. What adaptations have fish developed that enable them to live in water?
2. Explain why amphibians must live near water or in a moist environment.
3. Describe the life cycle of a frog.
4. What adaptations have reptiles developed that allow them to live successfully on land?
5. Hypothesize why those vertebrates that reproduce by internal fertilization tend to produce fewer eggs than do those animals that reproduce by external fertilization.

Warmblooded Vertebrates

11

CHAPTER OBJECTIVES

After completing this chapter, you will be able to

11–1 Define warmblooded.

11–1 Identify the main characteristics of birds.

11–1 Identify the adaptations that enable birds to fly.

11–2 Describe the main characteristics of mammals.

11–3 Describe the characteristics of monotremes and marsupials.

11–4 Describe the characteristics of placental mammals.

11–4 Classify ten groups of placental mammals.

On a beautiful, sunny day, a woman went swimming off a deserted beach on the Florida coast. She had swum some distance away from the shore before she realized that she was being pulled under by the current. She remembered being terrified and swallowing water. She was beginning to lose consciousness when suddenly she felt a push from below and from the side. Something was pushing her toward the shore!

Eventually she landed on the beach. But she was so tired that she could not even turn around to thank whoever had saved her. When she finally did look up, no one was there. All she saw were some dolphins swimming a couple of meters away. Then a man came running over and told her that he had witnessed the end of her adventure. A dolphin had saved her life!

Though this story is true, scientists do not know if the dolphin actually "thought" about rescuing the woman. It has been reported that dolphins will also push a soaked mattress to shore. However, dolphins have been known to come to the aid of other dolphins that are sick or wounded. They have even been known to help members of other animal species. And, scientists believe that dolphins are among the most intelligent animals on the earth.

Although they live in the water, dolphins are not fish. They belong to the same class of warmblooded vertebrates that you do—mammals. In the following chapter, you will learn about many other fascinating warmblooded vertebrates.

There are many cases on record of dolphins assisting people in distress.

11–1 Birds

Sharpen Your Skills

Bird Watching

Many people find bird watching a relaxing, enjoyable activity. Obtain a field guide to birds from the library. See if you can borrow a pair of binoculars. Then early in the morning, see how many kinds of birds you can identify in your neighborhood. Make a list of those you spot. Check your list in different seasons to see if it changes.

For many people birds are the most fascinating animals on earth. Many birds are brilliantly colored, some sing beautiful songs, and most have the ability to fly. With the exception of bats, birds are the only members of the chordate phylum that can fly. **Birds are warmblooded vertebrates that have wings and a body covered with feathers.**

Birds are **warmblooded** vertebrates. Warmblooded organisms can maintain a constant body temperature despite the temperature of their environment. See Figure 11–1. In general, birds have a higher body temperature than other vertebrates. To maintain a high body temperature, birds must use up a great deal of food. What is the relationship between food and body temperature?

Birds can be divided into four main groups: perching birds, water birds, birds of prey, and flightless birds. The perching birds are perhaps the most familiar. These birds have feet that are adapted for perching. Their feet can easily grasp a branch. Cardinals, robins, and sparrows are perching birds. The beaks of perching birds are adapted for cracking seeds, catching insects or worms, or reaching deep into flowers for nectar.

Figure 11–1 *Birds are warmblooded vertebrates. The eastern rosella (right) lives in tropical areas. Penguins (left) inhabit the Antarctic. What other characteristics do birds have?*

Water birds, such as ducks and geese, have paddlelike feet for swimming. Other water birds, such as flamingos, have long legs and toes for wading.

Birds of prey, such as eagles, are excellent fliers and also have keen eyesight. Soaring high in the air, they can spot prey on the ground or in the water far below them. Birds of prey eat small animals including fish, reptiles, mammals, and other birds. In addition, these birds have sharp claws called **talons** on their toes. Talons enable the birds to grasp and hold their prey. Birds of prey also have strong curved beaks that are used to tear their prey into pieces small enough to be swallowed.

The flightless birds include ostriches, rheas, and penguins. These birds have small wings relative to the size of their bodies. Except for the penguins, flightless birds have strong leg muscles that enable them to run quickly from their enemies. And, if that does not work, these birds can also use their legs to kick at any enemy who challenges them.

Adaptations for Flight

The body of a bird is adapted for flying. For example, birds have light, hollow bones, which cut down on their body weight. And birds have strong muscles in their wings, which help them fly.

Feathers also help birds fly. There are two main types of feathers. The feathers on the wings and most of the bird's body feathers are called **contour feathers.** These feathers are the largest and most

Figure 11–2 *The blue jay (left) has feet that allow it to easily grasp a tree branch. The paddlelike feet of the king eider (center) enable it to swim in water. The sharp-billed barn owl (top right) has keen eyesight. It uses its talons, or sharp claws, to grasp its prey. A rhea (bottom right) has strong legs that help it run quickly. What are the four main groups of birds?*

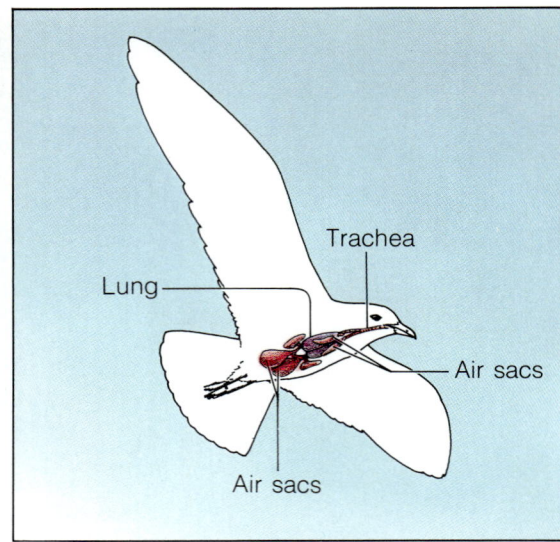

Figure 11–3 *Birds, such as this ring-billed gull, need large amounts of oxygen during flight. Oxygen is taken in by the lungs. In the diagram, you can see the internal structures of a bird. What is the function of the air sacs?*

familiar feathers. They give birds their streamlined shape. Other feathers, called **down feathers,** are short and fuzzy and act as insulation. Most birds have down feathers on their breasts. Down feathers are also found covering young birds after they hatch, but before contour feathers have grown in.

Birds need large amounts of oxygen during flight. Like all warmblooded vertebrates, birds take in oxygen through their lungs. But the lungs of a bird are connected to hollow structures called **air sacs.** See Figure 11–3. These air sacs increase the amount of air available to the lungs and help provide the bird with a constant supply of oxygen during flight. The air sacs also help cool the bird's body during flight by bringing cool fresh air close to internal body organs.

Behavior and Development

Many birds have complicated behaviors. For example, bird songs are used to establish a **territory,** or an area where an animal lives. Establishing a territory is important so that there will not be too many birds competing for food and living space in the same area.

Birds use a variety of methods to attract mates. Some male birds have bright feathers to attract females. Other birds attract mates by constructing large and colorful nesting sites. The male penguin presents his intended mate with a pebble to indicate that he is ready to breed and care for a young bird.

Most birds build nests. These nests can be little more than a space hollowed out in the ground, or they can be quite complex. See Figure 11–4. All nests are designed to protect the eggs and the young birds as they develop.

As in reptiles, fertilization in birds is internal. However, unlike a reptile's egg, a bird's egg is encased within a hard, strong shell. The shell protects the developing bird and contains food for the bird. The shell, which seems to be quite solid, allows oxygen to pass into the egg and carbon dioxide to pass out of the egg.

Most birds **incubate,** or warm their eggs by sitting on them. The eggs must be kept at a certain temperature in order for the young birds to develop. Often only one parent has this job. But in some species, both parents take turns incubating the eggs. After the eggs hatch, most young birds are quite helpless. They cannot fly to look for food. Their parents must bring them food and water until they are old enough to take care of themselves.

Migration

Many birds **migrate,** or move to a new environment, during the course of a year. Birds migrate for many reasons, but perhaps the most important reason is to follow seasonal food supplies.

Birds have developed extremely accurate mechanisms for migrating. Scientists have learned that some birds navigate, or find their way, by observing the sun and other stars. Other birds follow natural

Figure 11–4 *Birds are the most famous nest-builders. An osprey (left) builds its nest atop tall trees near large bodies of water. A weaver bird (right) uses its feet and bill to weave its enclosed nest, which hangs from a tree branch.*

Figure 11–5 *This 10-day-old flamingo is not yet able to fly and must depend upon its parents to bring it food.*

Figure 11–6 *Many birds migrate, or move to a new environment, during the course of a year. Why do birds migrate?*

formations such as coastlines and mountain ranges. Still other birds are believed to have a magnetic center in their brain. This center acts like a compass to help the bird find its way.

SECTION REVIEW

1. What are three characteristics of birds?
2. List the four main groups of birds and give an example of each.
3. How do air sacs help birds during flight?
4. Why is incubation important for the development of a bird's egg?
5. The feathers of many male birds are bright and colorful only during the breeding season. Why do you think having bright colors all year would be a disadvantage?

11–2 Characteristics of Mammals

You, along with lions, bats, and walruses, belong to a class of chordates called mammals. There are about 4000 species of mammals on the earth and many of them look very different from one another. But all mammals have certain characteristics that set them apart from all other living things.

Mammals are warmblooded vertebrates that have hair and feed their young with milk produced in mammary glands. In fact, the word *mammal* comes from the term **mammary gland.** A mammary gland is a structure in a female mammal that produces milk. Most mammals also give their young more care and protection than do other animals.

At one time during their lives, all mammals possess fur or hair. The fur or hair, if it is thick enough, acts as insulation and enables some mammals to survive in very cold parts of the world. Mammals also can survive in harsh climates because, like birds, mammals are warmblooded. The body temperature of mammals remains almost unchanged despite the temperature of their surroundings.

Mammals are believed to be the most intelligent animals on the earth. This intelligence comes from a

brain that is better developed than that of any other group of animals.

As in the birds, fertilization in mammals is internal. But their young develop in different ways. This difference is used to place mammals in three groups: monotremes, marsupials, and placental mammals.

Figure 11–7 *As with most mammals, a thick coat of fur covers the body and provides warmth. What other characteristics do the moose (left) and polar bear (right) share with other mammals?*

SECTION REVIEW

1. What are the main characteristics of mammals?
2. Into which animal phylum are mammals classified?
3. Compare birds and mammals.

11–3 Monotremes and Marsupials

Section Objective

To compare the marsupials with the monotremes

One of the strangest mammals in the world lives in Australia. It has fur like a beaver, feet like a frog, a bill like a bird, and it lays eggs!

This mammal is the duckbill platypus and it belongs to the group of egg-laying mammals called **monotremes** (MAHN-uh-treemz). There is only one other living monotreme—the spiny anteater. See Figure 11–8 on page 254. Like all mammals, monotremes feed their young with milk produced in the mother's mammary glands.

Figure 11–8 *Unlike other kinds of mammals, the duckbill platypus (left) and spiny anteater (right) lay eggs that have a leathery shell. To which group of mammals do these animals belong?*

Figure 11–9 *Marsupials are pouched mammals. Notice the way in which the baby kangaroo, or joey, (left) and the baby koala (right) are carried around by their mothers.*

Unlike the monotremes, the marsupials, or pouched mammals, do not lay eggs. The young of the **marsupials** (mahr-soo-pee-uhlz) are born when they are only partially developed. They complete their development in their mothers' pouch where they feed on her milk.

The marsupial with which you are probably most familiar is the kangaroo. When a kangaroo is born, it is only 2 centimeters long and can neither see nor hear. However, it manages to crawl as far as 30 centimeters to its mother's pouch.

The koala (koh-AH-luh) is a marsupial that lives in Australia and feeds only on eucalyptus leaves. Oils from these leaves are put into cough drops. So it is no wonder that koalas actually smell like cough drops!

The only marsupial that is found in North America is the opossum. Female opossums may give

birth to as many as 24 young at one time. And, the newborn opossums are so tiny that all 24 of them could fit into a single teaspoon!

SECTION REVIEW

1. What are the similarities and differences between marsupials and monotremes?
2. Give an example of each type of mammal.
3. Which group of mammals, the marsupials or the monotremes, is more similar to birds? Explain your answer.

11–4 Placental Mammals

Unlike egg-laying and pouched mammals, the young of placental (pluh-SEHN-tuhl) mammals develop totally within the female. The females in this group of mammals have **placentas** (pluh-SEHN-tuhz). The placenta is a structure through which the developing young receive food and oxygen while in the mother. The placenta also removes wastes from the developing young. After the young are born, female placental mammals, like all mammals, supply their young with milk from mammary glands.

There are more than 20 groups of placental mammals. They are grouped according to how they eat, how they move, or where they live. Ten groups are discussed in the remainder of this section.

Insect-Eating Mammals

What has a nose with 22 tentacles and spends half its time in water? If you have given up, the answer is the star-nosed mole. The star-nosed mole gets its name from the ring of 22 tentacles on the end of its nose. See Figure 11–10. No other mammal has this structure. Each tentacle has very sensitive feelers, which enable the mole to find insects to eat and to feel its way around.

In addition to moles, hedgehogs and shrews are also included among the insect-eating mammals. One type of shrew, the pygmy shrew, is the smallest mammal in the world. It has a mass of only 1.5 to 2 grams—as an adult!

Figure 11–10 *Insect-eating mammals include the star-nosed mole (top) and pygmy shrew (bottom). The star-nosed mole lives in moist or muddy soil in eastern parts of North America. The pygmy shrew must eat twice its weight in insects each day to stay alive.*

Flying Mammals

Figure 11–11 *Bats are the only flying mammals. What are the two types of bats?*

Bats look like mice and are the only flying mammals. In fact, the German word for bat is *Fledermaus,* which means "flying mouse." Bats are able to fly because they have skin stretched over their arms and fingers, which forms wings. How are the wings of bats different from the wings of birds?

Although a bat's eyesight is poor, its hearing is excellent. While flying, bats give off high-pitched squeaks that people cannot hear. These squeaks bounce off nearby objects and return to the bat as echoes. By listening to these echoes, the bat knows where objects are.

There are two types of bats: fruit eaters and insect eaters. Fruit-eating bats are found in tropical areas, such as Africa, Australia, India, and the Orient. Insect-eating bats live almost everywhere.

Flesh-Eating Mammals

Flesh-eating mammals are called **carnivores** (KAHR-nuh-vorz). Most carnivores have sharp pointed teeth called **canines** (KAY-nighnz) that they use for tearing and shredding meat.

CAREER   *Animal Technician*

HELP WANTED: ANIMAL TECHNICIAN
Completion of a two-year animal technology program required. Experience handling animals and knowledge of laboratory procedures desirable.

A veterinarian and an assistant are called to a farm to treat a sick horse. The vet's assistant is an **animal technician,** or someone who has had training in assisting vets and working in laboratories and animal research. Animal technicians work on farms, in kennels, hospitals, and laboratories.

Animal technicians must work under the supervision and instruction of a veterinarian. An animal technician's duties include record keeping, specimen collection, laboratory work, and wound dressing. They also help with animals and equipment during surgery. To receive more information about this career, write to the American Veterinary Medical Association, 930 North Meacham Road, Schaumburg, IL 60196.

Figure 11–12 *Flesh eaters, such as these walruses, have large pointed teeth that tear and shred meat. What are these teeth called?*

You also have canines; two in the top set of teeth and two on the bottom. To locate your upper canines, look at your top set of teeth in a mirror. Find your **incisors** (ihn-SIGH-zuhrs). They are your four front teeth, which are used for biting. On each side of the incisors are teeth that come to a point. These teeth are your canines.

Carnivores are predators. Most land-living carnivores include any members of the dog, cat, and bear families. Sea-living carnivores include otters, sea lions, walruses, and seals. Most carnivores, such as lions, wolves, and bears, have very muscular legs that help them chase other animals. Carnivores also have sharp claws on their toes to help them hold their prey.

"Toothless" Mammals

Although "true" anteaters belong to this group of mammals, they are the only members of the group that actually have no teeth. The other members, the armadillos and the sloths, have poorly developed teeth.

Unlike the spiny anteaters mentioned earlier in this chapter, the true anteater does not lay eggs. The young remain inside the female until they are fully developed. However, both types of anteaters have something in common—a long, sticky tongue that is used to catch insects.

The second group of "toothless" mammals is the armadillos. These mammals eat plants, insects, and small animals. The most striking feature of the armadillo is its protective, armorlike coat. In fact, the

Figure 11–13 *The armor-plated armadillo (top) and three-toed sloth (bottom) are examples of toothless mammals. What other type of mammal is a member of this group?*

257

Figure 11–14 *Elephants are the largest land animals. The larger ears of the African elephant (top) distinguish it from the Asiatic elephant (bottom). To which group of mammals do elephants belong?*

word *armadillo* comes from a Spanish word, meaning "armored."

Sloths are the third type of "toothless" mammals. There are two kinds of sloths: the two-toed sloth and the three-toed sloth. They eat leaves and fruits and are slow-moving creatures. Sloths spend most of their lives hanging upside down in trees.

Trunk-Nosed Mammals

Holding its trunk high in the air, the elephant moves clumsily into the deep river. Little by little, the water seems to creep up the elephant's body. Will it drown? The answer comes a few seconds later as the huge animal actually begins to swim!

To an observer on the shore, nothing can be seen of the elephant except its trunk, through which the animal breathes in air. The trunk is the distinguishing feature of all elephants. It is powerful enough to tear large branches from trees. Yet, at the same time, elephant trunks are capable of such delicate movements as picking up a single peanut thrown by a child at a zoo.

Elephants are the largest land animals. There are two kinds of elephants, African and Asiatic. Although there are a number of differences between the two, the most obvious one is ear size. The ears of the African elephants are much larger than those of their Asiatic cousins.

Hoofed Mammals

What do pigs, camels, horses, and rhinoceroses have in common? Not much at first glance. They could not look more different. Yet look again—down where these animals meet the ground—and you see a common characteristic. The feet of these animals end in hoofs.

One kind of hoof has an even number of toes and belongs to such mammals as pigs, camels, goats, cows, and the tallest of all mammals, the giraffes. The other kind of hoof has an odd number of toes and belongs to mammals such as horses, rhinoceroses, zebras, and tapirs.

Hoofed animals are among the most important "partners" of human beings and have been so for

thousands of years. People eat their meat, drink their milk, wear their skins, ride on them, and use them to pull devices used in farming. Most of the hoofed mammals are **herbivores** (HER-buh-vorz). Herbivores are organisms that feed only on plants.

Gnawing Mammals

Hardly a day goes by that people in the country or even in cities do not see a gnawing mammal. There are more of these mammals than any other mammals on earth. Gnawing mammals are commonly known as rodents.

Among the rodents are such animals as squirrels, beavers, chipmunks, rats, mice, and porcupines. As you might guess, what they have in common has something to do with the way they eat—by gnawing.

Figure 11–15 *A rhinoceros (left) has an odd number of toes. A giraffe (right) has an even number of toes. Why are hoofed mammals important to people?*

Figure 11–16 *The most numerous of all mammals are the gnawing mammals. The tender tips of tree branches are a special delicacy for the porcupine (left). Berries provide the mouse (right) with a tasty treat. What is another name for gnawing mammals?*

Figure 11–17 *A jack rabbit is actually a hare and not a rabbit. To which group of mammals does the jack rabbit belong?*

The common characteristics are four special incisors that are used for gnawing. These teeth are chisellike and constantly grow for as long as the animal lives. Because rodents gnaw or chew on hard objects such as wood, nuts, and grain, their teeth are worn down as the teeth grow. If this were not the case, a rodent's incisors would grow so long that the animal could not open its mouth wide enough to eat.

Some rodents, especially rats and mice, compete with human beings for food. They eat the seeds of plants and many other foods used by people.

Rodentlike Mammals

Rabbits, hares, and pikas (PIGH-kuhz) belong to the group of rodentlike mammals. These mammals have gnawing teeth, similar to rodents. But unlike rodents, they have a small pair of grinding teeth behind their gnawing teeth. Another difference between these two groups of mammals is that the rodentlike mammals move their jaws from side to side as they chew their food, while rodents move their jaws from front to back as they chew.

Water-Dwelling Mammals

Like the dolphin that you read about earlier, whales, porpoises, dugongs, and manatees are water-dwelling mammals. Although they live in water most or all of the time, they breathe air. They have hair and feed their young with milk. And, they are intelligent.

Figure 11–18 *Two groups of mammals live in water. One group includes whales, such as the killer whale (left). The other group includes mammals called manatees (right). What are some other examples of water-dwelling mammals?*

Whales, dolphins, and porpoises spend their entire lives in the ocean and cannot survive on land. Dugongs and manatees live in shallow water, often in rivers and canals. Because of their large size, it is difficult for these animals to move around on land. However, they do so for short periods of time when they become stranded.

Primates

On a visit to your local zoo, you see a family of chimpanzees entertaining the crowd by running and tumbling around their cage. The baby chimpanzee comes to the front of the cage and extends its hand to you. You are amazed to see how much the chimpanzee's hand looks like yours. It is no wonder they are similar. After all, the chimpanzee along with the gibbon (GIHB-uhn), orangutan (o-RANG-oo-tan), and gorilla are the closest mammals, in structure, to human beings. These mammals, along with baboons, monkeys, and human beings, belong to the same group of mammals—the **primates.**

All primates have eyes that face forward, enabling the animal to see depth. The primates also have five fingers on each hand and five toes on each foot. The fingers are capable of very complicated movements, especially grasping objects.

Primates also have large brains and are the most intelligent of all mammals. There is evidence that chimpanzees can be taught to communicate with people by using a kind of sign language. Some scientists have reported that chimpanzees can *use* tools, such as twigs, to remove insects from a log. Human beings, on the other hand, are the only primates that can *make* their own tools.

Figure 11–19 *The gorilla (top) is the largest primate. It may grow to a height of 1.8 meters. The lemur (bottom), on the other hand, may be only a few centimeters tall. What are two characteristics of all primates?*

SECTION REVIEW

1. How are placental mammals different from monotremes and marsupial mammals?
2. What characteristics are used to classify placental mammals?
3. List ten groups of mammals and give an example of each.
4. Compare a carnivore to a herbivore.
5. Why do whales usually come to the surface of the ocean several times an hour?

Classifying Vertebrate Bones in Owl Pellets

Problem

What small vertebrates are eaten by an owl?

Materials *(per group)*

owl pellet   magnifying glass
dissecting needle   small metric ruler

Procedure

1. Observe the outside of a pellet coughed up by an owl and record your observations.
2. Gently break the pellet into two pieces.
3. Using a dissecting needle or small spatula, separate any undigested bones and fur from the pellet. **CAUTION:** *Be very careful when using a dissecting needle.* Also remove all fur from any skulls found in the pellet.
4. Group similar bones together in a pile. For example, put all skull bones in one group.

Observe the skulls. Record the length, number, shape, and color of their teeth.
5. Now try to fit together bones from the different piles to form complete skeletons.

Observations

1. Which parts of the skeleton did you find the most of?
2. Which parts of the skeleton were missing?
3. On the basis of your observations of the skulls, use the chart to identify the kinds of animals eaten by the owl.

Conclusions

1. Which animals appear to be eaten most frequently by the owl?
2. Which group of animals do owls generally eat?
3. To which group of birds do owls belong? How do you know?
4. Why do you think owls cough up bones in their pellets?

Shrew	Upper jaw has at least 18 teeth. Skull length is 23 mm or less. Teeth are brown.	
House Mouse	Upper jaw has 2 biting teeth. Upper jaw extends past lower jaw. Skull length is 22 mm or less.	
Meadow Vole	Upper jaw has 2 biting teeth. Upper jaw does not extend past lower jaw. Molar teeth are flat.	
Mole	Upper jaw has at least 18 teeth. Skull length is 23 mm or more.	
Rat	Upper jaw has 2 biting teeth. Upper jaw extends past lower jaw. Skull length is 22 mm or more.	

CHAPTER REVIEW

SUMMARY

11–1 Birds

❑ Birds are warmblooded vertebrates that have wings and are covered with feathers. Along with bats and insects, birds are the only animals that can fly.

❑ The four main groups of birds are perching birds, water birds, birds of prey, and flightless birds.

❑ The beaks of birds are adapted for the kinds of foods they eat.

❑ Light, hollow bones, feathers, strong wing muscles, and air sacs all help birds fly.

❑ Contour feathers give birds their streamlined appearance. Down feathers act as insulation.

❑ Many birds migrate long distances in search of food and warm climates.

11–2 Characteristics of Mammals

❑ Mammals are warmblooded vertebrates that have hair or fur.

❑ Female mammals feed their young milk from the mammary glands. Most mammals give their young more care and protection than do other animals.

11–3 Monotremes and Marsupials

❑ Like birds and reptiles, egg-laying mammals, or monotremes, lay eggs. The spiny anteater and duckbill platypus are examples of egg-laying mammals.

❑ The young of pouched mammals, or marsupials, are born only partially developed. They further develop in the pouch of the mother. The koala, kangaroo, and opossum are examples of marsupials.

11–4 Placental Mammals

❑ In placental mammals, the placenta provides food for the developing young inside the females.

❑ Insect-eating mammals include moles, hedgehogs, and shrews.

❑ Bats are the only flying mammals.

❑ Flesh-eating mammals, or carnivores, include sea-living animals such as walruses. The land-living carnivores include any member of the dog, cat, and bear families.

❑ Armadillos, anteaters, and sloths are "toothless" mammals.

❑ The only members of the trunk-nosed mammal group are the elephants.

❑ Hoofed mammals are divided into those with an even number of toes on each hoof and those with an odd number of toes.

❑ Gnawing mammals, such as beavers, chipmunks, rats, mice, and porcupines, have chisel-like incisors for chewing.

❑ Rabbits, hares, and pikas are examples of rodentlike mammals.

❑ Mammals, such as whales, dolphins, porpoises, dugongs, and manatees are water-dwelling mammals.

❑ Human beings, monkeys, and apes are known as primates.

VOCABULARY

Define each term in a complete sentence.

air sac	herbivore	migrate	territory
canine	incisor	monotreme	warmblooded
carnivore	incubate	placenta	
contour feather	mammary gland	primate	
down feather	marsupial	talon	

CONTENT REVIEW: MULTIPLE CHOICE

On a separate sheet of paper, write the letter of the answer that best completes each statement.

1. Birds
   a. are invertebrates.   b. have an exoskeleton.
   c. are coldblooded.   d. are chordates.
2. Which is *not* one of the four main groups of birds?
   a. perching birds   b. flightless birds
   c. singing birds   d. water birds
3. Which act as insulation for birds?
   a. down feathers   b. contour feathers
   c. hollow bones   d. air sacs
4. About how many species of mammals live on the earth?
   a. 20   b. 10   c. 400   d. 4000
5. Which is *not* one of the three basic mammal groups?
   a. monotremes   b. marsupial mammals
   c. primates   d. placental mammals
6. The duckbill platypus is a
   a. monotreme.   b. marsupial.
   c. placental mammal.   d. toothless mammal.
7. Marsupial mammals complete their development
   a. in eggs with hard shells.   b. in eggs with leathery shells.
   c. in their mother's pouch.   d. inside their mother's body.
8. Incisors are used for
   a. tearing.   b. shredding.   c. biting.   d. chewing.
9. The largest land animals are
   a. whales.   b. elephants.   c. human beings.   d. gorillas.
10. Which mammals are believed to be the most intelligent?
    a. monotremes   b. marsupials   c. water-dwelling mammals   d. primates

CONTENT REVIEW: COMPLETION

On a separate sheet of paper, write the word or words that best complete each statement.

1. Birds are members of the _____ phylum.
2. _____ animals can maintain a constant body temperature.
3. Bird songs are used to establish _____.
4. In birds and in mammals, fertilization is _____.
5. Many birds _____, or move to a new environment during the course of the year.
6. Mammals feed their young with milk produced in _____.
7. The opossum is the only _____ that lives in North America.
8. The internal structure through which the developing young of some mammals receive food and oxygen is the _____.
9. Animals that eat meat are _____.
10. Rats are examples of _____ mammals.

CONTENT REVIEW: TRUE OR FALSE

Determine whether each statement is true or false. Then on a separate sheet of paper, write "true" if it is true. If it is false, change the underlined word or words to make the statement true.

1. Flightless birds have feet that are adapted for swimming.
2. Contour feathers give birds their stream-lined shape.
3. Female birds often have bright colors to attract mates.
4. A bird's egg is encased in a hard shell.
5. Mammals are coldblooded vertebrates that have hair and feed their young with milk.
6. Mammals are placed into 20 groups based on how their young develop.
7. Monotremes are mammals that lay eggs.
8. The kangaroo is a placental mammal.
9. Sea otters are herbivores.
10. Gnawing mammals have four special incisor teeth.

CONCEPT REVIEW: SKILL BUILDING

Use the skills you have developed in the chapter to complete each activity.

1. **Making charts**   Prepare a chart in which you list ten groups of placental mammals, describe each group's major characteristic, and provide examples of animals in each group.
2. **Classifying objects**   Koalas are often called koala bears. Are koalas really bears? Explain your answer.
3. **Developing a hypothesis**   The Galapagos Islands are a group of small islands located off the west coast of South America. When people visited these islands they noticed distinct differences in the beaks of birds called finches that lived on different islands. Suggest a hypothesis to explain these differences.
4. **Relating concepts**   Many of the hoofed mammals feed in herds. Suggest a possible advantage of this behavior in the survival of these mammals.
5. **Relating concepts**   What is the relationship between the amount of care animals give to their young and how complex the animal is? Provide a logical explanation to support your answer.

CONCEPT REVIEW: ESSAY

Discuss each of the following in a brief paragraph.

1. Describe at least three adaptations that enable birds to fly.
2. Explain three possible ways in which birds may navigate across great distances.
3. Describe the similarities and the differences between birds and mammals.
4. What features make carnivores good predators?
5. What does the tiny pygmy shrew have in common with the 140,000-kilogram blue whale?
6. What are the three groups of mammals? How do they differ from one another?
7. Name two pouched mammals. How are their young born?
8. Compare the structure and function of canines and incisors.

Adventures in Science

katharine payne & the "Language" of elephants

Sometimes Mrs. Payne follows the travels of an elephant herd in her land rover. Rumbling noises picked up by microphones near a water hole tell her that elephants are on the way.

Soon the elephants appear—roaring, trumpeting, snorting. Humans can easily hear these elephant sounds. However, Mrs. Payne, a biologist working for Cornell University, has found that elephants make some sounds that human ears cannot hear.

These newly discovered elephant sounds can travel over a much greater distance than an elephant's trumpet or roar. Mrs. Payne wondered whether elephants use these sounds to locate each other in the vast plains and forests of Africa. In 1986, Mrs. Payne went to Africa to find out.

A chance discovery in the fall of 1984 gave her the idea for her trip. While observing elephants in a zoo, Mrs. Payne felt the air around her throb, as if thunder were rolling in from some far-off place. "Only after a week," Mrs. Payne recalls, "did I think that these might actually be very low frequency sounds that I couldn't hear."

Later, she returned to the zoo with two other researchers from Cornell. Using sensitive recording equipment, they discovered that the elephants were indeed making sounds with a frequency too low for human ears to hear.

You may know that sounds have different frequencies. The sounds made by a flute have a very high frequency. The sounds made by a tuba have a very low frequency. Human ears can hear sounds over a wide range of frequencies. But they cannot hear sounds of a very high frequency, such as the squeaks made by a bat flying through the night air. And, Mrs. Payne found out, humans cannot hear the low-frequency sounds made by elephants. Suddenly, researchers had a clue to explain one long-standing mystery of elephant behavior.

Wild elephants constantly move from place to place in well-organized herds and family groups. But sometimes individual elephants become separated from their group as they wander over many kilometers of grassland or forest. Yet, without a signal that can be detected by humans, the separated elephants come together. "It's always been mysterious," says Mrs. Payne. "African elephant researchers have always said that there must be some kind of ESP between the animals. And there is. It's extrasensory as far as human beings are concerned, but ordinary as far as elephants are concerned."

Mrs. Payne and the Cornell team think that elephant calls are the key to elephant communication. Most of the elephants' calls are too low for human ears to hear. But these low-frequency calls can travel over great distances, and elephants have no trouble hearing them.

While in Africa, Mrs. Payne observed that elephants' foreheads flutter and their ears flap when they make low-frequency calls as well as calls humans can hear. The forehead flutterings most likely occur when sounds made by the elephants' vocal cords cause the animal's forehead to vibrate.

Some of the elephant calls bring calves running to their mothers. Other calls cause elephants to respond over a great distance. Sometimes the sounds can be heard by human observers, says Mrs. Payne, as "soft, puttering, furry rumbles." But, Mrs. Payne points out that the sounds humans can hear are only a small part of the calls made by elephants.

During the second part of Mrs. Payne's research, recorded calls will be played back to the elephants to see how they react. In the past, researchers have placed electronic collars on some wild elephants. The collars are used to keep track of the animals as they wander over great distances. Mrs. Payne would like to put microphones on some of the collared elephants so that the sounds they make as they move can be recorded. "Our ultimate hope is that our work may increase the elephant's chances of survival," says Mrs. Payne.

Elephants have few natural enemies. Why are they in danger of becoming extinct?

Issues in Science

Wildlife Conservation
Which Animals Should Be Saved?

Thirty years ago thousands of dusky sparrows nested in the marshes near Titusville, Florida. Today, not one sparrow remains. Due to the destruction of its habitat, the dusky sparrow is extinct. But it was not a natural disaster, such as a fire or an earthquake that destroyed the birds' habitat. It was people building condominiums near the marshland that were responsible for the fate of the dusky sparrow.

People are largely responsible for causing the extinction of nearly 10,000 species each year. As the human population continues to grow, people need to find places to live and grow food. Many species are threatened with extinction because people destroy wetlands, tropical rain forests, coral reefs, and other natural areas. People also use natural resources by logging, poaching, and, as in the case of the dusky sparrow, by building homes. Scientists predict that in the next few decades, 20 percent of the earth's species will become extinct. But even though they are the biggest threat to other species, humans are also their greatest hope for survival.

Conservationists around the world are working to protect endangered species. This important job requires careful planning, good management techniques, and wise regulations.

The last dusky sparrow, which lived out its life in a protected cage, died in 1988.

Conservationists use a process called *triage* as part of their planning. Triage is a French word that means to pick or sift. During World War I, physicians used a triage process to separate and classify wounded soldiers. They first treated those wounded soldiers who seemed most likely to survive.

Today, some scientists use a similar selection process to decide which of the thousands of endangered species to try to save. It is not possible for scientists to save all of the world's endangered species.

Therefore, they must choose endangered species much like the doctors who decided which patients should receive immediate care. Of course these choices are not easy to make. How can a scientist decide which species are more deserving or more valuable than others? How can a scientist decide which species ought to be saved and which species ought to be permitted to become extinct?

Some conservationists want to focus their efforts on the "charismatic or beautiful animals." These animals include California condors, Florida panthers, pandas, tigers, grizzly bears, and black-footed ferrets. These conservationists hope that these well-loved, appealing animals will call attention to the plight of all endangered species. But this method of choosing—based on attractiveness—has met with criticism. Many people argue that it is not fair to judge a species' value solely by its appearance.

Another viewpoint currently circulating among scientists is based on the usefulness of the species. Some scientists want to concentrate on saving species that will most benefit people and other living things. For example, the periwinkle, a tropical plant, contains an ingredient that has been found useful in treating some cancers. The venom of some snakes can be used to prevent blood clots that cause heart attacks. Some conservationists believe that these useful species should be protected immediately.

The process of deciding which endangered species to protect is under considerable study. Almost all scientists now understand the urgent need to slow down the rate of species extinction. Conservationists realize that it is up to all people to preserve the diverse life forms on earth and to prevent species from becoming extinct. With that in mind, how would you decide which species to protect?

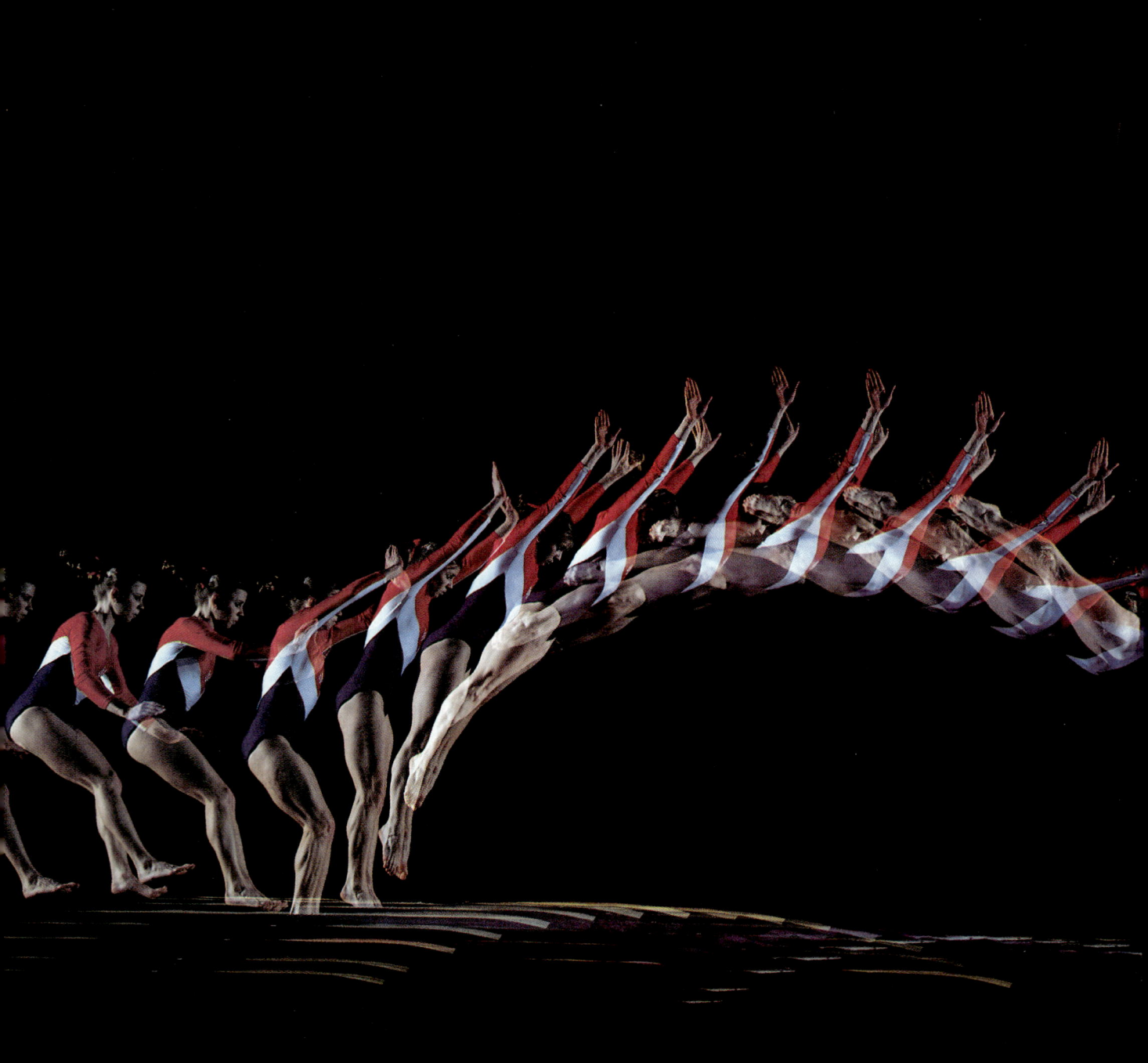

Human Biology

The gymnast begins her exercise. She leaps through the air, twisting and turning as if she were a leaf blowing in the wind. The audience gasps in wonder. But what the audience sees is only part of the performance. Within the gymnast's body, another fascinating performance is occurring—a performance of grace and balance. If the gymnast is to win a medal, all the parts of her body must work in perfect coordination. Nerves must carry information to her brain, telling the body exactly what movements must come next. Muscles must move and pull on bones so that each leap and twist occurs at precisely the right moment. Chemicals flowing in her blood inform certain parts of her body to speed up while other parts slow down.

If all the parts of her body are working in harmony, then the rhythms of her body and the rhythms of the music will seem as one. And, while you may never compete as a gymnast in the Olympic Games, at this very moment your own body is also working in a kind of harmony—the harmony of life. In this unit, you will discover how your body works.

CHAPTERS

Olympic gymnast Kathy Johnson performs a forward flip.

Skeletal and Muscular Systems 12

CHAPTER OBJECTIVES

After completing this chapter, you will be able to

12–1 Classify the four basic types of tissue.

12–2 Describe the functions of the skeletal system.

12–2 Describe the characteristics and structure of bones.

12–2 Compare the three types of movable joints.

12–3 Describe the function of the muscular system.

12–3 Classify the three types of muscles.

First-time visitors to East Africa are often astonished by the sight of women walking great distances with heavy loads on their heads. These women can carry up to 70 percent of their body mass. Yet they seem to glide effortlessly with their heads erect under enormous burdens.

In 1986, scientists tried to find out more about the art of head carrying. The scientists knew that the amount of oxygen people take in is a good indicator of how much energy they are using. So the scientists measured the amount of oxygen the East African women used to carry loads of different sizes. To the scientists' astonishment, the women could carry up to 20 percent of their body mass with no increase in their use of oxygen. Scientists are still not certain how the women can perform such feats of strength and use so little energy. But they suspect it has something to do with the women's straight posture and smooth streamlined walk.

The head-carrying skill of the East African women is proof of the efficiency of the human body. The body's strong bones and powerful muscles enable it to spring into action in a fraction of a second. And, each movement seems effortless and almost unplanned. Now use your own power and grace to quickly turn this page. When you do, you will discover a marvelous relationship between your body's skeletal and muscular systems.

This photograph shows some East African women carrying heavy loads on their heads.

12–1 The Human Body

Shake hands with a new robot arm that was invented by an electronics company. The hand is sensitive enough to hold an egg without breaking it, yet strong enough to lift objects that have a mass of 2 kilograms. Yet, the robot hand does not in any way compare to a similar-looking machine that can perform much more complicated tasks and lift much heavier masses—the human hand.

For hundreds of years, people have tried to build machines that could work as efficiently as their own bodies do. But so far, no one has been able to invent a machine that can grow or repair its own parts, find its own fuel, or move from place to place without outside directions. Of course, that is no real surprise. For the human body is much more than a machine. It is a living masterpiece of timing and organization.

Types of Body Tissue

Your body is made up of more than 50 trillion cells. These cells are organized into **tissues.** A tissue is a group of similar cells that perform the same function. **There are four main types of tissue in the human body: muscle, connective, nerve, and epithelial (ehp-uh-THEE-lee-uhl).** If you were to ob-

Figure 12–1 *Scientists have developed a robot hand that is sensitive enough to hold an egg without breaking it. However, even this robot hand cannot perform the complicated tasks that a human hand can.*

serve these tissues under a microscope, you would see the many differences among the four tissues.

MUSCLE TISSUE   The only kind of tissue in your body that has the ability to contract, or shorten, is **muscle tissue.** By contracting and pulling on bones, one type of muscle tissue makes your body move. Another type of muscle tissue lines the walls of organs inside your body. These muscles do jobs such as moving food from your mouth to your stomach. The third type of body muscle tissue is found only in the heart and enables it to contract and pump blood to all parts of the body.

CONNECTIVE TISSUE   The tissue that provides support for your body and unites its parts is called **connective tissue.** Bone is an example of connective tissue. Without bone, you would have no skeleton. Your body would have no support and no definite shape. Another example of connective tissue is blood. Among other things this type of tissue brings food and oxygen to body parts and removes cell wastes from them. A third kind of connective tissue is fat. It keeps the body warm, cushions organs from the shock of a sudden blow, and stores some of the body's food.

NERVE TISSUE   Another type of tissue is **nerve tissue.** Nerve tissue carries messages back and forth between the brain and spinal cord to every part of the body. In the fraction of a second it takes for you to feel the cold after touching an ice cube, your nerve tissue can carry a message from your finger to your brain.

EPITHELIAL TISSUE   The fourth type of tissue is **epithelial tissue.** Epithelial tissue forms a protective surface on the outside of your body and lines the cavities, or hollow spaces, of the mouth, throat, ears, stomach, and other body parts. When you look in a mirror you are looking at one kind of epithelial tissue—in this case, skin!

Organs and Organ Systems

Just as cells join together to form tissues, so do the different types of tissue in your body combine to form **organs.** An organ is a group of different

Figure 12–2   *The human body is made up of trillions of cells, which group together to form tissues. The combined actions of these tissues allow people to perform complex activities like downhill skiing.*

HUMAN ORGAN SYSTEMS

System	Function	System	Function
Skeletal	Supports and protects the body	**Excretory**	Removes wastes from the body
Muscular	Supports and enables the body to move	**Nervous**	Carries messages throughout the body to aid in controlling body functions and in responding to the environment
Digestive	Breaks down food for use by the body		
Circulatory	Transports food, oxygen, and wastes throughout the body	**Endocrine**	Controls various body functions
		Reproductive	Produces sperm in males and eggs in females
Respiratory	Supplies oxygen to the body and gets rid of carbon dioxide	**Integumentary (skin)**	Protects the body

Figure 12–3 *This chart lists the ten organ systems of the body. What is the function of the excretory system?*

tissues that has a specific function. Your heart, stomach, and brain are all organs. The heart is an example of an organ made up of all four kinds of tissue. Although it is mostly muscle, it is covered by epithelial tissue and also contains connective and nerve tissue.

Groups of organs working together are called **organ systems.** The ten organ systems of the body and their functions are shown in Figure 12–3. Though each system performs a special function for the body, none of them acts alone. Each organ system contributes to the constant teamwork that allows all the parts of the human body to work together in perfect harmony.

SECTION REVIEW

1. What are the four types of tissues in the human body?
2. List three examples of connective tissue.
3. Name two places in the human body that contain epithelial tissue.
4. List the ten organ systems of the human body.
5. Provide two examples of body activities in which two or more organ systems work together.

12–2 The Skeletal System

Although you may not be aware of all your bones, your body actually contains about 206 bones of different shapes and sizes. All of these bones are arranged in a very orderly way. They are fastened together by stringy connective tissues called **ligaments.** Together, your bones and the ligaments that hold them in place help make up most of your skeletal system. **The skeletal system has five important functions: it provides shape, allows movement, protects organs, produces blood cells, and stores certain materials.**

Functions of the Skeleton

Without your skeletal system, your body would look like little more than a mass of jelly. Besides giving the body shape and support, the skeletal system also serves four other important functions. One of

Figure 12–4 *Your body contains about 206 bones. Working together, the bones and the muscles to which most of the bones are attached allow you to do many things, including play basketball like Magic Johnson. What is the name of the connective tissue that connects bone to bone?*

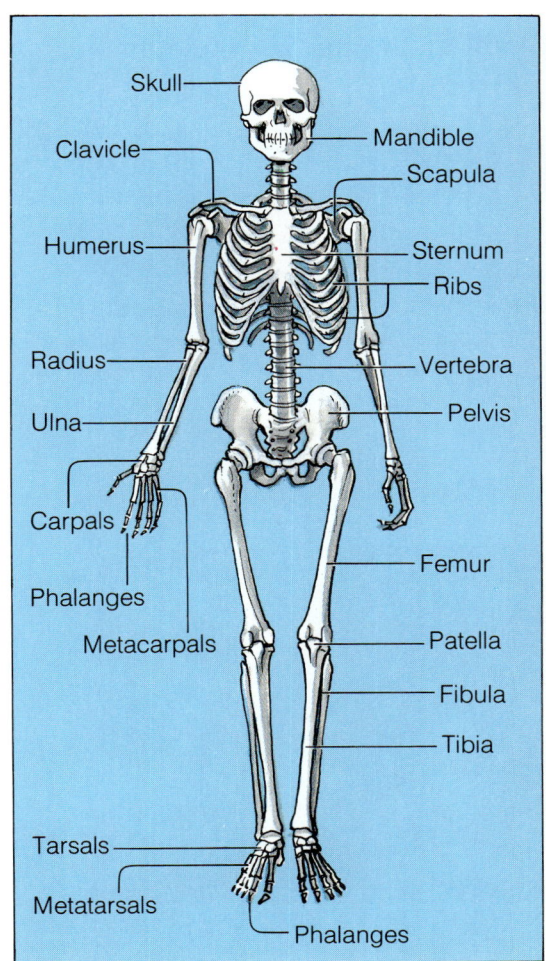

Figure 12–5 *This X-ray shows how the 19 bones in each of your hands allow you to unscrew a bottle. What are the five functions of the skeletal system?*

these is movement. Almost all of your bones are attached to muscles. Working together, bones and muscles allow you to walk, sit, stand, and even smile.

Bones also protect the organs of your body. Your brain is surrounded by a bony skull that keeps it from being injured. If you move your fingers along the center of your back, you can feel your backbone, or vertebral column. Your backbone protects your spinal cord, which contains nerve tissue. The spinal cord is the main message "cable" between the brain and other parts of the body. What other organs are protected by bones?

Bones store important minerals. For example, calcium and phosphorus are constantly deposited in bones for future use. The long bones in your body also produce many of the blood cells necessary for healthy blood.

Parts of the Skeleton

Suppose you were asked to make a life-sized model of a skeleton. Where would you start? You may begin by thinking of the human skeleton as

being divided into two major parts. The two major parts of the human skeleton are the axial (AK-see-uhl) skeleton and the appendicular (ap-uhn-DIHK-yuh-luhr) skeleton.

AXIAL SKELETON   The axial skeleton covers the area that runs from the top of your head down your body in a straight line to your hips. This part includes the skull, the chest with its ribs and other bones, and the vertebral column. The vertebral column consists of 33 separate bones. Each of these bones is called a **vertebra** (VER-tuh-bruh; plural: vertebrae).

APPENDICULAR SKELETON   The appendicular skeleton includes the bones that branch out from the axial skeleton. These bones include the pelvis, or hip bones, the bones of the arms and legs, the clavicle (KLAV-uh-kuhl), or collar bone, and the scapula (SKAP-yoo-luh), or shoulder blade.

Formation of Bones

Of all our body parts, bones are the strongest and the hardest. Perhaps this is the reason why many people tend to think of bone as permanent and unchanging. Actually, bone is made up partly of living tissue. Your bones grow and change just as your skin does. During childhood, bones are shaped and reshaped as the shape and weight of the body

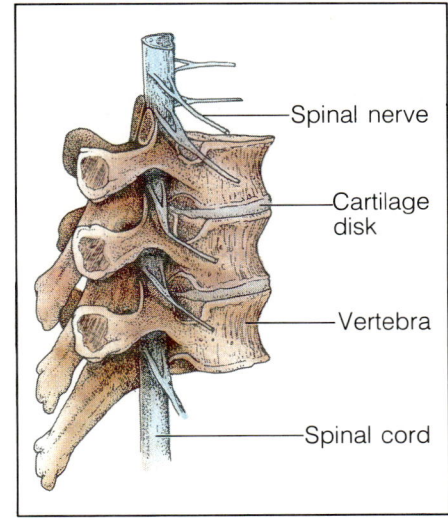

Figure 12–6   *The vertebral column consists of 33 vertebrae, or small bones, stacked one on top of the other. These bones not only protect the spinal cord, but they also form a strong support for the body. Is the vertebral column part of the axial or appendicular skeleton?*

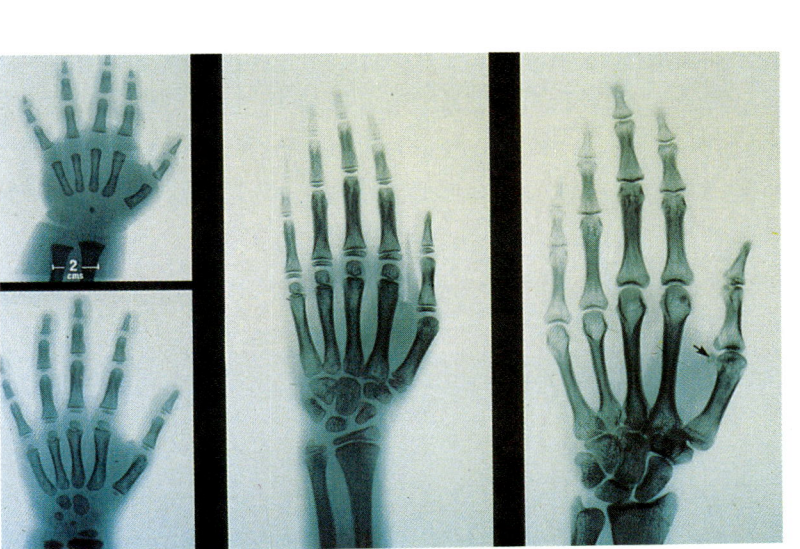

Figure 12–7   *X-rays of the hands of a two year old (top left) and a three year old (bottom left) show that bone has not yet replaced the cartilage in the wrist. In the X-ray of the 14-year-old hand (center), bone formation is almost as complete as it is in the X-ray of a 60-year-old's hand (right). What process changes cartilage to bone?*

changes through growth. Even in adulthood, bone is being continually made and removed, especially where it is under pressure.

With your index finger, touch the tip of your nose. Now gently move the tip of your nose from side to side. You are able to move the tip of your nose because it contains **cartilage** (KAHRT-uhl-ihj). Cartilage is a type of flexible connective tissue. It supports such structures as the nose and ears. Cartilage also connects the ribs to the sternum, or breast bone, and acts as a cushion between the bones of the vertebral column. This connective tissue also covers the ends of bones to keep them from rubbing against one another where they meet.

Most bone is made from cartilage. In the early stages before birth, the skeleton is composed almost entirely of this flexible tissue. Then, over several months, a great deal of cartilage disappears and is replaced by bone. This process is called **ossification** (ahs-uh-fuh-KAY-shuhn). Most of the bones in the skeleton are formed this way.

Structure of Bones

Bones are made up of living tissue. Yet in some ways they are similar to such nonliving things as rocks. Two obvious similarities are hardness and strength. Both bones and rocks owe their hardness and strength to chemical substances called minerals. Although rocks contain a wide variety of minerals, bones are made up mainly of mineral compounds that contain the elements calcium and phosphorus.

Figure 12–8 *The femur, which is the longest bone in the body, has two enlarged ends with a long shaft connecting them. The ends of the femur are made up of softer spongy bone, while the shaft is made up of compact bone. What is the tough membrane that surrounds the shaft called?*

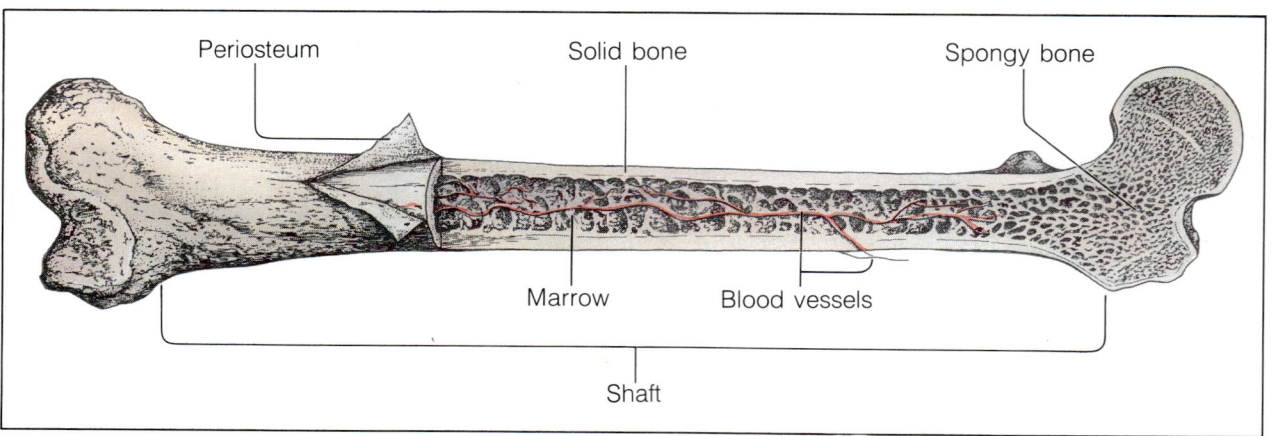

Periosteum   Solid bone   Spongy bone

Marrow   Blood vessels

Shaft

Dairy products are rich in calcium and phosphorus. Next time someone suggests that you drink milk "to keep your bones strong and healthy," you will know why this suggestion makes sense.

Let's take a close look at the longest bone in the body to see what it, and other bones, are made of. This bone, called the femur, links your hip to your knee. Perhaps the most obvious part of this bone is its long shaft, or column, which is shaped something like a hollow cylinder. The shaft contains solid bone and is surrounded by the **periosteum** (pehr-ih-AHS-tee-uhm). The periosteum is a tough membrane that contains bone-forming cells and blood vessels. The bone-forming cells aid in repairing injuries to the bone. The blood vessels supply food and oxygen to the bone's living tissue. Muscles are attached to the periosteum's surface. At each end of the shaft is an enlarged knob. These knobs are made of a softer type of bone called spongy bone.

Running through the middle of the thick bone is a system of pipelike passageways called the **Haversian** (huh-VER-shuhn) **canals.** They contain blood vessels, which bring food and oxygen to the living bone cells. These canals also contain nerves. The nerves send messages through the canals to living parts of the bone.

If you have ever broken a chicken bone in half, you may have noticed that it contains a hollow cavity, or space. Inside the cavity of every bone is a soft red or yellow material called **marrow.** Red marrow produces most of the body's blood cells, and yellow marrow contains fat and blood vessels.

Skeletal Joints

You walk slowly to the edge of the diving board and gaze nervously into the water. Then you raise both arms high and lower your head between them. Bending slightly at the knees, you leap into the air and kick your legs out behind you, meeting the surface of the water with your fingertips. None of these movements would be possible if your body was not equipped with structures called **joints.** A joint is any place where two bones meet.

There are several types of joints. Some joints, such as the gliding, ball-and-socket, hinge, and pivot

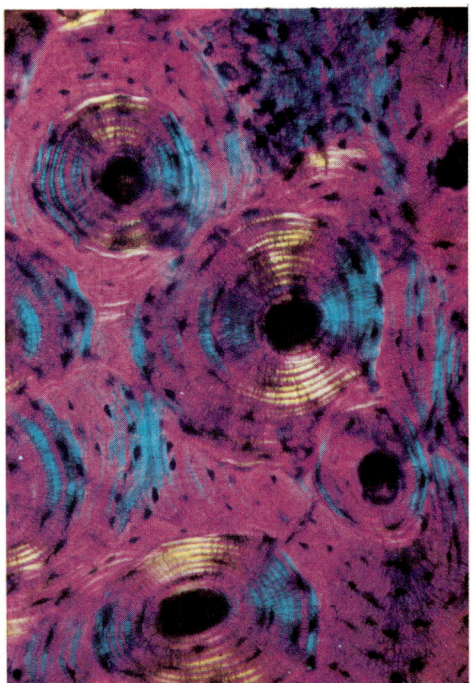

Figure 12–9 *The tiny openings in this photograph of thick, compact bone are the Haversian canals. What do these canals contain?*

joints, are called movable joints. Movable joints allow the bones they connect to move. Others, such as the joints in the skull do not permit movement. These joints are called immovable joints.

In the diving example, a gliding joint allows you to lower your head as you ready yourself for your dive into the water. The gliding joint allows the vertebrae in the neck to slide over one another.

When you raise your arms high, the ball-and-socket joints of your shoulders make the movements possible. The ball-and-socket joints allow you to swing your arms in a circle. These joints are made of a bone with a rounded head that fits into the cuplike pocket of another bone. Can you name another example of a ball-and-socket joint in the body?

As you swing your arms forward, you turn them palm side down. This movement is made possible by the pivot joint. The pivot joint permits a turning motion in which one bone rotates on a ring-shaped bone.

As you get ready to leap off the diving board, you bend your knees. The knee is another type of

Figure 12–10 *The actions involved in diving require the use of many types of joints. What movement does the ball-and-socket joint allow you to do?*

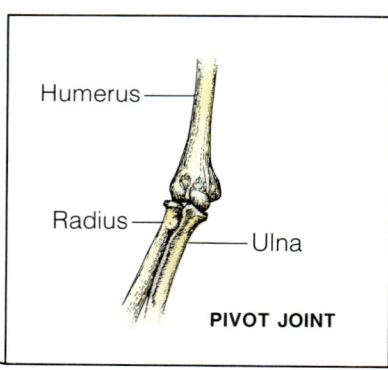

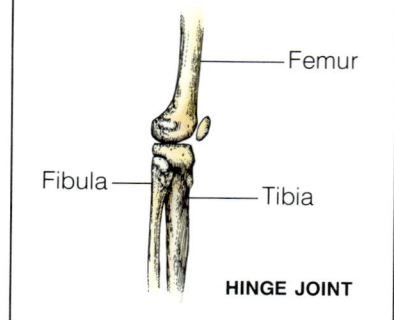

282

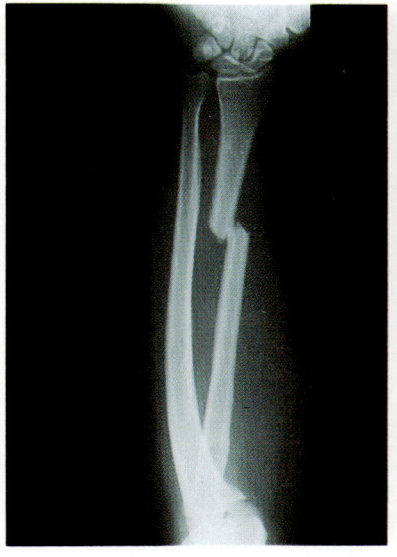
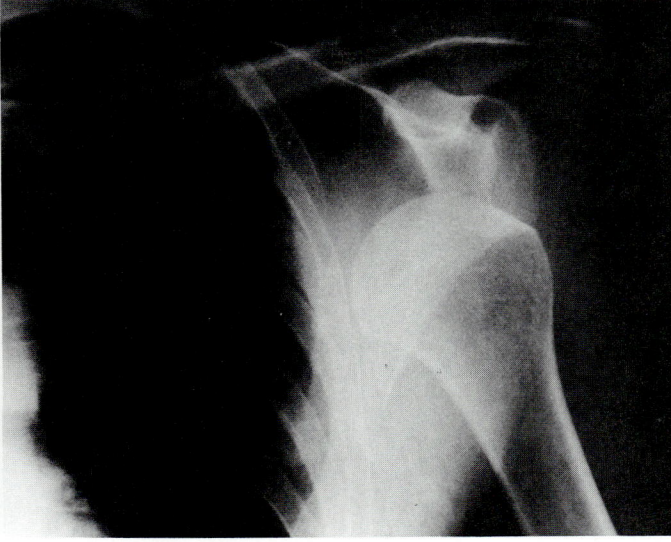

joint, a hinge joint. The hinge joint, also found at the elbow, allows for forward and backward movement. See Figure 12–10.

Figure 12–11 *The X-ray of a shoulder (right) shows how a bone can be forced out of a joint in a dislocation. In the X-ray of a lower arm (left), you can see a break in the radius. What is the other bone in the lower arm called?*

SECTION REVIEW

1. What are the five functions of the skeletal system?
2. What is a ligament?
3. List four places in the body where cartilage is found.
4. What is marrow?
5. Compare the movements of four types of movable joints.
6. Suggest a reason why the ribs are attached to the sternum, or breastbone, by cartilage.

12–3 The Muscular System

Section Objective

To describe the three types of muscles

It is 3 A.M. and you have been asleep for several hours. All day you walked, ran, and played, using your muscles in a variety of ways. Now that you are asleep, all the muscles in your body are also at rest. Or are they?

Without waking you, many of the more than 600 muscles in your body are still working to keep you alive. The muscles of your heart are contracting to pump blood throughout the body. Your chest muscles are working to help move air in and out of

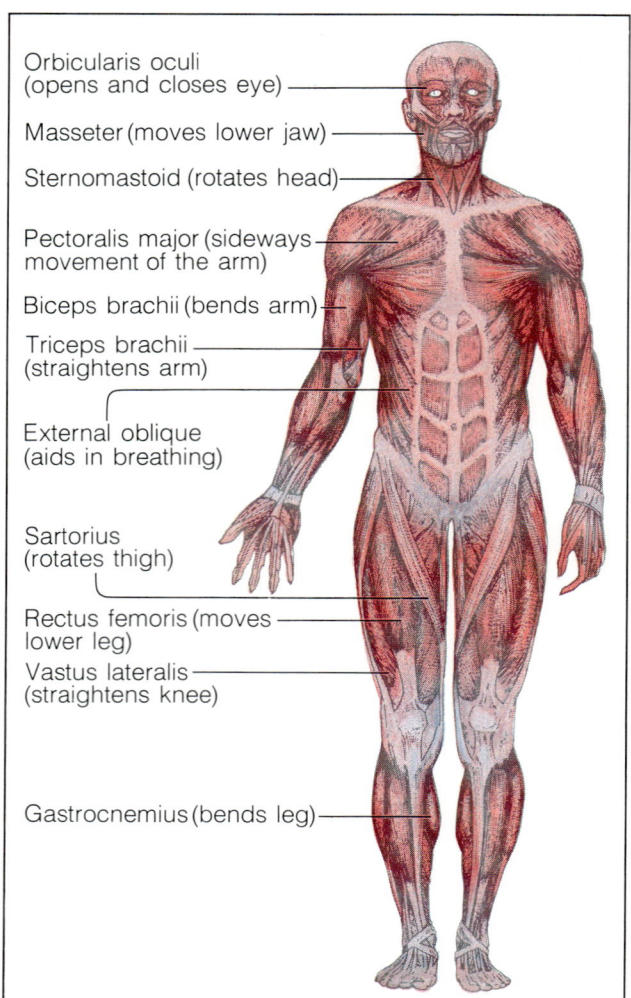

Orbicularis oculi
(opens and closes eye)

Masseter (moves lower jaw)

Sternomastoid (rotates head)

Pectoralis major (sideways
movement of the arm)

Biceps brachii (bends arm)

Triceps brachii
(straightens arm)

External oblique
(aids in breathing)

Sartorius
(rotates thigh)

Rectus femoris (moves
lower leg)

Vastus lateralis
(straightens knee)

Gastrocnemius (bends leg)

Figure 12–12 *The muscles of the body allow you to perform the movements needed to play soccer. Which muscle enables you to move your lower leg?*

your lungs. Perhaps the food that you had for dinner last night is still being moved through your body by muscles.

Your muscles are made of long, thin fibers. These tiny tissue fibers run beside, or parallel, to one another and are held together in bundles by connective tissue. These fibers give muscle tissue the special ability to shorten, or contract. As muscle tissue shortens, it causes movement in the body.

Types of Muscles

There are three types of muscle tissue: skeletal muscle, smooth muscle, and cardiac muscle. You are probably most familiar with the type of muscle tissue called **skeletal muscle.** Skeletal muscles are attached to bones. By contracting, they cause your arms, legs, head, and other body parts to move.

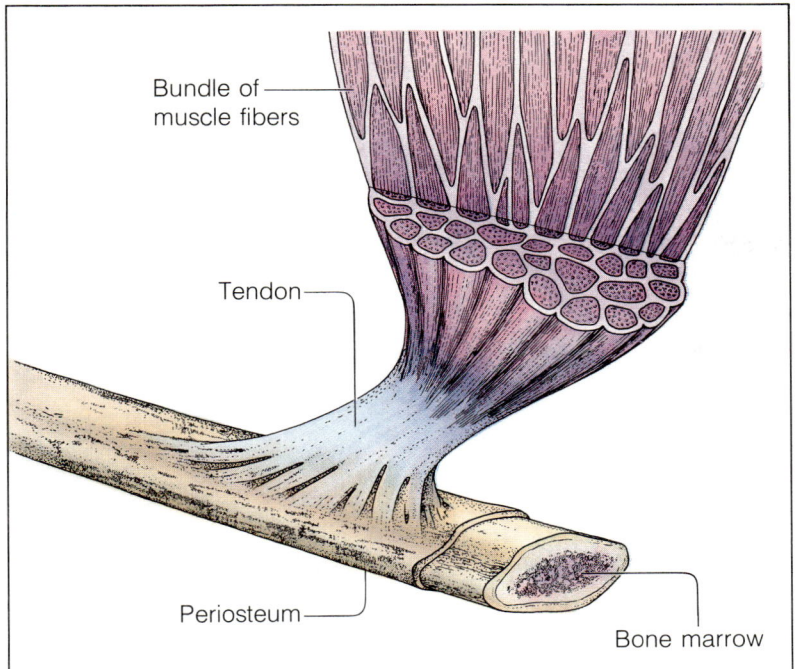

Bundle of muscle fibers

Tendon

Periosteum

Bone marrow

Figure 12–13 *Tendons are connective tissues that attach most skeletal muscles to bones.*

Each skeletal muscle is a separate organ because the muscle is surrounded by its own covering of connective tissue. The ends of the connective tissue form living cables called **tendons.** Tendons connect muscles to bone.

When viewed through a microscope, skeletal muscles are striated (STRIGH-ayt-uhd), or banded. For this reason, they are called striated muscles. See Figure 12–14. And, because skeletal muscles only move when *you* want them to, they are called voluntary muscles.

Think of the movements you make in order to write your name on a sheet of paper. The instant you want it to, your arm stretches out to pick up the paper and pencil. You grasp the pencil and lift it. Then, with the pencil, you press down on the paper. Your eyes move across the page as you write

Figure 12–14 *Skeletal muscle (left), smooth muscle (center), and cardiac muscle (right) are the three types of muscle tissue found in the body. Which muscle tissue is also known as involuntary muscle?*

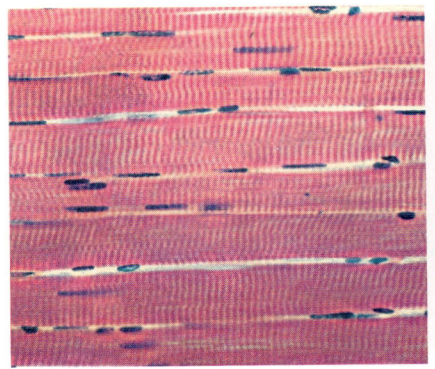
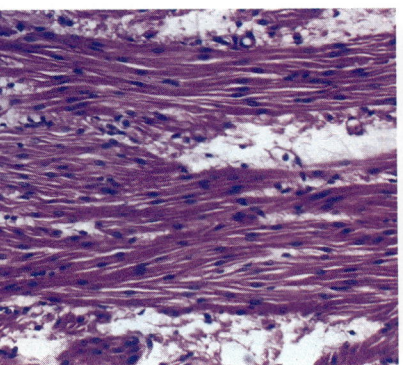
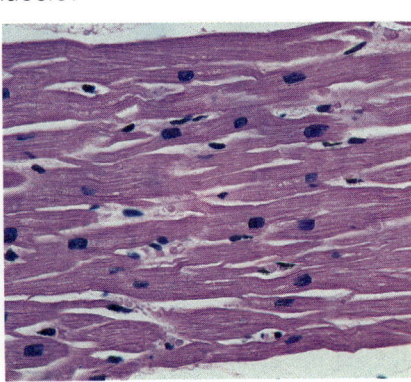

your name. You have to use more than 100 muscles to do all of this. Now, try writing your name 100 times. Do the muscles in your hand ache? They probably do. For, although skeletal muscles react very quickly when you want them to, they also tire very quickly.

The second type of muscle, **smooth muscle,** can contract without you consciously causing them to. They are called involuntary muscles. The involuntary muscles of the body help control breathing and the movements of the digestive system. The fibers of smooth muscle are organized into flat, thin sheets of tissue. Unlike skeletal muscles, smooth muscles react and tire slowly.

The third type of muscle, **cardiac muscle,** is found only in the heart. Branching out in many directions, cardiac muscle fibers weave a complicated mesh. The contractions of these muscle fibers make the heart beat. Like smooth muscles, cardiac muscles usually are involuntary. Heart muscle, as you may have guessed, does not normally tire.

Action of Muscles

Imagine that you could see through the skin of a weightlifter as she lifted a barbell. What would the muscles under her skin look like?

The muscle that makes her upper arm bulge is called the biceps. When the biceps contracts, the tendon at one end pulls the forearm upwards. At this point, the biceps bulge as the muscle contracts.

Muscles only do work by contracting. So for the weightlifter to move her arm and the barbell in one direction and then back to the original position requires the work of two muscles, or two groups of muscles. Put another way, muscles *always work in pairs*.

For example, to raise your own forearm at the elbow, your biceps must contract. At the same time, another muscle at the back of your upper arm, the triceps, must relax. But let's say you now want to straighten your arm. You want to lower your forearm at the elbow. To perform this simple feat, two actions must happen: Your triceps contracts and at the same time, your biceps relaxes. The two sets of muscles are working as a team. See Figure 12–15.

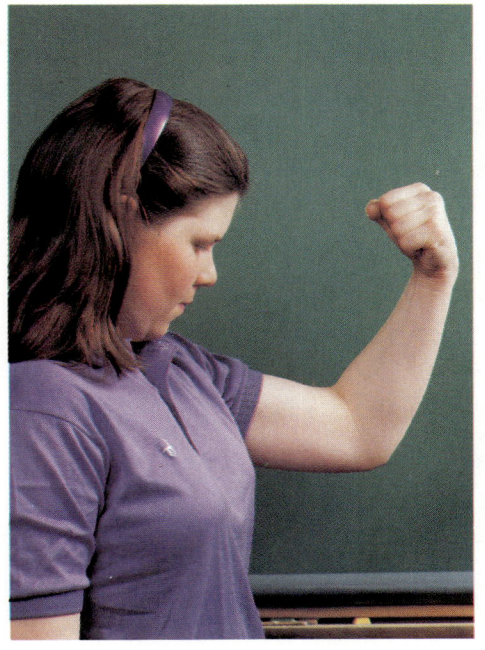

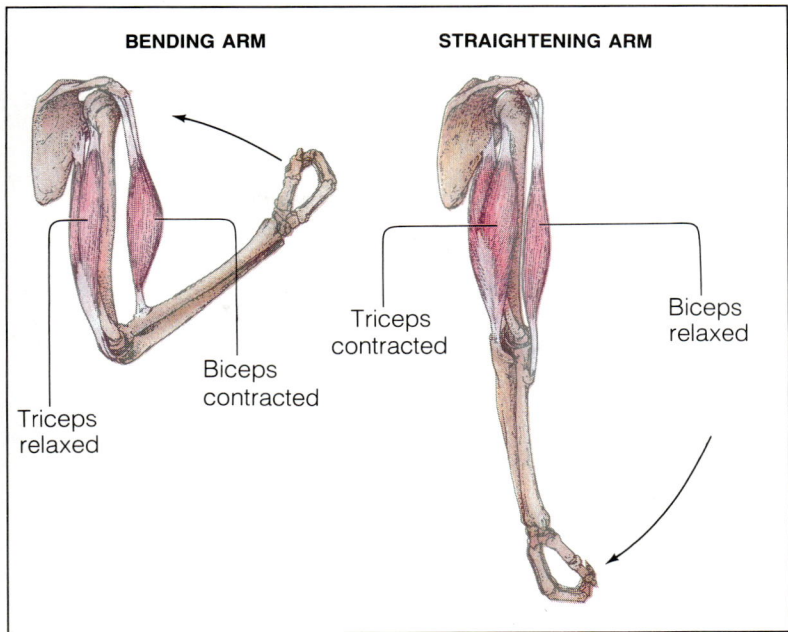

Triceps relaxed

Biceps contracted

Triceps contracted

Biceps relaxed

In order to contract, muscle fibers need energy and oxygen. The energy comes from the food you eat. In addition, muscles can only contract when they receive a message from a nerve to do so. These nerves carry electrical messages that signal the muscle to contract. The messages come from the brain and spinal cord.

There is no such thing as a weak or strong contraction of a muscle fiber. When a fiber receives a message to contract, it contracts as much as possible. The strength the muscle exerts depends on the number of fibers that have received the message to contract at one time. Strong muscle contractions, such as those that involve diving off a diving board, require the contractions of more muscle fibers than would be needed to open a book.

Figure 12–15 *When you "make a muscle" (left), the biceps and triceps muscles work together in pairs. In the drawing, you can see what happens to each of these muscles when you bend and straighten your arm. Which muscle contracts when you straighten your arm?*

SECTION REVIEW

1. Name the three types of muscle tissue.
2. What is the function of a tendon?
3. What is the difference between a voluntary and an involuntary muscle?
4. What happens to the biceps when the triceps contracts?
5. If your biceps were paralyzed, what movement would you be unable to do? Explain.

Sharpen Your Skills

Voluntary or Involuntary?

Voluntary muscles move when you want them to, while involuntary muscles move without you being aware.

1. Blink your eyes three times.
2. Then try not to blink. Time how long you were able not to blink. Record your data.
3. Repeat step 2. Determine the average time you can keep from blinking.

How does your average time compare with that of other students? Are your eye muscles voluntary or involuntary muscles? Explain.

Observing Bones and Muscles

Problem

What are the characteristics of bones and muscles?

Materials *(per group)*

2 chicken leg bones
tiny piece of raw, lean beef
2 dissecting needles
methylene blue stain

2 jars with lids	2 glass slides
knife	cover slip
vinegar	microscope
medicine dropper	laboratory apron
water	

Procedure
Part A

1. Place one chicken leg in an empty jar. Fill the jar about two-thirds full with vinegar.
2. Place the other leg in an empty jar. Cover both jars.
3. After five days, remove the bones from the jars. Rinse the bone from the vinegar jar with water.
4. Observe the texture and flexibility of the two bones.
5. With the knife, carefully cut each of the bones in half. **CAUTION:** *Be careful when using a knife.* Examine the inside of each.

Part B

1. Place the tiny piece of raw beef on one of the slides. With a medicine dropper, place a drop of water on top of the beef.
2. With the dissecting needles, carefully separate, or tease apart, the fibers of the beef. **CAUTION:** *Be careful when using dissecting needles.*
3. Transfer a few fibers to the second slide. Add a drop of methylene blue stain. Cover with a cover slip.
4. Examine the slide under the low and high powers of the microscope.
5. Draw diagrams of what you see.

Observations

1. How do the two bones differ in texture and flexibility?
2. Describe the appearance of the beef under the microscope?

Conclusions

1. What has happened to the minerals and the marrow within the bone that was submerged in vinegar?
2. Why was one bone put in the empty jar?
3. What type of muscle tissue did you observe under the microscope?
4. How does the structure of these muscles aid in their function?

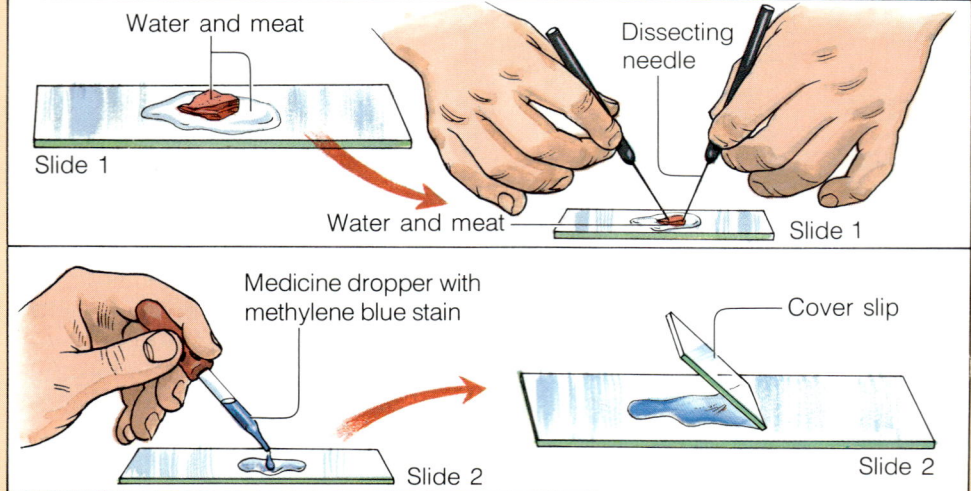

SUMMARY

12–1 The Human Body

❏ A tissue is a group of similar cells that perform the same function.

❏ The human body contains four types of tissue: muscle, connective, nerve, and epithelial.

❏ An organ is a group of different tissues that performs a specific function.

❏ The human body has ten organ systems that are made up of groups of organs working together.

12–2 The Skeletal System

❏ Bones are fastened together by connective tissues called ligaments.

❏ The skeleton gives the body shape and supports, moves, and protects it. The bones of the skeleton store minerals and produce blood cells.

❏ The human skeleton is divided into two parts: the axial skeleton and the appendicular skeleton.

❏ The axial skeleton consists of the skull, the chest, and the vertebral column.

❏ The appendicular skeleton is made up of the pelvis, the bones of the arms and legs, the clavicle, and the scapula.

❏ Cartilage is a type of flexible connective tissue that supports, connects, or cushions other skeletal parts.

❏ Most bone is made of cartilage. In the early stages before birth, the skeleton is made up of almost all cartilage. Over time, most of the cartilage disappears and is replaced by bone in a process called ossification.

❏ The periosteum is a tough membrane surrounding the bone shaft.

❏ The Haversian canals are a system of passageways through a bone. These canals contain blood vessels and nerves.

❏ Inside the bone cavity is a soft red or yellow material called marrow.

❏ A joint is a place where two bones meet. Gliding, ball-and-socket, hinge, and pivot joints are examples of movable joints. The joints in the skull are immovable.

12–3 The Muscular System

❏ Muscle tissue is made of fibers bundled together by connective tissue. Muscle tissue only moves by contracting, or shortening.

❏ Tendons are connective tissue that connect skeletal muscles to bone.

❏ Skeletal, or striated, muscles permit voluntary movements and are connected to bone. Smooth, or involuntary, muscles help control breathing and movements of the digestive system. Cardiac muscles make the heart beat.

❏ Skeletal muscles work in pairs. When one contracts, the other relaxes.

VOCABULARY

Define each term in a complete sentence.

cardiac muscle	Haversian canal	nerve tissue	smooth muscle
cartilage	joint	organ	tendon
connective tissue	ligament	organ system	tissue
epithelial tissue	marrow	ossification	vertebra
	muscle tissue	periosteum	
		skeletal muscle	

CONTENT REVIEW: MULTIPLE CHOICE

On a separate sheet of paper, write the word or words that best complete each statement.

1. A group of similar cells that perform the same function is called a (an)
   a. organ.   b. organ system.   c. tissue.   d. tendon.
2. A tissue that has the ability to contract is
   a. muscle tissue.   b. nerve tissue.
   c. connective tissue.   d. epithelial tissue.
3. A group of tissues that join together to perform a specific function is a (an)
   a. ligament.   b. muscle.   c. organ.   d. system.
4. Which type of tissue is blood?
   a. nerve   b. epithelial   c. muscle   d. connective
5. Bones are fastened together by
   a. tendons.   b. muscles.   c. ligaments.   d. other bones.
6. Two minerals that make up bone are
   a. sodium and chlorine.   b. magnesium and phosphorus.
   c. calcium and iron.   d. calcium and phosphorus.
7. A flexible, connective tissue found in the ear is
   a. tendon.   b. cartilage.   c. ligament.   d. muscle.
8. The soft material found inside the cavity of a bone is the
   a. marrow.   b. periosteum.   c. Haversian canals.   d. shaft.
9. The knee is an example of a (an)
   a. immovable joint.   b. pivot joint.
   c. hinge joint.   d. ball-and-socket joint.
10. Skeletal muscles are also known as
    a. involuntary muscles.   b. voluntary muscles.
    c. cardiac muscles.   d. smooth muscles.

CONTENT REVIEW: COMPLETION

On a separate sheet of paper, write the word or words that best complete each statement.

1. The tissue that carries electrical messages throughout the body is _____ tissue.
2. A tissue that provides support for the body and unites its parts is _____ tissue.
3. The 33 separate bones of the backbone are called _____.
4. The changing of cartilage into bone is called _____.
5. The tough membrane that surrounds the shaft of the bone is the _____.
6. The system of pipelike passageways within the shaft of the bone is called the _____.
7. The place where two bones meet is a (an) _____.
8. The connective tissue that joins muscle to bone is called a (an) _____.
9. Smooth muscles are also called _____ muscles.
10. When the triceps contracts, the biceps _____.

CONTENT REVIEW: TRUE OR FALSE

Determine whether each statement is true or false. Then on a separate sheet of paper, write "true" if it is true. If it is false, change the underlined word or words to make the statement true.

1. Bone is an example of <u>connective</u> tissue.
2. <u>Muscle</u> tissue carries electrical messages between the brain and the spinal cord.
3. An organ is a group of <u>tissues</u> that join together to perform a specific function.
4. <u>Ligaments</u> connect bones to bones.
5. The <u>femur</u> is another name for the breast bone.
6. Cartilage is a flexible <u>epithelial</u> tissue.
7. The periosteum contains <u>blood vessels</u> and bone-forming cells.
8. A <u>pivot</u> joint allows you to bend your knee.
9. <u>Skeletal</u> muscle is also called involuntary muscle.
10. <u>Voluntary</u> muscles help control breathing and blood pressure.

CONCEPT REVIEW: SKILL BUILDING

Use the skills you have developed in the chapter to complete each activity.

1. **Sequencing events** Arrange the following terms in order from the most complex to the least complex: cell, organism, tissue, organ system, organ. Give an example of each.
2. **Relating facts** Examine the hinge joint in Figure 12–10. Explain why this joint is often involved in athletic injuries.
3. **Relating facts** Explain why you feel pain when you fracture, or break, a bone.
4. **Relating concepts** Why do bones heal faster in children than they do in adults?
5. **Applying concepts** Some joints, such as the elbow and the knee, are covered by a fluid-filled sac. What is the advantage of this sac?
6. **Relating cause and effect** Osteoporosis is a disease that usually occurs in older women. It involves a loss and weakening of bone tissue. Doctors recommend that women over 45 eat more foods that contain calcium. How is this helpful in preventing osteoporosis?
7. **Making inferences** There are more joints in the hands and feet than in most other parts of the body. Suggest a reason for this.

CONCEPT REVIEW: ESSAY

Discuss each of the following in a brief paragraph.

1. List and describe each of the four types of tissue in the body.
2. What are the two major parts of the human skeletal system? Which bones belong to each part?
3. Compare the functions of the periosteum and the Haversian canals.
4. Using Figure 12–4, identify the following bones with their scientific names.

   a. jaw   c. kneecap   e. thigh
   b. toe   d. upper arm   f. collar bone

5. Name the type of joint that is found at each of the following locations.

   a. the elbow   c. the knee   e. the skull
   b. the finger   d. the neck

6. Explain the difference among ligaments, tendons, and cartilage.
7. List and describe the three types of muscles. Give the function of each and where each is located within the body.
8. What do muscle fibers need in order to contract?

Digestive System 13

CHAPTER OBJECTIVES

After completing this chapter, you will be able to

13–1 Name and describe the six kinds of nutrients.

13–1 Identify the nutrients in different types of food.

13–2 Describe how food is digested in the mouth, stomach, and small intestine.

13–2 Compare mechanical and chemical digestion.

13–2 Describe the role of the liver, pancreas, and gall bladder in digestion.

13–3 Explain how absorption occurs in the digestive system.

13–4 Apply weight control and exercise to maintaining good health.

Imagine swimming from an island in the Bahamas to the east coast of Florida without stopping! The distance is 97 kilometers. And the woman who broke the record by becoming the first person to accomplish such a feat is Diana Nyad.

The record-breaking swim took Nyad 27 and a half hours. But she would probably be the first to tell you that the final hours were not easy. She had to fight strong ocean currents pulling her underwater. Stinging jellyfish attacked her as she swam. And, as the kilometers passed by, she became increasingly tired and hungry.

But Nyad never stopped to rest or to eat. She was fed from a boat while she swam. As her body digested the food, it was able to replenish some of the enormous energy she was using to accomplish her task. Without food, the champion swimmer would never have made it to the other shore.

The human body requires a constant supply of energy in order to function. It obtains this energy from the chemicals in food. How food is converted into energy for the body is explained in the pages that follow.

Diana Nyad broke a world record when she swam 97 kilometers without stopping.

In a way, you and all human beings are chemical factories. Like any factory, you need raw materials for building new products, repairing old parts, and energy to keep the factory going. These needs are provided by the **nutrients** in the foods you eat. Nutrients are the parts of food your body can use.

To keep your body strong and healthy, you must eat a balanced diet. A balanced diet contains foods from four food groups. These groups are the milk group, the vegetable and fruit group, the meat group, and the bread and cereal group. **A balanced diet provides you with the six basic categories of nutrients that your body needs: proteins, carbohydrates, fats and oils, vitamins, minerals, and water.**

Proteins

Every living part of your body contains **protein**. Proteins are used to build and repair body parts. That is why you need a regular supply of the raw materials used to make proteins—**amino acids.**

Figure 13–1 *A balanced diet contains nutrients from each of the four basic food groups. These groups are the milk group (top left), the vegetable and fruit group (top right), the meat group (bottom left), and the bread and cereal group (bottom right). In addition to meat, what foods are found in the meat group?*

THE SIX BASIC NUTRIENTS

Substances	Sources	Needed For
Proteins	Soybeans, milk, eggs, lean meats, fish, beans, peas, cheese	Growth, maintenance, and repair of tissues Manufacture of enzymes, hormones, and antibodies
Carbohydrates	Cereals, breads, fruits, vegetables	Energy source Fiber or bulk in diet
Fats	Nuts, butter, vegetable oils, fatty meats, bacon, cheese	Energy source
Vitamins	Milk, butter, lean meats, leafy vegetables, fruits	Prevention of deficiency diseases Regulation of body processes Growth Efficient biochemical reactions
Mineral salts Calcium and phosphorus compounds	Whole-grain cereals, meats, milk, green leafy vegetables, vegetables, table salt	Strong bones and teeth Blood and other tissues
Iron compounds	Meats, liver, nuts, cereals	Hemoglobin formation
Iodine	Iodized salt, seafoods	Secretion by thyroid gland
Water	All foods	Dissolving substances Blood Tissue fluid Biochemical reactions

Figure 13–2 *In this chart, nutrients are grouped into six categories. Which nutrient is contained in all foods?*

All proteins are made up of chains of amino acids. For this reason amino acids are called the building blocks of proteins. When you eat a protein, your digestive system breaks it up into its amino acid parts.

Most animal proteins, such as meat, fish, poultry, and eggs, contain all the amino acids your body needs. Such proteins are called complete proteins. Other foods, such as rice, cereal, and vegetables contain only small amounts of one or another amino acid and are known as incomplete proteins.

Carbohydrates

Carbohydrates supply the body with its main source of energy. Fruits, vegetables, and grain products are good sources of carbohydrates.

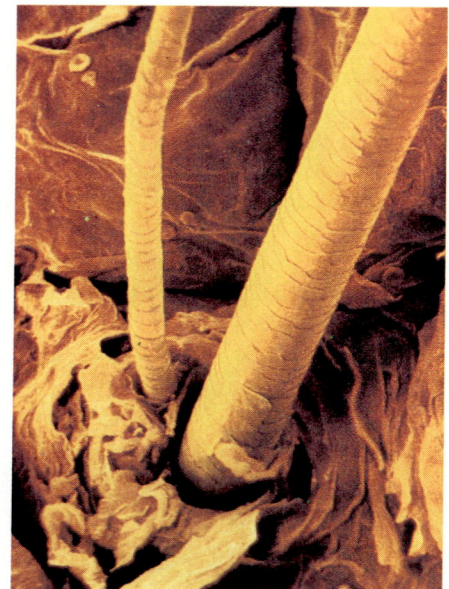

Figure 13–3 *This photograph shows two hairs penetrating the outer layer of skin cells. Both the hair and the skin are composed mainly of protein. What substances make up proteins?*

The energy value of foods, such as carbohydrates, is measured in units called **Calories.** A Calorie is equal to the amount of heat energy needed to raise the temperature of 1 kilogram of water by 1°C. The number of Calories that a person needs every day depends on the person's size, body build, occupation, type of activity, and age.

There are two types of carbohydrates—the sugars and the starches. A starch is a long chain of sugars. Sugars can be digested and used for fuel faster than starches. But if you eat more carbohydrates than you need, the excess is stored as starch in your muscles and liver. If these storage places become filled, the carbohydrates are then stored in other places in the body. As a result, a diet with too much carbohydrates can make a person overweight.

Fats and Oils

Like carbohydrates, **fats** and **oils** supply the body with energy. In fact, fats supply the body with twice as much energy as do equal amounts of proteins and carbohydrates. In addition to providing energy, fats help to support and cushion vital organs, protecting them from injury. Fats also insulate the body against loss of heat. Foods that are rich in fats come from both animals and plants. Some sources of fats are nuts, butter, vegetable oils, and cheeses.

Figure 13–4 *Like any activity, the high jump requires energy. Which group of nutrients supplies the body with its main source of energy?*

Vitamins and Minerals

In addition to substances that are used for building, repairing, and fueling your body, you also need **vitamins** and **minerals.** Vitamins help regulate growth and the normal functioning of your body.

Figure 13–5 *Some of the vitamins your body needs are listed in this chart. Which vitamins are contained in green or leafy vegetables?*

SOME IMPORTANT VITAMINS

Vitamins	Sources	Needed For
A (fat soluble)	Liver and kidney, fish-liver oils, eggs, butter, green and yellow vegetables, sweet potatoes, yellow fruit, tomatoes	Maintenance of skin, eyes, and mucous membranes Healthy bones and teeth Growth
D (fat soluble)	Fish-liver oils, liver, fortified milk, eggs, tuna, sunlight	Regulation of calcium and phosphorus metabolism Healthy bones and teeth Growth
E (fat soluble)	Milk, butter, vegetable oils	Maintenance of cell membranes
K (fat soluble)	Tomatoes, soybean oil, leafy vegetables	Blood clotting Normal liver functioning
Thiamine (B$_1$) (water soluble)	Meat, yeast, whole-grain cereals, green vegetables, nuts, peas, soybeans, seafood, milk	Carbohydrate metabolism Functioning of heart and nerves Growth
Riboflavin (B$_2$ or G) (water soluble)	Milk, cheese, fish, fowl, meat, green vegetables, liver, eggs, yeast	Healthy skin Growth Eye functioning Carbohydrate metabolism
Niacin (water soluble)	Yeast, lean meats, liver, fish, whole grains, peanut butter, potatoes, leafy vegetables	Growth Healthy skin Carbohydrate metabolism Functioning of stomach, intestines, and nerves
B$_{12}$ (water soluble)	Eggs, meats, milk, green vegetables, liver	Proper development of red blood cells
C (ascorbic acid) (water soluble)	Citrus and other fruits, tomatoes, potatoes, leafy vegetables	Healthy bones, teeth, and gums Growth Maintaining strength of blood vessels

Sharpen Your Skills

Vitamin-Deficiency Diseases

Use reference materials in the library to find out about a vitamin-deficiency disease, such as scurvy, pellagra, or rickets. How is the disease caused? What are its symptoms? How can the disease be prevented? Write a brief report on your findings.

Figure 13–6 *According to this chart, which minerals can be found in green or leafy vegetables?*

Because you only need them in small amounts, vitamins are sometimes called micronutrients. However when your body does not get enough of a certain vitamin, you could come down with a vitamin-deficiency disease.

There are two groups of vitamins. The fat-soluble vitamins are stored in body fat. So, high levels of fat-soluble vitamins can be harmful. Water-soluble vitamins are constantly washed out of the body and should be supplied in adequate amounts in the foods you eat each day. See Figure 13–5 on page 297.

Like vitamins, minerals help to maintain the normal functioning of your body. There are 16 essential minerals. Some of these minerals are listed in Figure 13–6.

SOME IMPORTANT MINERALS

Mineral	Sources	Needed For
Calcium	Milk products, green leafy vegetables, fish, eggs	Strong bones, teeth, and muscles, nerve function, blood clotting, enzyme activation
Iron	Liver, red meat, raisins, beans and peas, nuts, egg yolk, green leafy vegetables, whole grains	Carrying oxygen in red blood cells
Iodine	Iodized table salt, seafood	Controlling the rate at which nutrients are used
Magnesium	Milk products, leafy green vegetables, beans and peas, meat, potatoes, whole grains	Regulating body temperature, making proteins, nerve and muscle function
Phosphorus	Milk, cheese, meat, whole grains	Strong bones and teeth, cell wall formation, muscle function
Potassium	Bananas, vegetables, meat, nuts, avocados	Maintaining fluid balance, nerve and muscle function
Sodium	Table salt, cured meat, cheese	Maintaining fluid balance, nerve function
Sulfur	Milk, eggs, meat, beans and peas	Making amino acids
Zinc	Meat, eggs, seafood, cereals, nuts, green vegetables, beans and peas	Making enzymes, healing wounds, tissue growth

Water

Although you can survive many days without eating, several days without water can be fatal. Water is important because all the chemical reactions in the body take place in water. And water carries nutrients and other substances to and from body organs through the bloodstream. Water also helps your body stay at the right temperature, 37°C.

On the average your body is 55 to 75 percent water. Under normal conditions, you need about 2.4 to 2.8 liters of water daily. You get this water from the fluids you drink and the foods you eat.

SECTION REVIEW

1. List the six basic categories of nutrients.
2. Name the four food groups and give an example of each.
3. What are proteins composed of?
4. What is the difference between a fat-soluble and a water-soluble vitamin?
5. Explain how it is possible for a person to be overweight and suffer from improper nutrition at the same time.

Figure 13-7 *Marathon runners must drink water "on the run" in order to replace the water that they lose during the race.*

13-2 Digestion

Section Objective

To explain what happens to food during the process of digestion

Unfortunately, most foods that you eat cannot be used directly by your body. **Food must be broken down into nutrients in a process called digestion.** The breaking down of food into simpler substances for use by the body is the work of the **digestive system.** Once these simpler substances are broken down, or **digested,** they are carried to all the cells of the body by the blood. In the following sections, you will follow a mouthful of food on its path through the digestive system.

The Mouth

Close your eyes and imagine your favorite food. Your mouth probably watered. This response occurs because the mouth contains salivary (SAL-uh-vair-ee)

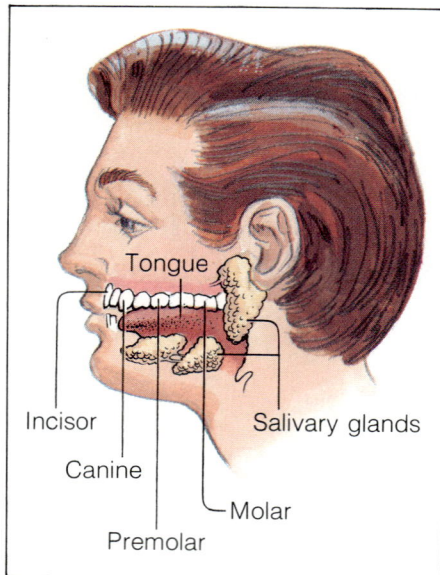

Figure 13–8 *Salivary glands in the mouth produce saliva. Saliva contains an enzyme that begins the chemical digestion of food. What is this enzyme called?*

glands. Salivary glands produce and release a liquid known as **saliva.** Seeing, smelling, or even thinking about food can increase the flow of saliva.

CHEMICAL DIGESTION Saliva, as you might expect, helps moisten food. It also contains a chemical substance called **ptyalin** (TIGH-uh-lihn). Ptyalin chemically breaks down some of the starches in your food into sugars.

Ptyalin is one of many **enzymes** in your body. Enzymes control a wide variety of chemical reactions, including the breaking down of food into simpler substances. The digestion of foods by enzymes is called **chemical digestion.**

MECHANICAL DIGESTION You bite into your food with your **incisors** (ihn-SIGH-zuhrz), or front teeth. Then you pull the food into your mouth with your lips and use your tongue to push it to the back of

CAREER

Food Inspector

HELP WANTED: FOOD INSPECTOR
Requires three years of experience taking care of animals or working with meat products. College courses in biology, chemistry, zoology, and agriculture may be substituted for experience. Must pass written test and be physically fit. Contact local Federal Job Information Center for application instructions.

Many people get the protein they need each day from meats and poultry. It is the responsibility of **food inspectors** to make sure that these protein sources are carefully prepared and processed and are safe to eat before they are sold.

Food inspectors look for any unusual conditions in live animals in slaughtering plants. In a processing plant, inspectors make certain that manufacturers follow all public health rules and regulations.

Food inspectors receive on-the-job training. During this time, they are taught the regulations, methods, and duties involved in food inspection. Food inspectors monitor all stages of food processing to make sure that food manufacturers follow all rules concerned with the

kinds and levels of food additives, correct product labeling, and cleanliness.

If you are interested in becoming a food inspector, write to the United States Department of Agriculture, Food Safety and Inspection Service, Rm. 3438-South, 14th and Independence Avenue, SW, Washington, DC 20250.

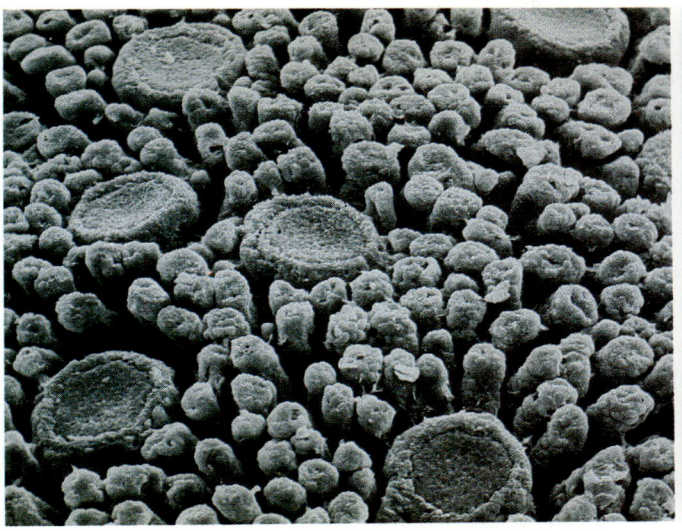

Figure 13–9 *The hills, valleys, and plateaus that you see in the photograph on the left are found on the surface of the tongue. The photograph on the right shows the bulb-shaped structure of a taste bud. Name the four types of taste buds.*

your mouth. Here, the flat-headed premolars and **molars,** or back teeth, grind and crush the food into small pieces. At the same time, the **canines** (KAY-nighnz), or eyeteeth, tear and shred the food. The physical action of breaking food into smaller parts is called **mechanical digestion.**

TASTE BUDS   The food in your mouth tastes good or bad to you because your tongue has taste buds. See Figure 13–9. There are four types of taste buds. Each reacts to a different group of chemicals in food to produce a taste. The four tastes are sweet, sour, bitter, and salty. But the flavor of food does not come from taste alone. Flavor is a mixture of taste, texture, and odor.

THE EPIGLOTTIS   After you chew your food a few times, you swallow. When you swallow, smooth muscles near the back of your throat begin to force the food downward. At the same time a small flap of tissue called the **epiglottis** (ehp-uh-GLAHT-ihs) automatically closes over your windpipe. The windpipe is the tube through which the air you breathe reaches your lungs. When the epiglottis moves over the windpipe, it prevents food or water from moving into the windpipe or "down the wrong pipe." After swallowing, the epiglottis moves back into place to allow air into the windpipe.

The Esophagus

After you swallow, smooth muscles force the food into a pipe-shaped tube called the **esophagus** (ih-SAHF-uh-guhs). The word *esophagus* comes from a Greek word meaning "to carry what is eaten." And that is exactly what this tube does as it transports food down to your **stomach.** The stomach is a J-shaped, muscular organ connected to the end of the esophagus.

The esophagus, like most of your digestive organs, is lined with slippery mucus. This mucus helps food travel through your esophagus to your stomach, which takes about 7 seconds. As soon as food enters the esophagus, powerful waves of muscle contractions push the food downward. This wavelike motion is called **peristalsis** (pair-uh-STAHL-sihs).

Digestion in the Stomach

As food enters the stomach, cells in the stomach release gastric juice. This gastric juice contains the enzyme **pepsin,** hydrochloric acid, and thick slippery mucus. The mucus coats and protects the stomach wall. Why do you think such protection is necessary?

With the help of the hydrochloric acid, pepsin breaks down some of the complex proteins in your food into simpler proteins. This is a form of chemi-

Figure 13–10 *The diagram (right) shows food being pushed down through the esophagus to the stomach. In the X-ray (left), you can actually see a ball of food in the esophagus. What is the name for the wavelike motion that moves food through the digestive system?*

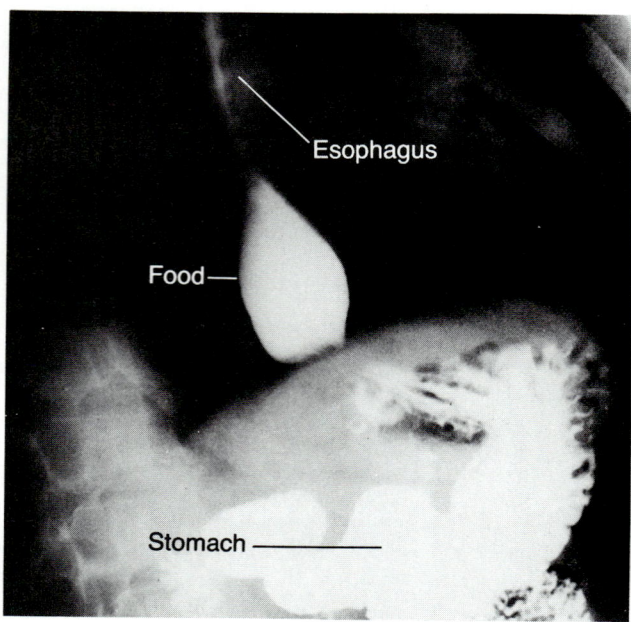

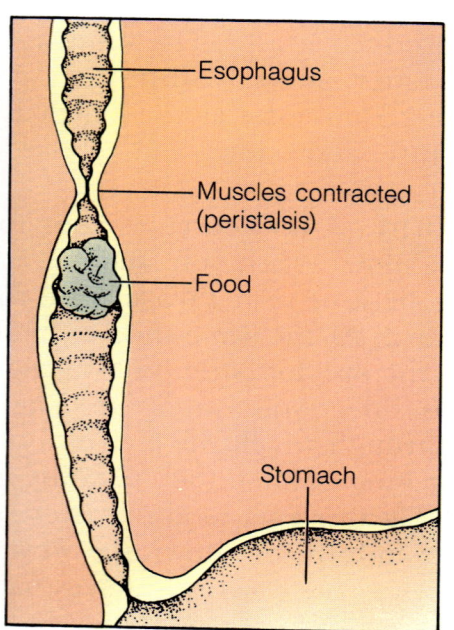

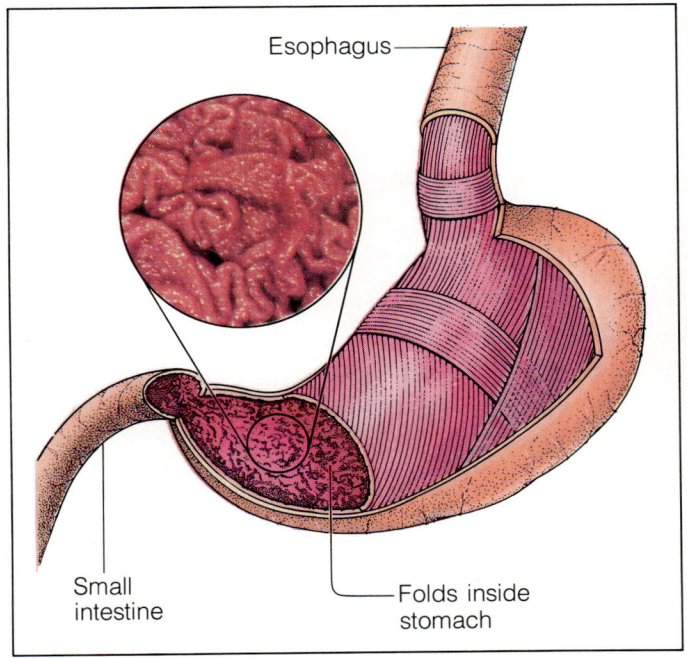

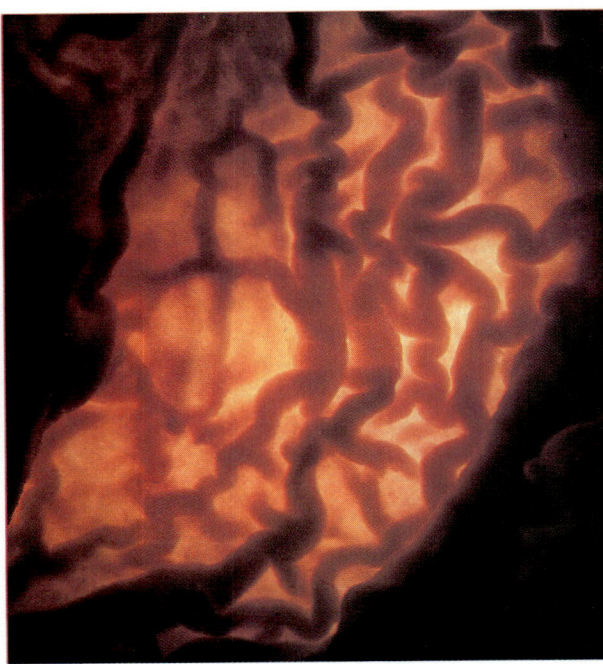

Esophagus

Small
intestine

Folds inside
stomach

Figure 13–11 *The stomach is made up of several layers of muscle. Notice that the inside wall of the stomach contains many folds (left). Today doctors can use an instrument called a gastroscope to examine the inside of the stomach. This remarkable photograph of the inner lining of the stomach (right) was taken using a gastroscope.*

cal digestion. Food also undergoes a kind of mechanical digestion in the stomach. Muscle contractions mix the food with gastric juice as peristalsis pushes the food towards the stomach's exit.

Digestion in the Small Intestine

Food moving out of your stomach has changed quite a bit since it was first placed in your mouth. After a few hours in your stomach, muscle contractions and enzymes have changed your food into a soft, watery substance. Now it is ready to move slowly from your stomach into another organ of the digestive system—the **small intestine.** Although this organ is only 2.5 centimeters in diameter, it is over 6 meters long. As in the esophagus and the stomach, food moves through the small intestine by peristalsis.

Although some chemical and mechanical digestion has already taken place in the mouth and the stomach, most digestion takes place in the small intestine. The cells that line the walls of the small intestine release intestinal juice containing enzymes.

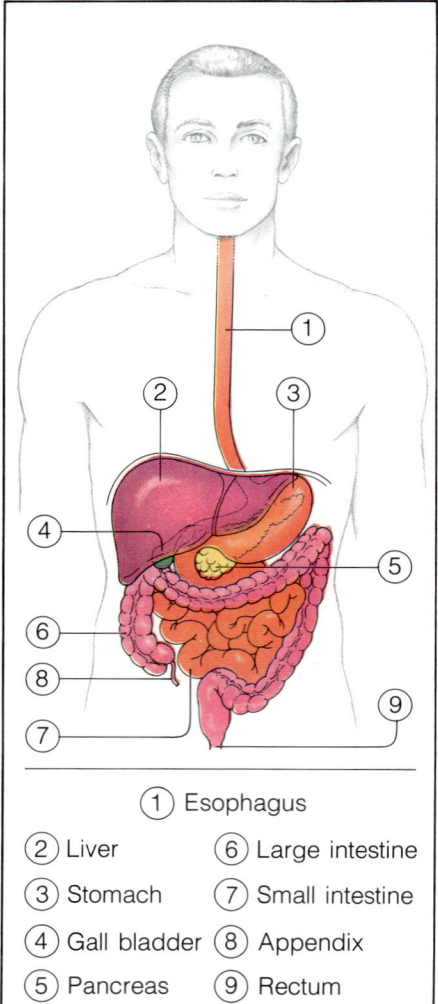

Figure 13–12 *The breaking down of food into simpler substances that can be used by the body is the work of the digestive system. Use this diagram to trace the path of food as it passes through the digestive system.*

1 Esophagus

2 Liver   6 Large intestine

3 Stomach   7 Small intestine

4 Gall bladder   8 Appendix

5 Pancreas   9 Rectum

Most chemical digestion that occurs in the small intestine takes place within 0.3 meter of the beginning of the small intestine. Here, intestinal juices help break down food arriving from the stomach. These juices do not work alone. They are helped by juices that are produced by two organs near the small intestine. These organs are the **liver** and the **pancreas** (PAN-kree-uhs). Because no food passes through the liver and the pancreas, they are considered to be digestive helpers.

The Liver

Located to the right of the stomach is the liver, the body's largest and heaviest internal organ. One of its many important functions is to aid digestion by producing a substance called **bile.** Once the liver produces bile, the bile moves into the **gall bladder.** The gall bladder is an organ that stores bile. As food moves into the small intestine from the stomach, the gall bladder releases bile through a duct, or tube, into the small intestine. Bile breaks up large fat molecules into smaller ones. Which type of digestion is this?

Figure 13–13 *Each enzyme in the digestive system has a specific function. What is the function of lipase?*

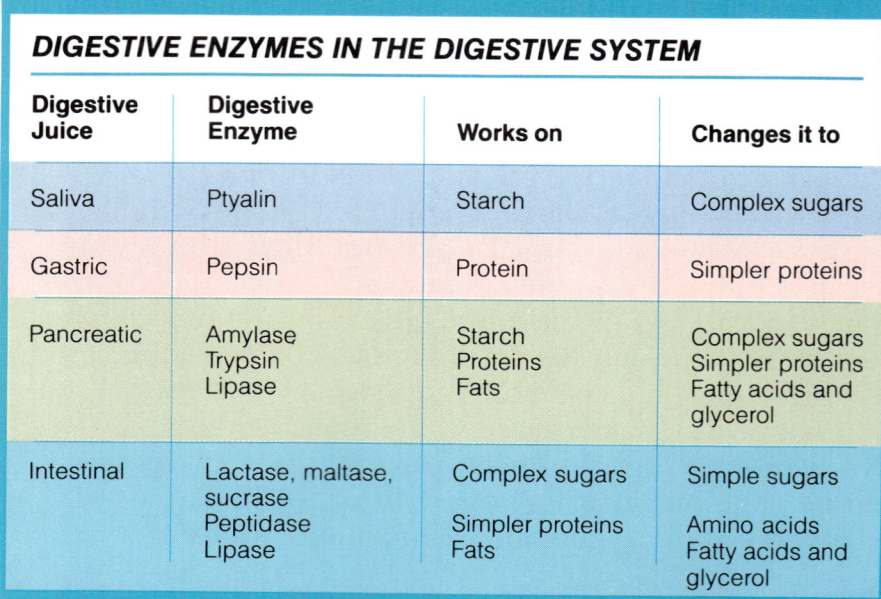

DIGESTIVE ENZYMES IN THE DIGESTIVE SYSTEM

Digestive Juice	Digestive Enzyme	Works on	Changes it to
Saliva	Ptyalin	Starch	Complex sugars
Gastric	Pepsin	Protein	Simpler proteins
Pancreatic	Amylase Trypsin Lipase	Starch Proteins Fats	Complex sugars Simpler proteins Fatty acids and glycerol
Intestinal	Lactase, maltase, sucrase Peptidase Lipase	Complex sugars Simpler proteins Fats	Simple sugars Amino acids Fatty acids and glycerol

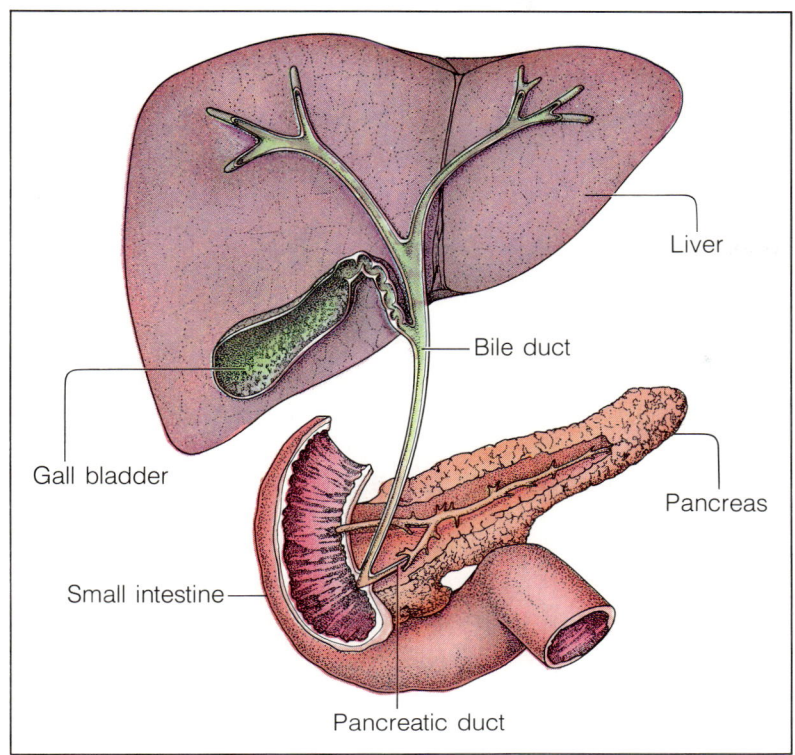

Figure 13–14 *The liver, pancreas, and gall bladder produce and store chemicals that are released into the small intestine to aid in the digestion of nutrients. What substance is produced by the liver?*

Liver

Bile duct

Gall bladder

Pancreas

Small intestine

Pancreatic duct

The Pancreas

The pancreas is a soft, triangular organ located between the stomach and the small intestine. The pancreas produces a substance called pancreatic juice, which is a mixture of several enzymes. These enzymes move into the small intestine at the same time bile does and help break down proteins, starches, and fats.

The pancreas also produces a substance called insulin, which is important in controlling the body's use of sugar. You will read more about the pancreas and insulin in Chapter 16.

SECTION REVIEW

1. Describe the process of digestion.
2. Explain the difference between chemical digestion and mechanical digestion.
3. Define peristalsis.
4. Where does most of the digestion of food occur?
5. Why are the liver and pancreas called digestive helpers rather than digestive organs?
6. Explain why food seems to have less flavor when you have a stuffy nose.

Eventually most of the food in the small intestine is digested. Proteins are broken down into individual amino acids. Carbohydrates are broken down into simple sugar molecules. And fats are broken down into fatty acids and glycerol. However, before these nutrients can be used for energy, they must first pass into the blood through the walls of the small intestine.

Absorption in the Small Intestine

The small intestine has an inner lining that looks something like wet velvet. This is because the inner walls of the small intestine are covered with millions of tiny fingerlike projections called **villi** (VIHL-igh, singular: villus).

Digested food is absorbed through the villi of the small intestine into a network of blood vessels that carries the nutrients to all parts of the body. If there were no ridges and no villi on the wall of the small intestine, nutrients would move through it too quickly. Nutrients would pass out of the body before they could be absorbed. The ridges and villi increase the surface area of the small intestine. This increased surface allows the maximum amount of nutrients to be absorbed.

Although most of the material in the small intestine is digested, there are some substances in food that cannot be digested. Cellulose, a part of fruits and vegetables, is one such substance.

Absorption in the Large Intestine

After most of the nutrients have been absorbed, undigested food along with water, mucus, bile salts, and scraped-off intestinal cells makes its way to the **large intestine.** During the time it takes for these remains to move through the large intestine, most of the water is absorbed. Also, helpful bacteria living in the large intestine feed on the leftovers and make certain vitamins, such as vitamin K and the B complexes, which are absorbed by the body.

The large intestine is shaped like a horseshoe that fits over the coils of the small intestine. It is

Figure 13–15 *These hills and ridges are not part of a mountain range on the earth's surface. They are tiny hairlike projections called villi that line the inside of the small intestine.*

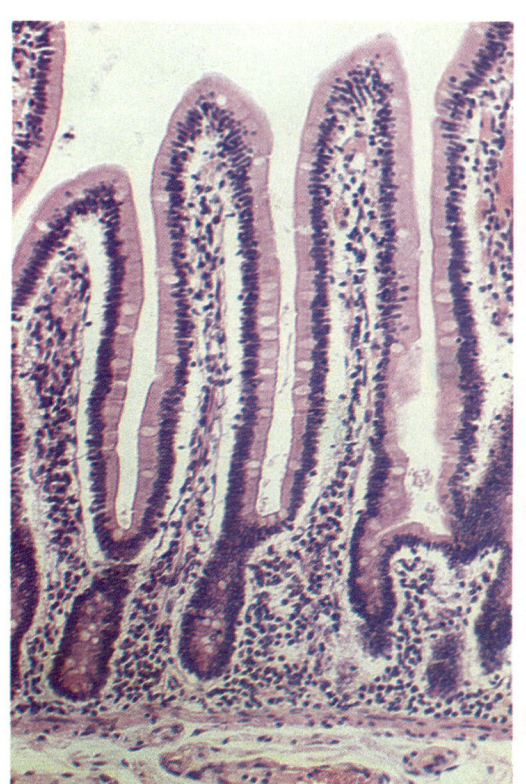
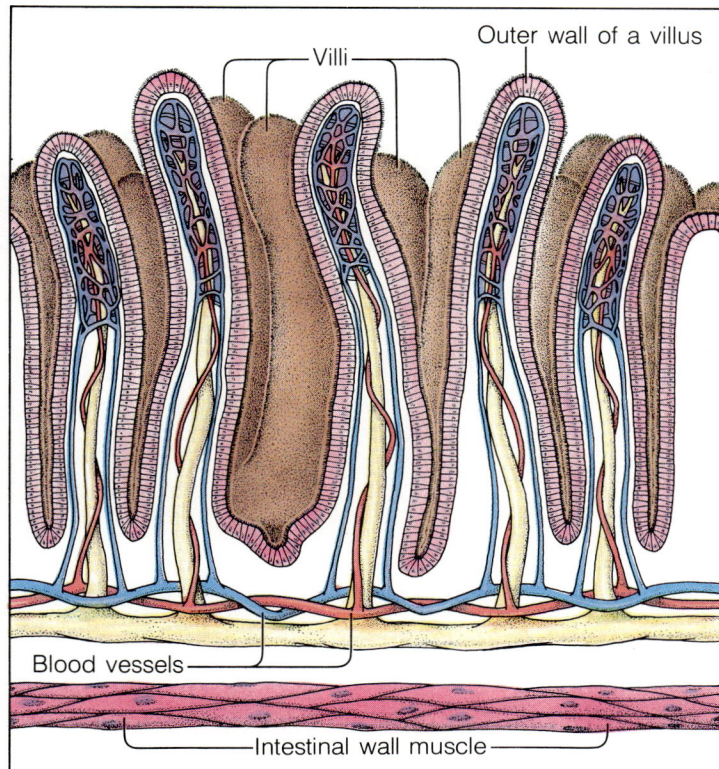

Figure 13–16 *The inner wall of the small intestine is covered with millions of villi. In the photograph, you can actually see the single layer of cells that covers each villus. The drawing shows a cross section of some villi. Notice the blood vessels. What is their function?*

about 6.5 centimeters in diameter but only 1.5 meters long. Why is it called the large intestine?

Materials that are not absorbed in the large intestine form a solid waste called **feces** (FEE-seez). At the end of the large intestine is a short tube called the rectum, which stores this waste. The rectum ends at an opening called the anus through which feces are eliminated from the body.

The appendix (uh-PEHN-diks) is a worm-shaped organ near the area where the small and large intestines meet. The appendix does nothing and leads nowhere. It sometimes becomes irritated, inflamed, or infected. An infection of the appendix is called appendicitis (uh-pehn-duh-SIGHT-ihs). If treated early, appendicitis usually is not dangerous.

Figure 13–17 *The horseshoe-shaped organ in this X-ray is the large intestine. What part of the skeleton is visible in the background?*

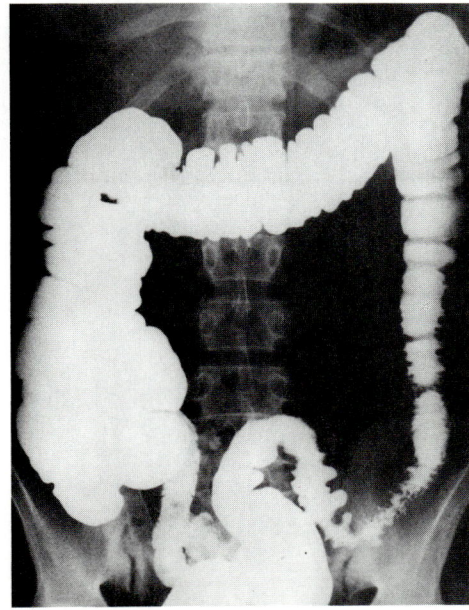

SECTION REVIEW

1. Describe the process of absorption.
2. What is the purpose of the villi in the small intestine?

3. Which substances are absorbed in the large intestine?
4. Appendicitis is usually treated with the removal of the appendix. Explain why this treatment does not interfere with the functioning of a person's digestive system.

13–4 Maintaining Good Health

You already know that by eating a balanced diet you supply your body with the materials it needs for growth, repair, and energy. **By controlling your weight and getting the proper amount of exercise, you can also keep your body healthy and running smoothly for many years.**

Exercise

Activities such as swimming, jogging, bicycling, hiking, calisthenics, or even walking briskly are good ways to exercise. Regular exercise helps to strengthen the heart. Exercise also gives you firmer muscles, better posture, greater strength, more endurance, and an improved sense of balance.

Figure 13–18 *If done regularly, bicycling helps maintain good health. What other activities are good forms of exercise?*

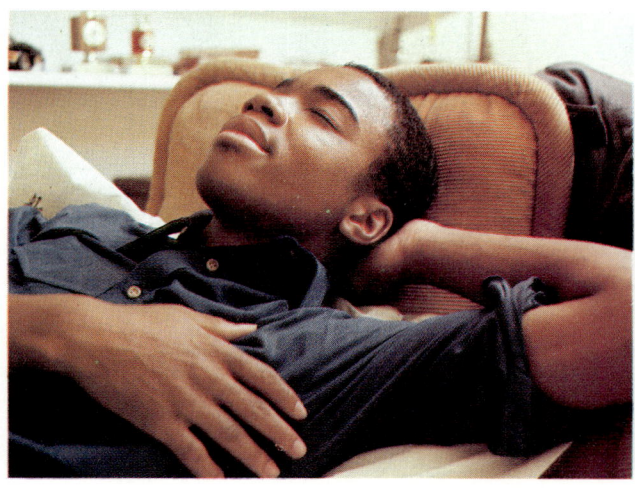 

Figure 13–19 *Everyone needs rest and sleep in order to stay healthy (left). These runners (right) need to rest after strenuous exercise.*

Weight Control

When the energy in the food you eat exceeds the work your body does, your body stores the excess energy in the form of fat. In order to get rid of this stored fat, you must use up more energy than is provided in the foods you eat. Then your body will break down the fat to release needed energy.

When a person is overweight, all of the organs of the body, including the heart, must work harder. Some people who are overweight may choose to go on a "diet." It is important that the diet chosen is a balanced one containing the proper amounts of nutrients needed by the body. Before starting any weight loss plan, a person should see a doctor. The doctor will give a complete checkup and discuss a sensible diet. Once a person loses the excess weight, he or she can maintain a steady weight by taking in only as many Calories in food as are used up.

SECTION REVIEW

1. Explain how exercise and weight control contribute to good health.
2. List five different activities that are good forms of exercise.
3. What must you be sure to include in any balanced diet?
4. How can you maintain a steady weight?
5. How does exercise help a person lose weight?

LABORATORY INVESTIGATION

Measuring Calories Used

Problem

How many Calories do you burn in a 24-hour period?

Materials *(per student)*

pencil and paper
scale

Procedure

1. Look over the chart of Calorie rates. It shows how various activities are related to the rates at which you burn Calories. The Calorie rate shown for each activity is actually the number of Calories used per hour for each kilogram of your body mass.

Activity	Average Calorie Rate *(Calories per hour per kilogram of body mass)*
Sleeping	1.1
Awake but at rest (sitting, reading, or eating)	1.5
Very slight exercise (bathing, dressing)	3.1
Slight exercise (walking quickly)	4.4
Strenuous exercise (dancing)	7.5
Very strenuous exercise (running, swimming rapidly)	10.5

2. Use a scale to record your mass in kilograms.
3. Classify all your activities for a given 24-hour period. Record the kind of activity, the Calorie rate, and the number of hours (or fractions of an hour) you were involved in that activity.
4. For each of your activities, multiply your mass by the Calorie rate shown in the chart. Then multiply the resulting number by the number of hours or fractions of hours you were involved in that activity. The result is the number of Calories you burned during that period of time. For example, if your mass is 50 kg and you exercised very strenuously, perhaps by running for a half hour, the Calories you burned during that activity would be equal to $50 \times 10.5 \times 0.5$.
5. Add together all the Calories you burned in the entire 24-hour period. The result is the total number of Calories you used during the 24 hours.

Observations

1. How many Calories did you use up in the 24-hour period?
2. How does the number of Calories you used compare to the number of Calories your classmates used?

Conclusions

1. Explain why the values for the Calorie rates of various activities are approximate rather than exact.
2. What factors could affect the number of Calories a person uses up during exercise?
3. Why do teenagers need to consume more Calories than adults?

Average Caloric Needs Chart					
	Age	*Calories*		*Age*	*Calories*
Boys	9–12 12–15	2400 3000	**Girls**	9–12 12–15	2200 2500

CHAPTER REVIEW

SUMMARY

13-1 Nutrition

❑ Nutrients provide the body with raw materials for repair, energy, and building new products. The six categories of nutrients are proteins, carbohydrates, fats and oils, vitamins, minerals, and water.

13-2 Digestion

❑ Digestion is the process of breaking down food into simpler substances.

❑ In the mouth, the salivary glands release saliva, which contains ptyalin. Ptyalin breaks down starches into simple sugars.

❑ The changing of foods into simple substances by enzymes is called chemical digestion.

❑ Mechanical digestion occurs when food is broken down by the chewing of the teeth and the churning movements of the digestive tract.

❑ As food enters the esophagus a wavelike motion called peristalsis pushes the food downward toward the stomach.

❑ Gastric juice contains the enzyme pepsin, which breaks down proteins into amino acids.

❑ After leaving the stomach, food enters the small intestine where intestinal juice is released and digests proteins, starches, and fats.

❑ Bile produced by the liver and stored in the gall bladder is carried to the small intestine. There bile aids in the digestion of fats.

❑ Pancreatic juice produced by the pancreas travels to the small intestine and digests proteins, starches, and fats.

13-3 Absorption

❑ Nutrients are absorbed into the bloodstream through villi in the small intestine.

❑ The large intestine absorbs water, as well as forming and absorbing some vitamins that are made by helpful bacteria.

❑ Undigested food substances, or feces, are eliminated out of the body through the anus.

13-4 Maintaining Good Health

❑ By controlling your weight and exercising regularly, you can help keep your body healthy.

VOCABULARY

Define each term in a complete sentence.

amino acid	esophagus	oil
bile	fat	pancreas
Calorie	feces	pepsin
canine	gall bladder	peristalsis
carbohydrate	incisor	protein
chemical digestion	large intestine	ptyalin
digest	liver	saliva
digestive system	mechanical digestion	small intestine
enzyme	mineral	stomach
epiglottis	molar	villus
	nutrient	vitamin

CONTENT REVIEW: MULTIPLE CHOICE

On a separate sheet of paper, write the letter of the answer that best completes each statement.

1. Which nutrient is used to build and repair body parts?
   a. proteins   b. minerals
   c. carbohydrates   d. vitamins
2. Which is *not* found in the mouth?
   a. pepsin   b. saliva   c. ptyalin   d. mucus
3. Mechanical digestion occurs when food is broken down by
   a. teeth.   b. enzymes.
   c. taste buds.   d. ptyalin.
4. Food moves from the mouth to the stomach through the
   a. pancreas.   b. small intestine.
   c. epiglottis.   d. esophagus.
5. In the digestive system, starches are broken down into
   a. glycerol.   b. simple sugars.
   c. fatty acids.   d. amino acids.
6. Gastric juice contains the enzyme
   a. bile.   b. ptyalin.   c. pepsin.   d. mucus.
7. Bile is stored in the
   a. stomach.   b. esophagus.
   c. gall bladder.   d. liver.
8. Pancreatic juice contains enzymes that break down
   a. starches.   b. proteins.   c. fats.   d. all of these.
9. Water is absorbed in the
   a. small intestine.   b. large intestine.   c. pancreas.   d. rectum.
10. Regular exercise helps a person to have
    a. good posture.   b. firm muscles.
    c. a stronger heart.   d. all of these.

CONTENT REVIEW: COMPLETION

On a separate sheet of paper, write the word or words that best complete each statement.

1. _____ are the usable parts of food.
2. Sugars and starches are _____.
3. The energy value of foods is measured in _____.
4. _____ digestion occurs when food is broken down by enzymes.
5. The teeth used for tearing and shredding food are _____.
6. The _____ is the flap of tissue that closes over the windpipe during swallowing.
7. In the digestive system, proteins are broken down into _____.
8. Bile helps to break up _____.
9. Digested food is absorbed in the _____.
10. The body stores excess energy in the form of _____.

CONTENT REVIEW: TRUE OR FALSE

Determine whether each statement is true or false. Then on a separate sheet of paper, write "true" if it is true. If it is false, change the underlined word or words to make the statement true.

1. The nutrients that supply the greatest amount of energy are the <u>fats and oils</u>.
2. There are <u>six</u> basic food groups.
3. Vitamins are needed by the body in <u>small</u> amounts.
4. <u>Enzymes</u> are chemicals that help control reactions in the body.
5. The physical action of breaking food into smaller parts is called <u>chemical</u> digestion.
6. <u>Ptyalin</u> helps break down proteins.
7. Food moves through the digestive system by a wavelike motion called <u>peristalsis</u>.
8. Most of the digestive organs are lined with a slippery substance called <u>saliva</u>.
9. Nutrients are absorbed through the walls of the <u>small intestine</u>.
10. In order to <u>gain</u> weight, you must use up more energy than is provided in the foods you eat.

CONCEPT REVIEW: SKILL BUILDING

Use the skills you have developed in the chapter to complete each activity.

1. **Making comparisons** Explain how the human digestive system can be compared to an "assembly line in reverse."
2. **Classifying objects** Construct a chart in which you classify each of the following as belonging to one of the four food groups: muffin, cottage cheese, nuts, yogurt, chicken, eggs, spaghetti, fish, cereal.
3. **Developing a model** Write menus for what you would consider to be ideal breakfasts, lunches, and dinners for four days. Explain why.
4. **Making predictions** How would your digestive system be affected if your doctor gave you an antibiotic that killed all of the bacteria in your body?
5. **Applying concepts** Explain why it is possible for astronauts to eat in space where there is no gravity.
6. **Making charts** Design a chart in which you list in order the organs of the digestive system through which food passes. Also indicate the roles of the liver, pancreas, and gall bladder in your chart.
7. **Developing a hypothesis** Suggest a hypothesis to explain why a meal high in fat makes a person feel fuller than a meal composed mainly of vegetables does.

CONCEPT REVIEW: ESSAY

Discuss each of the following in a brief paragraph.

1. Discuss the importance of proteins, carbohydrates, fats and oils, vitamins, and minerals in a daily diet. Include an example of a food that contains each.
2. Explain why it is helpful to chew your food thoroughly before swallowing it.
3. Describe the function of the epiglottis.
4. Compare the process of digestion to the process of absorption.
5. Describe the inner lining of the small intestine and explain why its structure is important in the absorption of nutrients.
6. You just ate a tuna fish sandwich with mayonnaise. What will happen to the bread (starch), tuna fish (protein), and mayonnaise (fat) during digestion?

Circulatory System 14

CHAPTER OBJECTIVES

After completing this chapter, you will be able to

14–1 Describe the function of the circulatory system.

14–2 Trace the flow of blood from the heart, through the body, and back to the heart.

14–2 Describe the structure and function of the heart and blood vessels.

14–3 Describe the parts of the blood.

14–3 Classify the four human blood groups.

14–4 Relate cardiovascular diseases to the circulatory system.

"Strap yourself in, we're taking off!" shouted the pilot over the roar of the helicopter. The chief flight nurse hopped into the cabin and buckled her seatbelt. She is a highly trained team member of Survival Flight, a special medical unit that helps people who are having heart attacks.

At 6:24 AM, the first call of the day came in. A 73-year-old man who lived 120 kilometers away was having severe chest pains. By 7:08 AM, the nurse was buckled into her seat. By 8:12 AM, she had helped give the patient an injection of a new drug called TPA (tissue plasminogen activator). TPA dissolves blockages in the blood vessels near the heart. Such blockages can prevent the heart from receiving an adequate amount of blood. By 8:23 AM the Survival Flight team was loading the patient onto the helicopter and preparing to rush him to a hospital.

The team is in a "race against time" to save heart tissue. When blockages clog the blood vessels near the heart, the muscle is deprived of oxygen. After about 30 minutes, tissue begins to die. After about six hours, there is little that can be done to save the heart.

The Survival Flight team has a helicopter supplied with a stretcher, medical instruments, and medicine. But the team's greatest advantage is the knowledge of its members. All of them understand how the heart works and know what to do when something goes wrong. Some of this special knowledge will be explained to you in the pages that follow.

Members of the Survival Flight team are specially trained to help people who are having heart attacks.

14–1 Circulation

The main task of the **circulatory system** in all organisms is transportation. **The circulatory system delivers food and oxygen to body cells and carries carbon dioxide and other waste products away from body cells.** The power behind this system is the heart. It pumps blood through a network of blood vessels. This network is so large that if it could ever be unraveled, it would wrap around the world more than twice.

One of the most important jobs of the circulatory system is delivering oxygen to the cells. As you know, oxygen combines with food inside your body cells to produce usable energy. Without energy, body cells would soon die. The cells that use the most oxygen—and the first to die without oxygen—are brain cells.

When cells combine oxygen and food to produce energy, they also produce a waste product called carbon dioxide. Removing carbon dioxide is another important job of the circulatory system.

Still another job of the circulatory system is to transport food to all body cells. At the same time, wastes produced by the cells are carried away by the blood. If the blood did not remove such wastes, the body would poison itself with its own waste products!

Figure 14–1 *The heart, like any muscle, contains lacy networks of blood vessels that deliver oxygen and other substances (right). The photograph on the left is a magnified view of such a network. What would happen to the heart if it did not have blood vessels?*

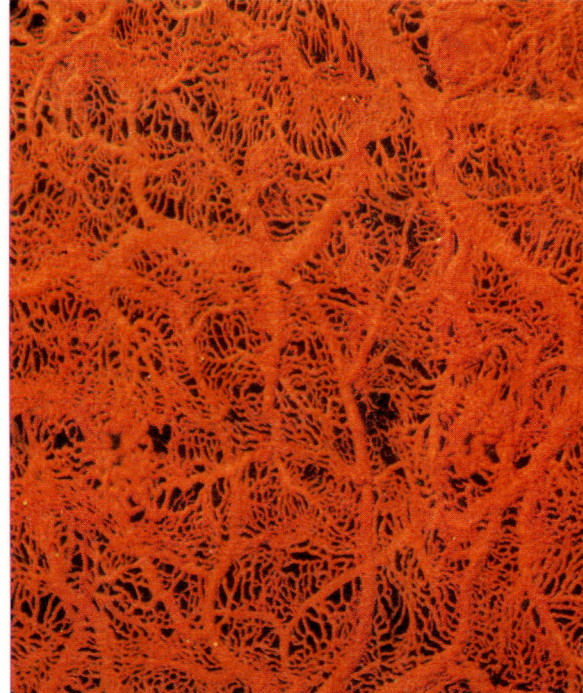
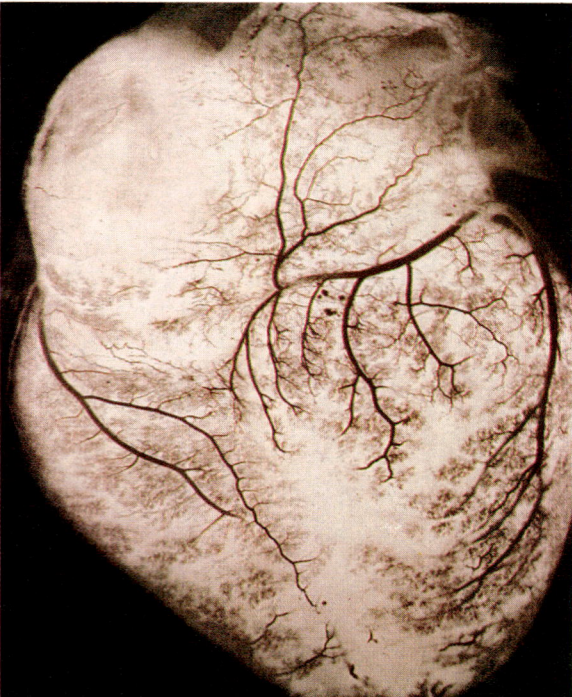

Figure 14–2 *The circulatory system transports materials to all parts of the body. What is the pumping power behind this transportation system?*

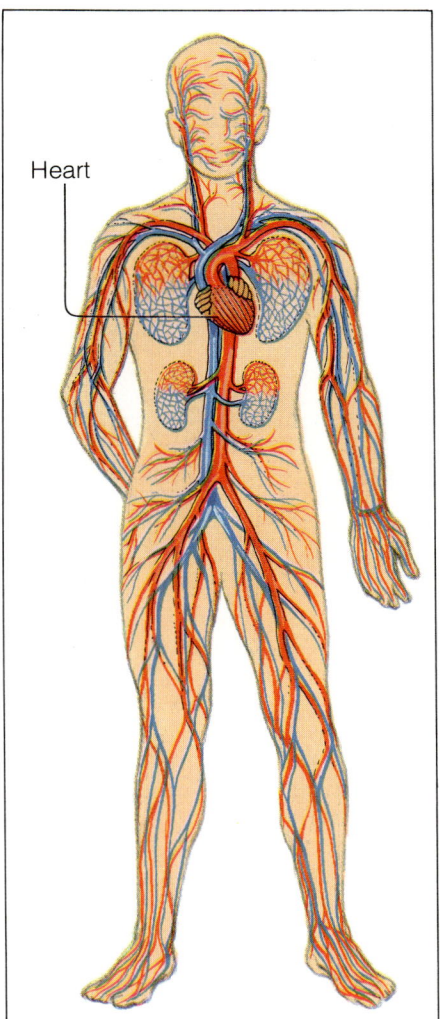

Heart

Sometimes the body comes under attack from microscopic organisms such as bacteria and viruses. At these times, another transporting function of the circulatory system comes into play—body defense. The blood rushes disease-fighting cells and chemicals to the area under attack.

The circulatory system transports other chemicals as well. These chemicals carry messages sent from one part of the body to another. For example, a chemical messenger from the pancreas is carried by the blood to the liver. Its message is "Too much sugar in the blood; remove some of the sugar and store it."

SECTION REVIEW

1. Describe the functions of the circulatory system.
2. What is the power behind the circulatory system?
3. When doctors perform operations on the heart, it is often necessary for them to cause the heart to temporarily stop beating. When they do so, they also lower the patient's body temperature. What is the purpose of this?

14–2 Circulation in the Body

Section Objective

To trace the flow of blood through the circulatory system

In a way, the entire circulatory system is like a vast maze that starts at the heart. Unlike most mazes, however, this one always leads back to the place it began. **Blood moves from the heart to the lungs and back to the heart. Then it travels to all the cells of the body and returns again to the heart.** In the next few pages, you will follow the blood on its journey through the circulatory maze. You will begin, of course, at the heart.

The heart is a muscle that rests only between beats. Even when you are asleep, at least 5 liters of blood are pumped through your body every minute. Not much larger than a fist, the heart is located slightly to the left of the center of your chest.

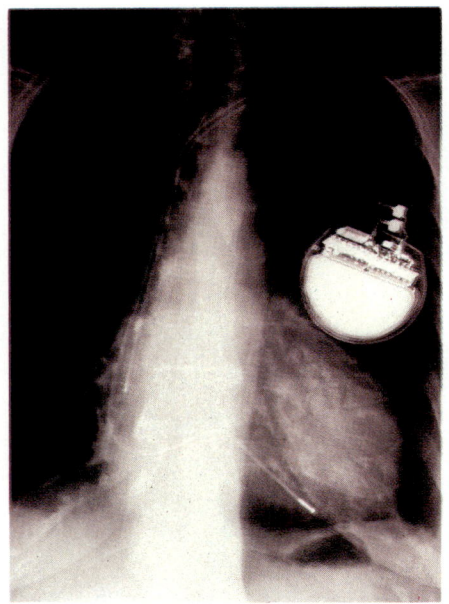

Figure 14–3 *The mechanical device in this X-ray of a human chest is an artificial pacemaker. When the body's original pacemaker becomes damaged, it can often be replaced by such a device. What is the function of a pacemaker?*

Heartbeat, or the heart's rhythm, is controlled by an area of special tissue in the heart known as the **pacemaker.** Located in the upper right side of the heart, the pacemaker sends out signals to heart muscle that control its contractions. If the body's original pacemaker becomes damaged, it can often be replaced by an artificial pacemaker.

The Right Side of the Heart

The human heart is made up of four hollow chambers. A thick wall of tissue called the **septum** separates the heart into right and left sides, with two chambers on each side. Your journey through the circulatory maze will begin in the right upper chamber, called the right **atrium** (plural: atria).

Inside the atrium, you find yourself swirling in a dark sea of blood. A great many **red blood cells** surround you. Red blood cells carry oxygen throughout the body. When a red blood cell hooks up to an oxygen molecule, the blood turns bright

Figure 14–4 *The heart has two upper chambers called atria and two lower chambers called ventricles. The atria receive blood from the body. What is the function of the ventricles?*

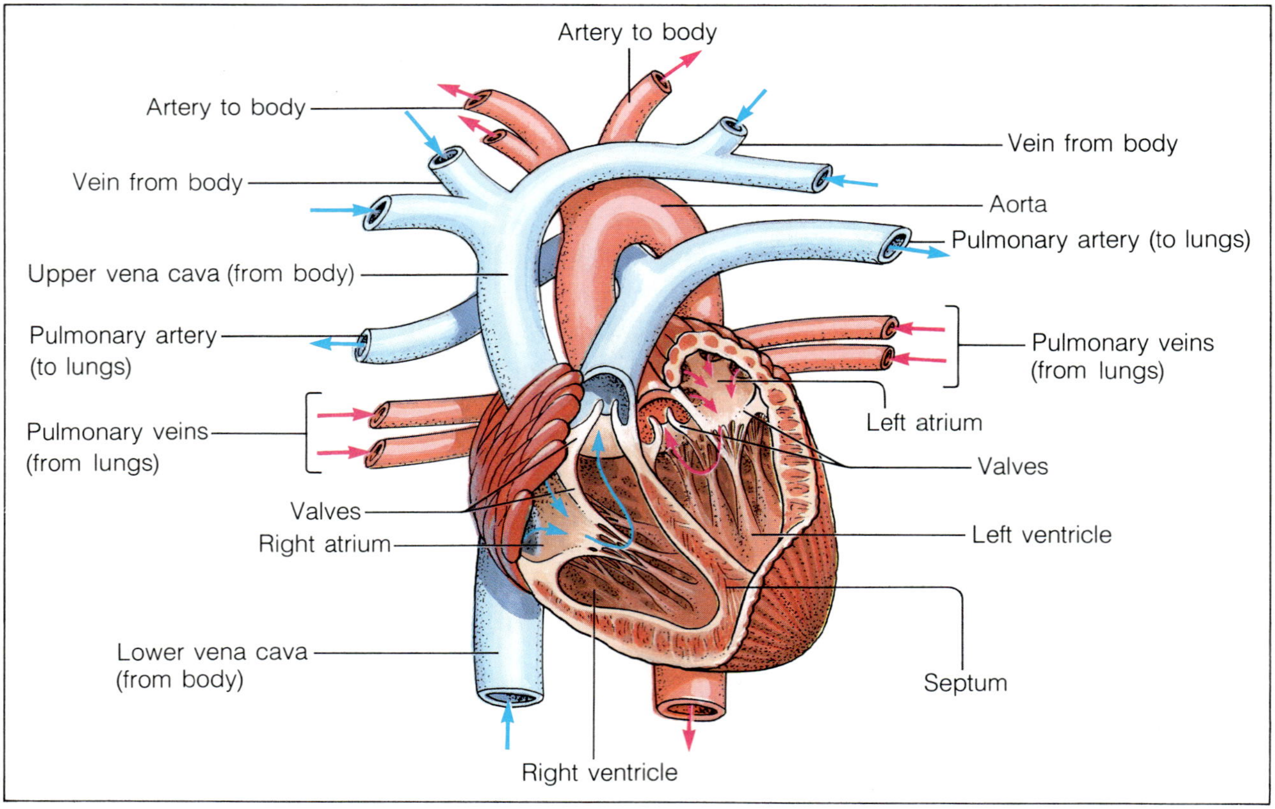

Artery to body

Artery to body

Vein from body

Vein from body

Aorta

Upper vena cava (from body)

Pulmonary artery (to lungs)

Pulmonary artery (to lungs)

Pulmonary veins (from lungs)

Pulmonary veins (from lungs)

Left atrium

Valves

Valves

Left ventricle

Right atrium

Lower vena cava (from body)

Septum

Right ventricle

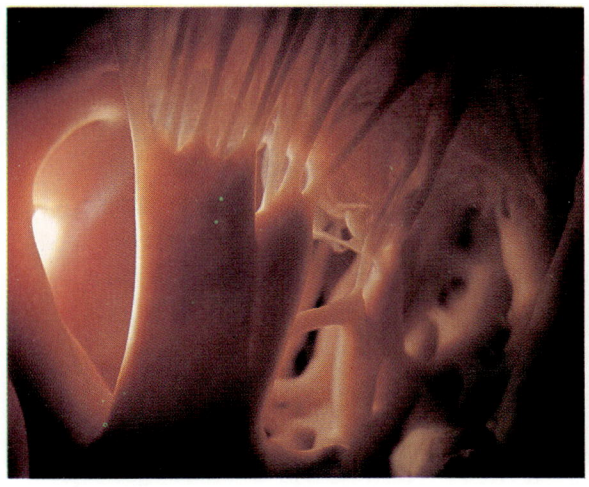

Figure 14–5 *This remarkable photograph shows a valve in a human heart (left). The valve is a small flap of tissue that prevents the backflow of blood. Sometimes a heart valve does not work properly and has to be replaced with an artificial valve (right).*

red. Such blood is said to be oxygenated. However, the blood in which you are swimming in the right atrium is dark red, not bright at all. This can only mean that these red blood cells are not carrying oxygen, but rather the waste carbon dioxide. This blood, then, is deoxygenated. And that makes sense. For the right atrium is a collecting chamber for blood returning from its trip through the body. Along the way, the red blood cells have dropped off their oxygen and picked up carbon dioxide.

Suddenly the blood begins to churn, and you feel yourself falling downward. You are about to enter the heart's right lower chamber, called the right **ventricle.** Before you do, you must pass through a small flap of tissue called a heart **valve.** The valve opens to allow blood to go from the upper chamber to the lower chamber. Then it closes immediately to prevent blood from backing up.

You have fallen into the muscular right ventricle. Your stay will be quite short. The ventricles, unlike the atria, are pumping chambers. Before you know it, you will feel the power of a heartbeat as the ventricle contracts and blood is forced out of the heart through a large blood vessel.

To the Lungs and Back Again

Now your journey has really begun. Since you are surrounded by dark, deoxygenated blood, your first stop should be obvious. The right ventricle has pumped you toward the lungs. It is a short trip. Soon red blood cells are dropping off the waste carbon

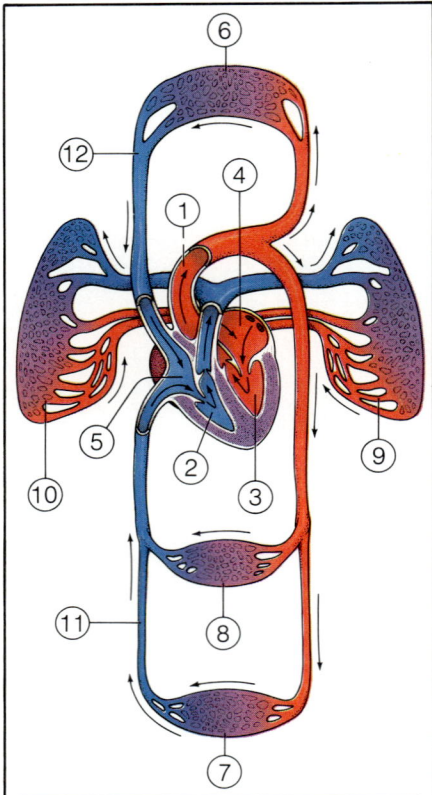

1. Aorta   2. Right ventricle
3. Left ventricle   4. Left atrium
5. Right atrium
6. Capillaries of head and arms
7. Capillaries of legs and feet
8. Capillaries of internal organs
9. Capillaries of left lung
10. Capillaries of right lung
11. Lower vena cava
12. Upper vena cava

Figure 14–6 *This diagram illustrates the path of the circulatory system. The path of oxygenated blood is shown in red, while that of deoxygenated blood is shown in blue. Where does deoxygenated blood enter the heart?*

dioxide. The carbon dioxide enters the lungs and is immediately exhaled. At the same time, the red blood cells are busy picking up oxygen. What color is the blood at this point?

From the lungs, you might expect the oxygenated blood to travel all over the body. But the first stop for the blood after leaving the lungs is the heart. This time you enter the left atrium.

The Left Side of the Heart

The left atrium, like the right atrium, is a collecting chamber for blood returning to the heart. However, the left chamber collects oxygenated blood as it returns from the lungs. Once again, the blood quickly flows downward through a valve and enters the left ventricle. The left ventricle has a lot more work to do than the right side. The right ventricle only pumps blood a short distance to the lungs. The left ventricle has to pump blood to every part of the body. In fact, the left side works about six times as hard as the right. That is one reason you feel your heartbeat on the left side of your chest. And it is also the reason that most heart attacks occur in the left side of the heart.

Arteries: Pipelines From the Heart

As the left ventricle pumps you and the blood you are floating in out of the heart, the blood passes through the largest blood vessel in the body, the **aorta** (ay-OR-tuh). The aorta is an **artery,** or a pipeline that carries blood *away from* the heart.

Like all arteries, the aorta is a thick-walled but flexible tube. Much of its flexibility comes from an elastic outer layer. The aorta soon branches into other, smaller arteries. Some return immediately to the heart, feeding heart muscle with food and oxygen. Other arteries branch and branch again, much like the branches of a tree. These branching arteries form a network leading to all parts of the body.

As you pass through the aorta and enter a smaller artery, you notice that the inner wall of the artery is quite smooth. The smooth wall allows blood to flow freely.

Outside the smooth layer in the artery is a middle layer that is made up mainly of smooth muscle tissue. When the muscle in the artery expands and contracts, it helps control the flow of blood. In this way, some arteries can expand to send large amounts of blood to one area, while other arteries can contract to lessen the amount of blood flowing somewhere else. Naturally, some blood must go to every part of the body—from the head to the toes.

From the aorta, your path leads you down one of the many branching arteries. Where you go from here depends on many factors. If food has been eaten recently, for example, much of the blood will be directed toward the intestines to pick up food. If the body is exercising, the blood supply to the muscles will probably be increased. If there are a great many wastes in the blood, you may be sent to the liver, where some wastes are changed into substances that are not poisonous to the body. Or you may travel to one of the kidneys, where other wastes are removed from the blood. However, whether it is thinking very hard or not, the brain always gets top priority ahead of any other part of the body.

Capillaries: The Unseen Pipelines

The artery network carries blood all over the body. But arteries cannot drop off or pick up any materials from body cells. Their walls are too thick. Oxygen and food, for example, cannot pass through the artery walls. In order to do its main job—transporting materials—the blood must pass from

Figure 14–7 *Although arteries, capillaries, and veins are all blood vessels, each has a unique structure. Their structures enable them to perform specialized functions. What is the function of each type of blood vessel?*

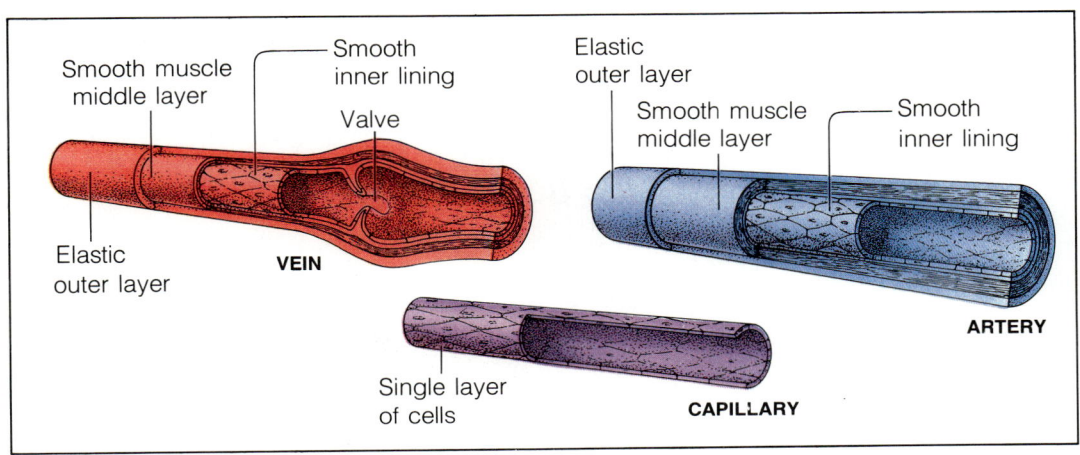

321

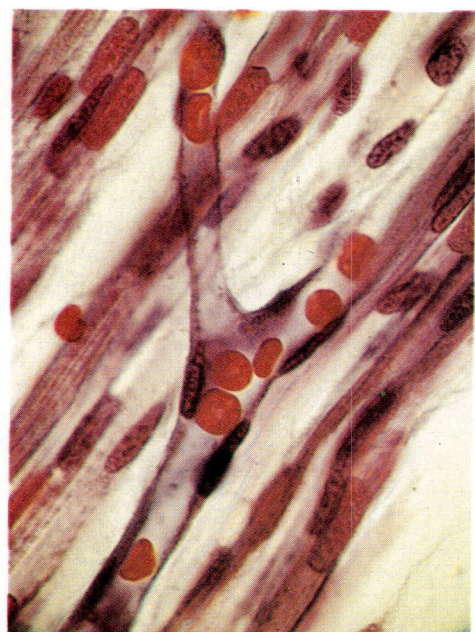

Figure 14–8 *Capillaries are so tiny that red blood cells can only pass through them in single file. What materials pass from blood to body cells through the capillary walls?*

the thick arteries into very thin-walled blood vessels called **capillaries.**

In the capillary in which you now find yourself, you must squeeze very tightly to pass through. In fact, in most capillaries there is only enough room for red blood cells to pass through in single file. Here in the capillaries, the real work of the blood is carried out. Food and oxygen pass through the thin walls and enter the cells. Wastes pass out of the cells and enter the blood. Other materials transported by the blood can leave and enter body cells at this time. What other materials are transported by the blood?

Veins: Pipelines to the Heart

Once the work is done in the capillary, you are ready to return to the heart. Your trip through the circulatory maze is just about over. The blood has given up its oxygen and is dark red again. So the blood leaves the capillaries and enters blood vessels called **veins.** Veins carry blood from the body *back to* the heart.

Although veins are much larger than capillaries, their walls are thinner than those of arteries. Unlike arteries, veins have tiny one-way valves. These valves help keep the blood from flowing backward. So blood always flows back to the heart.

SECTION REVIEW

1. To which organ does blood flow after it leaves the lungs?
2. List the four chambers of the heart.
3. What are three types of blood vessels?
4. Why are the walls of the ventricles much thicker than the walls of the atria?

14–3 Blood

In Chapter 12, you learned that blood is a type of connective tissue. That may have been a difficult concept to understand at the time. But now you can see how a fluid tissue—a tissue that does not stay in one place—can connect different parts of the body. Through the blood, the circulatory system transports many different substances. In this sense, blood "connects" all of the body systems.

Blood is a mixture of plasma, red blood cells, white blood cells, and platelets. In the following pages you will learn about the function of each of these components.

Plasma

Most people have between 4 and 6 liters of blood. About 55 percent of the blood is a yellowish liquid called **plasma.** Plasma is 92 percent water and so is easily and quickly replaced by the body. In the plasma, dissolved nutrients, enzymes, wastes, and other materials are transported through the body.

Red Blood Cells

The most numerous of the cells in whole blood are red blood cells. Under a microscope, they look like round hats with thickened brims—almost like tiny berets. The centers of these flexible red disks are so thin that they seem clear. Red blood cells can bend almost in two, a useful trick when trying to squeeze through a narrow capillary.

Have you ever heard people complain they have "iron-poor" blood? That phrase refers to a shortage of an iron-containing protein called **hemoglobin** (hee-muh-GLOH-bihn) that is found in red blood cells. Hemoglobin binds to oxygen in the lungs and carries the oxygen to body cells. Hemoglobin also helps carry carbon dioxide wastes back to the lungs.

Sharpen Your Skills

Pumping Power

The heart of an average person pumps about 9000 L of blood daily. How much blood will be pumped in an hour? In a year?

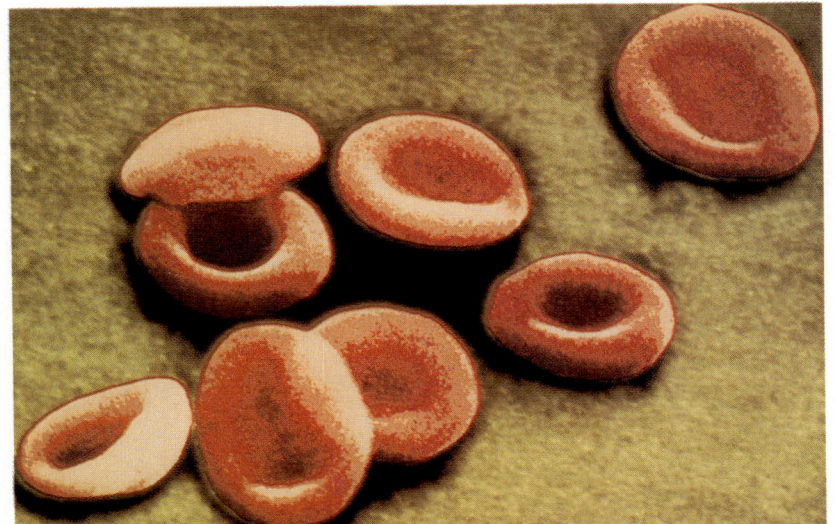

Figure 14–9 *These beret-shaped objects are red blood cells as seen through a scanning electron microscope. What is the function of red blood cells?*

Red blood cell production begins in bone marrow. An immature red blood cell, like all living body cells, contains a nucleus. However, as the cell matures, its nucleus grows smaller and smaller until it vanishes. Red blood cells pay a price for life without a nucleus. They are very delicate and have a life span of only 120 days. When a red blood cell becomes old or damaged, it is broken down in the liver and the spleen, an organ just to the left of the stomach. In fact, so many red blood cells are destroyed in the spleen each day that it has been called the "cemetery" of red blood cells.

White Blood Cells

The second kind of cell in whole blood is the **white blood cell.** White blood cells are outnumbered by red blood cells 700 to 1. But what they lack in number they make up for in size and life span. Some white blood cells are twice the size of red blood cells. And although certain kinds of white blood cells stay in the circulation for only a few hours, others can last for months or even years!

Like red blood cells, most white blood cells develop in bone marrow. However, white blood cells do not lose their nuclei when they mature. White blood cells are among the most important defense systems of the body. Carried by blood to areas under attack by tiny invaders such as bacteria, some white blood cells surround and digest the invaders. Other kinds of white blood cells produce special chemicals to attack viruses and various other microscopic invaders. You will learn more about white blood cells and body defense in Chapter 18.

Figure 14–10 *The round yellow objects in this photograph are the smallest type of white blood cells. What are the red objects?*

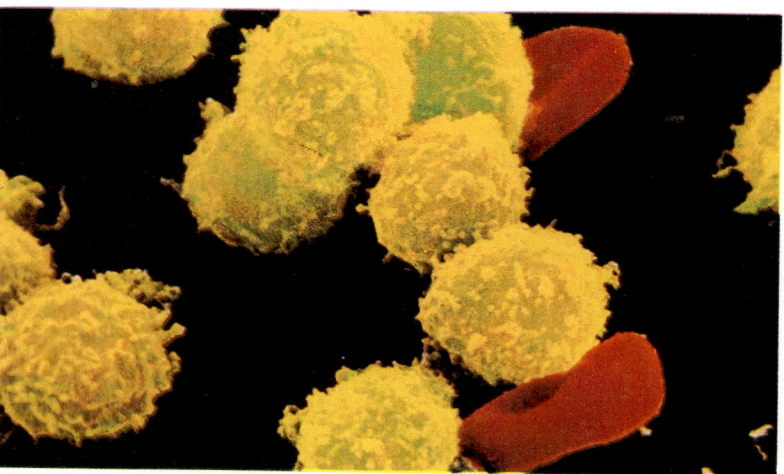

Platelets

The third kind of cell in whole blood is the **platelet** (PLAYT-liht). Actually, platelets are really only bits of cells. They have no nucleus or color. These tiny cell fragments break away from bone marrow and enter the bloodstream. There they last no more than ten days. But in that short time they may save a life!

Have you ever wondered what happens inside a blood vessel when you cut yourself? Why does all your blood not ooze out of your body? As soon as a blood vessel has been cut, platelets begin to collect around the cut. They then release an enzyme that acts on other substances in the blood to create **fibrin** (FIGH-bruhn). Fibrin gets its name from the tiny fibers it weaves across the cut in the blood vessel wall. Blood cells and plasma are trapped by fibrin. See Figure 14–11. The plasma will set and harden and form a clot. You have probably seen this process of clotting on the surface of your skin when a scab formed over a cut. A scab is a surface blood clot.

The Lymphatic System

As blood moves through the capillaries, some of the plasma leaks out of the blood and through the capillary walls. This fluid, called **lymph,** surrounds and bathes body cells. Eventually the lymph collects

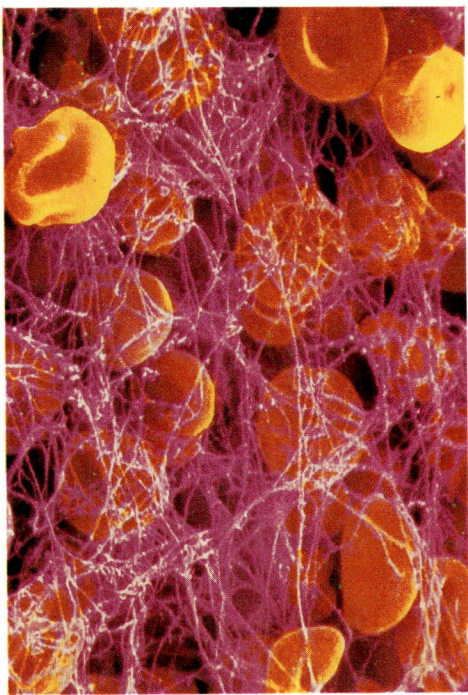

Figure 14–11 *When you cut yourself, a network of fibrin threads forms over the injured area. This unusual photograph shows how the threads trap red blood cells and stop bleeding.*

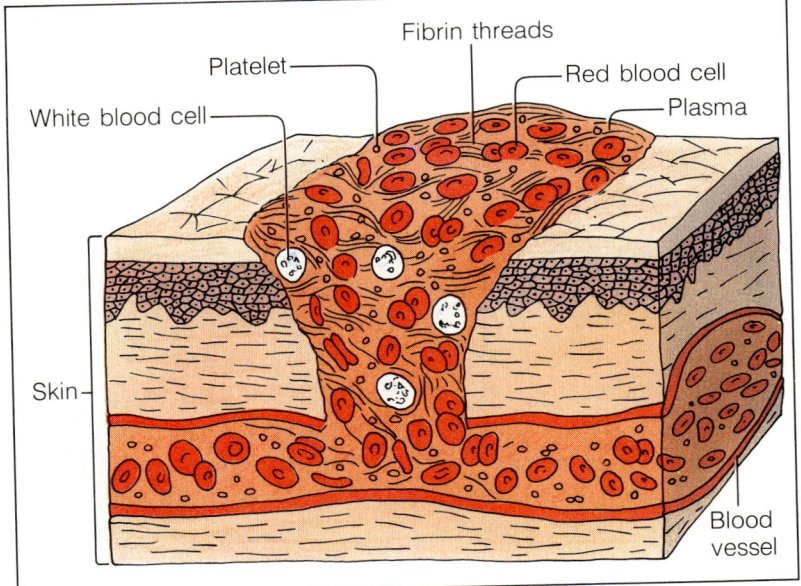

White blood cell — Platelet — Fibrin threads — Red blood cell — Plasma — Skin — Blood vessel

Figure 14–12 *This drawing shows a cross section of a cut and what happens inside the body to stop the bleeding. What are the fibers that trap blood cells and plasma called?*

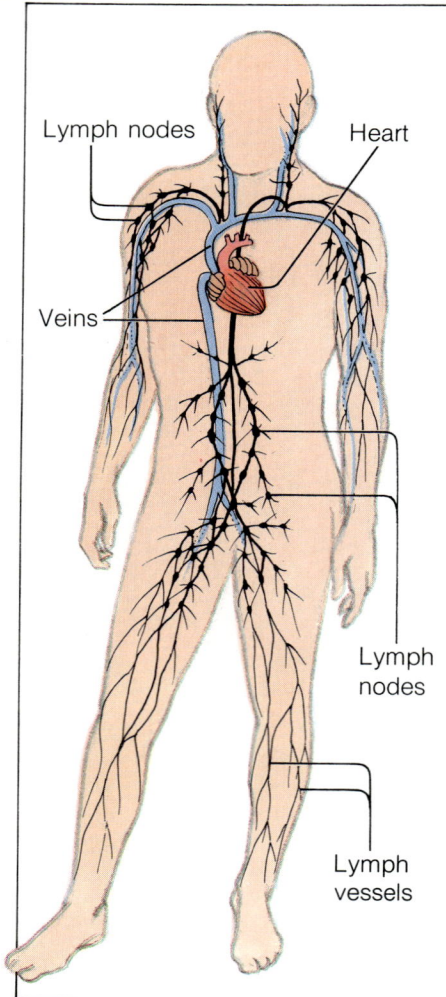

Figure 14–13 *This diagram illustrates the path lymph travels on its way through the lymphatic system. What is lymph?*

Labels on figure: Lymph nodes, Heart, Veins, Lymph nodes, Lymph vessels

in a system of tiny vessels. These vessels make up the lymphatic system.

From the lymph vessels, lymph flows into two veins in the neck. In this way, the lymphatic system transports fluid lost from the blood back into the bloodstream.

On its way through the lymphatic system, lymph passes through masses of tissues called lymph nodes. Special cells in these nodes remove harmful materials, such as bacteria, from the lymph.

Blood Groups

In 1900, Karl Landsteiner, an American scientist, was able to classify human blood into four basic groups, or types: A, B, AB, and O. Everyone is born with a certain type of blood, which stays the same for a lifetime. People with group A blood have an A protein attached to the outer coat of every red blood cell. People with group B blood have a B protein. Group O people have neither protein. What proteins do you think group AB people have?

The process of transferring blood from one body to another is called a **transfusion.** But there is one problem with transfusions. The problem is that people with group A blood also produce a chemical called anti-B. Anti-B causes red blood cells with group B protein to clump together. Such a clump in the bloodstream can cause death. In much the same way, people with group B blood produce anti-A. Anti-A causes red blood cells with group A protein to clump together. Because of the anti-A and anti-B clumping chemicals, blood groups must be carefully matched before a transfusion. In Figure 14–14, you can see the four groups of blood and how they can be mixed.

Rh Factor

Several years after the discovery of the ABO blood groupings, Landsteiner and another American scientist, Alexander S. Wiener, discovered a group of proteins that existed in the blood of rhesus monkeys as well as in the blood of many humans. These proteins were named the Rh factor after the first two letters in the word "rhesus." People who have the Rh factor are said to be Rh positive, or Rh$^+$.

ABO BLOOD SYSTEMS

Blood Group	Proteins on Red Blood Cells	Clumping Chemicals in Plasma	Can Accept Transfusions From Group(s)
A	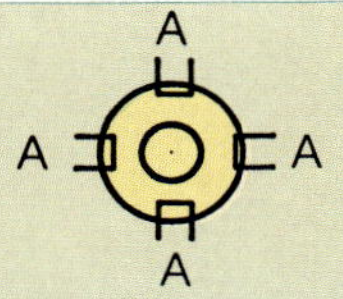 (A, A, A, A proteins shown)	Anti-B	A, O
B	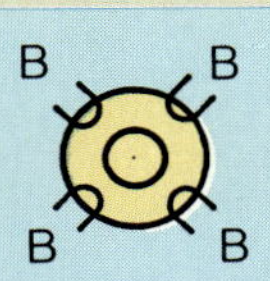 (B, B, B, B proteins shown)	Anti-A	B, O
AB	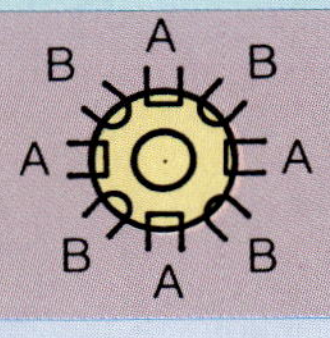 (A, B proteins shown)	None	A, B, AB, O
O	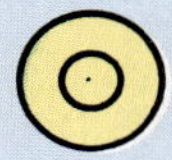 (no proteins shown)	Anti-A Anti-B	O

Those who do not have this factor in their blood are Rh negative, or Rh⁻.

Figure 14–14 *Each blood group is characterized by the type of protein on the red blood cells and the clumping chemicals in the plasma. What protein does blood group O have?*

SECTION REVIEW

1. List the four main components of whole blood.
2. What is the liquid portion of blood called?
3. Which blood cells contain hemoglobin?
4. Which blood cells help fight disease?
5. Certain people are able to donate blood to anyone. Which group of blood do these people have?

14–4 Cardiovascular Disease

Section Objective

To relate atherosclerosis and high blood pressure to the circulatory system

People are living longer now than they ever have. And, as people's life expectancy has increased, so has the number of people who suffer from **chronic disorders.** In fact, chronic disorders are a major health problem in the United States. Chronic disorders are lingering, or lasting, illnesses. Usually developing over a long time, chronic diseases also

HELP WANTED: BIOMEDICAL ENGINEER Ph.D. in biomedical engineering desired, but master's degree with experience acceptable. Apply to NASA.

A patient with a serious heart disease needs a heart transplant to survive. While waiting for a human donor to supply a heart, doctors might consider implanting a temporary artificial heart in the patient's body. The scientists who design and build artificial body parts are **biomedical engineers.**

The body part replacements created by biomedical engineers include pacemakers, heart valves, artificial kidneys, joints, and limbs. Much of the work of these scientists is done in a university laboratory or private research institute. They use their knowledge of engineering and technology to solve problems in biology and medicine. The biomedical engineers then publish and present papers at scientific meetings about their findings.

To learn more about biomedical engineering, write to the Alliance for Engineering in Medicine and Biology, Suite 402, 4405 East-West Highway, Bethesda, MD 20814.

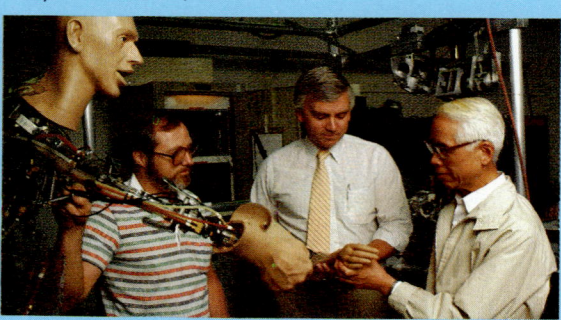

Figure 14-15 *In this photograph, you can see crystals of cholesterol magnified 250 times. Everyone has some cholesterol in his or her body, but the amount can be increased through diet. What foods are high in cholesterol?*

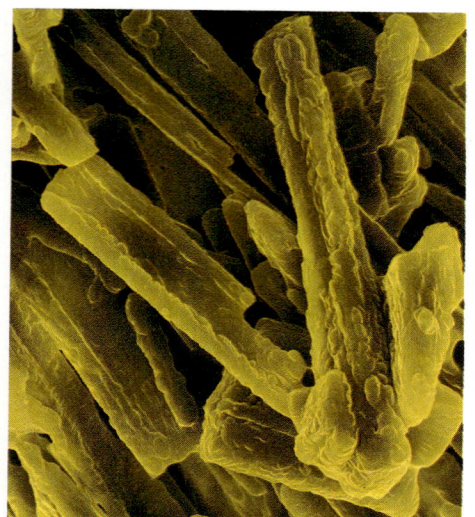

last a long time. Today, **cardiovascular** (kahr-dee-oh-VAS-kyoo-luhr) **diseases** are the most serious of all chronic disorders, in terms of numbers of people affected. **Cardiovascular diseases, such as atherosclerosis (ath-uhr-oh-skluh-ROH-sihs) and high blood pressure, affect the heart and blood vessels.**

Atherosclerosis

The most common cause of heart attacks in the United States today is **atherosclerosis.** Atherosclerosis is the thickening of the inner lining of the arteries. This thickening begins when certain fatty substances, such as **cholesterol,** enter a damaged area in the artery's lining. These fatty substances in the blood slowly collect on the inner lining of the arteries. As this happens, the space for blood to flow inside the artery becomes narrower and narrower. So the normal circulation of the blood through the arteries is lessened or even blocked. When blood flow is reduced, the cells served by the blood may die. If this occurs in the arteries and the cells of the heart, heart cells begin to die. A heart attack may then occur. A similar blockage can happen in the brain's arteries, killing some brain cells. This is called a stroke.

Atherosclerosis is normally thought of as a chronic disease that affects senior citizens. But it is a condition that begins much earlier in life. By the age of 20, in fact, most people have some atherosclerosis, which is often called "hardening of the arteries." For this reason, it is very important to begin proper health habits and care of the heart before atherosclerosis becomes severe.

Many doctors suggest avoiding foods rich in fatty substances such as cholesterol. These foods include animal fats, meats, and dairy products. These foods do not have to be eliminated. But doctors do believe that meals very high in cholesterol should not be the major part of a person's daily diet. Doctors also suggest a moderate amount of exercise to keep the heart and blood vessels healthy.

Figure 14–16 *These diagrams show how fatty substances in the blood build up on the inside wall of arteries, causing atherosclerosis. The photographs actually show the stages of atherosclerosis in human arteries.*

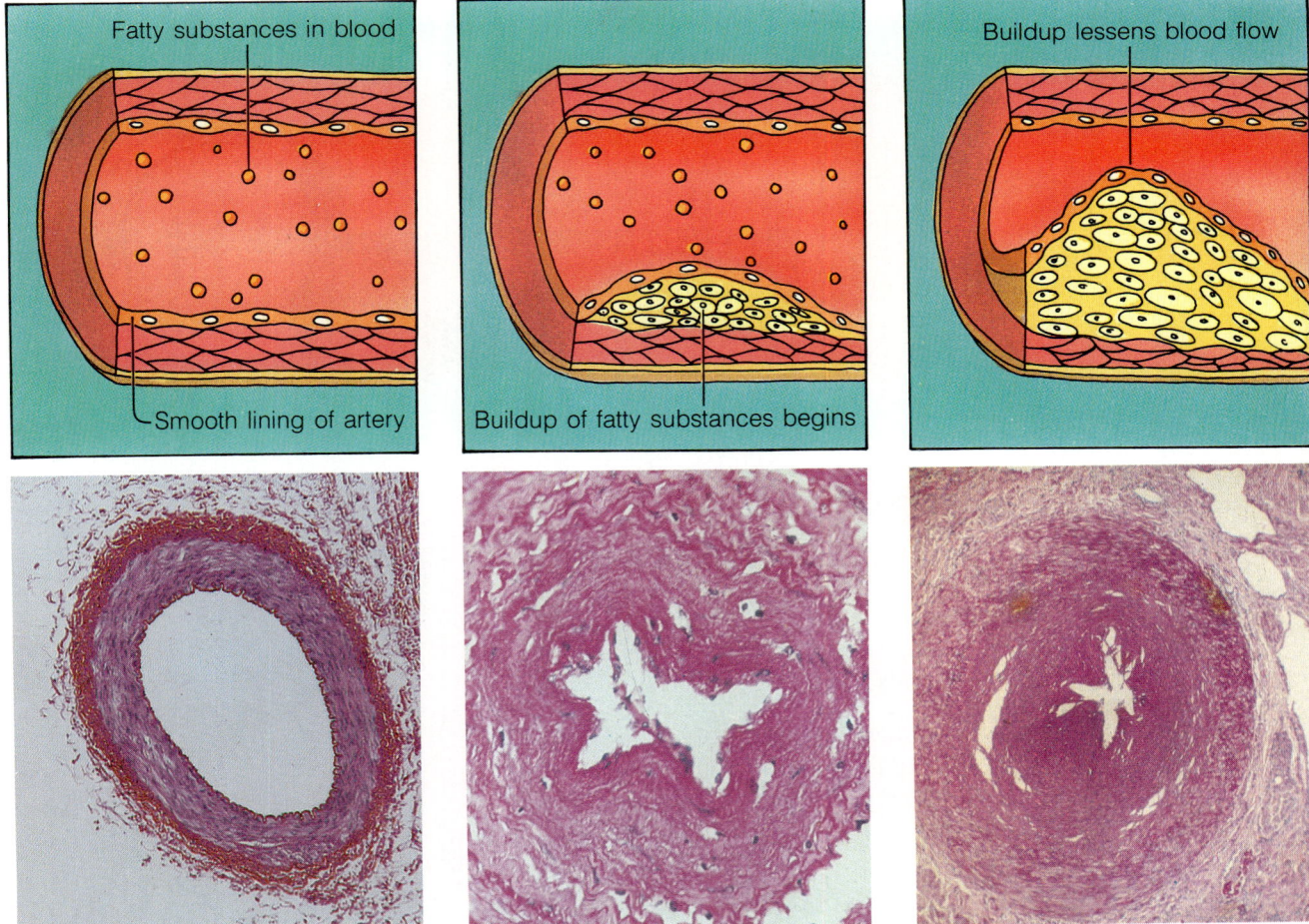

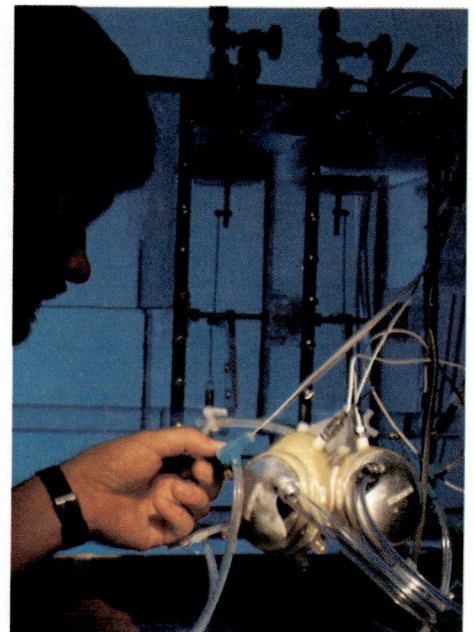

Figure 14–17 *This man is working on the Jarvik-7, an artificial heart. In the future, such hearts may routinely replace damaged human hearts. But today there are still many problems with artificial hearts.*

Despite all precautions, no one can guarantee that atherosclerosis and the heart problems it can cause can be avoided totally. Fortunately, in no area of medicine has more research and progress occurred than in the treatment of cardiovascular diseases. For example, heart bypass operations have become commonplace. In this operation, a healthy blood vessel from the leg is removed and used to bypass damaged arteries serving the heart. In this way, blood flow to heart cells is increased.

High Blood Pressure

If you have ever used a garden hose, you know that water pressure inside the hose can be increased in two ways. When you turn up the flow of water at the faucet, pressure increases in the hose. The water flows out of it with more force. Another way to increase the water pressure is to narrow the opening of the nozzle with your fingers. So the water pressure in the hose is a result of both the force of the water rushing through the hose and the resistance of the walls of the hose to that flow of water.

Figure 14–18 *High blood pressure occurs when too much pressure builds up in the arteries. If this condition continues and is left untreated, damage to the walls of the arteries, to the heart, and to other organs may occur. What is another term for high blood pressure?*

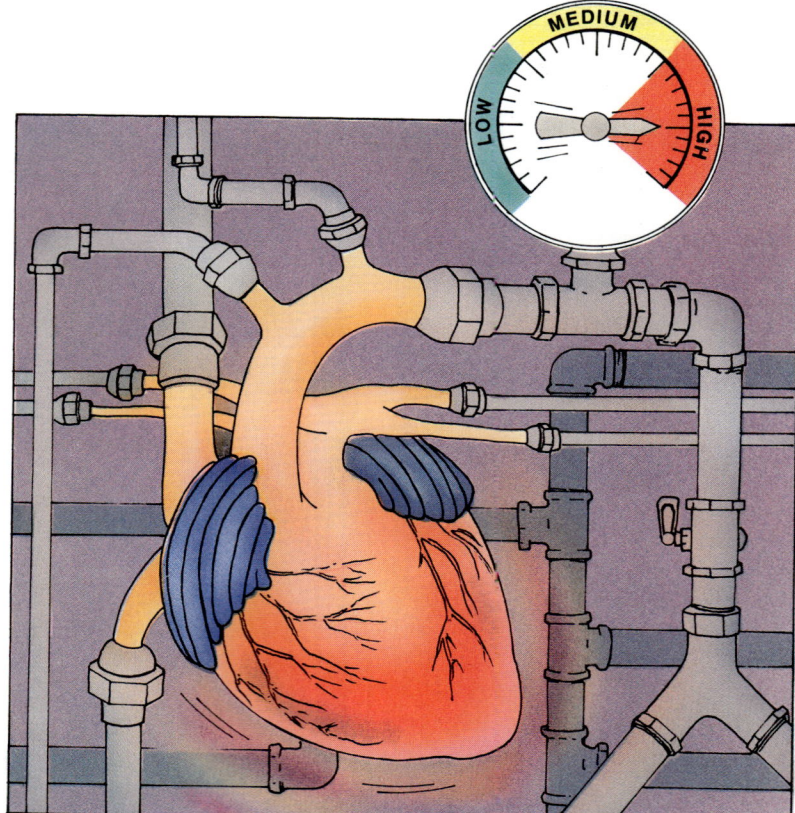

Blood pressure can be compared to the pressure of water in a hose. The force of the blood is caused by the pumping of the heart. Blood vessel walls are similar to the walls of a hose. Blood pressure is produced by the force of blood and the resistance of the blood vessel walls. Blood pressure rises each time the heart contracts to pump blood through blood vessels. Blood pressure falls each time the heart relaxes between beats.

Blood pressure can be measured with an instrument called a **sphygmomanometer** (sfihg-moh-muh-NAHM-uh-tuhr). See Figure 14–19. Measurements of blood pressure are indicated by two numbers. A normal blood pressure reading, for example, is 120/80. The first number is the pressure of the blood when the heart muscle contracts. The second number is the pressure of the blood when the heart relaxes between beats.

Normally blood pressure rises and falls from day to day and hour to hour. Sometimes though, blood pressure goes up and remains above normal. This condition is called **hypertension,** or high blood pressure. The increased pressure can damage blood vessels and make the heart work harder. As a result, hypertension can cause strokes and contribute to heart attacks.

Because people with hypertension often have no obvious symptoms to warn them, hypertension is often called the "silent killer." That is why it is very important to have blood pressure checked at least once a year. A person with high blood pressure should control weight gain, reduce the amount of salt in the diet, and exercise regularly. Sometimes medicines must also be taken.

Figure 14–19 *This woman is having her blood pressure taken. What is the name of the instrument that measures blood pressure?*

SECTION REVIEW

1. What is a cardiovascular disease?
2. What is another name for hypertension?
3. Name a fatty substance associated with heart disease.
4. How does atherosclerosis affect the circulatory system?
5. Predict two ways in which a regular exercise program could help a person with cardiovascular disease.

Sharpen Your Skills

Blood Pressure

Visit the library and look up information on what the two numbers of a blood pressure reading, such as 120/80, mean. In a written report, explain how a sphygmomanometer works. Explain what is meant by systole and diastole.

LABORATORY INVESTIGATION

Measuring Your Pulse Rate

Problem

What are the effects of activity on pulse rate?

Materials *(per group)*

clock or watch with a sweep second
  hand
graph paper

Procedure

1. On a separate sheet of paper, construct a data table similar to the one below.
2. Place the index and middle fingers of one hand on the other wrist where your wrist joins the base of your thumb. Move the two fingers slightly until you locate your pulse.
3. To determine your pulse rate, have a classmate observe a clock or watch with a sweep second hand as you count the number of beats that you feel in 60 seconds. Record the result.
4. Walk in place for one minute. Then measure your pulse and record the result.

5. Run in place for one minute. Again, measure your pulse and record the result.
6. Sit down and rest. Measure your pulse after you have been resting for one minute and again after three minutes. Enter these results in your data table.
7. Use the data table to construct a bar graph, comparing each activity and the pulse rate you measured.

Observations

1. What was the pulse rate you recorded in step 3? This is called your pulse rate at rest. How does your rate compare with those of your classmates? (Do not be alarmed if your pulse rate is somewhat different from those of other students. Rates may vary.)
2. What effect did walking have on your pulse rate? Running?
3. What effect did resting after running have on your pulse rate?

Conclusions

1. What conclusion can you draw from your data?
2. How is pulse rate related to heartbeat?
3. What happens to the blood supply to the muscles during exercise? How is this related to the change in pulse rate?
4. Did your pulse rate return to normal after exercising? If so, how long did it take? If not, how long do you think it would have taken?
5. What do you think would happen to your pulse rate if you were to run in place for 10 minutes?

Activity	Resting	Walking	Running	Resting After Exercise *(1 min.)*	Resting After Exercise *(3 min.)*
Pulse Rate					

CHAPTER REVIEW

SUMMARY

14–1 Circulation

❏ The main task of the circulatory system is to transport materials throughout the body.

❏ Among the materials carried by the blood are oxygen and carbon dioxide, food and wastes, disease-fighting cells, and chemical messengers.

14–2 Circulation in the Body

❏ The heart's pacemaker regulates the contraction of the heart muscle.

❏ The circulatory system is like a maze that begins and ends with the heart.

❏ A wall of tissue called the septum divides the heart into a right and left side.

❏ The two upper, collecting chambers of the heart are called atria. The two lower, pumping chambers of the heart are called ventricles.

❏ Valves between the atria and ventricles keep blood from flowing backward.

❏ Arteries carry blood away from the heart.

❏ Veins carry blood back to the heart.

❏ Capillaries connect the arteries and veins.

❏ Materials leave and enter the blood through the walls of the capillaries.

14–3 Blood

❏ Plasma, which is mainly water, is the yellowish fluid portion of whole blood.

❏ Red blood cells, white blood cells, and platelets make up the solid portion of whole blood.

❏ Hemoglobin binds to oxygen in the lungs and carries oxygen to body cells.

❏ White blood cells are part of the body's defense against invading bacteria, viruses, and other microscopic organisms.

❏ Platelets help the body form blood clots to stop the flow of blood in a cut blood vessel.

❏ Red blood cells, white blood cells, and platelets are all formed in bone marrow.

❏ All people are born with one of four blood groups: A, B, AB, or O.

❏ People whose blood contains the Rh factor are Rh positive. People without the factor are Rh negative.

❏ The lymphatic system transports fluid lost from the blood back into the bloodstream.

14–4 Cardiovascular Disease

❏ Cardiovascular diseases affect the heart and blood vessels.

❏ The thickening of the inner lining of the arteries is called atherosclerosis. If atherosclerosis affects the arteries of the heart, a heart attack may occur. If atherosclerosis affects the brain arteries, a stroke may occur.

❏ High blood pressure can damage blood vessels and make the heart work harder.

VOCABULARY

Define each term in a complete sentence.

aorta	cholesterol	hypertension	sphygmomanometer
artery	chronic disorder	lymph	transfusion
atherosclerosis	circulatory system	pacemaker	valve
atrium	fibrin	plasma	vein
capillary	hemoglobin	platelet	ventricle
cardiovascular disease		red blood cell	white blood cell
		septum	

CONTENT REVIEW: MULTIPLE CHOICE

On a separate sheet of paper, write the letter of the answer that best completes each statement.

1. The heart is separated into right and left sides by a wall of tissue called a (an)
   a. ventricle.   b. atrium.   c. septum.   d. valve.
2. Oxygenated blood from the lungs enters the heart through the
   a. left atrium.   b. right atrium.
   c. left ventricle.   d. right ventricle.
3. The first stop for blood pumped from the right atrium is the
   a. brain.   b. lungs.   c. right ventricle.   d. capillary network.
4. The part of the heart that works hardest is the
   a. right atrium.   b. right ventricle.
   c. left atrium.   d. left ventricle.
5. The blood vessels that carry blood back to the heart are the
   a. arteries.   b. veins.   c. capillaries.   d. ventricles.
6. The cells that contain hemoglobin are the
   a. plasma.   b. platelets.
   c. white blood cells.   d. red blood cells.
7. Red blood cells are produced in the
   a. heart.   b. liver.
   c. spleen.   d. bone marrow.
8. Platelets help the body to
   a. control bleeding.   b. fight infection.
   c. carry oxygen.   d. do all of these.
9. People with group AB blood have
   a. both A and B proteins.   b. neither A nor B proteins.
   c. both anti-A and anti-B clumping chemicals.   d. none of these.
10. Cholesterol is a fatty substance associated with
    a. hemoglobin.   b. hypertension.   c. atherosclerosis.   d. salt.

CONTENT REVIEW: COMPLETION

On a separate sheet of paper, write the word or words that best complete each statement.

1. The two lower heart chambers are called _____.
2. Deoxygenated blood enters the heart through the _____.
3. The _____ are the thinnest blood vessels.
4. The blood vessels that contain valves are the _____.
5. The special tissue that controls the heartbeat is the _____.
6. An iron-containing substance in red blood cells is _____.
7. Worn-out or damaged red blood cells are broken down in the _____ and the liver.
8. The type of blood cell that fights infection is the _____ cell.
9. Plasma that leaks out of capillary walls is _____.
10. Blood pressure can be measured by an instrument called a (an) _____.

CONTENT REVIEW: TRUE OR FALSE

Determine whether each statement is true or false. Then on a separate sheet of paper, write "true" if it is true. If it is false, change the underlined word or words to make the statement true.

1. The <u>atrium</u> separates the left side of the heart from the right side.
2. The two upper chambers of the heart are called <u>ventricles</u>.
3. Blood carries food and <u>carbon dioxide</u> to the cells.
4. Red blood cells pick up <u>oxygen</u> in the lungs.
5. The <u>aorta</u> is the largest blood vessel in the body.
6. <u>Arteries</u> are blood vessels that carry blood to the heart.
7. The <u>capillaries</u> are blood vessels that connect arteries to veins.
8. <u>Platelets</u> help to create fibrin.
9. Those people who have the Rh factor in their blood are said to be <u>Rh$^+$</u>.
10. <u>Hypertension</u> is sometimes called "hardening of the arteries."

CONCEPT REVIEW: SKILL BUILDING

Use the skills you have developed in the chapter to complete each activity.

1. **Sequencing events** Do all arteries carry oxygenated blood? Do all veins carry deoxygenated blood? Explain.
2. **Relating facts** On a separate sheet of paper add another column to Figure 14–14. Call it "Can Donate Blood to Group(s)." Fill in this column.
3. **Making predictions** Explain why chronic diseases are more of a problem today than they were 200 years ago. Predict how great a problem they will be in the future.
4. **Applying concepts** To determine whether a person has an infection, doctors often do blood tests in which they count white blood cells. Explain why such a count is useful.
5. **Making generalizations** To determine whether a person has abnormally high blood pressure, at least three blood pressure measurements should be taken on three separate days at three different times. Explain why.
6. **Making inferences** An artificial heart actually replaces only the ventricles of a natural heart. Suggest a reason why replacing the atria is not necessary.
7. **Making inferences** Sometimes lymph nodes in the body, such as those in your neck, become swollen. Why do you think this happens?

CONCEPT REVIEW: ESSAY

Discuss each of the following in a brief paragraph.

1. List and describe the four components of blood.
2. Explain why blood must be crossmatched before a blood transfusion can be performed.
3. Construct a chart in which you list the three types of blood vessels, their structures, and their functions.
4. Starting in the right atrium, trace the path of blood through the body.
5. Why is blood considered a connective tissue?
6. Explain why, in human beings, oxygenated and deoxygenated blood never mix.

Respiratory and Excretory Systems 15

CHAPTER OBJECTIVES

After completing this chapter, you will be able to

15–1 Describe the function of the respiratory system.

15–1 Describe the structures and functions of the respiratory organs.

15–1 Trace the flow of air during breathing.

15–2 Describe the function of the excretory system.

15–2 Describe the structures and functions of the excretory organs.

Standing almost 6 kilometers above sea level, these mountain climbers have just climbed one of the highest mountains in the world—Mount McKinley. They have walked across snow, ice, and rugged terrain that would have kept most people from reaching the top of this mountain. As you might expect, this feat requires strong muscles, incredible stamina, and a great deal of courage.

Mountain climbing also requires healthy lungs and powerful chest muscles to provide the body with enough oxygen to reach the top. If you were with the mountain climbers, you would soon discover that the amount of oxygen in the air decreases as the altitude increases. As a result, you would have to breathe harder and deeper to supply your body with enough oxygen. If you were not used to reduced oxygen at high altitudes, you might grow tired or dizzy. Even experienced mountain climbers must bring oxygen equipment along on steep climbs.

Climbing Mount McKinley requires lungs that function at their best. But you do not have to be a mountain climber to need healthy lungs. Walking, running, eating, and even sleeping require oxygen. In fact, not a minute goes by that your body does not need oxygen. Even if you try to hold your breath, at some point your body disregards what you are telling it and you automatically begin to breathe again.

Now turn the page and begin your journey of discovery. You will learn about the remarkable feats of the respiratory system. On the way, you will meet another body system—the excretory system.

Mountain climbers atop Mount McKinley in Alaska

15–1 The Respiratory System

Try this: Breathe in. Easy, right? Now breathe out. Again, no problem. You can breathe in or out whenever you think about it. But what about when you are asleep? When you sleep, you do not have to worry that your body will not breathe. You breathe automatically. But why breathe at all?

In Chapter 13, you learned that the digestive system breaks down food into small molecules that are transported by the blood to your body cells. In Chapter 14, you read how oxygen is carried by the blood to body cells. In the cells, the oxygen is combined with food to produce energy that the body can use. This process is called **respiration.** You may also recall that the waste carbon dioxide is produced during respiration.

The job of getting oxygen into the body and removing carbon dioxide and water is the main task of the respiratory system. This happens automatically when you breathe. Breathing in and breathing out are the first and last steps in the process of respiration.

The Nose and Throat

All the air that you use to breathe, sing, speak, or shout is usually first taken into the **respiratory system** through openings in the nose. These open-

Figure 15–1 *The main job of the respiratory system is to get oxygen into the body. Once inside the body, the oxygen combines with food to produce energy. This energy enables you to perform many activities, such as playing volleyball. What is the process called that combines oxygen with food?*

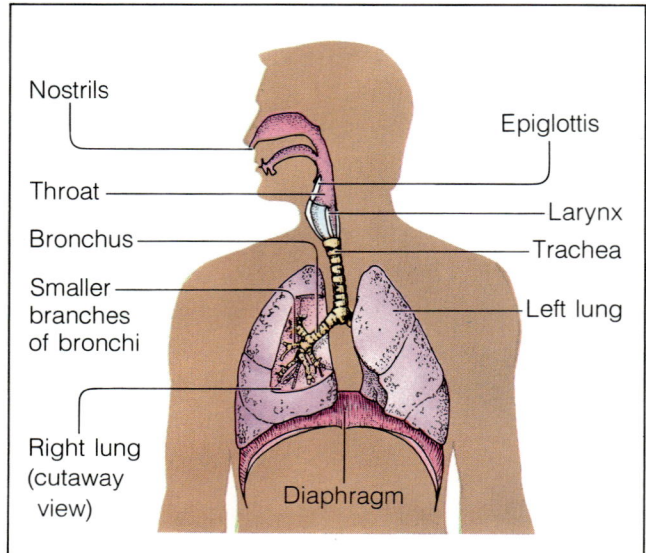

Nostrils
Throat
Bronchus
Smaller branches of bronchi
Right lung (cutaway view)
Diaphragm
Epiglottis
Larynx
Trachea
Left lung

ings are called **nostrils.** If the air is cold, as it may be in winter, it is quickly heated by warm blood that flows through vessels near the inner lining of the nose. Meanwhile, mucus in the nose moistens the air. This keeps the delicate tissues of the respiratory system from drying out. In addition, large hairs and tiny hairs in the nose trap dust particles and microscopic organisms such as bacteria and keep them from going any farther into the respiratory system. If the nose becomes irritated by these trapped particles, your body responds by producing a little "explosion" to force the particles out. As you may have guessed, this explosion is a sneeze!

Because the nose warms, moistens, and filters the air coming into the body, it is healthier to take in air through the nose than through the mouth. But when the nose is blocked, such as when you have a cold, the mouth acts as a backup organ so that you can continue breathing.

From the nose, the moist, clean air moves into your throat. The air will come to a kind of fork in the road. One path leads to the digestive system. The other path leads deeper into the respiratory system. What directs the air down the respiratory path and objects such as food and water down the digestive path? The "traffic" is routed down the right path by a small flap of tissue called the **epiglottis.** The epiglottis cuts off the opening to your windpipe when you swallow and routes all food and water down to your digestive system.

The Trachea

Take a moment to gently run your finger up and down the front of your neck. You will feel alternating bands, or rings, of cartilage and smooth muscle. These rings form the protective wall of your **trachea** (TRAY-kee-uh), or windpipe. If the epiglottis is open, air passes from your throat into the trachea. As the air moves downward, tiny hairs lining the trachea trap dirt particles and bacteria that have managed to get through the nose. Like the nose, the trachea produces tiny explosions in response to irritations. These explosions are called coughs. During a cough, air is sometimes forced out of the trachea at speeds of up to 160 kilometers per hour!

Figure 15–2 *In order to shout cheers, these cheerleaders must control the flow of air through their respiratory systems.*

Figure 15–3 *In the drawing, you can see the position of the vocal cords at the top of the trachea. How are sounds produced by the vocal cords?*

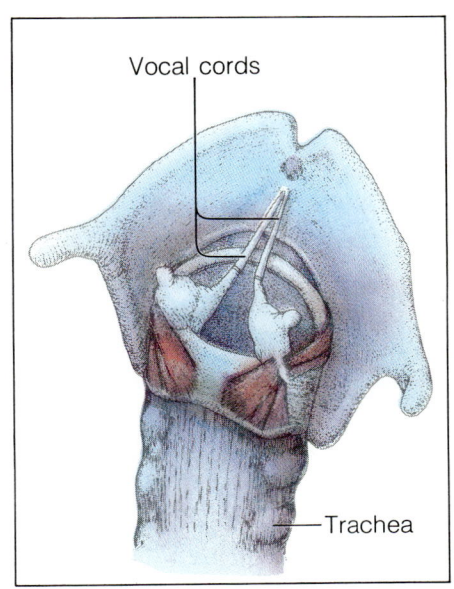

Vocal cords

Trachea

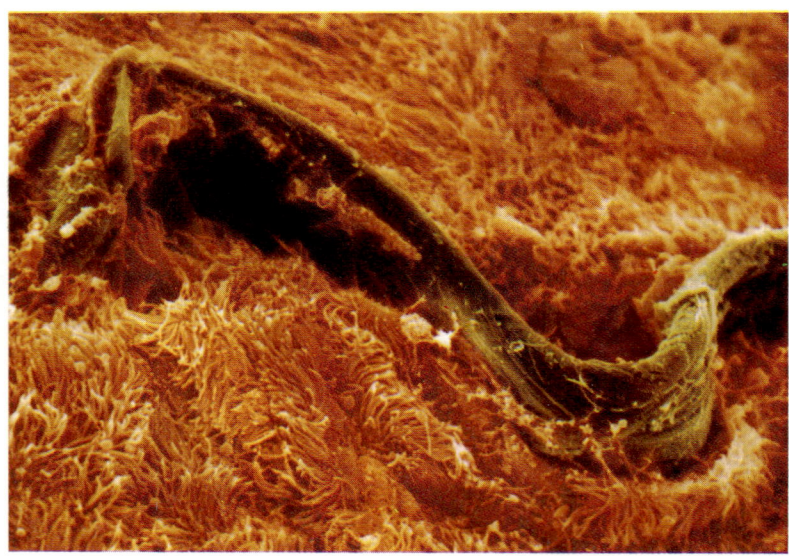

Figure 15–4 *The long, brown-colored object in this photograph is a dirt particle trapped in the tiny hairs in the bronchus. These hairs move constantly, carrying the dirt particle up through the windpipe to the throat where it is coughed out or swallowed. What is another name for the windpipe?*

Located at the top of the trachea is the **larynx** (LAR-ihngks), or voice box. The larynx is made of cartilage. Within the lining of the larynx are folds of tissue called **vocal cords.** As air passes out and past the vocal cords, they vibrate. These vibrations, together with the movements of the mouth and tongue, produce sounds.

CAREER   *Respiratory Therapy Technician*

HELP WANTED: RESPIRATORY THERAPY TECHNICIAN Completion of a one-year program in respiratory therapy, taken either at a hospital or technical school. Certification as a respiratory therapy technician (CRTT) is desirable. A strong desire to work with people is needed. Apply at the hospital personnel office.

Somewhere each day tragic events occur, such as drowning, drug poisoning, heart failure, stroke, head injury, or electrical shock that can cause a person to stop breathing. It is vital that the person be treated immediately for breathing problems. That is the job of the **respiratory therapy technician.**

In addition to working in emergency situations, these technicians work under the direction of doctors. With special equipment, the technicians help people with respiratory disorders such as asthma, emphysema, pneumonia, and bronchitis. Some respiratory therapy technicians work in laboratories where they measure lung volumes or give radioactive gas to patients. The radioactive gas leaves a "trail" that shows up on X-rays. This helps the doctor in determining whether there is a respiratory problem and, if so, its degree of seriousness.

For more information about this exciting and challenging field, write to the American Association for Respiratory Therapy, 1720 Regal Row, Dallas, TX 75235.

The Respiratory Tree

As you breathe, air passes down the throat and into the trachea. Soon the air reaches a place where the trachea branches into two tubes. Each of these tubes is called a **bronchus** (BRAHNG-kuhs; plural: bronchi). Each bronchus continues to branch into smaller and smaller tubes. The bronchi and their many smaller branches are often described as forming an upside-down tree—the respiratory tree. See Figure 15–5.

The thinnest branches of the respiratory tree lead to grapelike clusters of tiny "balloons" called **alveoli** (al-VEE-uh-ligh; singular: alveolus). The alveoli make up most of the tissue of the **lungs,** the main organs of respiration. It is the alveoli that make the lungs soft and spongy. Because of the hollow alveoli, your lungs are so light that they could float. Each thin-walled alveolus is surrounded by a network of capillaries. It is here that your blood picks up its cargo of oxygen from the air. Oxygen

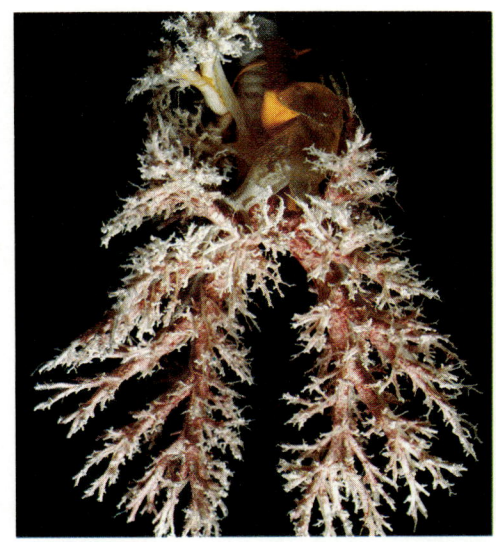

Figure 15–5 *Notice how the trachea, bronchi, smaller branches of the bronchi, and alveoli form a respiratory tree.*

Figure 15–6 *In this drawing, you can see how oxygen and carbon dioxide are exchanged in the lungs. Where in the lungs does the exchange of these gases occur?*

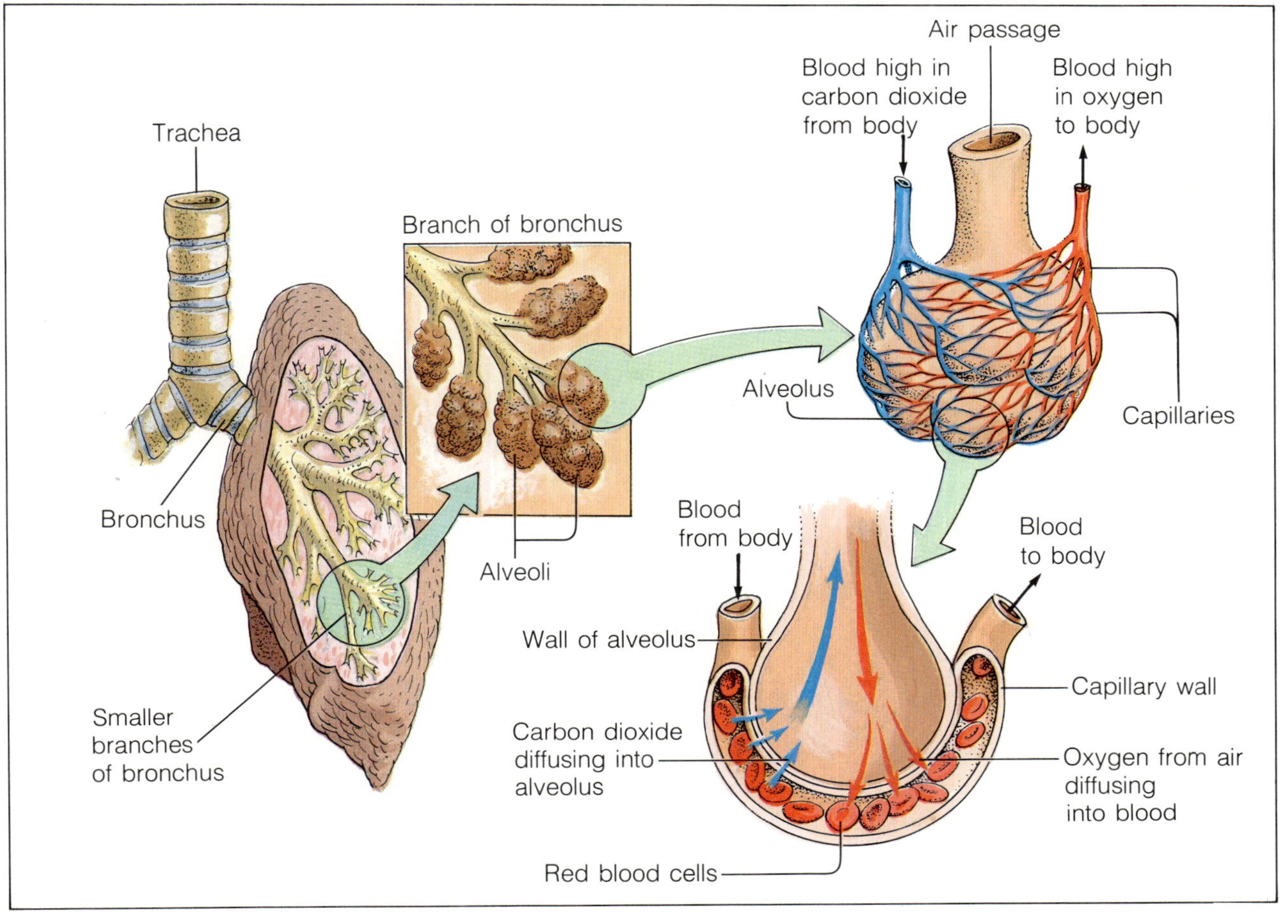

Trachea

Branch of bronchus

Air passage

Blood high in carbon dioxide from body

Blood high in oxygen to body

Alveolus

Capillaries

Bronchus

Alveoli

Blood from body

Blood to body

Wall of alveolus

Smaller branches of bronchus

Capillary wall

Carbon dioxide diffusing into alveolus

Oxygen from air diffusing into blood

Red blood cells

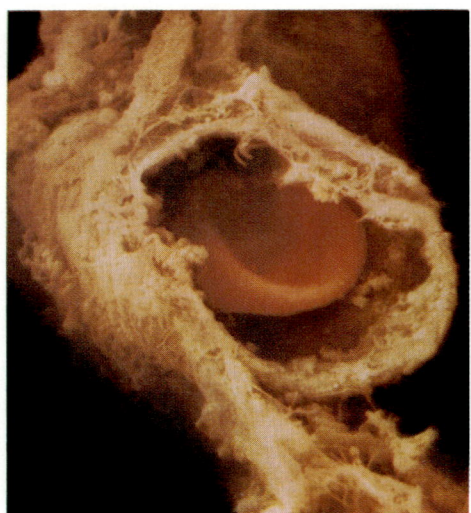

Figure 15–7 *The capillaries that surround each alveolus are so tiny that red blood cells must squeeze through one at a time. What gas is picked up by the red blood cells in the alveoli?*

from the alveoli passes into the blood flowing in the capillaries. Afterward, the oxygen-rich blood will be pumped back to the heart and sent through the arteries to all the tissues of the body. Meanwhile, the waste gas carbon dioxide passes from the blood into the alveoli. Within seconds, you breathe out the waste gas carbon dioxide.

Mechanics of Breathing

Look down at your chest as you breathe in and out. What do you see? As you take in a breath, your chest expands. As you let the air out, your chest becomes smaller. Why?

You may be quick to reply that air rushing into your lungs makes your chest expand. And that air rushing out makes your chest shrink. But that is not the way it happens at all. Something else happens first, before the air moves in either direction. That something has to do with some sets of muscles in your chest. Here is how they work.

When you are about to take a breath, muscles attached to your ribs contract and pull upward and outward. At the bottom of your chest, a muscle called the **diaphragm** (DIGH-uh-fram) contracts and pulls down the bottom of your chest. Both of these actions make the chest expand. Suddenly there is more space in it.

When your chest expands, there is more room for air in your lungs. So the same amount of air is in a larger space. This causes the air pressure in your lungs to decrease. As a result, the air pressure in your lungs becomes lower than the air pressure outside your body. The difference in pressure

Figure 15–8 *As the diaphragm contracts (left), air pressure decreases in the chest, causing the lungs to inflate. When the diaphragm relaxes (right), air pressure increases in the chest, causing the lungs to deflate. What is this process called?*

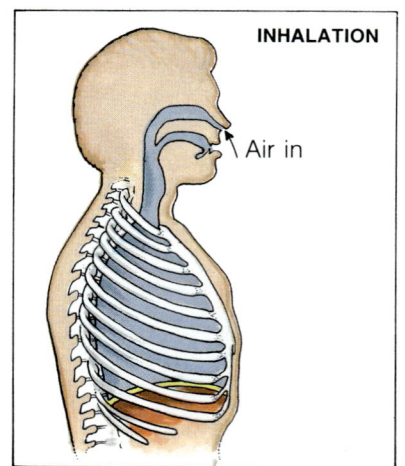
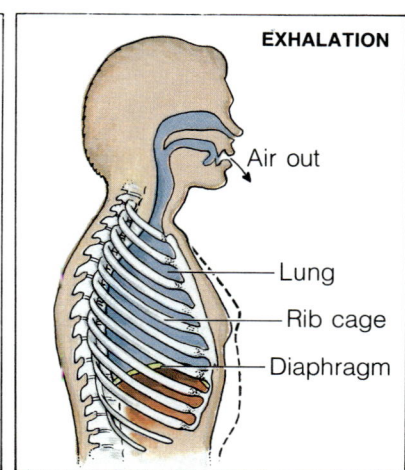

342

forces air to rush into your lungs. That is why your chest must expand before you **inhale,** or breathe in.

The reverse happens when you breathe out, or **exhale.** Your chest muscles relax. The space in your chest becomes smaller. The air pressure becomes greater inside than outside. The result? Air rushes out of your lungs as it would out of a squeezed balloon. How many breaths do you take in one minute?

SECTION REVIEW

1. What are the structures and function of the respiratory system?
2. What is respiration?
3. In which part of the lungs does the exchange of oxygen and carbon dioxide take place?
4. Explain why you should not talk while you are eating.

15–2 The Excretory System

The lungs, as you just learned, are part of the respiratory system. But did you know that they are also part of the **excretory** (EHKS-kruh-tor-ee) **system?**

The job of the excretory system is to remove various wastes produced by the body. The removal process is known as **excretion.** Because the lungs remove carbon dioxide and water, they are considered organs of excretion as well as organs of respiration. Of course, excretion does not refer simply to removing carbon dioxide. Excretion includes the removal of excess water, salts, and certain nitrogen wastes. Nitrogen wastes are produced when excess amino acids, which are the building blocks of proteins, are broken down in the body. Nitrogen wastes, as you might expect, contain the element nitrogen. Excretion also includes the removal of drugs and certain poisons that are taken into the body and absorbed by the blood.

Although you probably do not often think about it, excretion is just as important to your body as breathing and eating. During certain activities, the body produces **toxic** (TAHK-sihk), or poisonous,

Figure 15–9 *In the skin, wastes are excreted through tiny openings called pores. The green circular objects in this pore are bacteria. What are some wastes that are excreted through the pores?*

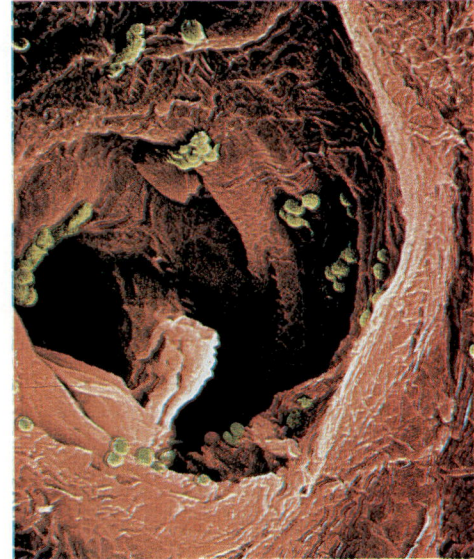

Figure 15–10 *The photograph shows the large number of blood vessels in the kidney. In the drawings, some of the structures in the kidney are labeled, and the structure of the microscopic nephron is magnified. In which section of the kidney are the nephrons found?*

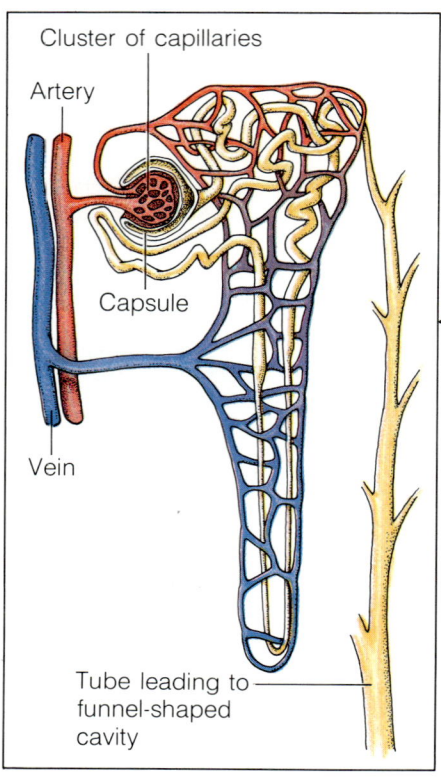

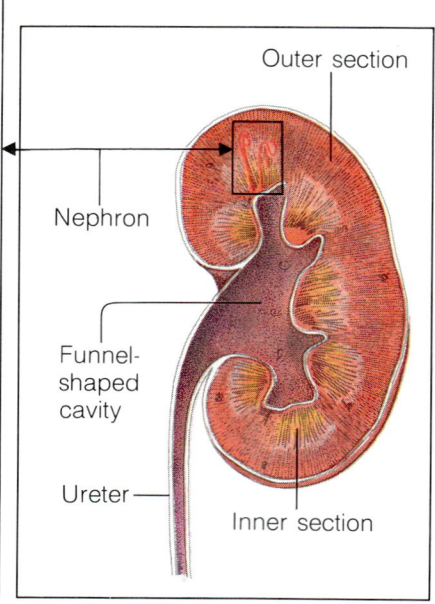

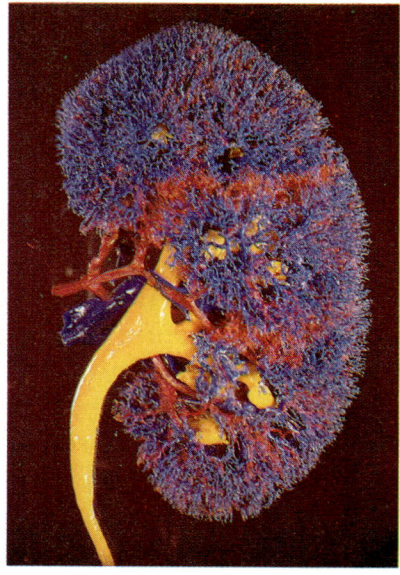

wastes. Without excretion, the toxic wastes would build up in the body. Such a buildup could lead to severe illness and death.

The Kidneys

Here is a riddle for you. What do you and an ocean have in common? Although on the outside your body appears to be solid, inside you are flooded with fluids. About 60 percent of your body weight is water, over half of which is inside body cells and tissues. The rest of the water bathes all the body's cells. This water is mixed with salts, making it almost like the sea water in the ocean.

The amounts of various salts in body water, particularly in the watery blood tissue, is very important. Too much or too little salt can lead to problems. So it is necessary that the salts in the body's "sea water" be kept at exactly the right concentrations. That task is the job of the two **kidneys,** the most important excretory organs.

Place the palms of your hands over your lowest ribs near your spine. Your palms should now cover the areas in which your kidneys are located inside your body. Each bean-shaped kidney is divided into an inner section and an outer section. The outer section contains millions of microscopic chemical filtering factories called **nephrons** (NEHF-rahnz). See Figure 15–10.

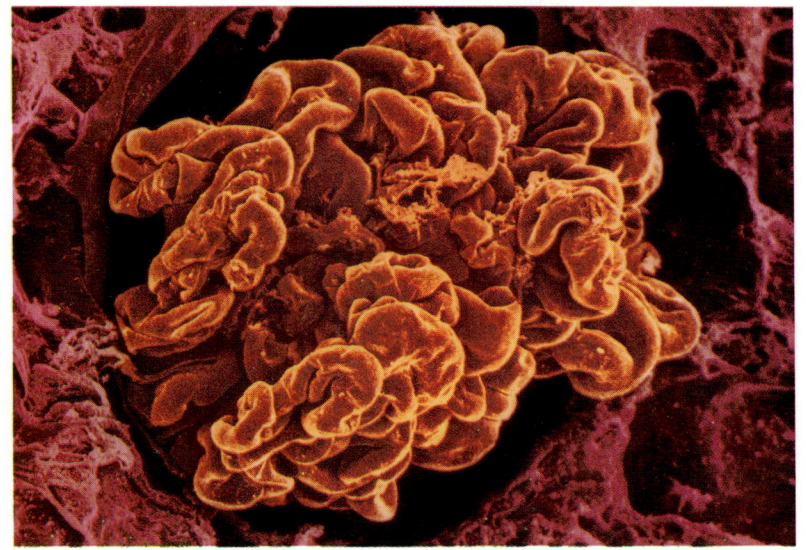

Figure 15–11 *The ball-shaped structure in the middle of this rare photograph is a cluster of capillaries in a human nephron. Wastes leave the bloodstream by passing through the capillary cells. Which part of the nephron surrounds this cluster of capillaries?*

Blood is carried by arteries to the kidneys. Once inside the kidneys, the blood travels through smaller and smaller arteries. Finally, it enters a cluster of capillaries. Various substances in the blood including water, salts, digested food particles, and other materials are filtered out of the blood here. A nitrogen waste called **urea** (yoo-REE-uh) is also filtered out of the blood. These substances pass into a cup-shaped part of the nephron called the **capsule.** The capsule leads to a complicated set of tiny tubes in the nephron.

As the filtered material moves through pipelike passages in the nephron, much of the water and digested food that was filtered out of the blood is reabsorbed into the bloodstream. If this reabsorption process did not take place, the body would soon lose most of the fluid and nutrients it needs to survive.

The liquid that is left after reabsorption is called urine. What materials make up urine? In the inner section of the kidney, the urine passes through cone-shaped areas into a funnel-shaped cavity. This cavity is connected to a tube called a **ureter** (yoo-REET-uhr), which conducts the urine to the **urinary bladder.** The urinary bladder is a muscular sac of tissue that stores urine. The urinary bladder has the ability to expand as it fills with urine. Eventually the urine passes out of the body through a tube at the bottom of the bladder called the **urethra** (yoo-REE-thruh). Why do you think the body must rid itself of liquid wastes?

Figure 15–12 *In the urinary system, urine travels from the kidneys through the ureters to the urinary bladder and out of the body through the urethra. What is the function of the urinary bladder?*

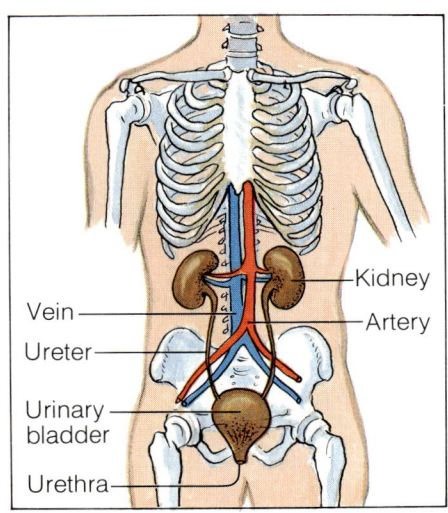

Vein
Ureter
Urinary bladder
Urethra
Kidney
Artery

Close at Hand

The skin is an excretory organ.

1. Using a hand lens, examine the skin on your hand.

2. Identify the epidermis and pores, or small openings, on the ridges of the skin.

3. Place a plastic glove on your hand and take it off after five minutes.

Explain what happened to your hand. If you put your hand against a chalkboard, what would you see when you removed it?

Figure 15–13 *The skin, an excretory organ, is composed of two layers. The upper layer is called the epidermis. The lower layer is called the dermis. What structures are found in the dermis?*

Other Excretory Organs

Aside from the kidneys and the lungs, two other organs of the body play a major role in excretion. These organs are the **liver** and the **skin.**

LIVER The liver acts as a filter for the blood passing through it. Among other things, the liver removes amino acids not needed by the body. The excess amino acids are broken down to form the urea that is excreted in urine. The liver also changes hemoglobin from worn-out red blood cells into substances that the body can use. In addition, the liver can turn some poisons that have collected in the blood into harmless substances.

SKIN The other excretory organ, the skin, is sometimes thought of as the largest organ of the human body. The skin covers an area of 1.5 to 2 square meters in an average person. The skin is about 4 millimeters deep. It is composed of two major layers: the **epidermis** and the **dermis.**

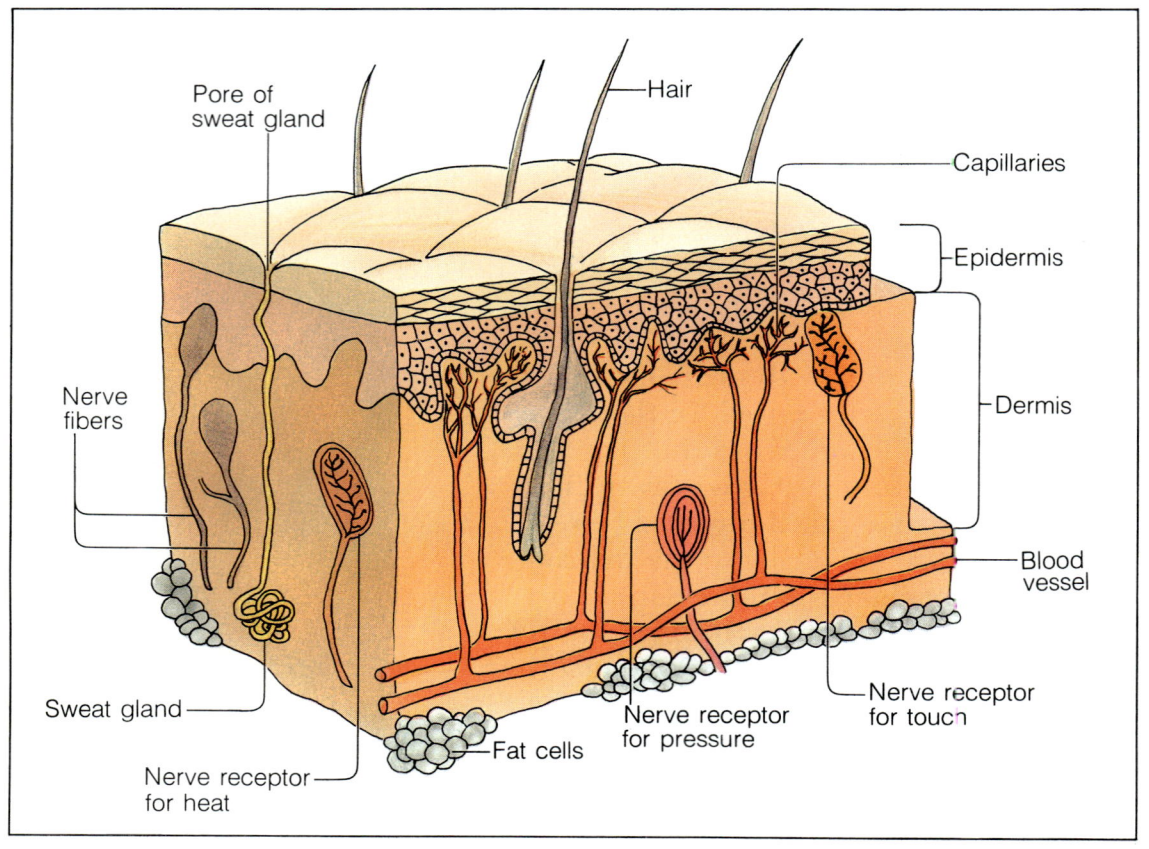

The epidermis is the outer layer of the skin. This layer contains flat cells that are connected like bricks in a wall. As new cells are produced in the bottom layer of the epidermis, the old cells are pushed upward. Because the cells at the outer layer of the epidermis are far away from a food supply, they soon die. Later, these dead cells are shed and replaced.

The dermis, or inner layer of the skin, is thicker than the epidermis. The dermis is rich in blood vessels and connective tissue. The upper layer of the dermis contains small fingerlike projections similar to the villi in the small intestine. Because the epidermis is built on top of these projections, it has an irregular outline that forms ridges. In turn, these ridges form patterns on the fingertips, on the palms of the hands, and on the soles of the feet. On the fingers, these patterns are known as fingerprints.

The skin also contains hair, nails, and oil and sweat glands. The hair and nails are forms of dead skin cells. The sweat glands are coiled tubes in the dermis that connect to pores, or openings, in the surface of the skin.

Sweat glands help the body get rid of excess water, salts, and wastes such as urea. These materials form a liquid called perspiration, or sweat. When your body perspires, it helps you not only to get rid of wastes but also to regulate your body temperature. This happens because, as perspiration evaporates from the skin, the body is cooled.

The oil glands give off oil, which keeps the hair from becoming brittle and dry. The oil also keeps the skin soft.

The skin also acts as a sense organ. Nerve endings in the skin make you sensitive to pain, cold, heat, and pressure.

SECTION REVIEW

1. What are the structures and function of the excretory system?
2. What is urea? How is it formed?
3. Name and describe the two layers of the skin.
4. Suppose all of the nerve endings in your skin were damaged. How might this affect your ability to respond to your surroundings?

Figure 15–14 *Fingerprints are formed by layers of cells that are built on tiny projections on the upper layer of the dermis. These projections form ridges. The tiny depressions on the ridges are actually the sweat gland pores. What is the function of the sweat glands?*

Figure 15–15 *The skin, the body's largest organ, is covered with coarse (left) and fine (right) hairs. What kinds of skin cells form hair?*

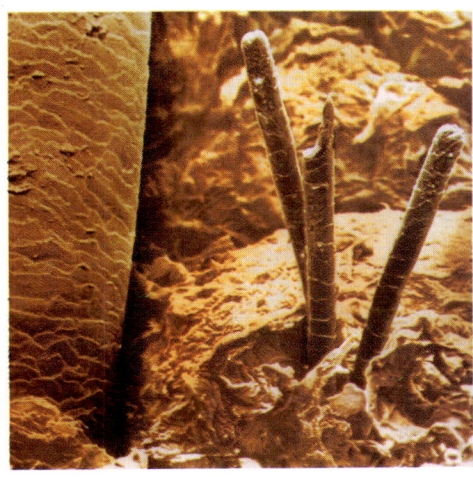

Measuring the Volume of Exhaled Air

Problem

What is the volume of air exhaled?

Materials *(per group)*

glass-marking pencil	spirometer
red vegetable coloring	paper towel
water	

Procedure

1. Obtain a spirometer. A spirometer is an instrument used to measure the volume of air that the lungs can hold.
2. Fill the plastic bottle four-fifths full of water. Add several drops of vegetable coloring to the water. With the glass-marking pencil, mark the level of the water.
3. Reattach the rubber tubing as shown in the diagram.
4. Cover the opening of the shorter length of rubber tubing with a paper towel and after inhaling normally, exhale normally into the rubber tubing.

5. The exhaled air will cause an equal volume of water to move through the other length of tubing into the graduated cylinder. Record the volume of this water in mL in a data table like the one shown.
6. Pour the colored water from the cylinder into the 2-L plastic bottle.
7. Repeat steps 3 to 6 two more times. Record the results in a data table. Calculate the average of the three readings.
8. Run in place for two minutes and exhale into the rubber tubing. Record the volume of the water in the graduated cylinder.
9. Rest for a few minutes until your breathing returns to normal. Then repeat step 8 two times and record the results. Calculate the average of the three readings.

Observations

Describe what happens in the plastic bottle as you exhale into the rubber tubing.

Conclusions

1. Why is it important to measure the volume of exhaled air three times before and after exercise?
2. How does your average volume of exhaled air before exercise compare to your average volume of exhaled air after exercise?
3. What effect does exercise have on the volume of exhaled air? Explain.

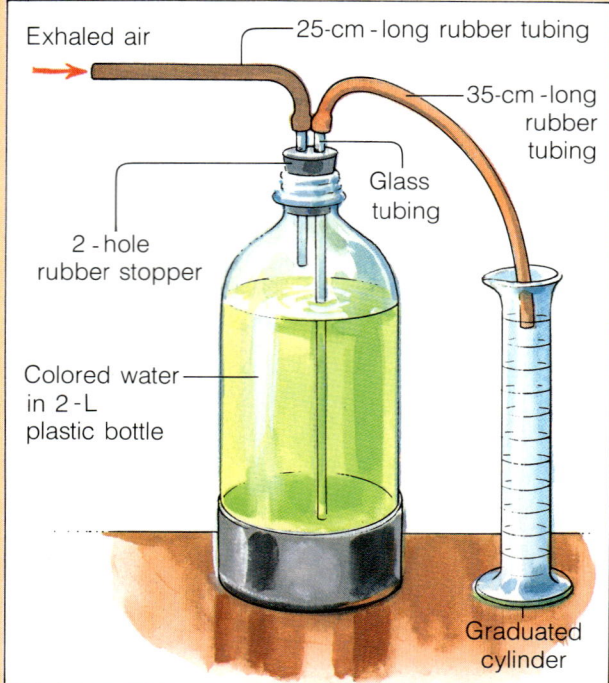

Exhaled air — 25-cm-long rubber tubing
35-cm-long rubber tubing
Glass tubing
2-hole rubber stopper
Colored water in 2-L plastic bottle
Graduated cylinder

	Volume of Exhaled Air Before Exercise	Volume of Exhaled Air After Exercise
Trial 1		
Trial 2		
Trial 3		
Average		

CHAPTER REVIEW

SUMMARY

15–1 The Respiratory System

❑ Respiration is the combining of oxygen and food in the body to produce usable energy.

❑ The main task of the respiratory system is getting oxygen into the body and getting carbon dioxide out.

❑ The respiratory system consists of the nostrils, throat, larynx, trachea, bronchi, and lungs. The bronchi divide into smaller tubes, which end inside the lungs in grapelike clusters called alveoli.

❑ The exchange of oxygen and carbon dioxide occurs in the alveoli, which are surrounded by a network of capillaries.

❑ The two stages of breathing involve inhaling and exhaling.

15–2 The Excretory System

❑ Excretion is the process by which wastes produced in the body are removed.

❑ The principal organs of the excretory system are the kidneys. They control the salt and fluid balance in the body.

❑ The nephrons are microscopic chemical filtration factories.

❑ Within the nephron, water, salt, urea, and other substances are filtered from the blood into the tubes of the nephron. Much of the water and digested food are reabsorbed into the bloodstream.

❑ The material that remains in the tubes of the nephron is called urine. The urine travels through the ureter to the urinary bladder and then is removed through the urethra.

❑ The lungs remove carbon dioxide.

❑ The liver removes excess amino acids and breaks them down into urea.

❑ The skin, another excretory organ, has two main layers: the epidermis and the dermis.

❑ Hair and nails are derived from the skin.

❑ The oil glands keep the skin and hair soft.

❑ The sweat glands excrete excess water, salt, and urea.

❑ Sense organs in the skin respond to pain, cold, heat, and pressure.

VOCABULARY

Define each term in a complete sentence.

alveolus	excretion	lung	trachea
bronchus	excretory system	nephron	urea
capsule	exhale	nostril	ureter
dermis	inhale	respiration	urethra
diaphragm	kidney	respiratory system	urinary bladder
epidermis	larynx	skin	vocal cord
epiglottis	liver	toxic	

CONTENT REVIEW: MULTIPLE CHOICE

On a separate sheet of paper, write the letter of the answer that best completes each statement.

1. In the body cells, food and oxygen combine to produce usable energy in a process called
   a. digestion. b. respiration. c. elimination. d. excretion.

2. The lungs, nostrils, and trachea are all part of the
   a. skeletal system.   b. digestive system.
   c. respiratory system.   d. circulatory system.
3. Air enters the body through the
   a. lungs.   b. nostrils.   c. larynx.   d. trachea.
4. Alternating bands of cartilage and smooth muscle help form the
   a. alveolus.   b. larynx.   c. epiglottis.   d. trachea.
5. The voice box is also known as the
   a. larynx.   b. windpipe.   c. trachea.   d. alveoli.
6. The trachea divides into two tubes called
   a. alveoli.   b. air sacs.   c. bronchi.   d. ureters.
7. A term that means ''breathe out'' is
   a. inhale.   b. exhale.   c. excrete.   d. breathe.
8. Which of these is eliminated from the body during excretion?
   a. excess water   b. urea   c. excess salt   d. all of these
9. Urine is stored in the
   a. urinary bladder.   b. urethra.   c. nephron.   d. kidney.
10. The inner layer of the skin is the
   a. epidermis.   b. dermis.   c. sweat gland.   d. epiglottis.

CONTENT REVIEW: COMPLETION

On a separate sheet of paper, write the word or words that best complete each statement.

1. The waste carbon dioxide is produced during _____.

2. In the nose, _____ moistens incoming air.

3. The trachea is also known as the _____.

4. The vocal cords are located in the _____.

5. The _____ are the main organs of respiration.

6. The dome-shaped muscle that aids in breathing is the _____.

7. The process by which wastes are eliminated from the body is called _____.

8. Microscopic kidney structures that act as filtration factories are called _____.

9. The _____ carries urine from each kidney into the urinary bladder.

10. The _____ is the outer layer of skin.

CONTENT REVIEW: TRUE OR FALSE

Determine whether each statement is true or false. Then on a separate sheet of paper, write ''true'' if it is true. If it is false, change the underlined word or words to make the statement true.

1. The job of getting oxygen into the body and getting carbon dioxide out is the main task of the respiratory system.

2. Dust particles in the incoming air are filtered by blood vessels in the nose.

3. The flap of tissue that covers the trachea whenever food is swallowed is the larynx.

4. The clusters of air sacs in the lungs are called alveoli.

5. Nitrogen wastes come from the breakdown of amino acids.
6. The organs that regulate the concentration of salt in body water are the kidneys.
7. The ureter is the tube through which urine leaves the body.
8. The lungs are excretory organs.
9. The sweat glands in the skin help to keep the skin soft.
10. Perspiration is composed of excess water, salts, and urea.

CONCEPT REVIEW: SKILL BUILDING

Use the skills you have developed in the chapter to complete each activity.

1. **Sequencing events** Trace the path of oxygen through the respiratory system.
2. **Applying concepts** What do you think is the advantage of having two kidneys?
3. **Making comparisons** How are the substances in the capsule of the nephron different from those in the urine that leaves the kidney?
4. **Relating cause and effect** Emphysema is a disease in which the alveoli are damaged. How would this affect a person's ability to breathe?
5. **Making comparisons** How are respiration and the burning of fuel similar? How are they different?
6. **Relating concepts** Would urine contain more or less water on a hot day? Explain.
7. **Relating concepts** How do respiration and excretion relate to the process of homeostasis?
8. **Designing an experiment** Design an experiment to show that the lungs excrete water.
9. **Interpreting data** Use your knowledge of the respiratory system to interpret the data table. What are the differences between inhaled air and exhaled air? How do you account for these differences?

	Nitrogen	Oxygen	Carbon Dioxide
Inhaled air	79%	21%	0.04%
Exhaled air	79%	16%	4.00%

CONCEPT REVIEW: ESSAY

Discuss each of the following in a brief paragraph.

1. What is the difference between respiration and breathing?
2. Explain why it is better to breathe through your nose than through your mouth.
3. What role do the rib muscles and diaphragm play in breathing?
4. Why are the lungs considered to be both respiratory and excretory organs?
5. How are nitrogen wastes produced in the body?
6. Explain the structure and function of a nephron.
7. Why is the skin classified as an excretory organ?
8. How is the kidney like a filter?
9. What changes occur in the lungs when you inhale? What changes occur in the lungs when you exhale?
10. Name and describe four organs of excretion.
11. Trace the path of nitrogen wastes from the nephron to outside the body.
12. What is perspiration? What two functions does perspiration serve in the body?

Nervous and Endocrine Systems 16

CHAPTER OBJECTIVES

After completing this chapter, you will be able to

16–1 Describe the function of the nervous system.

16–1 Identify the structures of a neuron.

16–1 Describe a nerve impulse.

16–2 List the structures and functions of the central nervous system.

16–3 Describe the peripheral and autonomic nervous systems.

16–4 Describe the functions of the five sense organs.

16–5 Describe the functions of the endocrine glands.

16–6 Explain the feedback mechanism.

On Saturday afternoon, you and your friends decide to go to your neighborhood theater to see the movie that is playing. Once inside, you head for the seats in the front row. The lights dim and the music begins. You settle back in your seat, munching popcorn. Images of people appear on the screen.

After a few minutes, the mood of the movie changes. The beat of the music quickens and it begins to get louder. The screen shows the inside of an old house. It is dark inside, very dark.

You slowly start to breathe deeper. Your heart pounds within your chest. You grab onto the arms of your chair. Beads of perspiration appear on your forehead.

Suddenly, a terrible creature jumps out of the darkness. You scream and jump halfway out of your seat, spilling your popcorn. Even though you know the creature is not real, your whole body seems to be ready to launch you out of the theater. Or, if you are trapped, to fight that oncoming "thing."

Perhaps you do not understand what is happening inside your body. Unknown to you, your nervous system and the endocrine system are at work. Together they regulate your body's activities. Parts of your nervous system—the brain, nerves, and senses—obtain information from the outside world. They, in turn, alert the endocrine system to flood your bloodstream with a fountain of chemicals. The chemicals cannot tell that the threat is not real. But one day when you are faced with a real danger, the same chemicals may save your life. How? Do not be afraid to turn the next few pages. You will find nothing there but . . . the answer.

An artist's conception of a sea monster

16–1 Structure and Function of the Nervous System

You have been wrestling for over an hour with a math problem that seems to be unsolvable. Then suddenly, while you are daydreaming about your summer vacation, the solution finally comes to you! This example is one of the many remarkable and often mysterious ways your **nervous system** functions. **The nervous system controls all of the activities of the body. It is made up of the brain, spinal cord, and nerves that are found throughout the body.**

The nervous system also allows you to react to **stimuli** (singular: stimulus). A stimulus is a change in the environment. At their simplest, these reactions are involuntary, or automatic. If an insect or other object zooms toward your eye, you blink without thinking. Your body reacts quickly and automatically to avoid damage to the eye. Such a response, or action caused by the stimulus, is controlled by your nervous system.

Figure 16–1 *The human nervous system controls most of the activities of the body, such as those that are needed to juggle objects. What are the two parts of the human nervous system?*

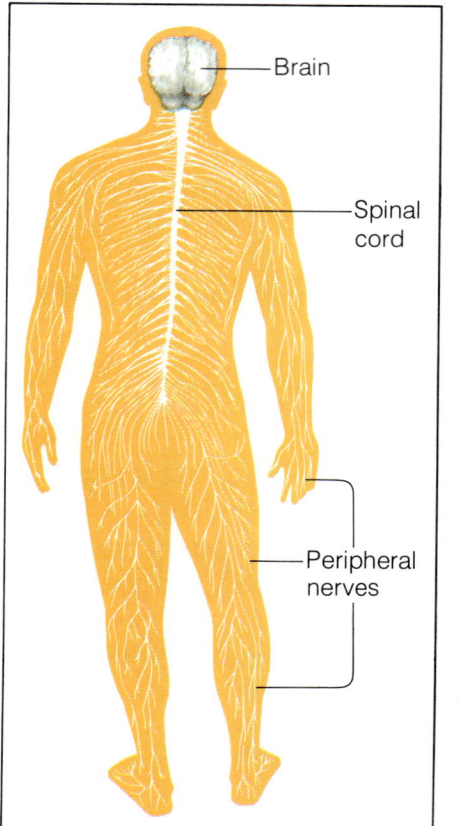

- Brain
- Spinal cord
- Peripheral nerves

Although some responses to stimuli are involuntary, such as blinking your eye, many responses of the nervous system are far more complex. For example, leaving a football game because it begins to rain is a voluntary reaction. It is a conscious choice that involves the feelings of the moment, the memory of what happened the last time you stayed out in the rain, and the ability to reason.

Because the human nervous system controls reactions that involve emotion, reason, and habit, it often has been compared to a computer. The circuits of this human computer are located throughout the body. From the second you are born to the second you die, this computer controls your emotions, your thoughts, and every movement you make. Without it, you could not feel pain, move, or think. You also would not be able to enjoy the taste of food.

Attempts at understanding the human nervous system usually begin by dividing it into two parts. One part of the human nervous system is the **central nervous system.** It is made up of the brain and the spinal cord. The central nervous system is the control center of the body. All information about what is happening in the outside world or within the body itself is brought here.

The other part of the human nervous system branches out from the central nervous system. It is a network of nerves and sense organs, which makes up the **peripheral** (puh-RIHF-uh-ruhl) **nervous system.** *Peripheral* means "outer." Included in the peripheral nervous system are all the nerves that connect the central nervous system to other parts of the body. A division of the peripheral nervous system controls all involuntary body processes, such as heartbeat and peristalsis. This division is known as the **autonomic nervous system.**

Neuron

A nerve is actually a bundle of **neurons,** or nerve cells. Neurons are the basic units of structure and function in the nervous system. Although the design of neurons may vary, they are all built to do one basic task—to carry messages throughout your body.

All neurons have a cell body that contains a nucleus. See Figure 16–2 on page 356. You can think

Sharpen Your Skills

Reaction Time

Reaction time is the length of time that passes between your seeing a change in your environment and your reacting to that change.

1. Have a partner hold a ruler vertically above a table.

2. Position your thumb and forefinger around the zero end of the ruler but not touching it.

3. Your partner may drop the ruler whenever he or she chooses. Moving only your thumb and forefinger, not your hand, you must catch the ruler as soon as it falls. Record the distance, in centimeters, from zero that the ruler falls until it is caught.

4. Repeat steps 1 to 3 four more times.

5. To obtain an average distance, add the five distances together and divide this sum by five. Record the average distance.

Why does measuring the distance the ruler falls give a relative measure of reaction time? Compare the reaction times of all the students. What can you conclude?

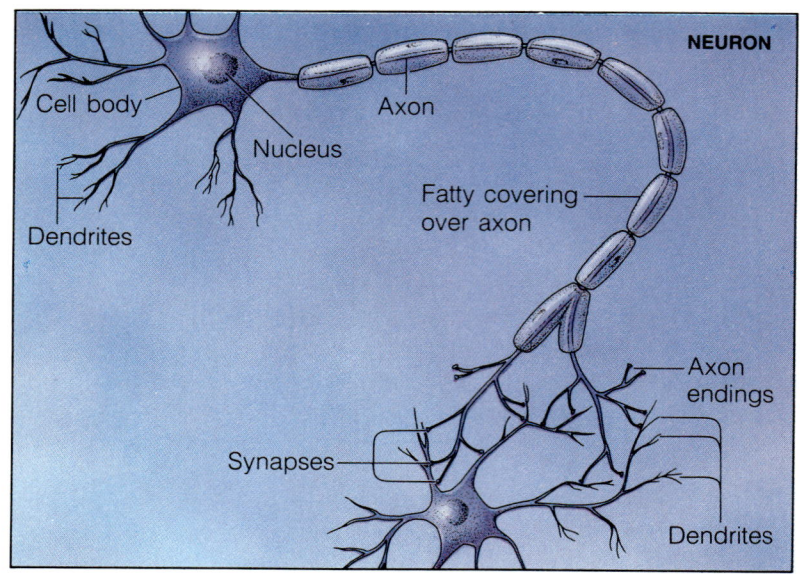

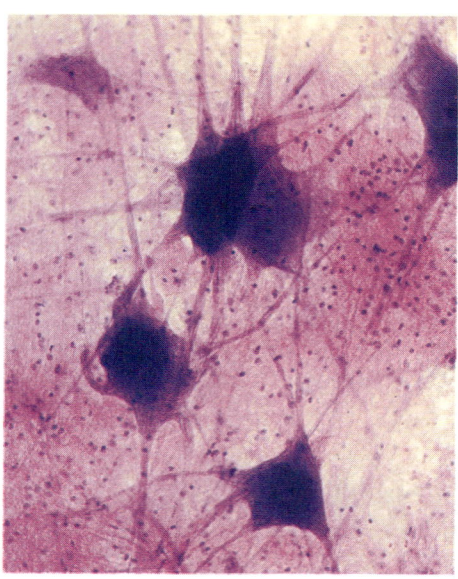

Figure 16–2 *The drawing shows the dendrites, cell body, axon, and axon endings of a typical neuron. The dark branched structures in the photograph are neurons. What is a neuron?*

Figure 16–3 *The neuron in this photograph is magnified 100 times. Notice the ropelike axon at the bottom of the photograph. What is the function of the axon?*

of the cell body as the switchboard of the message-carrying neuron. Running into this switchboard are one or more branches of short fibers called **dendrites.** The dendrites carry messages from other neurons toward the cell body. A longer fiber, called an **axon,** carries messages away from the cell body. Although microscopically thin, an axon can be as much as a meter in length.

There are three main types of neurons in the human nervous system. One type of neuron is the **sensory neuron.** A sensory neuron carries messages from special **receptors** to the central nervous system. Receptors are the part of the nervous system that respond to stimuli. For instance, receptors in your eye respond to light reflected off this page. Messages are carried through sensory neurons from your eyes to your brain. Your brain interprets these messages and you can read this sentence.

When you finish reading the next page, you may turn the page to continue. To do so, another type of neuron, a **motor neuron** carries messages from your central nervous system to **effectors.** Effectors are the parts of the body that carry out the instructions of the nervous system. In this example, the effectors are the muscles in your fingers, hand, and arm. Some glands in the body also act as effectors when they receive messages from the nervous system. A gland is an organ that produces and releases chemicals that affect other organs.

Sensory neurons carry messages *to* the central nervous system while motor neurons carry messages *away* from it. Obviously, if the activities of sensory and motor neurons are to be coordinated, a third kind of neuron must connect them. These neurons are called **interneurons.** Interneurons connect sensory and motor neurons.

Nerve Impulse

You are just about to get up from your desk when wham—you hit your "funny bone" against the back of your chair. The "shock" that travels up your spinal cord to your brain is called a **nerve impulse.** Nerve impulses are messages carried throughout your body by nerves. In the case of the "funny bone," the nerve runs the entire length of your arm. However, this nerve is well protected everywhere except at a spot at the back of the elbow. When you strike that particular spot you feel the shock of hitting your "funny bone."

Nerve impulses enter the cell body of a neuron through the dendrites. The impulse leaves the cell body through the long axon. You can think of the nerve impulse traveling down the axon of a nerve in much the same way electricity travels through a wire. Of course, nerve impulses move more slowly than electricity. Nevertheless, they can travel through an axon at speeds as high as 120 meters per second!

When the nerve impulse reaches the end of an axon, it can go no further. This is because the axon of one neuron is not directly attached to the dendrite of the next neuron. There is a tiny gap called a **synapse** (sih-NAPS) between an axon and a dendrite. An impulse cannot leap across that gap. Does the impulse stop? No.

The tip of the axon produces certain chemicals in tiny bubbles. When an impulse reaches the end of an axon, a chemical-containing bubble bursts and the chemicals move across the synapse. On the other side, the chemicals trigger a new impulse in a dendrite. The dendrite sends the impulse to its cell body, which leads to its axon, and then to a synapse, and so forth. Because only the axon ending can produce the triggering chemicals, impulses in a

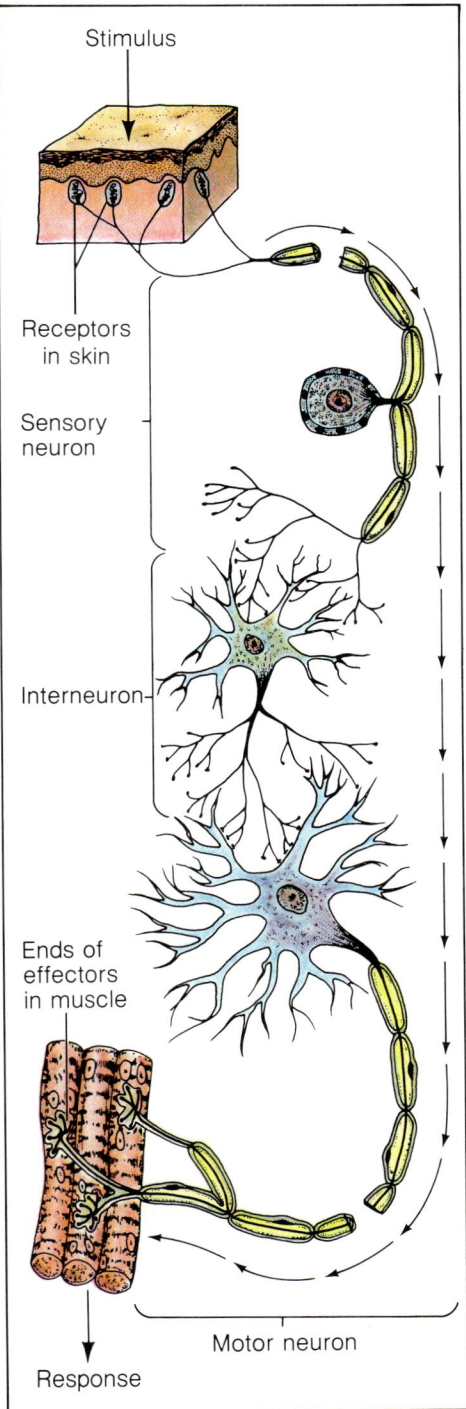

Figure 16–4 *There are three main types of neurons in the nervous system. Sensory neurons carry messages from receptors to the central nervous system. Motor neurons carry messages from the central nervous system to effectors. Interneurons connect sensory and motor neurons.*

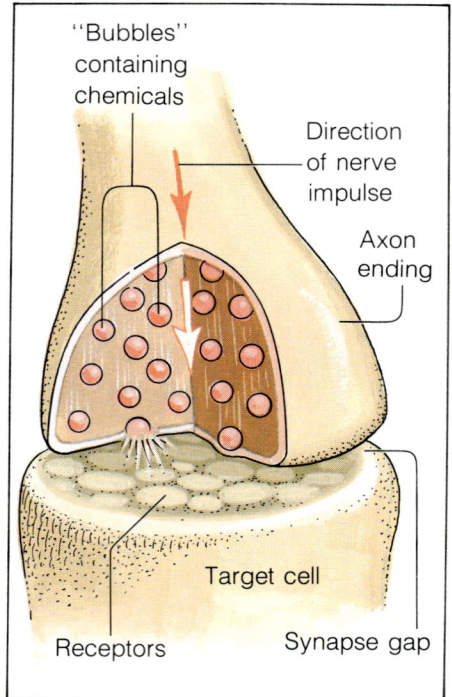

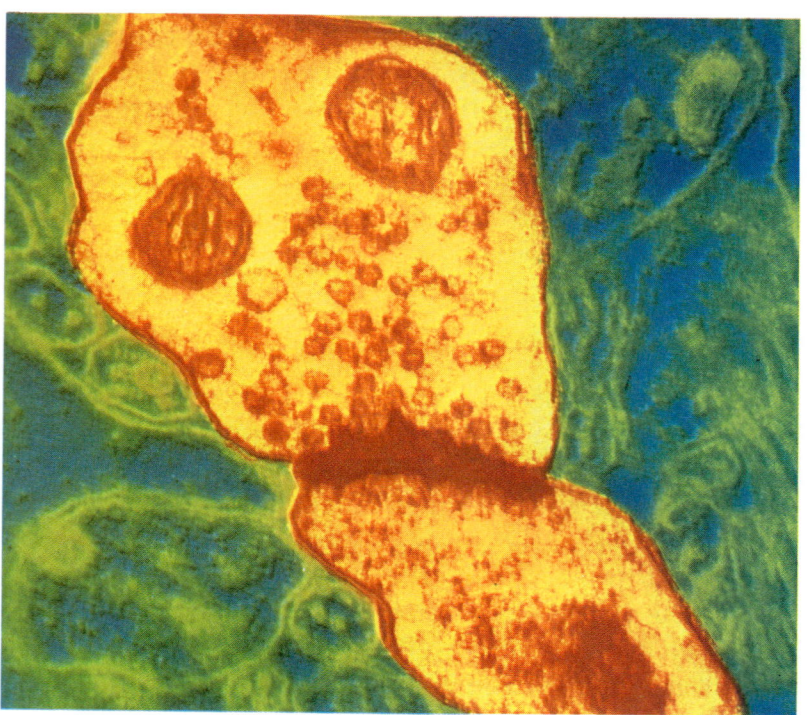

Figure 16–5 *This remarkable photograph, magnified 17,600 times, shows the synapse between two neurons. The small reddish-yellow circles are the chemical-containing bubbles in the axon ending. In the drawing, you can see how the bubbles burst and the chemicals move across the synapse, triggering an impulse in the second neuron. In what direction do nerve impulses travel?*

nerve can only travel in one direction—from dendrite to axon endings.

SECTION REVIEW

1. Compare the functions of the different parts of the nervous system.
2. Define neuron. Describe the three types of neurons.
3. What is a nerve impulse?
4. Why is it an advantage for a nerve impulse to travel in one direction?

Section Objective

To describe the structures and function of the central nervous system

16–2 The Central Nervous System

If you were asked to write a list of your ten favorite rock stars and name all fifty states out loud at the same time, you would probably say you were being asked to do the impossible. It is obvious that the nervous system cannot always control certain thinking functions while it is busy controlling others.

What is less obvious is how many functions the brain can control at one time. For example, even an action as simple as sitting quietly in a movie theater requires several mental operations.

Many kinds of messages travel between your central nervous system and other parts of your body as you watch a movie. Some nerve impulses control the focus of your eyes and the amount of light that enters them. Others control your understanding of what you see and hear. At the same time, other nerve impulses regulate a variety of body activities such as breathing and blood circulation. What is more, all these impulses may be interrelated. For example, if you are frightened by a scene, your breathing and heart rate will likely increase. If you are bored, such rates may decrease. You might even fall asleep.

As you can see, the workings of the central nervous system are very complicated. **The central nervous system, which regulates body activities, is made up of the brain and the spinal cord.** The brain is the main control center of the central nervous system. It transmits and receives messages through the **spinal cord.** The spinal cord is the part of the central nervous system that connects the brain and the rest of the nervous system.

The Brain

More than 10 billion neurons make up the human brain. No one has yet been able to unravel all the interconnections of these neurons or unlock all of the brain's hidden activities. As one scientist said, "Every time you look, you find more."

Despite its many neurons, the brain has a mass of only about 1.5 kilograms. The brain is made of spongy nerve tissue, which is surrounded by three membranes. These membranes nourish and protect the brain. The inner membrane clings to the surface of the brain and follows its many folds. Between the inner membrane and the middle membrane is a watery fluid. The brain floats in this fluid, which cushions it against shock. The outer membrane is thicker and tougher than the other two and makes contact with the inside of the bony skull. As you will discover, the brain is divided into three parts.

Figure 16–6 *This is an X-ray of a human skull. A photograph of the brain is placed in its proper position over the X-ray. To what part of the human nervous system does the brain belong?*

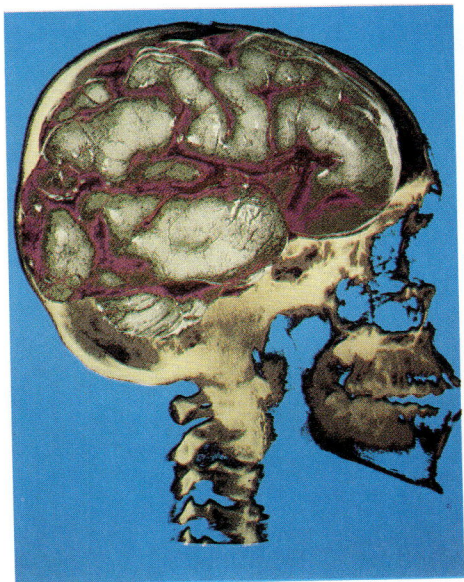

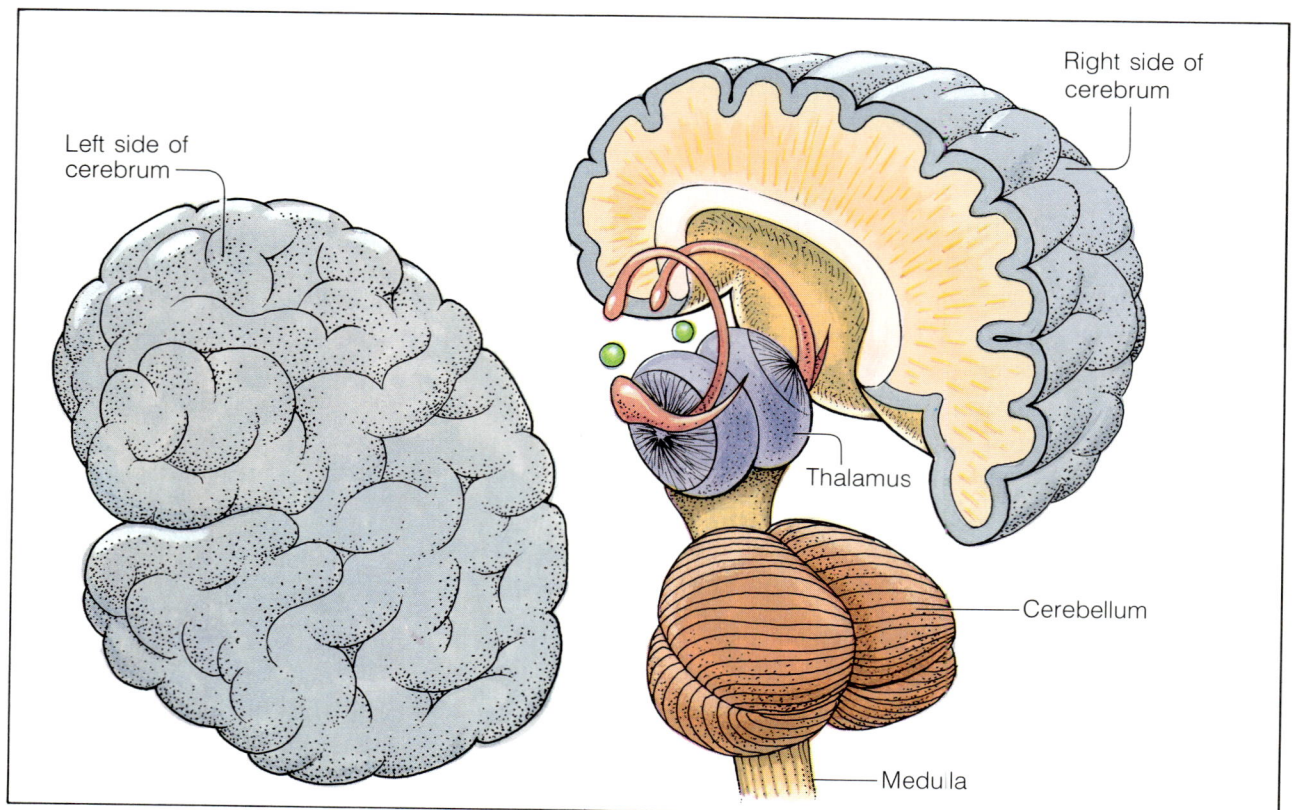

Left side of
cerebrum

Right side of
cerebrum

Thalamus

Cerebellum

Medulla

Figure 16–7 *The brain consists of the cerebrum, the cerebellum, and the brain stem, which includes the medulla. The thalamus is the relay center for sensory and motor impulses. What is the largest part of the brain called?*

THE CEREBRUM Most of the mental activities that make human beings different from other animals take place in the **cerebrum** (SAIR-uh-bruhm). The cerebrum is the largest part of the human brain. Located here are the nerve impulses that allow you to think, remember, and speak. The cerebrum also controls most voluntary muscle contractions and identifies the information gathered by your senses. Why is this important?

Looking like an oversized walnut without a shell, the cerebrum is lined with deep, wrinkled grooves that greatly increase the surface area. See Figure 16–7. Increased surface area means increased thinking ability. The gray outer surface of the cerebrum is composed of the cell bodies of neurons. Because of its color, this tissue is known as gray matter. Below the gray matter is the white matter. The white matter is tissue composed of the axons of neurons and fatty connective tissue. It is here in the gray and white matter of the cerebrum that your attitudes, your emotions, and even your personality are shaped.

The cerebrum actually is composed of two halves. Each of these halves controls different kinds of mental activity. And each controls movement and records sensations on the side of the body opposite it. Fortunately, the two halves are able to communicate their activities to each other by a series of nerve paths running between them.

THE CEREBELLUM   Below and to the rear of the cerebrum is the **cerebellum** (sair-uh-BEHL-uhm), the second largest part of the human brain. Motor nerve impulses beginning in the cerebrum pass through the cerebellum. The cerebellum adjusts the motor impulses so that the body movements they produce are smooth rather than jerky. The cerebellum also sends out impulses that help control balance and posture.

THE BRAIN STEM   Bundles of nerves passing from the cerebrum and cerebellum form a thick stalk called the brain stem at the base of the brain. The lowest part of this stem is the **medulla** (mih-DUHL-uh), which joins the brain to the spinal cord. Nerve impulses from the medulla control many of your automatic body processes, such as heartbeat, breathing, and blood pressure. Can you name some other types of automatic body processses?

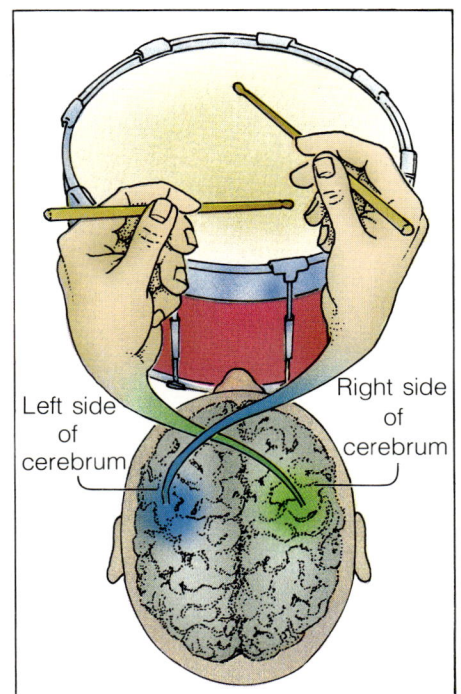

Figure 16–8   *In this drawing, you can see that the left side of the cerebrum controls the right side of the body and the right side of the cerebrum controls the left side of the body.*

Figure 16–9   *The brain directs and coordinates all of the body's activities. What is the function of the medulla?*

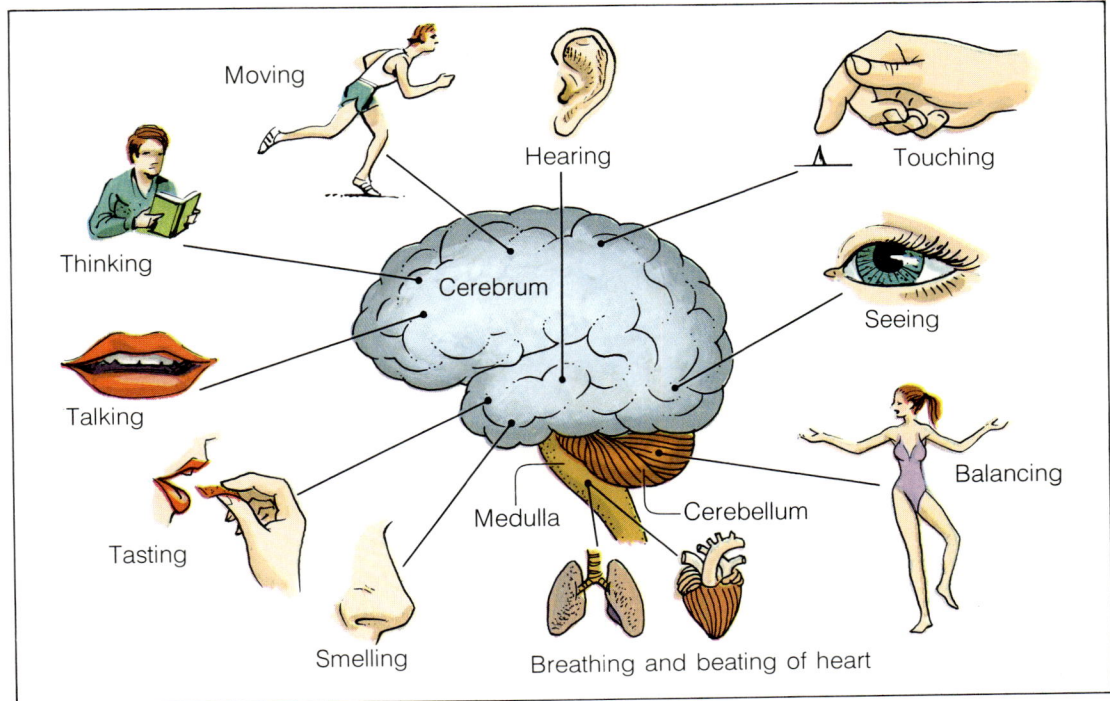

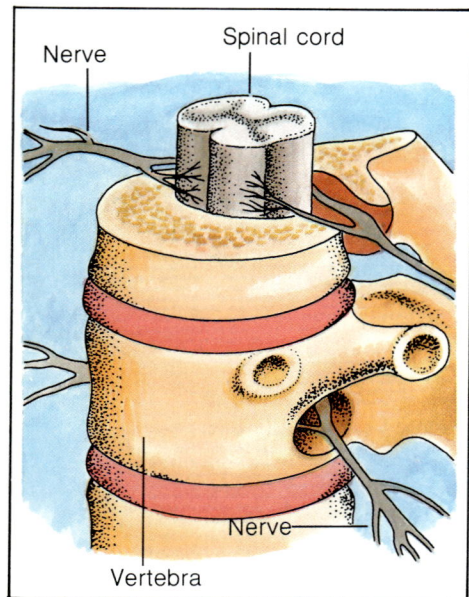

Figure 16–10 *The nerves running through the spinal cord are protected by a series of interlocking bones called vertebrae. To which organ system do the vertebrae belong?*

The Spinal Cord

If you bend forward slightly and run your thumb down the center of your back, you can feel the vertebrae through your skin. The vertebrae are a series of interlocking bones that protect the nerves running through the spinal cord. The spinal cord runs the entire length of the neck and back. It connects the brain with the rest of the nervous system through a series of 31 pairs of nerves.

SECTION REVIEW

1. Name the structures of the central nervous system and describe their functions.
2. Compare the functions of the three main parts of the brain.
3. What structures protect the spinal cord?
4. If a person's cerebellum is injured in an automobile accident, how might the person be affected?

16–3 The Peripheral Nervous System

Section Objective

To describe the function of the peripheral nervous system

The central nervous system is made up of the brain and the spinal cord. To function properly, however, the central nervous system must be connected to the rest of the body. This is the job of the peripheral nervous system. **The peripheral nervous system carries messages between the central nervous system and the rest of the body.**

Some activities are controlled automatically by the peripheral nervous system and the spinal cord without involving the brain. Read on to find out more about these activities.

Reflexes

Hopefully, you are so interested in what this chapter has to say about the nervous system that you would not even notice a fly circling in the air near your face. Perhaps even if it landed on your hair and started to crawl down the side of your face, you still would not notice it. But if the fly's

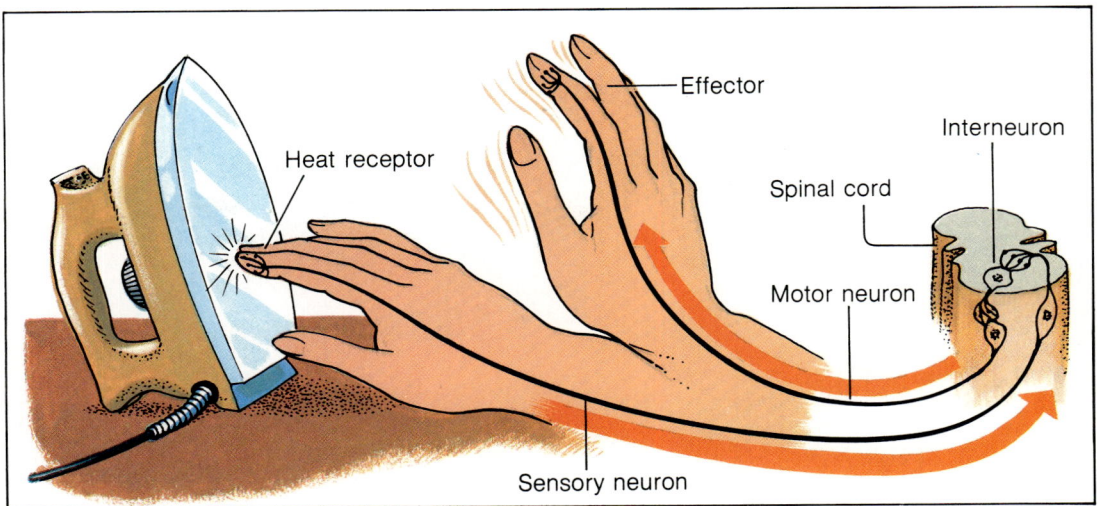

Heat receptor

Effector

Interneuron

Spinal cord

Motor neuron

Sensory neuron

legs happened to touch your eyelash, your eye would automatically blink shut.

Such an automatic reaction to the environment is called a **reflex.** In this case, the reflex begins as soon as the fly's legs bend an eyelash. This bending stimulates the sensory nerves that are in contact with the root of the eyelash. This action sends an impulse through the sensory nerves toward the spinal cord. The impulse is immediately transferred across a synapse to the dendrites of a motor nerve in the spinal cord. Motor nerves then relay the nerve impulse from the spinal cord to the muscles of the eyelid, causing them to contract and the eye to blink shut. The entire process takes less than a second. Reflexes occur quickly and automatically because the nerve impulses that cause them do not first have to pass through the brain. Of course, the brain is quickly alerted to what is going on. So a moment after you blink, your brain knows that you have done so and why.

There are many different reflexes your spinal cord can control without first notifying the brain. The knee jerk reflex, for example, occurs when a doctor checks your reflexes by striking your knee in a certain place.

Autonomic Nervous System

Many of the nerves in the peripheral nervous system are under the direct control of your conscious mind. When you "tell" your leg to move, for

Figure 16–11 *If you touch a hot iron, you will pull away from it quickly. This reaction is an example of a simple reflex. Describe the sequence of events during this simple reflex, beginning with the finger touching the hot iron.*

363

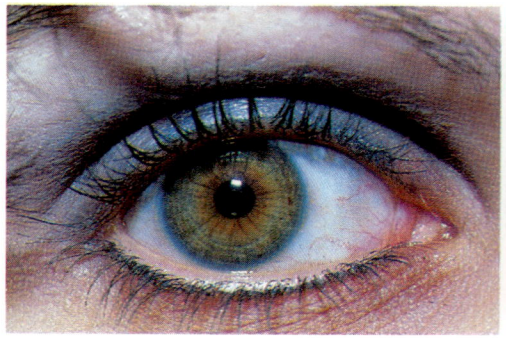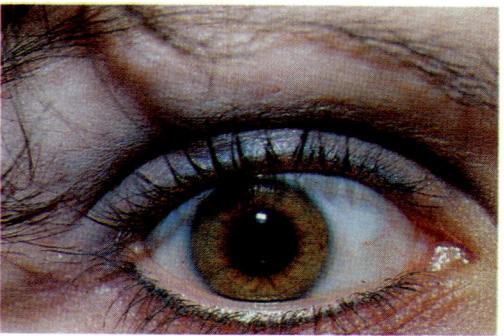

Figure 16–12 *The size of the eye's pupil, or circular opening, changes in response to changes in the intensity of light. The photograph on the left shows the size of the pupil in bright light. The photograph on the right shows the pupil's size in dim light. Why does the pupil enlarge in dim light?*

example, a message travels from your brain down your spinal cord and through a peripheral nerve to your leg. However, there is one important part of the peripheral nervous system that normally acts without you thinking about it. This part is called the autonomic nervous system.

The autonomic nervous system controls nerves that lead to smooth muscles of organs such as your small intestine. That is why you do not have to think about the peristalsis motion that pushes food through your small intestine. Your autonomic nervous system controls this muscle action. Other autonomic nerves lead to organs.

The nerves of the autonomic nervous system work together in sets of two. In fact, if one set of nerves triggers an action by a muscle or gland then a second set of nerves can slow down or stop that action. So the nerves of the autonomic nervous system work against each other to keep body activities in perfect balance. For example, if you are frightened, nerves leading to various muscles and glands are activated. They increase your breathing rate and heartbeat. Such an increase may be necessary when extra energy and strength are needed. But when the frightening situation is over, you need the second set of autonomic nerves to bring your breathing and heartbeat back to normal.

SECTION REVIEW

1. Describe the function of the peripheral nervous system.
2. Define reflex.
3. Explain why it is an advantage to you that your reflexes respond quickly and automatically.

16–4 The Senses

To compare the functions of the sense organs

You know what is going on inside your body and around you because of special sense receptors. Many of these receptors are found in sense organs. Sense organs are structures that carry messages about your surroundings to the central nervous system. **Sense organs respond to light, sound, heat, pressure, and chemicals and detect changes in the position of your body. The eyes, ears, nose, mouth, and skin are examples of sense organs.**

Most sense organs respond to stimuli from your body's external environment. Other kinds of sense organs keep track of the environment inside your body. Without your being aware of it, these sense organs send messages to the central nervous system about body temperature, carbon dioxide and oxygen levels in your blood, and the amount of light entering your eyes.

Vision

Seeing is not possible with your eyes alone. People whose vision center in the brain is damaged cannot see. Your eyes are designed to focus light rays to produce images of objects. But your eyes are useless without a brain to interpret these images.

Your eyes are made of three layers of tissue. The outer protective layer is called the sclera (SKLIHR-uh). This is the "white" of your eyes. At the center front of the eyeball, the sclera is transparent. The transparent tissue forms a protective shield called the **cornea.** Beneath the sclera is the choroid (KOR-oid), or middle layer of the eye. It contains nourishing blood vessels. The choroid layer also includes the circular, colored portion of the eye called the **iris.** When people say someone's eyes are blue, brown, or hazel, they are actually describing the color of the iris.

At the center of the iris is a circular opening called the **pupil.** The size of this opening is controlled by muscles in the iris. They relax or contract to make the pupil larger or smaller.

Watch your pupils change size by looking at them in a mirror as you vary the amount of light in

Figure 16–13 *In this photograph, it appears that the boy on the right is bigger than the girl on the left. Actually the rear walls of this special room are slanted backward making the girl appear smaller. This is called an optical illusion because the eyes are tricked into seeing something that actually is not correct.*

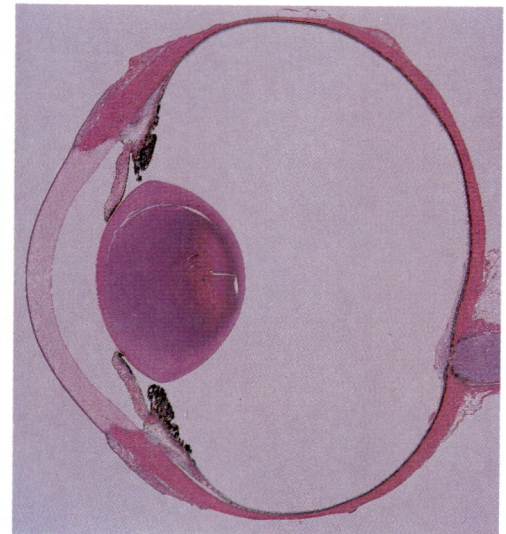

Figure 16–14 *Using the diagram, see if you can identify some of the eye structures in the photograph of the human eye. What is the name of the fluid that gives the eyeball its roundish shape?*

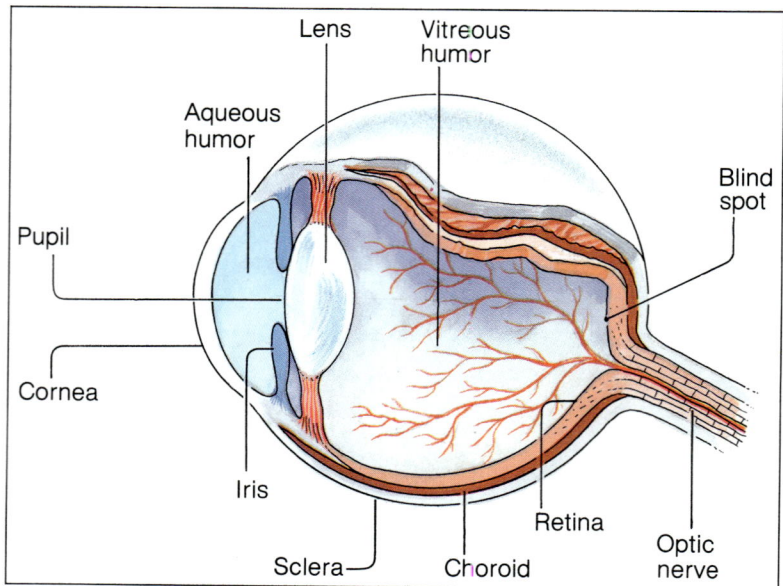

Figure 16–15 *This photograph of the retina was taken through an instrument called an opthalmoscope, which is used to examine the eye. Notice the image of a girl on the retina. Why is the image upside-down?*

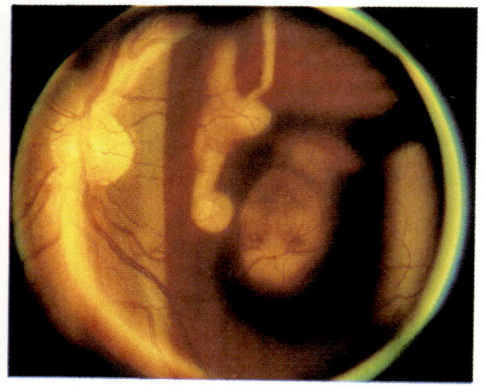

the room. Your pupils open in dim light and close as the light gets brighter. Pupils narrow in bright light to prevent light damage to the inside of the eye. They widen in dim light to let more light in.

As the light passes through the pupil, it travels through the aqueous (AY-kwee-uhs) humor. The aqueous humor is a watery fluid that is found between the cornea and **lens.** The lens focuses the light rays coming into the eye. A human eye lens is different from a camera lens in that the human eye lens adjusts its focus by actually changing shape. The focus of a camera lens is adjusted by moving the lens forward or backward.

A human lens focuses light on the back surface of the eyeball, an area known as the **retina** (REHT-uhn-uh). The retina is the eye's third layer of tissue. It contains light-sensitive cells called rods and cones. See Figure 16–16. Rods react to dim light, while cones react to colors and to bright light. Both produce nerve impulses that travel along the optic nerve. The optic nerve carries these impulses to the vision center of the brain. Between the lens and retina is a large compartment that contains a fluid called the vitreous (VIHT-ree-uhs) humor. This fluid gives the eyeball its roundish shape.

Vision, of course, does not end at the retina. The nerve impulses passing through the optic nerve from the retina have to be interpreted by the brain.

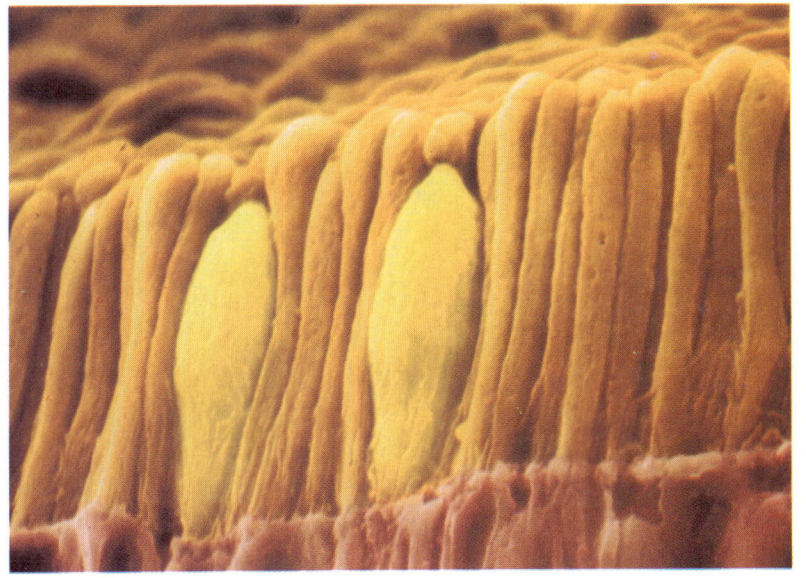

Figure 16–16 *In this photograph, you can see the light-sensitive cells called rods and cones magnified 10,000 times. The long rods respond to dim light while the shorter cones respond to colors and bright light. Which layer of the eye contains these light-sensitive cells?*

Because of the way the lens bends light rays, the brain receives images from the retina upside down and must automatically turn them right side up. The brain must also combine the two slightly different images provided by each eye into one three-dimensional image.

Hearing

When someone laughs or the telephone rings, the air around the source of the sound vibrates. These vibrations move through the air in waves. Hearing actually begins when some of the sound waves enter the external ear. The external ear is probably the part of the ear with which you are most familiar. Made mostly of cartilage covered with skin, it acts as a funnel to gather sound waves. These waves pass through the ear canal. The ear canal is a tubelike structure that is found between the external ear and the **eardrum.** The eardrum is a tightly stretched membrane that separates the ear canal from the middle ear. As the sound waves strike the eardrum, it vibrates in much the same way as the surface of a drum vibrates when it is struck.

Vibrations from the eardrum enter the middle ear, which is composed of the three smallest bones in the body. The hammer, the first of these bones,

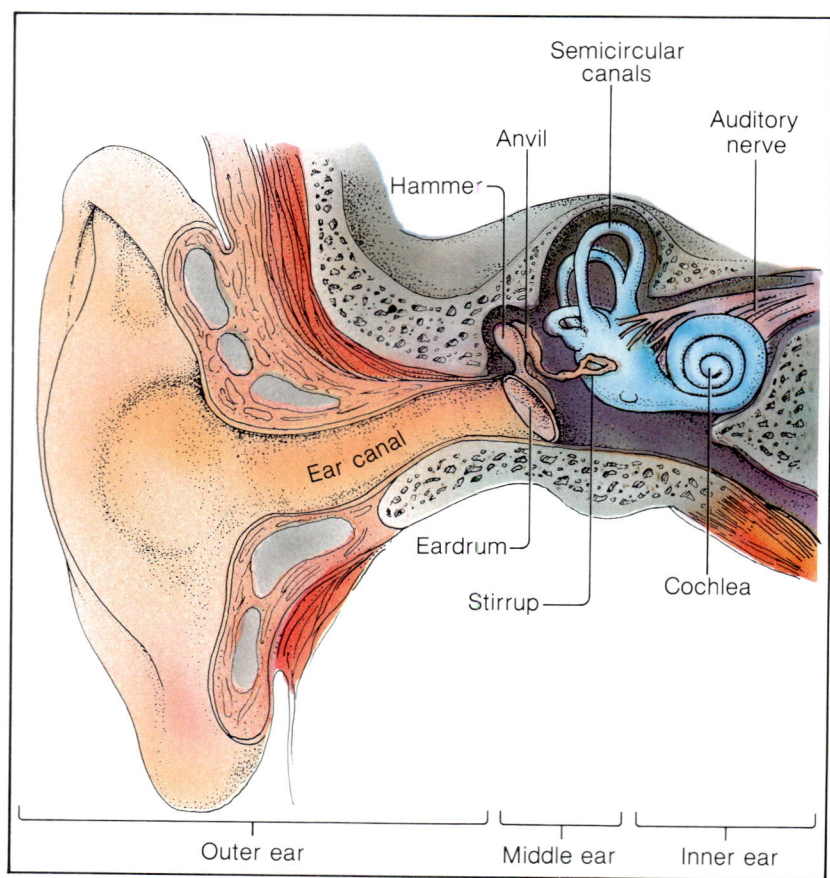

Figure 16–17 *The ear is made up of an outer, middle, and inner part. Sound vibrations entering the ear travel through these parts to the auditory nerve, which carries the nerve impulse to the brain. What is the spiral shell-like structure called?*

Figure 16–18 *The yellow structure in this photograph is the eardrum, which is connected to the cochlea by three tiny bones. Try to identify each bone in the photograph.*

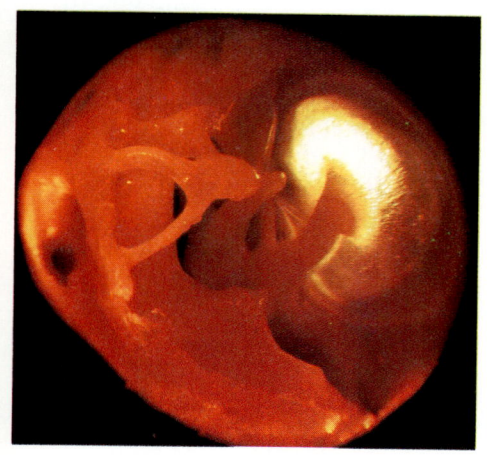

picks up the vibrations from the eardrum. These vibrations are passed along to the anvil, or second bone. Finally, the vibrations go to the stirrup, the last bone. The stirrup vibrates against another membrane. This membrane transfers the vibrations of sound into the fluid-filled inner ear.

Vibrations in the inner ear pass through the fluid and are channeled into a spiraling tube shaped like a snail shell. This tube is called the **cochlea** (KAHK-lee-uh). The cochlea contains nerve endings that react to the vibrations from a wide variety of sounds and range of loudness. Nerve impulses beginning in the cochlea are carried by the auditory nerve to the brain, where they are interpreted.

Balance

If you were asked what do lifting a spoonful of food to your mouth and balancing on a tightrope have in common, you would probably have to think

Figure 16–19 *Within the inner ear, there are three curved tubes called the semicircular canals, as shown in this photograph. These canals alert the brain to a change in body position. One of the canals responds to up-and-down movement, one to side-to-side movement, and one to swaying movement.*

a long time. Actually these two actions involve two important processes—balance and muscle coordination.

Three curved tubes called **semicircular canals** found in your inner ear are responsible for your sense of balance. These tubes are filled with a fluid that moves whenever your head changes position. As the fluid shifts, it presses against tiny hairs on the inner surface of the curved tubes. The hairs bend, producing nerve impulses. These impulses travel to the cerebellum and provide it with information about which way your head is moving and

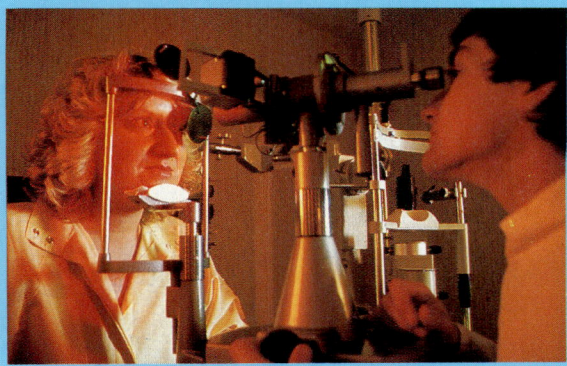

Touch Receptors

The sense of touch involves the reaction of many receptors in the skin.

1. Place a nickel, a dime, a penny, and a quarter in a small covered shoe box.

2. With your eyes closed, reach into the shoe box and remove one coin.

3. Identify the coin by holding it in your hand and touching it.

4. After you have identified the coin, open your eyes to see what coin it actually is.

5. Repeat steps 2 to 4 with each of the other coins. Record your observations.

6. Then set the coins flat on a table.

7. Close your eyes and with the tip of your index finger, touch the top of each coin. *Note: Do not pick up the coins.*

8. Identify each coin. Record your observations.

Was it easier to identify the coins by picking them up and touching them or by touching their tops with your index finger? In which case are you actually using only touch receptors to identify the coins? What other receptors are you using?

which way is up or down. Your brain responds by coordinating the movements of many muscles, which enables you to keep your balance.

Even the most ordinary actions, such as walking, jogging, jumping, swimming, and skipping, call for smooth coordination of muscles with sight, balance, and touch. After much training and practice, your brain can learn to coordinate balance and eye and hand movements with such speed that you could balance on a tightrope as easily as you could lift a spoon to your mouth!

Touch

The sense of touch is actually several different senses, all located in the skin. Near the surface of the skin are touch receptors that allow you to feel the textures of objects. Very little force is needed to produce impulses in these nerves. When you run your fingers gently across a piece of wood so that you can feel the grain, you are stimulating this sense.

Deeper down in the skin are the receptors that respond to pressure. The sense of pressure differs as much from the sense of texture as the experience

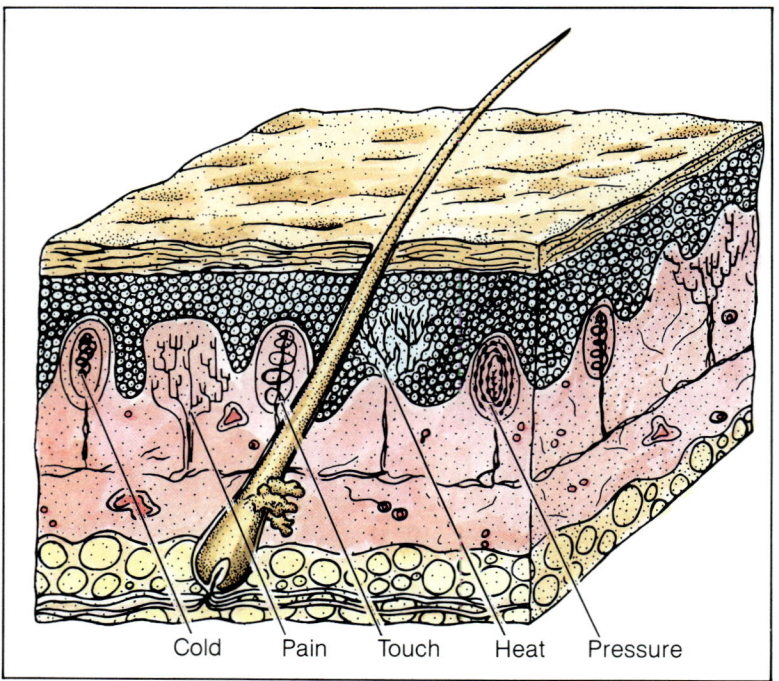

Figure 16–20 *Human skin contains five types of sense receptors. To which stimuli do these receptors respond?*

Cold   Pain   Touch   Heat   Pressure

of pressing your hand firmly against a piece of wood differs from feeling it with your fingertips. Other types of sense receptors in the skin respond to heat, cold, and pain. See Figure 16–20. Nerves that carry messages of pain to the brain are important survival tools because pain is often an indication that the body is in danger.

Smell and Taste

What makes smell and taste unusual senses is that they do not respond to physical stimuli such as light and sound vibrations. What do these senses respond to? A common experience in your family kitchen may provide a clue to the answer.

You are standing near the oven. Suddenly, you smell the aroma of a baking cake. But you cannot see, hear, or feel the cake. What then tells you there is a cake baking in the oven? Your sense of smell, of course, which is being stimulated by invisible "messengers" crossing from the oven to sense receptors in your nose.

The receptors in the nose are sensitive to chemicals carried through the air. So your sense of smell is a chemical sense. Your sense of taste is also a chemical sense. In the case of taste, the chemicals are not carried through the air but in liquids in your mouth. The receptors for taste are located in the taste buds on the tongue. Although taste buds produce only four basic kinds of tastes—sweet, sour, bitter, and salty—there are at least 80 basic odors. Taken together, tastes and odors produce flavors.

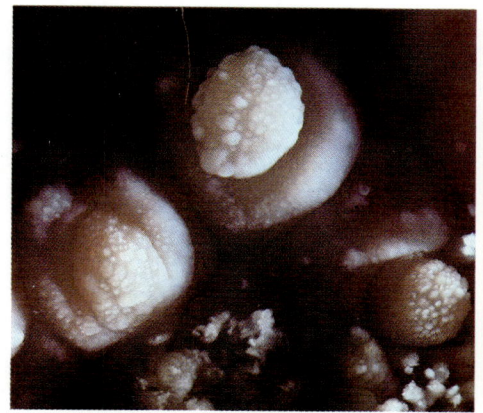

Figure 16–21 *The cluster of tiny bumps in this photograph contains taste buds. Each of the 10,000 or so taste buds responds to one of four tastes: bitter, salty, sour, and sweet. What causes flavor?*

SECTION REVIEW

1. Compare the functions of the five sense organs.
2. What are the functions of the three layers of tissue in the eye?
3. Trace the path of sound from the external ear to the auditory nerve.
4. Explain the role of the ear in maintaining balance.
5. What are the four basic tastes?
6. Design an experiment to show that your sense of smell is important in determining the flavor of food.

16–5 The Endocrine System

The night is late and the house is dark. As you feel your way toward the light switch in the pitch-black room, something warm brushes against your leg. You might let out a piercing shriek, or perhaps only a gasp. Not until you realize that it is the cat do you breathe a sigh of relief. As your pounding heart begins to slow down and your stiffened muscles relax, you begin to feel calmer. Have you ever wondered what causes these split-second changes?

Hormones

Such changes are set in motion by the glands of the **endocrine** (EHN-duh-krihn) **system** and the nerves of the nervous system. **In the endocrine system, glands produce chemical messengers called hormones that help regulate certain body activities.** These chemicals act by turning on and off—or speeding up and slowing down—the activities of different organs and tissues. Together, the nervous system and endocrine system function to keep all the parts of the body running smoothly.

The rush of fear in a dark room is an example of how the nervous system and endocrine system work together. When you brush against an unknown object in the dark, your senses report this

Figure 16–22 *In the diagram, you can see the major glands of the endocrine system. The rush of fear and excitement that these people experience as they ride the roller coaster stimulates the release of hormones from the endocrine system. These hormones cause changes in heartbeat, breathing, and muscle tension. What is a hormone?*

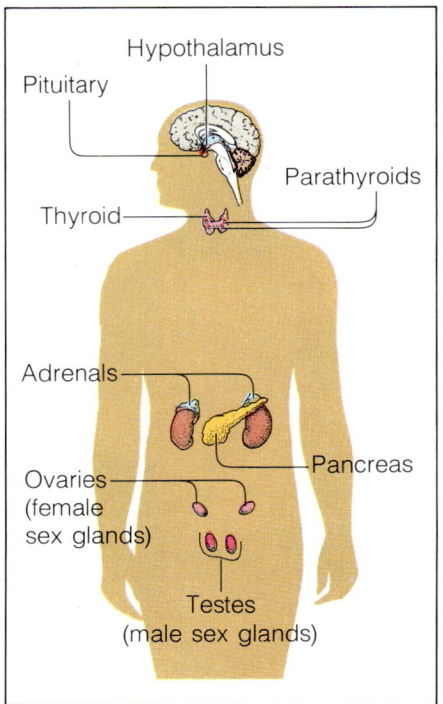

information to your brain. Because your brain interprets the information as a threat, it quickly sends impulses through selected nerves. These nerves trigger certain glands of the endocrine system. The selected glands then produce chemicals called **hormones** (HOR-mohnz). Hormones are chemical messengers that travel through the blood.

In the case of fear in the dark, the hormones that are produced cause your heartbeat to increase, your lungs to work harder, and prepare your muscles for immediate action. In this state, you are ready for fight or flight. Put another way, you are ready to defend yourself or to run. Your body stays prepared for trouble until your brain stops sending out danger signals. Then the glands slow their release of hormones, and your body calms down.

Fight-or-flight reactions are part of the many physical processes controlled by **endocrine glands.** All of these glands work by releasing their hormones directly into the bloodstream.

A second type of gland found in the body, known as an **exocrine** (EK-suh-krihn) **gland,** works by pumping its chemicals through ducts, or tubes, into a nearby organ. Exocrine glands do not produce hormones. The salivary glands, which produce saliva, are an example of an exocrine gland. Saliva, you may recall from Chapter 13, contains enzymes that help your body digest food in the mouth.

Endocrine glands use the circulatory system to deliver their hormones. So endocrine glands do not need to be near the organs they control. No matter where hormones enter the bloodstream, they always find their way through the nearly 100,000 kilometers of blood vessels to their intended target area. How? Body tissues have the ability to "recognize" the hormones that are made for them. Tissue cells are programmed to accept certain hormones and reject others. Hormones not meant for a particular type of tissue or organ will pass on until they come to the target tissue or organ.

The Hypothalamus

Have you ever seen photographs of people practicing yoga? For years, scientists were amazed by the ability of some of those who practiced yoga to lower

Figure 16–23 *Unlike endocrine glands, exocrine glands release their material through ducts, or tubes. What substance is released from the exocrine tear glands of this crying child?*

373

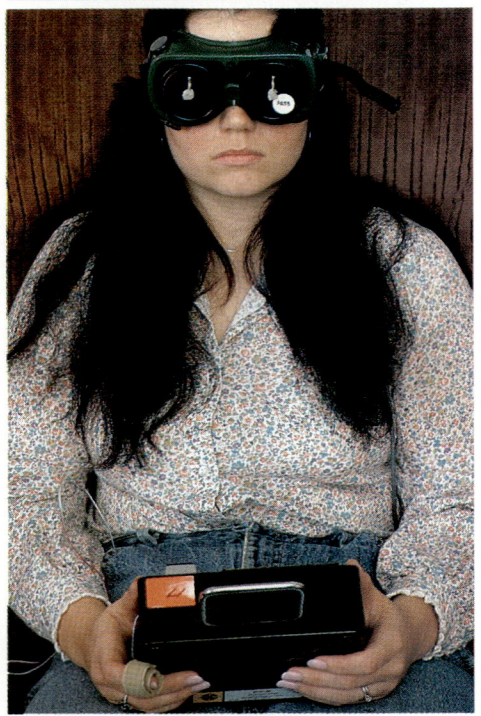

Figure 16–24 *By practicing yoga (top) and using a biofeedback machine (bottom), people can learn to control involuntary actions, such as heartbeat and body temperature. Such involuntary actions are normally under the control of the hypothalamus.*

their body temperature or slow down their heartbeat at will. Today, scientists suspect that this effect is accomplished in part by a conscious control of the **hypothalamus** (high-puh-THAL-uh-muhs). The hypothalamus is a tiny gland at the base of the brain.

The hypothalamus is the major link between the nervous system and the endocrine system. In fact, it is as much a part of one as of the other. As an organ of the nervous system, it is the control station of body temperature, water balance, appetite, and sleep.

The hypothalamus produces some very special hormones that help turn on and off the activities of other endocrine glands all over the body. So a yoga expert who can control the action of the hypothalamus can control many body processes, including heartbeat and body temperature.

The Pituitary

Most of the hormones of the hypothalamus trigger action in the **pituitary** (pih-TOO-uh-tair-ee). The pituitary is a small pea-shaped gland located at the base of the brain just below the hypothalamus. Do not let its size fool you. Although the pituitary is little more than a centimeter in diameter, it is able to manufacture several different hormones. These pituitary hormones play an important role in controlling body processes.

The pituitary is divided into two lobes (lohbz), or rounded sections. Each lobe stores and produces different types of hormones. The back lobe is made mostly of nerve tissue and sometimes is considered to be a part of the brain. This back lobe releases several hormones. One hormone helps control blood pressure. Another hormone causes the powerful muscle contractions during childbirth. Still another hormone from the back lobe of the pituitary regulates the way the kidneys control water balance in the body.

The front lobe of the pituitary also manufactures hormones. Many of these pituitary hormones act directly on other endocrine glands, turning them on or off. In fact, it is mainly through the pituitary that the hypothalamus controls the activities of the other endocrine glands.

Not all pituitary hormones act on other glands. The front lobe of the pituitary produces a hormone called human growth hormone, or HGH. In order for a person to grow, the pituitary must release enough HGH during the person's childhood years. Do not get the idea that a small person does not produce enough HGH. Such a person's pituitary produces the right amount of HGH for that person. Similarly, a tall person's pituitary produces the right amount for that person as well. In almost every person, the pituitary produces the right amount of HGH at the proper time.

Sometimes, however, the pituitary does not produce the correct amount of HGH. Whenever there is an underproduction of HGH, bones and tissues grow very slowly. This results in a condition called pituitary dwarfism. Although body growth is stunted, the body parts are in the proper proportion to one another. Mental development is also quite normal. Recently, scientists were able to engineer certain bacteria to produce HGH. The HGH is collected and can be given to children who do not produce enough HGH of their own. Such children now have a chance to grow normally.

Not only can glands produce too little of a hormone, they can also produce too much of a

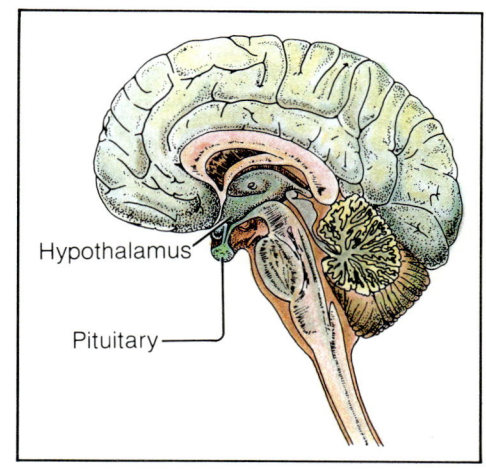

Figure 16–25 *The pituitary is a small pea-shaped gland located at the base of the brain just below the hypothalamus. What are the two rounded sections of the pituitary called?*

Figure 16–26 *One of the functions of the pituitary is to produce human growth hormone, or HGH, which controls body growth. Although these students are all different heights, the pituitary of each one of them has produced the right amount of HGH for that person. What condition occurs if there is an underproduction of HGH?*

hormone. An overproduction of HGH causes a condition called giantism. A person affected with giantism can grow as tall as three meters!

The Thyroid

The **thyroid** (THIGH-roid) is located at the base of the neck slightly below the larynx, or voice box. The larynx contains the vocal cords. In fact, before the mid-1800s, doctors thought that the function of the thyroid was to lubricate and protect the vocal cords. It was not until 1859 that the true functions of the thyroid were discovered. The thyroid actually produces two hormones. The first hormone, called calcitonin (kal-sih-TOH-nihn), controls the level of the minerals calcium and phosphorus in the body. The second hormone is called thyroxine (thigh-RAHK-suhn). Thyroxine helps control metabolism. Metabolism is the total of all the chemical reactions that keep an organism alive, including the reactions that help produce usable energy from food. So you can see the proper level of thyroxine is important.

Sometimes the thyroid does not produce the proper amount of thyroxine. For example, an overactive thyroid produces too much thyroxine. The result is an increased metabolism. Such a rapid

Figure 16–27 *The parathyroids are four pea-shaped glands located on, or sometimes in, the thyroid. What hormones does the thyroid produce?*

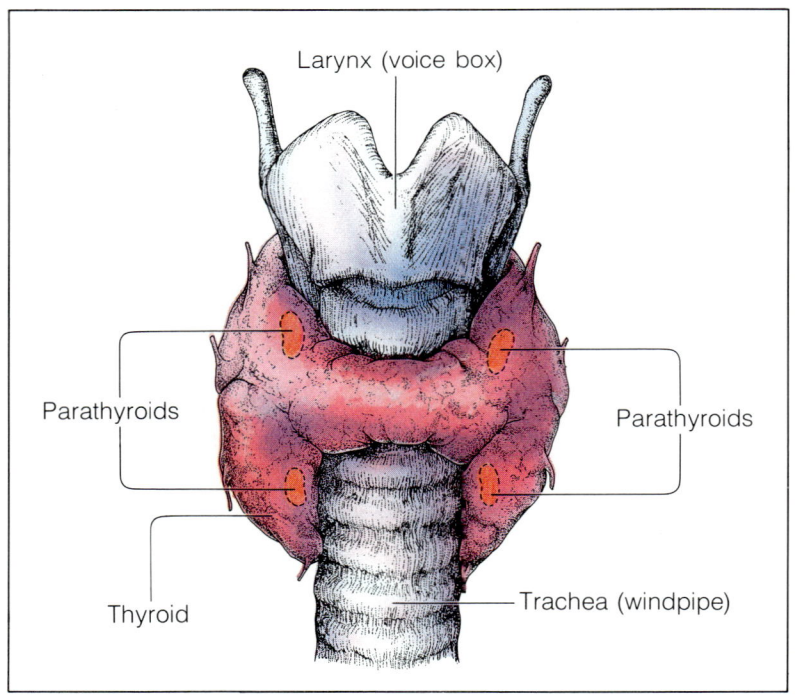

Larynx (voice box)

Parathyroids

Parathyroids

Thyroid

Trachea (windpipe)

SOME ENDOCRINE GLANDS

Gland	Location	Hormone Produced	Functions
Pituitary Front lobe	Base of brain	Human Growth Hormone (HGH)	Stimulates body skeleton growth
		Gonadotropic Hormone	Stimulates development of male and female sex organs
		Lactogenic Hormone	Stimulates production of milk
		Thyrotropic Hormone	Proper functioning of thyroid
		Adrenocorticotrophic Hormone (ACTH)	Proper functioning of adrenals
Back lobe		Oxytocin	Regulates blood pressure and stimulates smooth muscles; stimulates the birth process
		Vasopressin	Increases rate of water reabsorption in the kidneys
Thyroid	Neck	Thyroxine	Increases rate of metabolism
		Calcitonin	Maintains the level of calcium and phosphorus in the blood
Parathyroids	Behind thyroid lobes	Parathyroid Hormone	Regulates the level of calcium and phosphorus
Adrenals Inner tissue	Above kidneys	Adrenaline	Increased heart rate; elevated blood pressure; rise in blood sugar; increased breathing rate; decreased digestive activity
Outer tissue		Mineralocorticoids	Maintains balance of salt and water in the kidneys
		Glucocorticoids — Cortisone	Breaks down stored proteins to amino acids; aids in breakdown of fat tissue; promotes increase in blood sugar
		Sex Hormones	Supplements sex hormones produced by sex glands; promotes development of sexual characteristics
Pancreas Islets of Langerhans	Abdomen, near stomach	Insulin	Enables liver to store sugar; regulates sugar breakdown in tissues; decreases blood sugar level
		Glucagon	Increases blood sugar level
Ovaries	Pelvic area	Estrogen	Produces female secondary sex characteristics
		Progesterone	Growth of lining of uterus
Testes	Scrotum	Testosterone	Produces male secondary sex characteristics

Figure 16–28   *This chart shows the location of some endocrine glands as well as the hormones they produce and their functions. Which are the sex glands?*

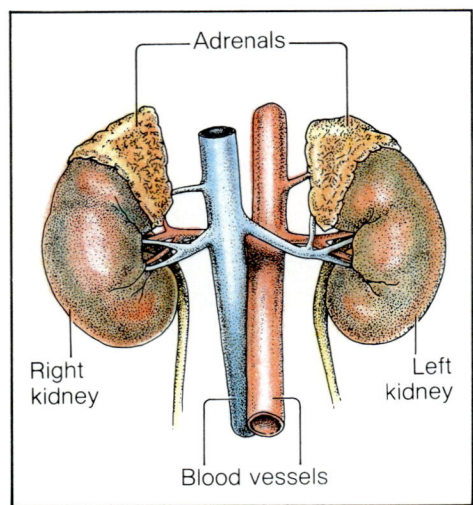

Figure 16–29 *There are two adrenal glands, one atop each kidney. What hormone secreted by the adrenals causes the heart to beat faster?*

metabolism can make a person feel nervous and unable to sleep. Because food is "burned" up so quickly, an overactive thyroid can cause a serious loss of weight. An underactive thyroid, on the other hand, produces the opposite effect. It slows down metabolism and may lead to weight gain and a lack of energy. Both overactive and underactive thyroid conditions usually can be treated by a change in diet or with medicine.

The Parathyroids

Located on, or sometimes in, the thyroid are four small, pea-shaped glands known as the **parathyroids.** A hormone produced by the parathyroid controls the level of calcium in the blood. It also helps maintain a balance of phosphorus. Both of these substances play a part in the healthy functioning of bones, muscles, and nerves.

Parathyroid hormone increases the calcium level of blood by releasing calcium from bone tissue. If the parathyroids are too active, the bones are drained of large amounts of calcium. In time, the bones become soft and spongy. If too little parathyroid hormone is released, the muscles go into uncontrolled contractions. This condition produces painful muscle cramps throughout the body.

The Adrenals

What could give you the power to perform deeds of almost superhuman strength, such as the ability to run faster than you have ever run before? What substance once allowed a young boy to lift a car when his father was trapped in an accident? It is not spinach, or a new miracle vitamin, but simply a hormone produced by the **adrenals** (uh-DREE-nuhlz).

In moments of sudden stress, the adrenal glands, one of which is located atop each kidney, release a hormone called adrenaline (uh-DREHN-uh-lihn). This hormone causes the heart to beat faster and steps up the rate of other body processes. It also allows for increased physical strength and the ability to think and move quicker than is normally possible. Although the effects of adrenaline are not lasting, the rush of energy it gives can save lives.

Adrenaline is produced and released by the inner layer of the adrenals. The outer layer of these glands releases several other important hormones. Some of these adrenal hormones work with the kidneys, the salivary glands, and the sweat glands to maintain a proper balance of salt and water in the body. Others keep tissues from swelling up too much after they are injured.

The Pancreas

Below the stomach is the **pancreas,** an organ you read about in Chapter 13. The pancreas is both an exocrine gland and an endocrine gland. In its role as an exocrine gland, the pancreas acts as a digestive organ, releasing its digestive enzymes through a duct into the small intestine. As an endocrine gland, it contains special cells that release two important hormones.

One of the two endocrine hormones of the pancreas is insulin (IHN-suh-lihn). Insulin reduces the level of sugar in the bloodstream by helping body cells absorb the sugar and use it for energy. After you eat, your blood sugar level is at its highest and the release of insulin into the bloodstream increases. As the level of sugar in the blood drops, the release of insulin slowly decreases.

The second hormone produced by the pancreas, glucagon (GLOO-kuh-gahn), causes an effect directly opposite to that of insulin. Between meals, the level of sugar in the bloodstream falls. This causes the pancreas to produce glucagon, which causes a rise

Figure 16–30 *Located near the beginning of the small intestine is the pancreas. It contains cells organized into groups called islets of Langerhans. What two hormones are produced by the islets of Langerhans in the pancreas?*

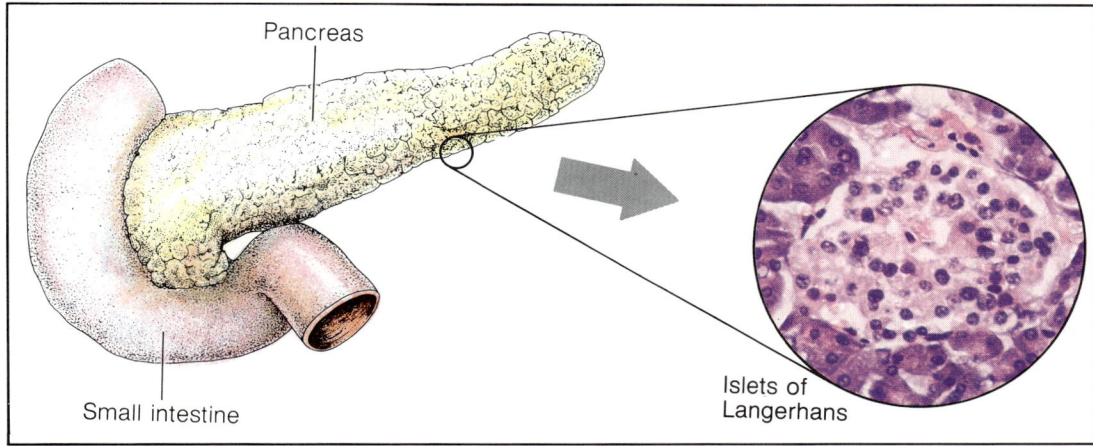

Pancreas

Small intestine

Islets of Langerhans

in the level of sugar in the blood. It does this by changing glycogen, or animal starch, that is stored in the liver into sugar.

Sometimes the pancreas does not produce enough insulin. Without the insulin, sugar cannot be absorbed into the body cells. As a result, the blood sugar level increases. This causes the condition known as diabetes mellitus (digh-uh-BEET-ihs muh-LIGHT-ihs). Some symptoms of diabetes include thirst, hunger, and loss of weight. In many cases, diabetes can be controlled through a person's diet. In other cases, diabetes is treated with regular injections of insulin or drugs that stimulate the pancreas to produce insulin.

Sex Glands

The **ovaries** (OH-vuhr-eez) are the female sex glands. The **testes** (TEHS-teez; singular: testis) are the male sex glands. The sex glands begin to produce sex hormones during puberty. Puberty is a stage during which boys and girls mature physically and sexually.

The ovaries produce two hormones, estrogen (EHS-truh-juhn) and progesterone (proh-JEHS-tuh-rohn). Estrogen stimulates the development of the female secondary sex characteristics, such as the development of the breasts and the broadening of the hips. Together with estrogen, progesterone regulates the menstrual (MEHN-struhl) cycle, which is discussed in Chapter 17.

The testes produce a hormone called testosterone (tehs-TAHS-tuh-rohn). Testosterone stimulates the development of the male secondary sex characteristics. Some examples of male secondary sex characteristics are a deep voice and growth of a beard.

SECTION REVIEW

1. Name the endocrine glands and tell the function of each gland.
2. What is a hormone?
3. Compare endocrine and exocrine glands.
4. If a person has diabetes mellitus, would his or her production of glucagon be increased or decreased? Explain your answer.

16-6 The Feedback Mechanism

Section Objective

To relate the feedback mechanism to maintaining the balance of hormones in the body

In a way, the endocrine system works much like the thermostat in a heating system. When room temperature is too cool, the thermostat instructs the heating system to switch on. When the temperature rises, the thermostat again comes into operation. It switches the heat off. In a sense, the temperature controls the thermostat as much as the thermostat controls the temperature.

Through the action of the **feedback mechanism,** the endocrine system works in much the same way. **In the feedback mechanism, the production of a hormone is controlled by the concentration of another hormone.**

The actions of the pituitary and the hypothalamus are probably the best example of the feedback mechanism. For many years, the pituitary gland was called the "master gland" of the body. The hormones the pituitary produces control many of the activities of the other endocrine glands. Eventually, it was discovered that the pituitary itself was controlled by its own master—the hypothalamus. The nickname "master gland" gradually dropped out of use.

However, in reality, not even the hypothalamus is the master of the endocrine system. The activities of the hypothalamus are controlled, in turn, by the amount of pituitary hormone circulating in the blood at any given time. A decrease in the amount of pituitary hormone in the blood causes the hypothalamus to release hormones that activate the pituitary. Then, as the level of pituitary hormones increase, the hypothalamus reverses its function. This reversal releases another type of hormone that slows the pituitary down. In this way, the feedback mechanism helps keep an inner state of hormone balance.

Figure 16–31 *The levels of various hormones in the body are controlled by a feedback process. This diagram shows the feedback process between two glands. What happens when there is an increase of Hormone 1 in the blood?*

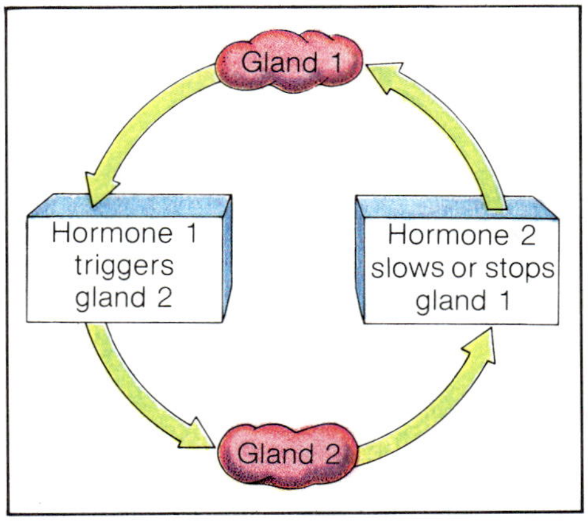

Gland 1

Hormone 1 triggers gland 2

Hormone 2 slows or stops gland 1

Gland 2

SECTION REVIEW

1. What is the feedback mechanism?
2. Explain how the feedback mechanism helps maintain a state of balance within the body.

Locating Touch Receptors

Problem

Where are the touch receptors located on the body?

Materials (per 2 students)

cardboard (6 × 10 cm)
scissors
metric ruler
9 straight pins
blindfold

Procedure

1. Using the scissors, cut the cardboard into five 6 × 2 cm-rectangles. **CAUTION:** *Be careful when using the scissors.*
2. Into one cardboard rectangle, insert two straight pins 5 mm apart. Into the second cardboard rectangle, insert two pins 1 cm apart. Insert two pins 2 cm apart into the third rectangle. Insert two pins 3 cm apart into the fourth rectangle. In the center of the remaining cardboard rectangle, insert one pin. Construct a data table.
3. Blindfold your partner.
4. Using the cardboard rectangle with the straight pins 5 mm apart, carefully touch the palm surface of your partner's fingertip, palm of the hand, back of the hand, back of the neck, and inside of the forearm. **CAUTION:** *Do not apply pressure when touching your partner's skin.*

5. If your partner feels two points, place the number 2 in the data table. If your partner feels only one point, place the number 1 in the data table. The more points felt, the more sensation there is.
6. Repeat steps 4 and 5, with the other cardboard rectangles.
7. Reverse roles with your partner and repeat the investigation.

Observations

1. On which part of the body that you tested did you feel the most sensation?
2. On which part of the body that you tested did you feel the least sensation?

Conclusions

1. Which part of the body that you tested had the most touch receptors? How do you know?
2. Which part of the body that you tested had the fewest touch receptors? How do you know?
3. What do the answers to questions 1 and 2 indicate about the distribution of touch receptors in the skin?
4. Using your data table, rank the body parts in order from the most to the least sensitive.

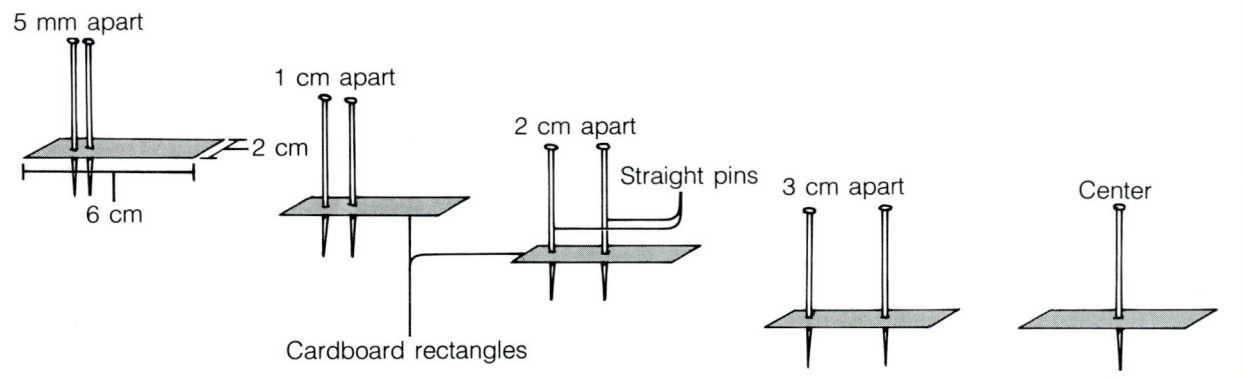

5 mm apart
1 cm apart
2 cm apart
Straight pins
3 cm apart
Center
2 cm
6 cm
Cardboard rectangles

CHAPTER REVIEW

SUMMARY

16–1 Structure and Function of the Nervous System

❏ The human nervous system is divided into the central nervous system and the peripheral nervous system. The autonomic nervous system is a part of the peripheral nervous system.

❏ The basic unit of structure and function of the nervous system is the neuron.

❏ Nerve impulses are signals that move from one neuron to another. They move across a space called a synapse.

16–2 The Central Nervous System

❏ The central nervous system is composed of the brain and the spinal cord.

❏ The brain is divided into three parts: the cerebrum, the cerebellum, and the medulla.

16–3 The Peripheral Nervous System

❏ The peripheral nervous system consists of all the nerves that connect to the central nervous system.

❏ The autonomic nervous system consists of two sets of nerves that control glands and smooth muscles of organs.

16–4 The Senses

❏ Light entering the eye passes through the cornea, aqueous humor, lens, and vitreous humor to the retina. The optic nerve carries the impulses to the brain.

❏ Sound enters the ear as vibrations and strikes the eardrum, causing the hammer, anvil, and stirrup to vibrate. These vibrations finally reach the cochlea. The auditory nerve carries the impulses to the brain.

❏ Smell and taste are chemical senses.

❏ The skin contains receptors for touch, pressure, pain, heat, and cold.

16–5 The Endocrine System

❏ The endocrine system is made up of glands that release hormones.

❏ The endocrine system includes the hypothalamus, pituitary, thyroid, parathyroids, adrenals, pancreas, ovaries, and testes.

16–6 The Feedback Mechanism

❏ The levels of various hormones in the body are controlled by a feedback mechanism that helps keep an inner state of hormone balance.

VOCABULARY

Define each term in a complete sentence.

adrenal

autonomic
  nervous
  system

axon

central
  nervous
  system

cerebellum

cerebrum

cochlea

cornea

dendrite

eardrum

effector

endocrine
  gland

endocrine
  system

exocrine gland

feedback
  mechanism

hormone

hypothalamus

interneuron

iris

lens

medulla

motor neuron

nerve impulse

nervous
  system

neuron

ovary

pancreas

parathyroid

peripheral
  nervous
  system

pituitary

pupil

receptor

reflex

retina

semicircular
  canal

sensory
  neuron

spinal cord

stimulus

synapse

testis

thyroid

CONTENT REVIEW: MULTIPLE CHOICE

On a separate sheet of paper, write the letter of the answer that best completes each statement.

1. The gap between two neurons is called the
   a. dendrite.   b. cell body.   c. synapse.   d. axon.
2. The short fibers that carry messages from neurons toward the cell body are the
   a. dendrites.   b. cell bodies.   c. synapses.   d. axons.
3. Muscles and glands are examples of
   a. synapses.   b. receptors.   c. neurons.   d. effectors.
4. The part of the brain that controls balance is the
   a. spinal cord.   b. cerebrum.   c. cerebellum.   d. medulla.
5. The largest part of the brain is the
   a. spinal cord.   b. cerebrum.   c. cerebellum.   d. medulla.
6. How many membranes nourish or protect the brain?
   a. 2   b. 4   c. 3   d. 5
7. Which of the following is part of the central nervous system?
   a. medulla   b. semicircular canals   c. retina   d. auditory nerve
8. Underproduction of the human growth hormone results in
   a. giantism.   b. dwarfism.   c. diabetes.   d. weight gain.
9. The pancreas produces the hormones insulin and
   a. thyroxine.   b. glucagon.   c. human growth hormone.   d. adrenaline.
10. The ovaries produce
   a. testosterone.   b. estrogen.   c. thyroxine.   d. adrenaline.

CONTENT REVIEW: COMPLETION

On a separate sheet of paper, write the word or words that best complete each statement.

1. The part of the neuron that contains the nucleus is the _____.
2. The part of the brain that controls thinking is the _____.
3. The knee jerk is an example of a (an) _____.
4. The colored portion of the eye is called the _____.
5. The structures in the ear that control balance are the _____.
6. The pituitary gland is an example of a (an) _____ gland.
7. The thyroid gland produces the hormone _____.
8. The gland that produces HGH is the _____.
9. The glands that are located above each kidney are the _____.
10. The female sex glands are called the _____.

CONTENT REVIEW: TRUE OR FALSE

Determine whether each statement is true or false. Then on a separate sheet of paper, write "true" if it is true. If it is false, change the underlined word or words to make the statement true.

1. The brain and the spinal cord make up the <u>peripheral</u> nervous system.
2. The <u>retina</u> is the watery fluid between the cornea and the lens of the eye.

3. The <u>lens</u> is the layer of the eye onto which an image is focused.
4. The <u>auditory</u> nerve carries impulses from the ear to the brain.
5. The outer layer of the eye is the <u>choroid</u>.
6. Touch receptors are found in the <u>ear</u>.
7. The pituitary produces <u>adrenaline</u>.

8. <u>Calcitonin</u> controls the level of calcium and phosphorus in the body.
9. <u>Metabolism</u> is the sum total of all the chemical reactions that keep an organism alive.
10. The feedback mechanism works much like a <u>thermostat</u>.

CONCEPT REVIEW: SKILL BUILDING

Use the skills you have developed in the chapter to complete each activity.

1. **Making comparisons** Draw and label a skin cell and a neuron. How are these cells similar? Different?
2. **Applying concepts** Explain why many people become dizzy after spinning around for a length of time.
3. **Relating concepts** A routine examination by a doctor usually includes the knee-jerk test. What is the purpose of this test? What could the absence of a response indicate?
4. **Applying concepts** Sometimes as a result of a cold, the middle ear becomes filled with fluid. Why do you think this can cause a temporary loss of hearing?
5. **Applying concepts** Suppose you enter a dark room and are surprised to see how colorful it is when the lights are turned on. Explain how this is possible.
6. **Interpreting graphs** This graph shows the levels of sugar in the blood of two people, during a five-hour period immediately after a typical meal. Which line represents an average person? Which line represents a person with diabetes? Explain.

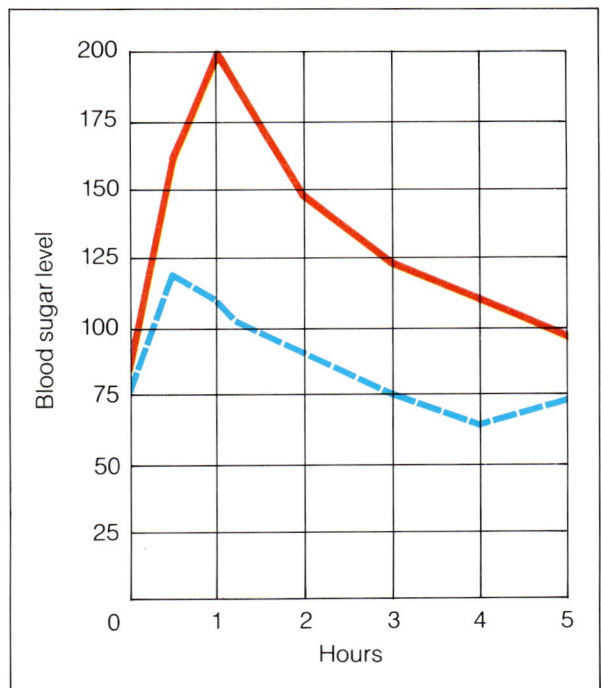

CONCEPT REVIEW: ESSAY

Discuss each of the following in a brief paragraph.

1. What is a stimulus? Give two examples.
2. What are the functions of the three types of nerves found in the nervous system?
3. Compare the functions of a receptor and an effector.
4. Explain how an impulse crosses a synapse.
5. Describe the role of the lens in vision.
6. What is diabetes mellitus? How is it treated?
7. What hormones are produced by the sex glands? What are their functions?
8. Explain what a feedback mechanism is. Give an example.

Reproduction and Development

17

CHAPTER OBJECTIVES

After completing this chapter, you will be able to

17–1 Define fertilization.

17–1 Compare the male and female reproductive systems.

17–2 Describe the changes that occur between fertilization and birth.

17–2 Describe the process of birth.

17–2 List and describe the stages of development after birth.

Is baby talk all nonsense? You are probably familiar with the screams, giggles, and other sounds babies make before they learn to talk. But did you know that these infant sounds might really make sense? Recently, scientists who have been studying baby sounds have come to some startling conclusions. It seems that these noises are actually a form of communication!

For example, the scientists discovered that well-fed, healthy babies cry at a different pitch, loudness, and rhythm than babies with certain health problems. Sometimes doctors can even figure out what is wrong with infants by listening to their cries.

Another surprising discovery is that babies are aware of the names of objects at a very early age. Scientists measured electrical impulses in the brains of 16-month-old babies at the moment they heard a familiar word. The scientists then compared these measurements to brain wave measurements taken at the moment the babies heard nonsense sounds. It was discovered that, even at the age of one month, babies have some understanding of the meaning of words.

The ability to speak and understand language is an important part of human development. It is, however, just one of the many changes that occur during development. For the most part, these changes make a person more independent. As you read this chapter, you will see that human development is an ongoing process and that there is almost no end to its changes.

The sounds a baby makes are actually a form of communication.

17–1 The Reproductive System

Section Objective

To compare the structures and functions of the male and female reproductive systems

You may find this unbelievable, but you began life as a single cell! This single cell was produced by the joining of two other cells. These other cells are the **sperm** and the **egg,** or ovum (OH-vuhm; plural: ova). The sperm is the male sex cell and the egg is the female sex cell. The joining of a sperm cell with an egg cell is called **fertilization.** Fertilization is an important part of **reproduction,** the process by which organisms produce more of their own kind. **Sperm are produced in the male reproductive system and eggs are produced in the female reproductive system.**

A sperm cell has a head, a middle part, and a tail. See Figure 17–1. Compared to the sperm, the egg is enormous. It is one of the largest cells in the body. The egg is so large that it can be seen with the unaided eye. Although the egg's shape and size differ from the sperm's, the cells share an important feature. Both cells contain thick, rodlike structures called **chromosomes** (KROH-muh-sohmz). Chromosomes pass on inherited characteristics from one generation of cells to the next.

Each cell of the human body, except the sex cells, contains 46 chromosomes. Each sex cell contains half this number, or 23 chromosomes. After fertilization, the fertilized egg contains 46 chromosomes. So a fertilized egg gets 23 chromosomes from the sperm and 23 chromosomes from the egg. The fertilized egg contains all of the information needed to produce a complete new human being.

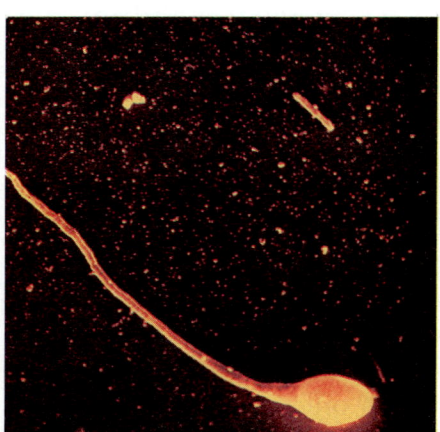
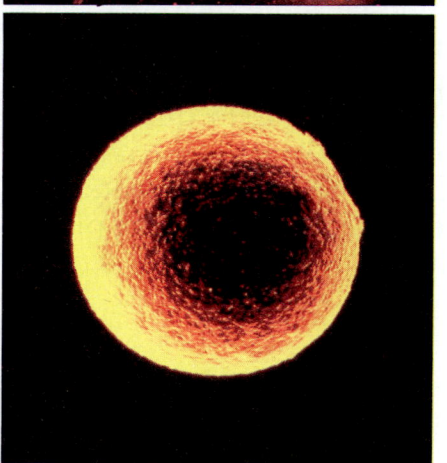

Figure 17–1 *The sperm (top) is the male sex cell. The egg, or ovum (bottom), is the female sex cell. In humans, how many chromosomes does each of these cells have?*

The Male Reproductive System

The male **reproductive system** has two **testes** (TEHS-teez; singular: testis). They are oval-shaped objects found inside an external sac called the **scrotum** (SKROHT-uhm). The testes are the organs that produce sperm. Sperm swim from each testis through many small tubes to a larger tube called the urethra (yoo-REE-thruh). As you learned in Chapter 15, the urethra is also the tube through which urine leaves the body. The urethra runs through an organ

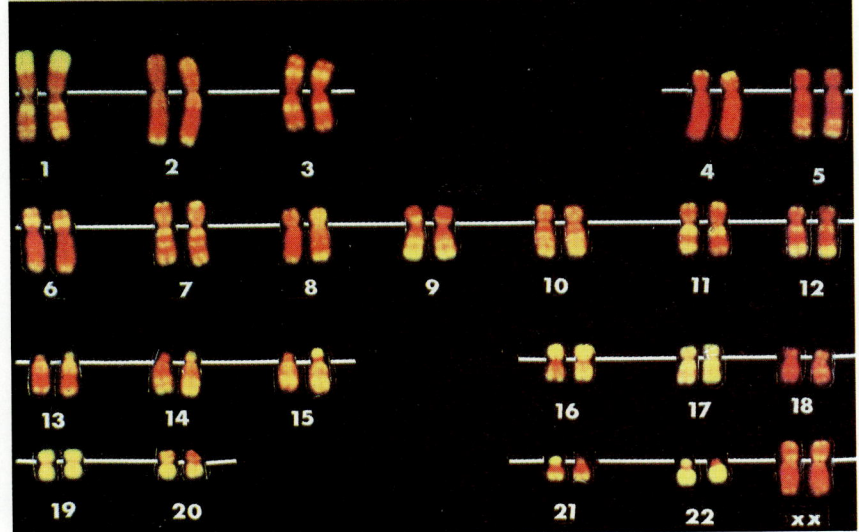

Figure 17–2 *This photograph shows the karyotype for a human cell. A karyotype illustrates the size, number, and pairs of chromosomes. How many chromosomes are found in a human skin cell?*

called the penis. Sperm is transferred to a female's body through the penis.

In addition to producing sperm, the testes produce a hormone called **testosterone** (tehs-TAHS-tuh-rohn). Testosterone is responsible for the growth of facial and body hair, broadening of the shoulders, and deepening of the voice in males.

The Female Reproductive System

Unlike the male reproductive system, all the parts of the female reproductive system are within the female's body. Two structures called **ovaries** (OH-vuh-reez) are located at about hip level, one on each side of a female's body. The almond-shaped ovaries produce the eggs. Like the testes, the ovaries produce hormones. One of these hormones, called **estrogen,** triggers the broadening of the hips in females. Estrogen also starts the maturation of egg cells in the ovaries.

Located near each ovary, but not directly connected to it, is a **Fallopian** (fuh-LOH-pee-uhn) **tube.** An egg travels through this tube from the ovary. Another name for a Fallopian tube is an **oviduct.**

At the end of the Fallopian tube, the egg reaches a hollow, muscular organ called the **uterus** (YOOT-uhr-uhs), or womb. The uterus is a pear-shaped structure in which the early development of a baby takes place. At the lower end of the uterus is a narrow section called the cervix (SER-vihks). The cervix

Figure 17–3 *Use this diagram to review the structures of the male reproductive system. Notice the two testes, which produce sperm and the hormone testosterone. What is the function of testosterone?*

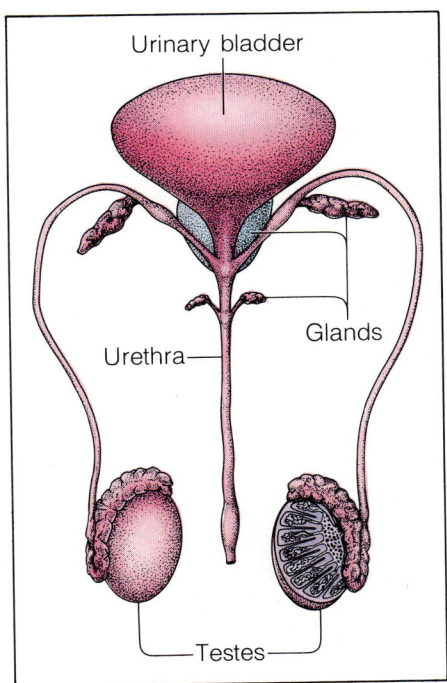

Urinary bladder

Urethra

Glands

Testes

Figure 17-4 *In the female reproductive system, the ovaries produce eggs as well as hormones. In which structure of the female reproductive system would the development of a baby take place?*

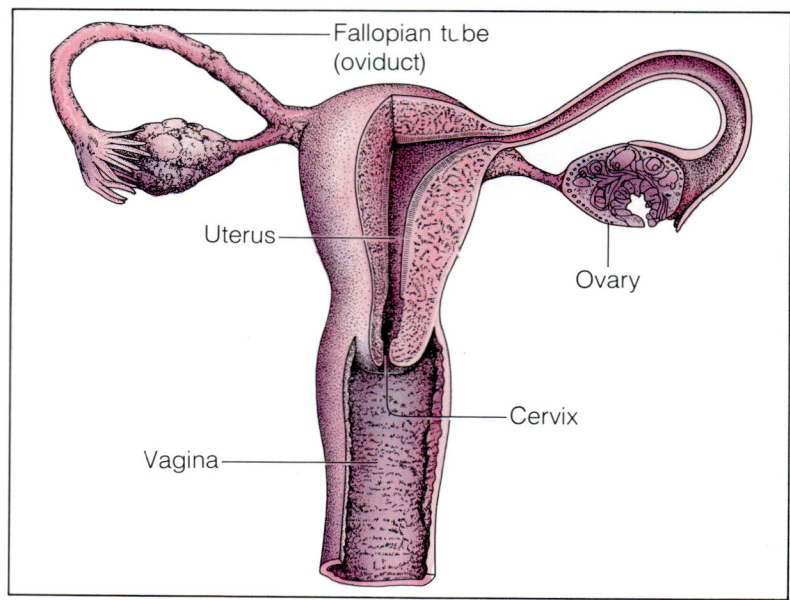

Sharpen Your Skills

Multiple Births

A multiple birth occurs when two or more children are born at one time to one mother. Using reference material in the library, look up information on twins and triplets.

In a written report, describe how twins and triplets occur. You may use diagrams to explain this. What is the difference between identical and fraternal twins? How are identical triplets produced?

opens into a wider channel called the vagina (vuh-JIGH-nuh), or birth canal. The vagina is the canal through which the baby passes during birth.

The Menstrual Cycle

The monthly cycle of change that occurs in the female reproductive system is called the **menstrual** (MEHN-struhl) **cycle.** This cycle has an average length of about 28 days, or almost a month. In fact, the word *menstrual* comes from the Latin word for *month*. During the menstrual cycle, an egg develops in an ovary. The mature egg is released into a Fallopian tube. This process is called **ovulation.**

Figure 17-5 *The monthly cycle of change that occurs in the female reproductive system is called the menstrual cycle. What is the average length of the menstrual cycle?*

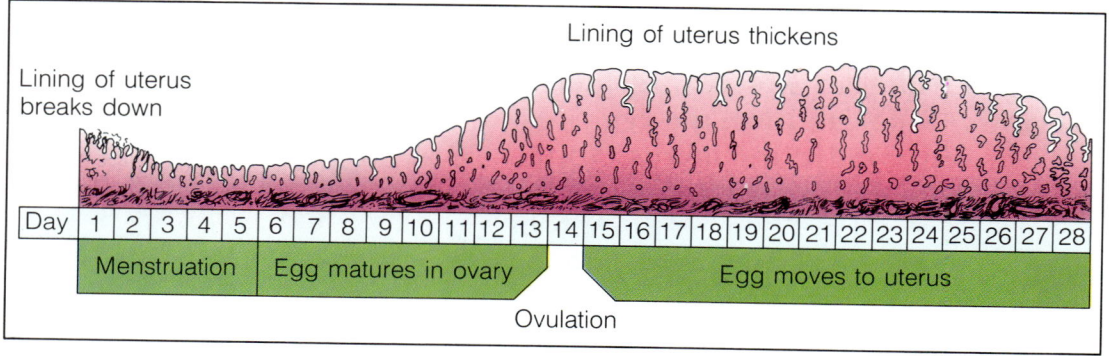

An egg meets a sperm and is fertilized in the Fallopian tube. During this time, not only are changes occurring in the ovary, but also in the uterus. For example, during most of the menstrual cycle, the lining of the uterus thickens. The lining is being prepared for a fertilized egg. If a fertilized egg reaches this lining, the egg attaches to the lining. There the egg grows and develops into a baby.

If the egg is not fertilized, both the egg and the lining of the uterus break down. The blood and tissue from the thickened lining of the uterus pass out of the body through the vagina. This process is called **menstruation** (mehn-STRAY-shuhn). On the average, menstruation lasts for five days. At the same time that menstruation occurs, a new egg is maturing in the ovary and the cycle continues.

An egg can only be fertilized during the short time of ovulation. Ovulation usually occurs about 14 days after the start of menstruation. After the egg is released into the Fallopian tube, the egg takes about one or two days to make its way along the tube to the uterus. Only if sperm are present at this time, can the egg be fertilized.

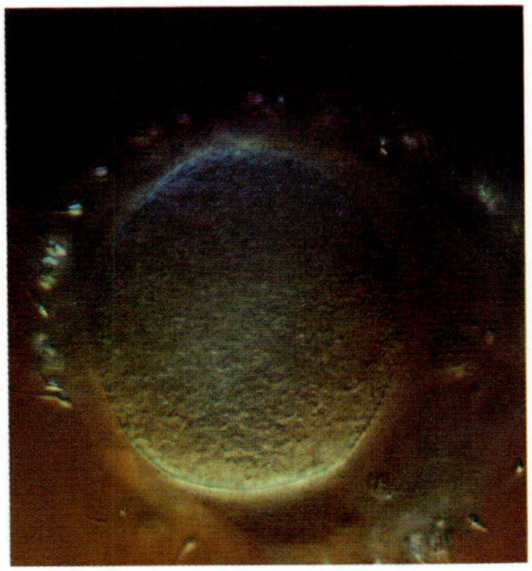

Figure 17–6 *The tiny tadpole-shaped objects clinging to the surface of the egg are sperm. Only one sperm will unite with the egg. What is this process called?*

SECTION REVIEW

1. Compare the functions of the male and female reproductive systems.
2. Define fertilization.
3. How many chromosomes do sex cells have?
4. Which organs produce sperm? Which produce eggs?
5. How does a sperm cell's shape help it function?

17–2 Stages of Development

Section Objective

To describe the stages of development that occur before and after birth

Almost immediately after a **zygote,** or fertilized egg, is formed, a new human being begins to develop. The zygote divides into two cells. Then each of these cells divides again and again, and so on. Change and growth continue for approximately nine months or 280 days until birth. This time between fertilization and birth is called pregnancy.

Human beings go through various stages of development before and after birth. Before birth, a single human cell develops into an embryo and then a fetus. After birth, humans pass through the stages of infancy, childhood, and adolescence to adulthood.

Development Before Birth

As the zygote first begins to divide, the cells remain close together and give the appearance of a bunch of berries. By the time this berry-shaped group of cells reaches the uterus, the cells have divided a few more times and look like a hollow ball of cells. During this early stage of development, and for the next eight weeks or so, the newly formed organism is called an **embryo** (EHM-bree-oh).

Once the embryo enters the uterus, it attaches itself to the wall of the uterus. In the uterus, several membranes form around the embryo. A clear membrane called the **amnion** forms a fluid-filled sac. This **amniotic sac** cushions and protects the developing baby. Another membrane forms the **placenta** (pluh-SEHN-tuh). The placenta is made partly from

Figure 17–7 *You can see in this illustration that blood vessels in the umbilical cord lead to and from the placenta and developing fetus. What substances pass between mother and baby through the placenta?*

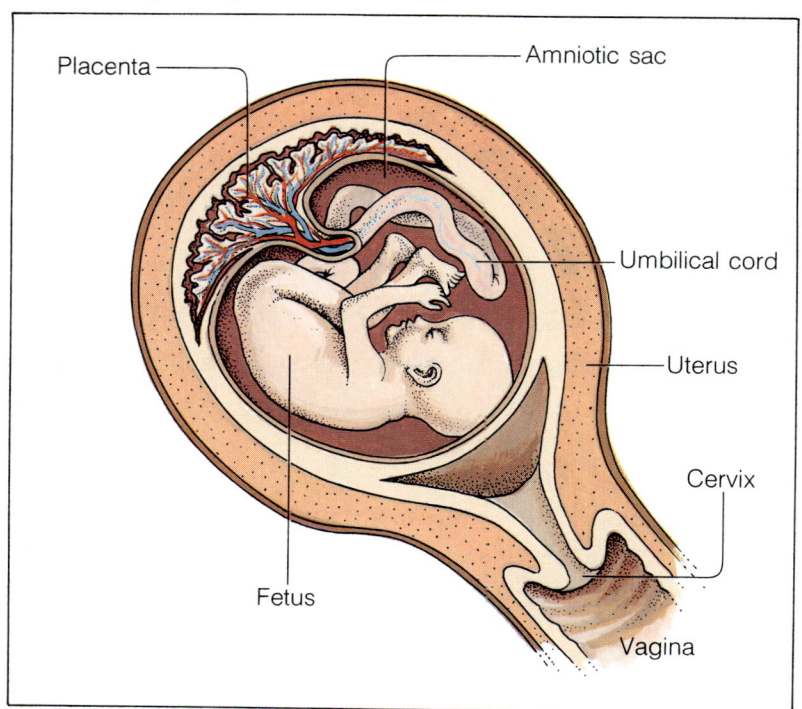

HELP WANTED: SCHOOL SOCIAL WORKER Bachelor's degree in social work required. Position involves working with students, parents, psychologists, and guidance counselors to evaluate problems and help students solve them. Submit résumé to school office.

During the school year, some teachers may discover students who have problems at school. People who have special training and experience to deal with such problems are **school social workers.**

School social workers are called upon for assistance by teachers, guidance counselors, administrators, parents, or the students themselves. After talking to those concerned, social workers try to learn the cause of the problem and find a solution. Sometimes they set up meetings with doctors or arrange for special tests to find the nature of a problem.

Those who wish to become school social workers should have a basic concern for stu-

dents and their problems. Social workers must be objective, sensitive, and emotionally mature in order to handle the responsibilities involved. In many states, it is necessary to become licensed or registered. This certification requires work experience and an examination. If you are interested in a career in social work, write to the National Association of Social Workers, Inc., 7981 Eastern Avenue, Silver Spring, MD 20910.

tissue that develops from the embryo and partly from tissue that makes up the wall of the uterus.

The placenta is a kind of transfer station between the embryo and its mother. Both mother and embryo have separate circulatory systems and separate blood supplies. However, oxygen obtained by the mother's lungs and food from her digestive system pass through the placenta to the embryo. Wastes pass out of the embryo through the placenta and are eliminated by the mother. An embryo is connected to the placenta by a cord called an **umbilical** (uhm-BIHL-ih-kuhl) **cord.** The umbilical cord contains blood vessels that transport food, oxygen, and wastes back and forth between the embryo and the placenta.

After eight weeks, the embryo is called a **fetus** (FEET-uhs). A fetus has developed eyes, ears, cheeks, arms, and legs. After three months, it is easy to identify the fetus as a male or a female. During months six through nine, the fetus grows rapidly.

In recent years, methods have been developed that allow doctors to study and even treat fetuses before they are born. One of these methods consists

Figure 17–8 *At about five to six months, a fetus is able to suck its thumb. The transparent veil behind the fetus is part of the membrane that surrounds and protects the fetus. What is this membrane called?*

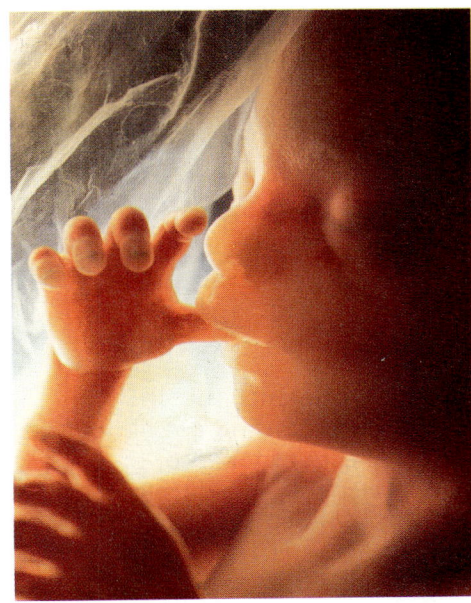

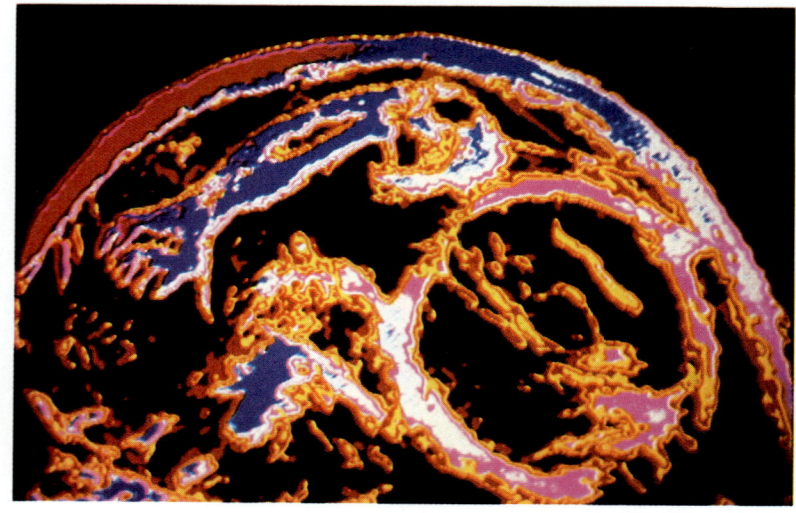

Figure 17–9 *You can see the outline of the fetus's head on the right and its arm and hand at the top of this photograph. The image was produced by sending sound waves through the mother's uterus.*

of drawing off a small amount of the fluid in which the fetus floats. The fluid contains cells from the fetus that can be studied to determine its health, as well as many of its inherited characteristics. In another process, sound waves that produce an image on a television screen are sent through the uterus of the mother. This procedure gives a detailed, moving image of the fetus. See Figure 17–9. Sound pictures have been used to safely diagnose and treat fetuses with medical problems before they are born.

Birth

After about nine months of development and growth inside the uterus, a baby is ready to be born. Strong muscular contractions of the uterus begin to push the baby out of the uterus into the vagina. This process is called labor. As labor progresses, the contractions of the uterus become stronger and occur more frequently. Eventually the baby, still connected to the placenta by the umbilical cord, is forced out of the mother.

Development After Birth

Within a few seconds after birth, a baby begins to cry. This action helps its lungs to expand and fill with air. The umbilical cord, which is still attached to the placenta, is cut close to the baby's body. This does not cause any pain to the baby. When the cut heals, it forms a scar. You know this scar as your navel. What is a common nickname for the navel?

Sharpen Your Skills

Tobacco, Drugs, and Alcohol

Use reference materials in the library to find out about the effects of tobacco, drugs, and alcohol on a developing fetus. Write a brief report on your findings.

INFANCY Have you ever watched a six-month-old infant for a few minutes? Its actions, or responses, are very simple. An infant can suck its thumb, grasp objects, yawn, stretch, blink, and sneeze. When lying in bed, an infant often curls up in a position much like the one it had in the uterus.

One of the most obvious changes during infancy is a rapid increase in size. The heads of young babies are very large in proportion to their bodies. As the infant gets older, the head grows more slowly and the body, legs, and arms begin to catch up.

Mental and muscular skills begin to develop in a fairly predictable order. However, the exact ages at which they occur vary from baby to baby. A newborn infant cannot even lift its own head. After about three months, it can hold its head up and can also reach for objects. In about two more months, the infant can grasp objects. At about seven months, most infants are able to move around by crawling. Somewhere between 10 and 14 months, most infants begin to walk by themselves.

CHILDHOOD Infancy ends and childhood begins around the age of 18 months. Childhood continues until the age of 13 years. During childhood, mental abilities increase. Memory is strengthened. Muscular skills develop. With practice, a small child becomes better and better at walking, holding a knife and fork, writing with a pencil, or playing sports. Over a period of several years, baby teeth are lost and replaced by permanent teeth.

Figure 17–10 *Most infants are able to move about by crawling at the age of seven months (left). At what age do infants often begin to walk by themselves (right)?*

Figure 17–11 *During childhood, children learn many skills by listening to, observing, and imitating adults (left). Muscular skills also develop during childhood (right). Between what ages is a person considered a child?*

Figure 17–12 *During adolescence, young people begin to socialize. What is the beginning of adolescence called?*

Little by little, young children develop language skills. All babies make basically the same babbling sounds. Then, as a child becomes aware of itself and others, these sounds are shaped into language. Language skills come from observing and imitating others. At first, a child uses only one word at a time, such as "ball." Soon the child uses an action word and produces a two-word sentence, for example, "Hit ball." By the age of four or five, the child is able to hold an adultlike conversation.

In addition to understanding and speaking a language, children can be taught to read and solve problems even before they attend school. During childhood, children learn a great deal about their environment. They also learn to behave in socially appropriate ways.

ADOLESCENCE In many cultures, adolescence is thought of as a passage from childhood to adulthood. The word *adolescence* comes from a Latin word that means "growing up." Adolescence begins after the thirteenth year and continues through the teenage years. The beginning of adolescence is called **puberty** (PYOO-buhr-tee). During puberty, the sex organs develop rapidly. Menstruation begins in girls and the production of sperm begins in boys.

In addition, a growth spurt also occurs during adolescence. A growth spurt is a sudden increase in height and weight. In females, this rapid growth occurs between the ages of 10 and 16. During these years, females may grow about 15 centimeters in height and gain about 16 kilograms or more. In males, the growth spurt occurs between the ages of 11 and 17. During this time, males may grow about 20 centimeters and gain 20 kilograms or more.

ADULTHOOD At about the age of 20, adulthood begins. All body systems, including the reproductive system, have become fully matured and full height has been reached. As a human being passes from infancy through adulthood, fat beneath the skin keeps moving farther and farther away from the surface. The round, padded, button-nosed face of the baby is slowly replaced by the leaner, more defined face of the adult. The nose and the ears continue to grow and take on a more individual shape.

After about 30 years, a process known as aging begins. This process becomes more noticeable between the ages of 40 and 65. At this time, the skin loses some of its elasticity, the hair sometimes loses its coloring, and muscle strength decreases. During this period, females go through a physical change known as **menopause.** After menopause, menstruation stops and ovulation no longer occurs. As men age, they may become bald. After age 65, the aging process continues, often leading to less efficient heart and lung action. But the effects of aging can be lessened if a person follows a good diet and exercise plan throughout life.

Figure 17–13 *If a person follows a good diet and exercise plan throughout life, the effects of aging can be lessened.*

SECTION REVIEW

1. Name and describe the four stages of development after birth.
2. What is a developing baby called during the first eight weeks of development? After the first eight weeks?
3. How does the amniotic sac protect a developing baby?
4. Through which structure would harmful substances, such as drugs, be transferred from a mother to her baby during pregnancy?

How Many Offspring?

Problem

How do the length of gestation, number of off-spring per birth, age of puberty, and life span of various mammals compare?

Materials (per group)

graph paper   colored pencils

Procedure

1. Study the chart, which shows the length of gestation, the average number of off-spring per birth, the average age of puberty, and the average life span of certain mammals.
2. Make a bar graph that shows the length of gestation for each of these mammals.
3. On another bar graph, show each mammal's average number of offspring.
4. Then make a third bar graph showing the life span of each animal. Color the part of the bar that shows the length of child-hood, or time from birth to puberty.

Observations

1. Which of the mammals listed has the longest gestation period? Do you know of any mammals with longer pregnancies?
2. Which of the mammals listed has the shortest life span?
3. Which of the mammals listed is most closely related to human beings?

Conclusions

1. What general conclusions can you draw after studying the graphs you have made? For instance, what seems to be the relationship between length of gestation and number of offspring, between length of gestation and size of the animal, and between age of puberty and life span?
2. If a mouse produces five litters per year, how many mice does the average female mouse produce in a lifetime?
3. Of all the mammals listed, which care for their young for the longest period of time after birth? Why do you think this is the case?

Mammal	Gestation Period (days)	Number of Offspring per Birth	Age at Puberty	Life Span (years)
Opossum	12	13	8 months	2
House mouse	20	6	2 months	3
Rabbit	30	4	4 months	5
Dog	61	7	7 months	15
Lion	108	3	2 years	23
Rhesus monkey	175	1	3 years	20
Human being	280	1	13 years	74
Horse	330	1	1.5 years	25

CHAPTER REVIEW

SUMMARY

17-1 The Reproductive System

❏ When a sperm, or male sex cell, joins with an egg, or female sex cell, fertilization occurs.

❏ Each sperm and each egg contain 23 chromosomes, which pass on inherited characteristics from one generation of cells to the next.

❏ The male reproductive system includes two testes, which produce sperm and a hormone called testosterone.

❏ The female reproductive system includes two ovaries, which produce eggs and hormones. One of these hormones is estrogen.

❏ Located near each ovary is an oviduct, or Fallopian tube, which leads to a hollow, muscular uterus.

❏ At the lower end of the uterus, there is a narrow cervix, which opens into the vagina, or birth canal.

❏ The monthly cycle of change that occurs in the female reproductive system is called the menstrual cycle.

❏ During the menstrual cycle, ovulation, or the release of an egg from an ovary, occurs. The lining of the uterus also thickens in preparation for the attachment of a fertilized egg.

❏ If fertilization does not take place, the egg and thickened lining of the uterus break down and pass out of the body. This process is called menstruation.

❏ If, at the time of ovulation, sperm are present in the Fallopian tube, the egg may become fertilized.

17-2 Stages of Development

❏ After an egg is fertilized, it begins to divide again and again, forming a ball-shaped structure of many cells.

❏ During the first eight weeks of its development, an organism is called an embryo.

❏ During the embryo's early period of growth, several types of membranes surround and protect it. One of these membranes, the placenta, provides the embryo with food and oxygen and eliminates its wastes.

❏ The embryo is connected to the placenta by the umbilical cord, which contains blood vessels.

❏ During the birth process, the fetus passes through the vagina and into the outside world.

❏ Human beings go through various stages of development during their lives. These stages are infancy, childhood, adolescence, and adulthood.

VOCABULARY

Define each term in a complete sentence.

amnion

amniotic sac

chromosome

egg

embryo

estrogen

Fallopian tube

fertilization

fetus

menopause

menstrual cycle

menstruation

ovary

oviduct

ovulation

placenta

puberty

reproduction

reproductive system

scrotum

sperm

testis

testosterone

umbilical cord

uterus

zygote

On a separate sheet of paper, write the letter of the answer that best completes each statement.

1. The female sex cell is the
   a. testis.   b. sperm.   c. egg.   d. ovary.
2. Sperm are produced in male sex organs called
   a. testes.   b. ovaries.   c. scrotums.   d. ureters.
3. In males, sperm and urine leave the body through the
   a. testes.   b. scrotum.   c. vagina.   d. urethra.
4. A hormone responsible for the deepening of the voice in males is
   a. estrogen.   b. sperm.   c. testosterone.   d. menstruation.
5. Eggs are produced in the
   a. scrotum.   b. ovaries.   c. cervix.   d. Fallopian tubes.
6. The thickened lining of the uterus passes out of the body during
   a. ovulation.   b. fertilization.   c. menstruation.   d. urination.
7. The structure in which a fertilized egg first divides is the
   a. ovary.   b. Fallopian tube.   c. uterus.   d. vagina.
8. A structure made up of tissue from both the embryo and the uterus is the
   a. ovum.   b. placenta.   c. umbilical cord.   d. fetus.
9. Adulthood begins at about the age of
   a. 13 years.   b. 1 year.   c. 20 years.   d. 30 years.
10. Sex organs develop rapidly during
    a. infancy.   b. childhood.   c. puberty.   d. adulthood.

CONTENT REVIEW: COMPLETION

On a separate sheet of paper, write the word or words that best complete each statement.

1. The male sex cell is the _____.
2. The number of chromosomes in a sex cell is _____.
3. Development of a baby takes place in its mother's _____.
4. The _____ is another name for the birth canal.
5. The release of a mature egg from an ovary is called _____.
6. Fertilization occurs in the _____.
7. The structure that provides cushioning for the developing baby is the _____.
8. After eight weeks, the embryo is called a (an) _____.
9. Most human beings learn to walk by themselves during the _____ stage.
10. _____ is the beginning of adolescence.

CONTENT REVIEW: TRUE OR FALSE

Determine whether each statement is true or false. Then on a separate sheet of paper, write "true" if it is true. If it is false, change the underlined word or words to make the statement true.

1. The joining of the sperm and egg is called fertilization.
2. The testes are found inside a sac called the scrotum.

3. To reach the uterus, an egg travels through the cervix.
4. Testosterone triggers the broadening of the hips in females.
5. The average length of the menstrual cycle is 40 days.
6. An egg can be fertilized only during ovulation.
7. If an egg is not fertilized, the lining of the uterus leaves the body through the urethra.
8. The structure that connects the embryo to the placenta is the uterus.
9. At two years of age, a young human being is considered to be a (an) infant.
10. After about 30 years, a person begins to go through a process called aging.

CONCEPT REVIEW: SKILL BUILDING

Use the skills you have developed in the chapter to complete each activity.

1. **Making comparisons** In what way is development during adolescence similar to development before birth?
2. **Relating cause and effect** Why is it dangerous for pregnant women to smoke, drink, or use drugs not prescribed by a doctor?
3. **Making inferences** The word *adolescence* means "growing up." Why is adolescence a good name for the teenage stage of life?
4. **Applying concepts** Explain why a proper diet and an adequate amount of exercise can lessen the effects of aging.
5. **Drawing conclusions** Why do broken bones heal more rapidly in young children than in elderly people?
6. **Making comparisons** In what way is amniotic fluid similar to the shock absorbers of a car?
7. **Making predictions** Explain how a child's environment can affect its development.
8. **Making graphs** Use the information in this table to construct a graph. What conclusions can you draw from the graph?

Age group in years	Average height in centimeters Female	Male
At birth	50	51
2	87	88
4	103	104
6	117	118
8	128	128
10	139	139
12	152	149
14	160	162
16	163	172
18	163	174

CONCEPT REVIEW: ESSAY

Discuss each of the following in a brief paragraph.

1. Describe the stages that occur during the menstrual cycle.
2. Describe the structure and function of the placenta.
3. Compare an embryo to a fetus.
4. What are two methods that are used by doctors to study fetuses?
5. Describe the birth process.

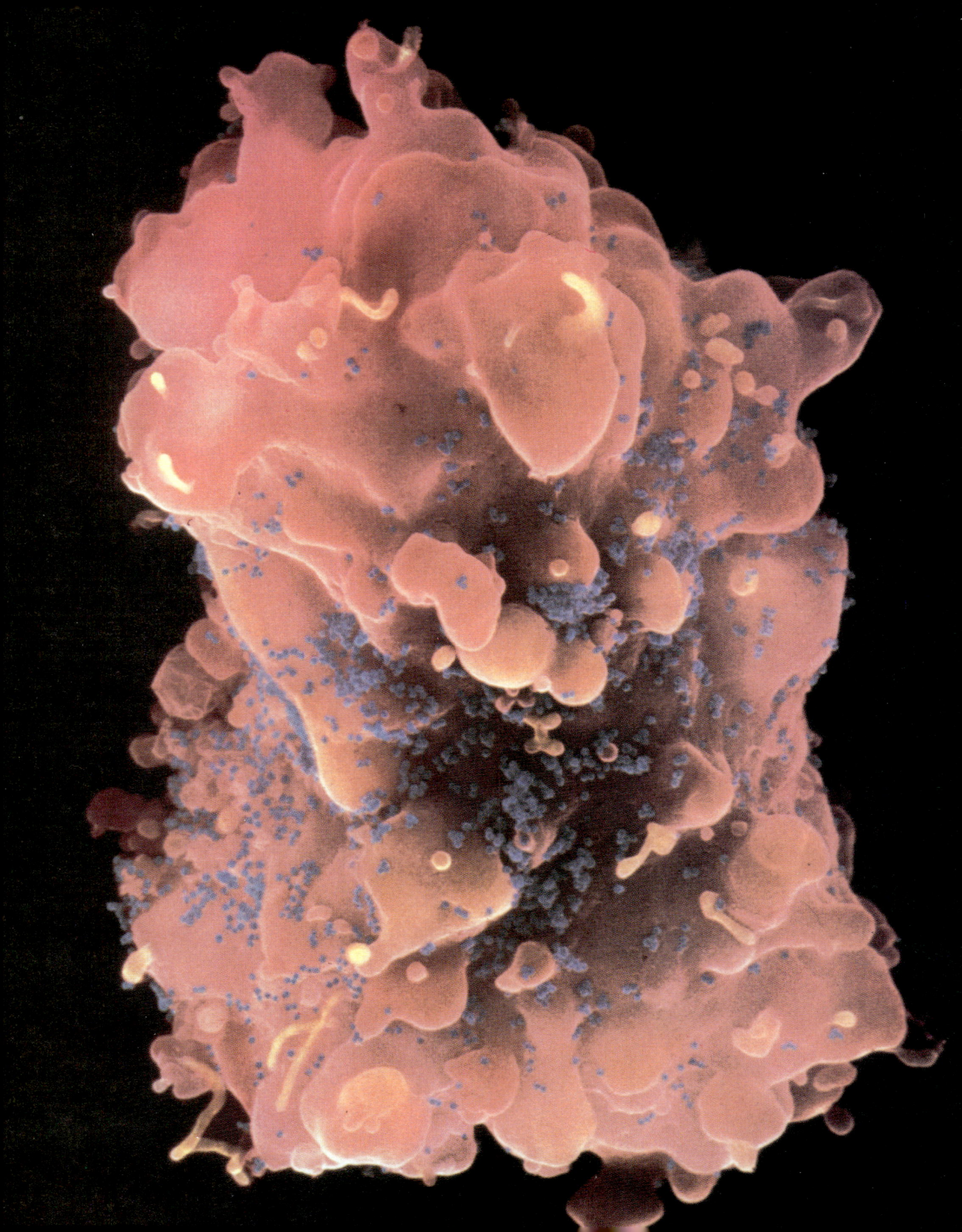

Infectious Disease and Chronic Disorders

18

CHAPTER OBJECTIVES

After completing this chapter, you will be able to

18–1 Define infectious disease and list some examples.

18–1 Describe how infectious disease may spread.

18–2 Describe the function of the immune system.

18–2 Compare the actions of the specific and nonspecific lines of defense.

18–2 Compare natural and acquired immunity.

18–3 Describe how vaccines can help prevent the spread of disease.

18–4 Define chronic disorder and list some examples.

Every minute of every day fierce battles are fought within your body. The attackers in these battles are very tiny. In fact, some are so small that more than 200 million of them would fit on the period at the end of this sentence. Yet they can destroy the body's much larger defender cells.

For the most part, you are never aware of these battles. Your defender cells act like scouts and keep a silent watch for invaders. These invaders include viruses and bacteria. Sometimes the defender cells mistake harmless invaders, such as pollen, for deadly enemies and cause you to sneeze. Sometimes the defender cells are "caught sleeping at their posts" and you develop the flu or a cold. At times, some of the body cells begin to grow uncontrollably and slip through the body's defenses. Fortunately, for every successful entry of invaders into the body, thousands of attempts are pushed back.

The photograph on the opposite page shows a defender cell under attack by the viruses that cause AIDS, or *A*cquired *I*mmune *D*eficiency *S*yndrome. The AIDS viruses appear as blue dots on a special white blood cell called a T cell. T cells are the body's main defenders against harmful invaders. When the AIDS viruses enter the body, they attack and kill T cells. Without T cells, the body's defense system is almost destroyed.

In this chapter, you will discover how the invaders enter the body. You will also learn how the body sets up a strong defense system to fight off its attackers.

The helper T cell in this photograph is under attack by AIDS viruses, seen as tiny blue dots.

You are not alone. At every moment of your life, you must share the food you eat, the water you drink, and the air you breathe with other forms of life. Most of these organisms are so tiny that they are microscopic, or invisible to the unaided eye. Many spend their entire lives in the lining of your nose and throat, in your mouth, and in your intestines. In these places, they take advantage of the warmth, food, and water in your body tissues. For these tiny organisms, you act as a **host.** A host is an organism in which another organism lives and gets nourishment and protection.

In most cases, the microscopic organisms within your body are not harmful to you. In your intestines, they may even help you digest the food you eat. And some produce vitamins important to your health.

Some microorganisms are harmful to people. They cause disease. An **infection** occurs when the body is successfully invaded by a disease-causing organism. **Diseases that are transmitted among people by harmful organisms such as viruses and bacteria are called infectious diseases.**

Few people, if any, can go through life without being infected by various organisms. Fortunately, the body has many kinds of defenses to protect itself. Let's first look at some of the organisms that cause **infectious diseases,** or diseases that can be transmitted from one organism to another. Let's also see how such organisms may be spread. Then you can discover some of the body's weapons against disease.

Diseases Caused by Viruses

Your head aches. Your nose is runny. Your eyes water. And you have a slight cough. No, you do not have some strange disease. More than likely, you suffer from the common cold. An achy feeling and a runny nose are among the **symptoms,** or signs, of a cold. Can you think of other cold symptoms?

The common cold is caused by perhaps the simplest and smallest disease-causing organism—a virus.

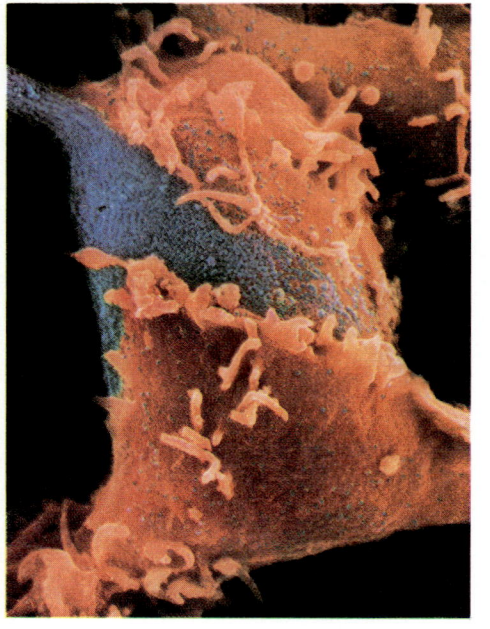

Figure 18–1 *This photograph shows how viruses, seen as tiny blue dots, are released by a human cell when it ruptures.*

SOME VIRAL DISEASES

Disease	Organism That Causes the Disease	Methods of Spreading the Disease
Common cold	Many viruses	Droplets in air, direct contact with infected person
Influenza	Two important types (A,B) of virus and many sub-types	Direct contact with an infected person, droplet infection, also may be airborne
Mumps	One virus	Droplet spread, direct contact with an infected person
Poliomyelitis	Poliovirus types 1, 2, and 3	Direct contact with an infected person
Rabies	One virus	Usually bite of mammal with rabies
Infectious mononucleosis	Probably one virus	Spread by droplets, may be spread by direct contact
Pneumonia (viral)	Several viruses	Droplets, oral contact with an infected person
Chicken pox	One virus	Droplets in air, direct contact with an infected person
Measles (rubeola)	One virus	Droplets in air, direct contact with secretions of infected person
German measles (rubella)	One virus	Droplet spread, direct contact with infected person

Figure 18–2 *This chart lists some common diseases caused by viruses. How is the virus that causes measles spread?*

There are hundreds of different kinds of viruses that can cause a cold.

In Figure 18–2, you will find some of the diseases that are caused by viruses. In all of these viral diseases, the actions of the virus in the body are similar. When a virus invades the body, it quickly enters a body cell. Inside the cell, the virus takes control. The virus uses the cell's food supplies. It also uses the cell's reproductive machinery. However, the virus does not cause the cell to reproduce. Actually it causes the cell to build more viruses. In time, the cell is full of thousands of viruses. Then the cell bursts open and dies. The viruses, however, are now free to invade many more body cells.

Diseases Caused by Bacteria

Your throat is very sore. Swallowing even ice cream is difficult. You visit your doctor to find out what is wrong. If your sore throat is caused by a bacterium, the doctor can prescribe an **antibiotic.** An antibiotic is a chemical substance that stops the growth of some microorganisms such as bacteria within the body. Most antibiotics are produced by certain microorganisms. Other antibiotics are chemicals produced in laboratories. If your sore throat is caused by a virus, no medicine is effective and the disease has to "run its course."

In Figure 18–3, you will find some of the most common diseases that are caused by bacteria. When disease-causing bacteria invade the body, they cause infection in one of two ways. Most bacteria attack body cells directly. However, some bacteria make you sick because they produce a **toxin,** or poison.

Figure 18–3 *This chart gives the names of some bacterial diseases, the shape of the bacteria, and how the disease is spread. What shape is the bacteria that causes tuberculosis?*

SOME BACTERIAL DISEASES

Disease	Shape of Organism	Methods of Spreading the Disease
Diphtheria	Rod-shaped	Contact with a patient or carrier, contaminated raw milk
Streptococcal sore throat	Sphere-shaped, occurring in chains	Droplets in air, direct contact with infected person or carrier, contaminated milk
Pneumonia (bacterial — many different types)	Sphere-shaped, occurring in pairs	Droplets in air, direct oral contact with infected person
Scarlet fever	Sphere-shaped, occurring in chains	Droplets in air, direct contact with infected person or carrier, contaminated milk
Meningitis	Sphere-shaped	Direct exposure to the organism
Tetanus (lockjaw)	Rod-shaped	Through a dirtied wound, usually a puncture wound
Tuberculosis	Rod-shaped	Droplets in air, contaminated milk and dairy products
Whooping cough	Rod-shaped	Droplets in air

For example, the reason you have a sore throat when streptococci invade your throat is that your body reacts to a toxin produced by the bacteria.

The Spread of Disease

Whenever a person develops an infectious disease, there is a possibility that the disease may spread to others. Infectious diseases are also called **communicable** (kuh-MYOO-nih-kuh-buhl) **diseases.**

As you might expect, the most common way infectious diseases spread is through contact with an infected person. For example, an infected person may cough or sneeze, spraying droplets of disease-causing organisms into the air. If another person inhales these droplets, the disease can spread. Infectious diseases such as the common cold and influenza, or the flu, are spread in this way.

Infectious disease may also be passed from person to person through sexual contact. These types of diseases are sexually transmitted diseases, or STDs. Some examples of STDs are gonorrhea (gahn-uh-REE-uh) and syphilis (SIHF-uh-lihs).

Another way infectious diseases can be spread is through contaminated food or water. A person with an infected cut, for example, may come into contact with some food product. If another person eats that food, he or she might take in some of the disease-causing organisms.

Other infectious diseases can be spread through contact with a dirty object. Tetanus (TEHT-uh-nuhs), a dangerous bacterial disease, is often spread when a person receives a puncture wound.

Still another way infectious diseases can be spread is through infected animals. Squirrels, for example, can get rabies, an extremely serious disease. The rabies virus then can be spread to a person if that person is bitten by the infected squirrel.

SECTION REVIEW

1. What is an infectious disease? Identify the causes of some infectious diseases.
2. What is an antibiotic?
3. List four ways in which diseases are spread.
4. Explain why there is no one cure for the common cold.

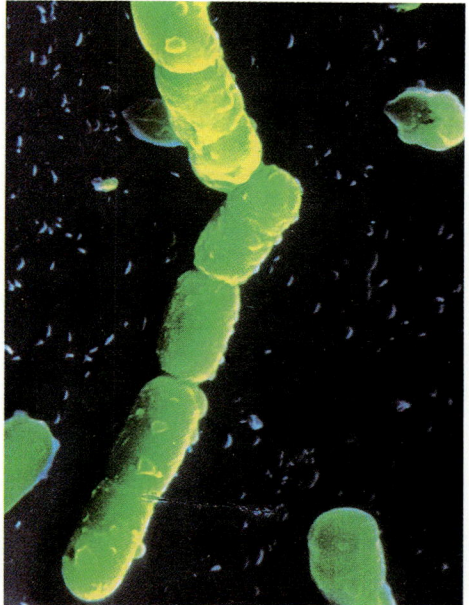

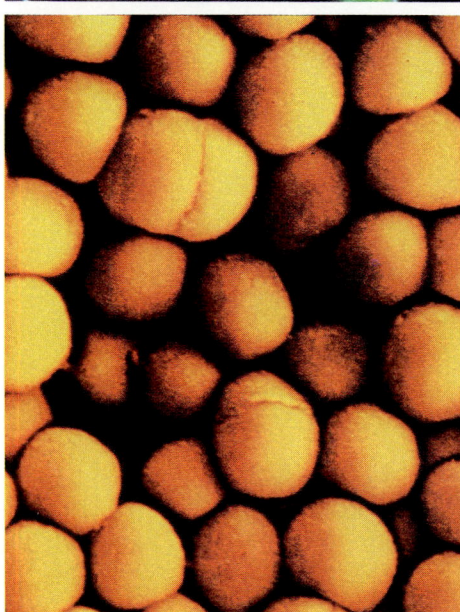

Figure 18–4 *Bacterial diseases can be caused by certain spherical bacteria (bottom) called cocci and by certain rod-shaped bacteria (top) called bacilli. What chemical substances stop the growth of bacteria?*

18–2 Body Defenses

The best way to keep from getting an infectious disease is to stop disease-causing organisms from entering the body. The body has two lines of defense against invading organisms. **The body's first line of defense against disease-causing organisms includes the skin, mucous membranes, and cilia. The second line of defense is the immune system.**

The body's first line of defense is made up of structures that form a barrier between the body and its surroundings. Unbroken skin forms an excellent barrier to disease-causing organisms. However, a cut in your skin is like an open door inviting these organisms to enter.

Not all disease-causing organisms enter the body through the skin; some might be inhaled from the air. In the process, these organisms meet the other first-line defenders—the mucous membranes and cilia. The membranes, or linings, of your nose, mouth, and respiratory, or breathing, system are coated with a sticky substance called mucus (MYOO-kuhs). Mucus helps trap organisms such as bacteria and prevents them from traveling farther into the body. Also, the lining of your nose, throat, and airways to the lungs contain cilia (SIHL-ee-uh). Cilia are tiny hairs that beat in a wavelike motion, sweeping

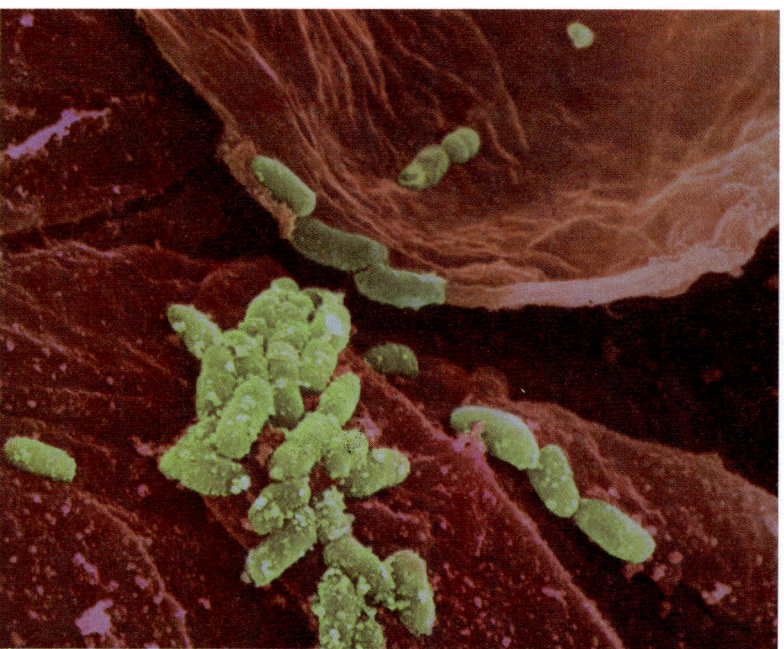

Figure 18–5 *The green objects in this photograph of the surface of the skin are bacteria. This photograph has been magnified 8000 times. The skin is one of the body's first lines of defense against disease-causing organisms. What are the body's other first line defenders?*

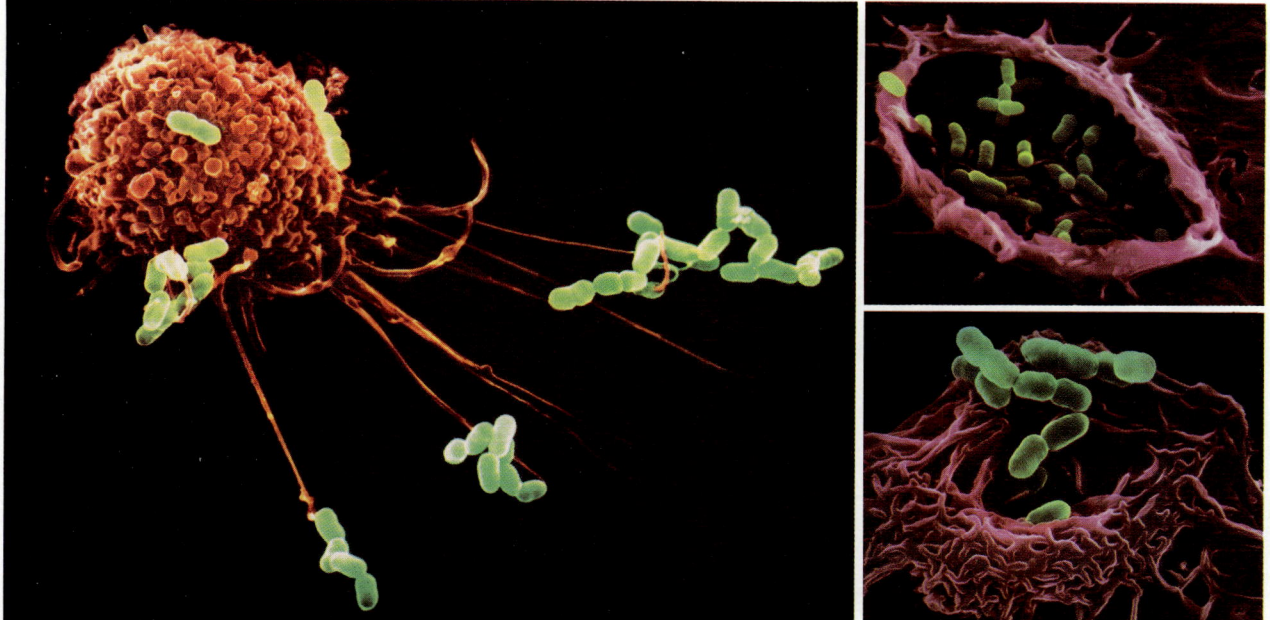

bacteria, dirt, and excess mucus away from your breathing tubes and lungs.

If the first line of defense fails and disease-causing organisms enter the body, the second line of defense—the **immune** (ih-MYOON) **system**—goes into action. The immune system consists of special cells and tissues that help resist invading organisms once they have entered the body. The actions of this system provide you with **immunity** (ih-MYOON-uh-tee). Immunity is the body's ability to fight off disease without becoming sick from the disease.

You can get a good idea of how your immune system works by studying what happens when disease-causing organisms such as bacteria enter a cut in your skin. The bacteria immediately attack nearby cells. But your body is quick to respond. The body's response to an attack on, or damage to, its cells is called an **inflammation** (ihn-fluh-MAY-shuhn). In the process of inflammation, one of the first things to occur is an increased blood supply to the infected area. This can cause the infected area to look red and swollen. If the infecting organism produces a toxin, the body often responds with a fever, or an increase in body temperature. Many scientists now believe that an increase in body temperature helps to stop the growth and reproduction of many disease-causing organisms. A fever, then, is a sign that your body is battling an infection.

Figure 18–6 *The function of white blood cells is to attack and destroy disease-causing organisms. In the photograph on the left, the large red object, which is a white blood cell, extends its "arms" to reach out and grab the tiny green objects, or bacteria. In the top right photograph, the bacteria are trapped within a white blood cell. Finally, in the photograph on the bottom right, you can see that the white blood cell has "gobbled up" the bacteria, one by one. Do these white blood cells form the body's first or second line of defense?*

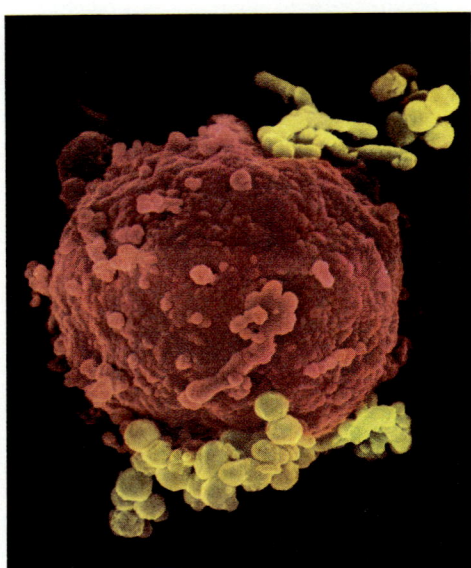

Figure 18-7 *This photograph, magnified 8500 times, shows a B cell covered with bacteria. B cells produce antibodies that attack particular kinds of invaders. What are these invaders called?*

Nonspecific Lines of Defense

Some lines of defense will attack any and all invading organisms. Other lines of defense will attack only a particular kind of invader. Because the skin, mucous membranes, and cilia attack all disease-causing organisms in the same way, they are called nonspecific lines of defense. These lines of defense set up a mechanical barrier against disease-causing organisms.

If disease-causing organisms do enter the body, another nonspecific line of defense, the white blood cells, takes over. These white blood cells are tiny. They constantly patrol the blood, ready to defend the body against disease-causing organisms. The white blood cells attack invading organisms, such as bacteria, by surrounding them. Then, the white blood cells "gobble up" and destroy the invaders. See Figure 18–6 on page 409.

Soon these tiny white blood cells become reinforced by much larger white blood cells. These larger cells are similar to heavy artillery, destroying almost any bacteria they attack. In time, the area around the infection becomes a battlefield. Dead bacteria, destroyed body cells, and dead white blood cells combine with fluids to form a white substance called **pus.** Pus in an infection shows that the body is battling disease-causing organisms.

Another nonspecific line of defense is a substance called **interferon.** Interferon is produced by body cells when they are attacked by viruses. As you read, viruses enter body cells and cause the cells to produce more viruses. Thus, when the body cells burst and release more viruses, interferon is also released. It is believed that this substance "interferes" with the actions of the released viruses by somehow warning nearby cells that a viral infection is occurring.

Specific Lines of Defense

Nonspecific lines of defense such as white blood cells are very important. White blood cells usually destroy the invaders. But sometimes invading organisms put up too strong a battle. Then it is time for more specific lines of defense to take over. Specific lines of defense only can attack a particular

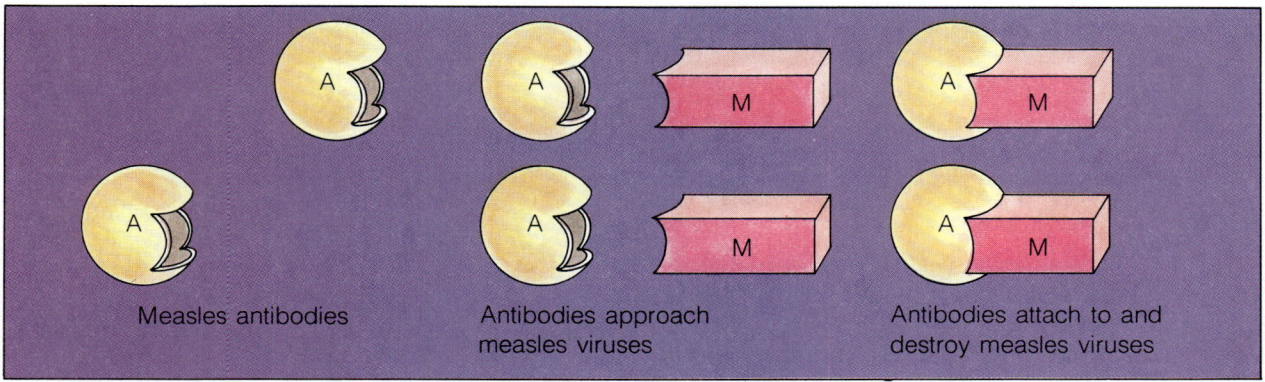

Measles antibodies

Antibodies approach measles viruses

Antibodies attach to and destroy measles viruses

Figure 18–8 *Antibodies are the main specific lines of defense in your body. In this drawing, you can see how an antibody joins to and eventually destroys a specific invading organism. What is an antibody?*

kind of invader. In a way, they are like the body's guided missiles, zeroing in on a particular target.

Substances called **antibodies** are the main specific line of defense in your body. Antibodies are proteins that are produced by certain kinds of white blood cells in response to an invasion by a particular organism. Any invading organism or substance is called an **antigen.** Antibodies are produced to fight specific *anti*body *gen*erators, or antigens.

Let's look at what happens. Once an antigen or organism invades the body, special white blood cells, called T cells and B cells, are alerted. T cells alert B cells to produce antibodies. If the body has never been invaded by this kind of antigen before, the B cells will take awhile to produce antibodies. During that time, the person is sick. Eventually, however, the cells produce antibodies. Antibodies destroy invading antigens by joining together like pieces of a jigsaw puzzle. Once joined to the invader, the antibody is able to destroy it. Each antibody, remember, is specific to a certain invader and will not work against other kinds of antigens.

With some invading antigens, the body has to take time to produce antibodies every time that antigen enters the body. Each time you are infected by that antigen, you become sick. However, with other antigens, the white blood cells "remember" how to produce antibodies. For example, when the body is infected with mumps virus for the first time, it can produce new antibodies almost immediately. So the next time the mumps virus enters the body, it is well prepared. The virus is destroyed quickly by antibodies. You now have immunity from mumps.

411

HELP WANTED: PHYSICAL THERA- PIST ASSISTANT Associate degree in physical therapy required. State license a must. Job involves working with the el- derly in a nursing home.

As people grow older, their chances of devel- oping arthritis increase. The word "arthritis" means inflammation of the joints. And in arthri- tis, the flexible cartilage between the joints wears down, causing pain and stiffness.

If arthritis is suspected, a doctor should be seen. Usually the doctor prescribes some type of medication to reduce the swelling. The doc- tor may suggest that the patient get help from a **physical therapist assistant.** A physical thera- pist assistant, under the supervision of a physi- cal therapist, teaches exercises to maintain or improve the movement of joints. In addition to helping patients with arthritis, physical therapist assistants also aid patients who are victims of accidents, strokes, nerve diseases, and birth defects.

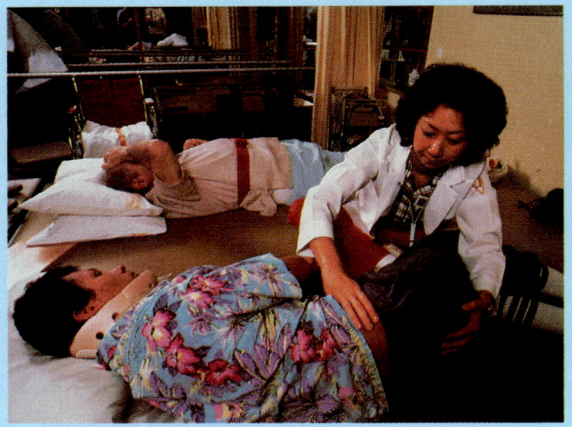

Physical therapy assistants work in hospitals, doctors' offices, clinics, nursing homes, and schools for the physically challenged. High school courses in health, biology, and physical education will help prepare for a career as a physical therapy assistant. Learn more from the American Physical Therapy Association, 1111 North Fairfax Street, Alexandria, VA 22314.

Immunity

People have two types of immunity: natural and acquired. **Natural immunity** is present at birth and protects people from some diseases that infect other types of organisms. For example, natural immunity prevents people from getting tobacco mosaic disease, a disease found in plants.

The other type of immunity is **acquired immunity.** Acquired immunity is an immunity that develops during a person's lifetime. There are two types of acquired immunity: passive and active.

PASSIVE IMMUNITY   In passive immunity, you re- ceive antibodies from another source rather than making antibodies yourself. These antibodies are usually given when you are exposed to a disease. For example, a person exposed to hepatitis, an in- flammation of the liver, would be given an injection of antibodies to help him or her fight the disease. These antibodies are produced by other people or animals that were exposed to a similar disease.

ACTIVE IMMUNITY You acquire active immunity to a disease when your body produces its own antibodies. You can acquire active immunity through vaccines. You will learn more about vaccines in the next section. Active immunity can also occur when you recover from an infection. In this case, you produce antibodies against that specific disease-causing organism.

Sometimes disease-causing organisms, such as the AIDS virus, attack the body's immune system. This action reduces the body's ability to fight other infections. As a result, people who have the AIDS virus easily get diseases such as certain cancers and pneumonias. AIDS can be transmitted through sexual contact with a person who carries the AIDS virus. It can also be transmitted through a blood transfusion that contains the AIDS virus. For this reason, all blood used in transfusions is tested to be sure that it does not contain the AIDS virus. AIDS can also be transmitted by a mother who has the disease to her unborn child. Presently, scientists are working on a vaccine against the AIDS virus. Researchers are confident that more effective treatments will be found within the next few years.

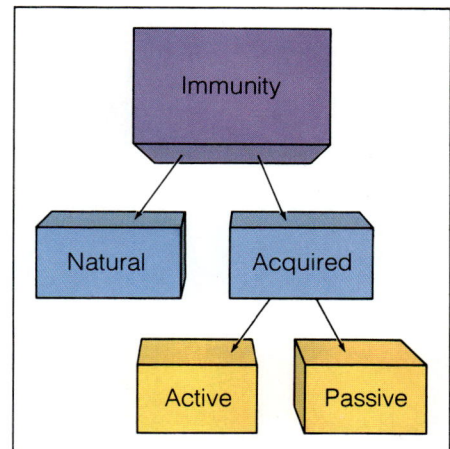

Figure 18–9 *There are two types of immunity: natural and acquired. Which immunity develops during a person's lifetime?*

SECTION REVIEW

1. What are the body's two lines of defense against disease-causing organisms?
2. What is an antibody? An antigen?
3. Compare natural immunity and acquired immunity.
4. Why might you get sick from chicken pox only once?

18–3 Vaccines

Section Objective

To describe how vaccines help control the spread of disease

As you read, infectious diseases are spread by direct contact with an infected organism, through contaminated water or food, by contact with dirty objects, and through infected animals. **To help control the spread of disease, certain substances called vaccines are used to increase the body's immunity to disease-causing organisms.**

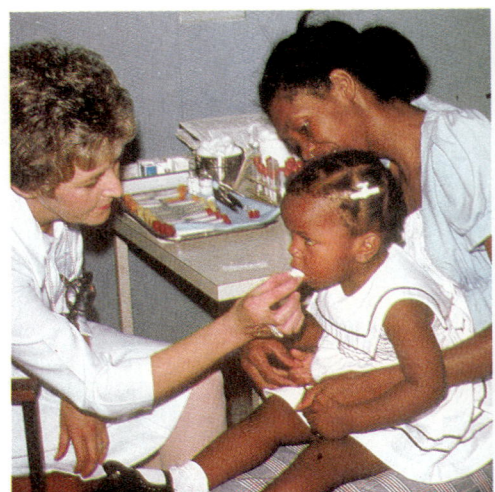

Figure 18–10 *Vaccines such as the poliomyelitis vaccine (left) are made from dead or weakened viruses. Recently scientists have engineered certain bacteria into producing a vaccine for the hepatitis B virus, which is seen on the television screen (right).*

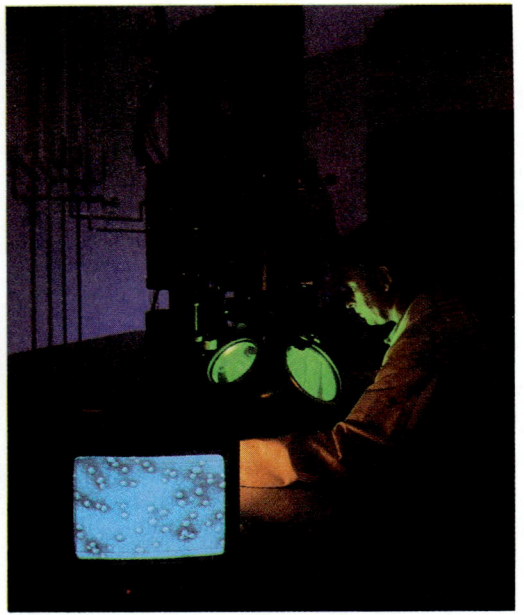

Remember that you can develop immunity from some diseases by catching the disease. In that way, your white blood cells "learn" how to produce antibodies against that disease. Catching a disease is the hard way to develop immunity. Can there be an easier way? Let's find out.

In the past, the virus that causes measles made almost all young children sick. Measles, in fact, came to be known as a childhood disease. Today, very few cases of measles are reported in the United States each year. In the future, measles may well become a forgotten disease. This dramatic change is the result of the development of a measles **vaccine.** A vaccine is a substance that increases an organism's immunity. Most vaccines are injected through the skin of an organism. Some may be taken orally.

A vaccine usually contains dead or weakened antigens. In their weakened state, these antigens usually do not make you sick. But they still cause the body's white blood cells to be alerted and to produce antibodies. Your body produces antibodies against the disease even though it does not really have the disease. Usually, once the antibodies are produced the first time, they can be made quickly any time that the disease-causing antigen invades the body again.

Suppose, for example, a child is given the measles vaccine. The body soon produces antibodies against the measles virus. Then, if the child's body is invaded by a measles virus at some other time, the child does not get sick. The body already has learned how to produce measles antibodies quickly. Of course, because measles antibodies cannot fight other invaders, the measles vaccine only allows a person to develop immunity from measles. It will not help the body against other diseases such as mumps.

Sometimes the body needs to be reminded how to produce antibodies, even after a vaccine. Booster shots help "boost" the production of antibodies in the blood. Some of the most common vaccines given in this country are vaccines against measles, German measles, poliomyelitis, mumps, whooping cough, diphtheria, and tetanus.

In the past, almost all vaccines were made from dead or weakened antigens. Today, however, scientists have developed some artificial vaccines. These vaccines contain specially designed chemicals that resemble antigens. They, too, cause the body to produce antibodies.

Sharpen Your Skills

Antiseptics and Disinfectants

Using reference material in the library, find the following information about antiseptics and disinfectants. How do these substances help control disease? What are three examples of each substance? Arrange this information in a report. Include the work of Joseph Lister and any other scientists who contributed to the discovery of these substances.

SECTION REVIEW

1. How do vaccines help control the spread of disease?
2. After receiving a vaccine, you may begin to develop mild symptoms of the disease. Explain why this might happen.

18–4 Chronic Disorders

Section Objective

To describe cancer and allergies as chronic disorders

Because modern medicine has found more and more ways to combat many infectious diseases, people have begun to live longer. And as people's life spans have increased, so have the number of people who suffer from **chronic disorders.** Chronic disorders are lingering, lasting illnesses. They have overtaken infectious disease as the major health problem in the United States. **Some of the more serious chronic disorders are cancer, allergies, cardiovascular diseases, and diabetes.**

This section will discuss two chronic disorders—cancer and allergies. Cardiovascular diseases, or dis-

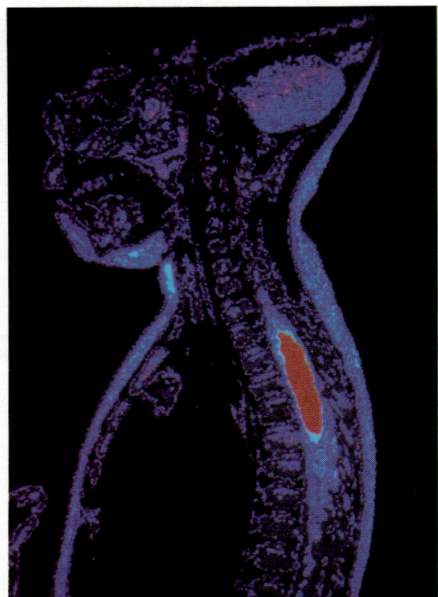

Figure 18–11 *This photograph shows a tumor, shaded red, growing on the spinal cord of a four-year-old girl. What is a tumor?*

eases that affect the heart and blood vessels, are discussed in Chapter 14. And in Chapter 16, you can read about how a lack of insulin causes diabetes.

Cancer

The chronic disorder second only to cardiovascular diseases in causing death is **cancer.** Cancer is the abnormal and uncontrolled reproduction of cells. As these cells multiply and grow, they form **tumors.** A tumor is a swelling of tissue that develops separately from the tissue that surrounds the tumor. Actually, most tumors are **benign,** or harmless. So most tumors are not cancers. Once benign tumors are removed surgically, they usually do not grow back.

Cancers, on the other hand, are *not* benign tumors. They are **malignant,** or life-threatening. The word "cancer" comes from the Latin word for crab because of cancer's crablike growth as it makes its way into healthy tissue. See Figure 18–12.

Cancer cells crowd out normal cells by robbing them of food and space. In time, some kinds of can-

Figure 18–12 *Notice the crablike growth of the large cancer cell in this photograph. The spherical objects are special white blood cells called T cells. They have surrounded the cancer cell and are preparing to attack and destroy it.*

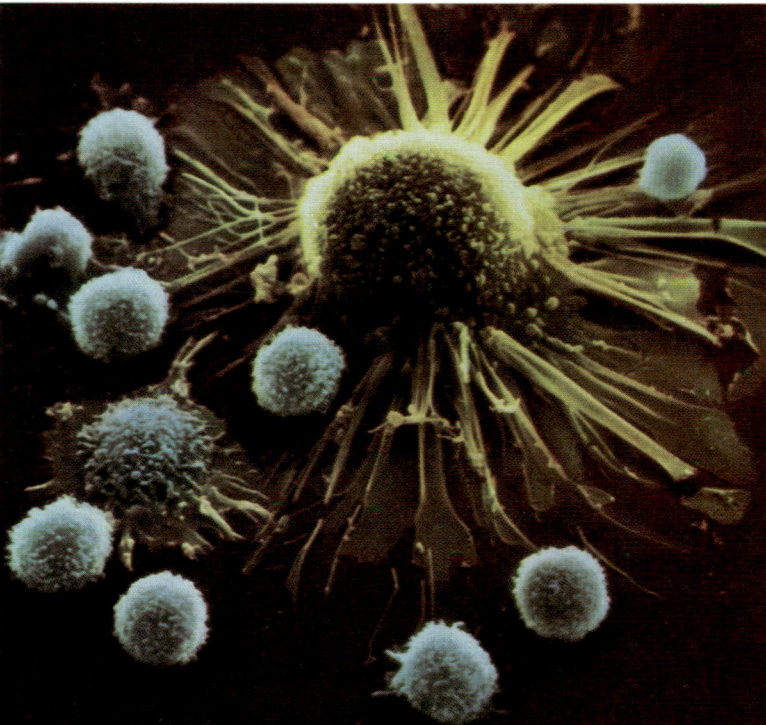

cer cells may break off and travel through the blood to other parts of the body. The cancer, in other words, may spread. How life-threatening a cancer is depends upon its location and stage of development.

In the beginning, cancers can develop without any signs or symptoms. In time, some signs or symptoms become apparent. Unfortunately, many people delay seeing a doctor when they first notice an unusual change in their bodies. Delaying treatment, however, is dangerous. During the delay, a cancer that could have been treated may spread.

CAUSES Because cancers are actually a group of more than 100 diseases, they have no single cause. In fact, the same type of cancer may have various causes in different people. Scientists have identified some materials as **carcinogens** (kahr-SIHN-uh-juhnz), or cancer-causing substances. Because there are a large number of carcinogens, it is difficult for scientists to find one cure for cancer.

In a few cases, scientists have found viruses that can cause cancer. Viruses, remember, are well equipped to take over a cell's reproductive processes. There are a few rare types of cancer in which heredity plays an important part. If a certain type of cancer seems to occur frequently in a family, then the members of that family should see a doctor regularly. In this sense, "heredity" means that some people may inherit the tendency toward a certain type of cancer. It does not mean in any way that such people will get cancer!

TREATMENT The major method of treating cancer in certain areas of the body is surgery. In many cases, surgery prevents the cancer from spreading. However, during the removal of cancerous tissue, some healthy tissue is also removed. Another method of treatment is the use of radiation, which destroys cancer cells. Unfortunately, the radiation kills normal cells as well as cancer cells.

Chemotherapy, or the use of specific chemicals, is also useful against some types of cancer. Like radiation, these chemicals not only destroy cancerous cells, but they also can injure normal cells. These injuries account for undesirable side effects such as nausea and high blood pressure that are felt during this treatment.

Figure 18–13 *Some carcinogens may make their way into people's water supply through leaking metal containers. What is a carcinogen?*

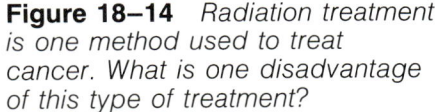

Figure 18–14 *Radiation treatment is one method used to treat cancer. What is one disadvantage of this type of treatment?*

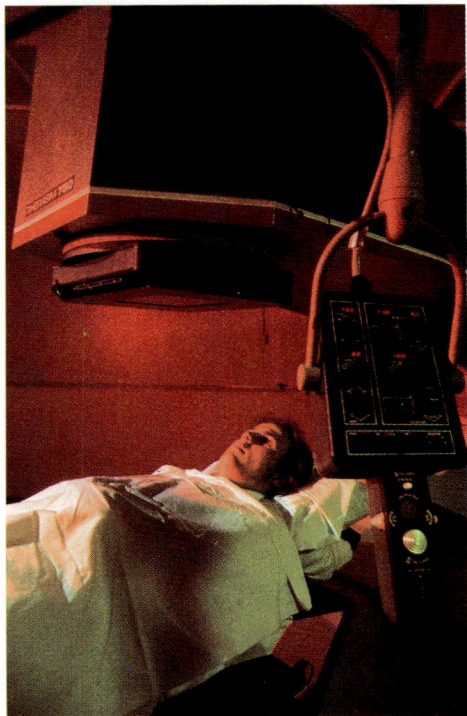

Recently, lasers have been used to treat cancer. Lasers are concentrated beams of light that can zero in on the cancer. Hence the number of normal cells destroyed during treatment is reduced.

Sometimes doctors use more than one of these methods to destroy cancer. The use of anticancer drugs after surgery is especially useful. This is because the drugs circulate to all parts of the body and can attack cancer cells that may have spread to other organs.

Scientists recently have been experimenting with drugs that can strengthen the body's immune system against cancer cells. The immune system is one of the body's systems of defense against disease. For example, scientists are trying to join cancer cells with the antibody-producing white blood cells. The joining of these cells produces **monoclonal** (MAHN-uh-kloh-nuhl) **antibodies.** These antibodies are similar to "guided missiles" because they seek out cancer targets. Scientists already have developed monoclonal antibodies that are effective against some types of flu virus and a type of hepatitis virus. In the future, monoclonal antibodies may become the most important tool in fighting cancers.

Allergies

Are the people in Figure 18–15 visitors from outer space? No. They are hay fever sufferers trying to avoid inhaling ragweed pollen. Actually, "hay fever" is neither a fever nor is it caused by hay. Hay fever is an **allergy.** An allergy occurs when the body is especially sensitive to certain substances called allergens (AL-uhr-juhnz). Allergens may be taken into the body by inhaling dusts, feathers, animal hairs, and pollens; by eating various foods; by coming in contact with certain substances; or by using certain drugs. In fact, you can become allergic to almost anything at anytime.

CAUSE When an allergen, such as pollen, enters the body, antibodies are produced. Unlike the antibodies that help fight infection, the antibodies produced by allergens release histamines (HIHS-tuh-meenz). Histamines are the chemicals that are responsible for the allergic symptoms. These

Figure 18–15 *The bubble helmets, hoses, and filters that these hayfever sufferers are wearing help them avoid inhaling pollen.*

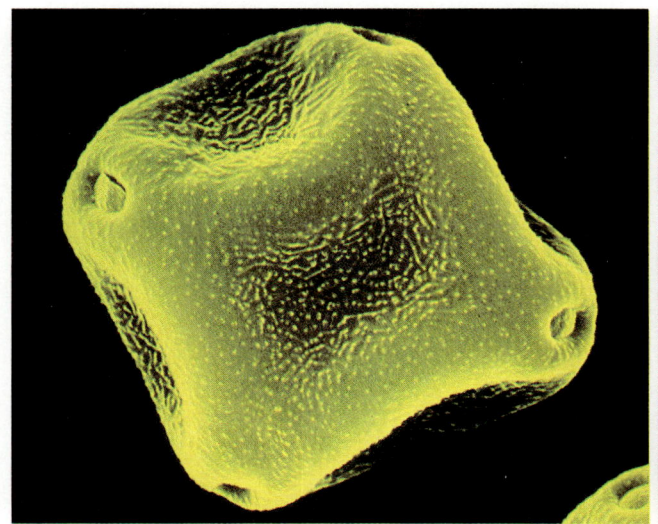
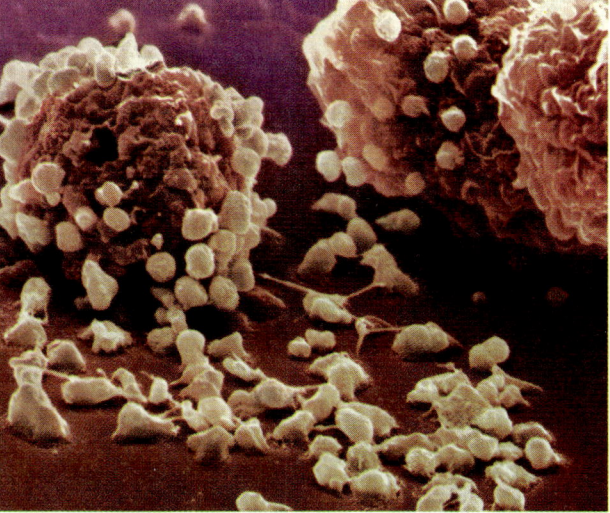

Figure 18–16 *When an allergen, such as pollen (left), enters the body, antibodies are produced by certain cells. Soon these cells explode and release tiny structures that contain histamines (right). What are histamines?*

symptoms include itchy, watery eyes, runny nose, and tickly throat.

Some allergies, such as asthma (AZ-muh) and hay fever, tend to run in families. For example, if both parents have an allergy, each of their children has a 75 percent chance of developing one. If one parent has an allergy, each child has a 50 percent chance of developing one.

TREATMENT There is no complete cure for an allergy. You may be able to avoid the symptoms of an allergy by avoiding the allergen that causes it. This may mean that you have to remove some foods that contain the allergen from your diet or you may have to find a new home for your pet. If you find it difficult to avoid the cause of the allergy, you may find some relief by obtaining medication from your doctor. Medication may include antihistamines. *Anti*histamines work *against* the effects of histamines.

SECTION REVIEW

1. What is cancer?
2. Define carcinogen.
3. Define allergy. List some allergens.
4. Why are chronic disorders more of a problem today than 200 years ago? What do you think will happen in the future?

Observing the Action of Alcohol on Microorganisms

Problem

What is the effect of alcohol on the growth of organisms?

Materials *(per group)*

glass-marking pencil
2 petri dishes with sterile nutrient agar
2 paper clips
2 thumbtacks
2 pennies
tape
graduated cylinder
alcohol
100-mL beaker
clean forceps

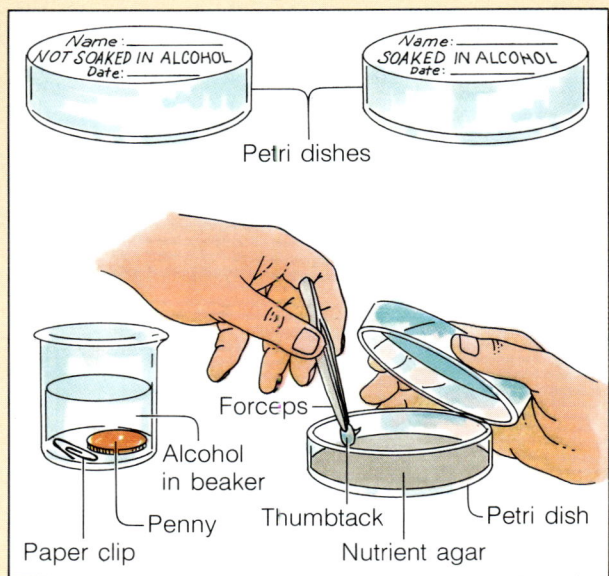

Petri dishes

Forceps

Alcohol in beaker

Penny — Thumbtack — Petri dish

Paper clip — Nutrient agar

Procedure

1. Obtain two petri dishes containing nutrient sterile agar.
2. Using a glass-marking pencil, label the lid of one dish "soaked in alcohol." Label the lid of the second dish "not soaked in alcohol." Write your name and the date on each lid. **Note:** *Be sure to keep the dishes closed while labeling them.*
3. Using a graduated cylinder, carefully pour 50 mL of alcohol into a beaker.
4. Place a paper clip, a thumbtack, and a penny into the alcohol in the beaker. Cover the beaker. Let these objects remain in the alcohol for 10 minutes.
5. Using clean forceps, place the other paper clip, thumbtack, and penny into the dish marked "not soaked in alcohol." Cover the dish and tape it closed.
6. Using clean forceps, remove the paper clip, thumbtack, and penny from the alcohol in the beaker. Place these objects into the dish marked "soaked in alcohol." Cover the dish and tape it closed.
7. Place the dishes in an undisturbed place for one week.

8. After one week, examine the two dishes. Make a sketch of what you observe.
9. Follow your teacher's instructions for the proper disposal of all materials.

Observations

1. After one week, what did you observe in the dish marked "not soaked in alcohol"?
2. After one week, what did you observe in the dish marked "soaked in alcohol"?
3. Compare the growth of organisms in these two dishes.

Conclusions

1. What effect did the alcohol have on the growth of organisms in the dish of objects soaked in alcohol?
2. Why did you use forceps, rather than your fingers, to place the objects in the dishes?
3. Why did you have to close the petri dishes immediately after adding the objects?
4. Why do doctors soak their instruments in alcohol?

CHAPTER REVIEW

SUMMARY

18–1 Infectious Disease

❏ The human body is the host for many microorganisms. A host is an organism in which another organism lives and gets nourishment and protection.

❏ When disease-causing microorganisms invade the body, an infection results. Diseases caused by viruses and bacteria are called infectious diseases.

❏ Diseases that can be spread from one infected organism to another are called communicable diseases.

❏ Other diseases can be spread through food or water, direct contact, contaminated objects, or through the air.

18–2 Body Defenses

❏ The skin, mucous membranes, cilia, and most white blood cells are the body's nonspecific lines of defense.

❏ The immune system is the body's specific line of defense against invading organisms.

❏ The ability to fight off disease is known as immunity.

❏ The body's response to an attack on its cells is called an inflammation.

❏ Within the blood, specific defenses called antibodies are produced. Antibodies are substances that are produced when there is an antigen, or invader, in the body.

❏ There are two types of immunity: natural and acquired. Natural immunity is present at birth. Acquired immunity develops during a person's lifetime.

18–3 Vaccines

❏ Vaccines are substances that increase an organism's immunity.

❏ Vaccines, which usually contain dead or weakened antigens, cause the white blood cells to produce antibodies.

18–4 Chronic Disorders

❏ Chronic disorders are lingering, lasting illnesses.

❏ Cancer is the abnormal and uncontrolled reproduction of cells.

❏ As cancer cells grow, they form tumors. Benign tumors are often harmless, while malignant tumors are cancerous.

❏ Some methods of treating cancer are surgery, radiation, and anticancer drugs.

❏ An allergy occurs when the body is especially sensitive to certain substances called allergens.

VOCABULARY

Define each term in a complete sentence.

acquired immunity	chronic disorder	infectious disease	pus
allergy	communicable disease	inflammation	symptom
antibiotic	host	interferon	toxin
antibody	immune system	malignant	tumor
antigen	immunity	monoclonal antibody	vaccine
benign	infection	natural immunity	
cancer			
carcinogen			

CONTENT REVIEW: MULTIPLE CHOICE

On a separate sheet of paper, write the letter of the answer that best completes each statement.

1. Which of the following would cause an infectious disease?
   a. bacteria   b. allergy
   c. antibiotics   d. antibodies
2. The itchy, watery eyes caused by a cold are an example of a (an)
   a. symptom.   b. antigen.
   c. antibody.   d. infection.
3. A chemical substance that is produced by some microorganisms and stops the growth of bacteria within the body is called a (an)
   a. allergy.   b. antigen.
   c. antibody.   d. antibiotic.
4. The poisons produced by bacteria are called
   a. antibodies.   b. antigens.   c. toxins.   d. antibiotics.
5. Which disease is caused by a virus?
   a. rabies   b. strep throat   c. tetanus   d. tuberculosis
6. Which is an example of a specific line of defense?
   a. skin   b. mucous membranes   c. cilia   d. antibodies
7. What substance increases an organism's immunity?
   a. antibiotics   b. antibodies   c. vaccines   d. antigens
8. Which of the following is *not* a chronic disorder?
   a. cancer   b. measles
   c. diabetes   d. allergy
9. Tumors that are not cancerous are said to be
   a. benign.   b. infectious.   c. malignant.   d. communicable.
10. The chronic disorder that occurs when the body is especially sensitive to certain substances is
    a. diabetes.   b. cancer.   c. influenza.   d. an allergy.

CONTENT REVIEW: COMPLETION

On a separate sheet of paper, write the word or words that best complete each statement.

1. The _____ is the organism in which another organism lives and gets its nourishment.
2. Infectious diseases are also called _____ diseases.
3. Measles is caused by a (an) _____.
4. The body's first line of defense includes the mucous membranes, cilia, and _____.
5. The body's response to an attack on, or damage to, its cells is called a (an) _____.
6. Antigens cause the body to produce _____.
7. Certain substances called _____ are used to increase the body's immunity to disease-causing organisms.
8. The second leading cause of death in the United States is _____.
9. _____ immunity is present at birth.
10. Hay fever is an example of a (an) _____.

CONTENT REVIEW: TRUE OR FALSE

Determine whether each statement is true or false. Then on a separate sheet of paper, write "true" if it is true. If it is false, change the underlined word or words to make the statement true.

1. Colds are caused by <u>bacteria</u>.
2. T cells are a <u>specific</u> line of defense.
3. Antibodies are produced to fight <u>antibiotics</u>.
4. A <u>vaccine</u> is a substance that increases an organism's immunity.
5. <u>Natural</u> immunity develops during a person's lifetime.
6. The leading cause of death in the United States is <u>cardiovascular disease</u>.
7. <u>Antibiotics</u> are cancer-causing substances.
8. Cancerous tumors that are life-threatening are called <u>malignant</u> tumors.
9. Pollen is an example of an <u>antibiotic</u>.
10. Asthma is an example of an <u>allergy</u>.

CONCEPT REVIEW: SKILL BUILDING

Use the skills you have developed in the chapter to complete each activity.

1. **Relating concepts** Why do you get mumps only once?
2. **Relating concepts** Why should you clean and bandage all cuts?
3. **Applying concepts** Explain why you should not go to school with the flu.
4. **Classifying types of immunity.** Classify each as natural immunity, passive acquired immunity, or active acquired immunity.
   a. a baby receives a DPT (diphtheria; pertussis, or whooping cough; and tetanus) shot
   b. a five-year-old child gets the measles
   c. a person comes in contact with Dutch elm disease
5. **Applying concepts** Explain why it is important for you to know what vaccines you have been given.
6. **Making predictions** Suppose all your cilia were destroyed. How would this affect your body?
7. **Relating facts** Are colds caught by sitting in a draft? Explain.
8. **Applying concepts** Suppose a person was born without a working immune system. What are some of the precautions that would have to be taken so that the person could survive?

CONCEPT REVIEW: ESSAY

Discuss each of the following in a brief paragraph.

1. Describe what happens during an inflammation.
2. Explain how antibodies are produced in a person's body.
3. Compare passive acquired immunity and active acquired immunity.
4. What is a benign tumor? A malignant tumor?
5. What are some of the methods of treating cancer?
6. What are monoclonal antibodies? How are they produced?
7. What is a communicable disease? List five communicable diseases and tell what kind of microorganism causes each disease. Explain how each of these diseases can be prevented or treated.
8. Describe what happens in the body of a person who has an allergy to dust.
9. Compare a histamine and an antihistamine.

The typical alcoholic American

young

old

male

female

black

white

rich

poor

employed

unemployed

executive

laborer

student

doctor

immigrant

native born

**There's no such thing as typical.
We have all kinds. Ten million alcoholic Americans.
It's our number one drug problem.**

NIAAA

NATIONAL
INSTITUTE
ON ALCOHOL
ABUSE AND
ALCOHOLISM

FOR INFORMATION OR FOR HELP, WRITE: NATIONAL CLEARINGHOUSE FOR ALCOHOL INFORMATION, BOX 2345, ROCKVILLE, MARYLAND 20852

U.S DEPARTMENT OF HEALTH AND HUMAN SERVICES · PUBLIC HEALTH SERVICE · ALCOHOL, DRUG ABUSE, AND MENTAL HEALTH ADMINISTRATION

Drugs, Alcohol, and Tobacco

19

CHAPTER OBJECTIVES

After completing this chapter, you will be able to

19–1 Define drug.

19–1 Discuss the dangers of drug abuse.

19–2 Describe the effects of alcohol abuse.

19–3 Relate cigarette smoking to certain diseases.

19–4 Describe the effects of smoking marijuana.

19–4 Classify depressants, stimulants, hallucinogens, and opiates.

"My name is Lisa, and I'm an alcoholic." The words slipped out as if someone else were saying them. Then Lisa, a high school student, took another deep breath and began to tell her story. The people at Alcoholics Anonymous listened quietly. Everyone in the room had been through what Lisa was feeling. They all knew how difficult it was to stand up and admit to the problem. And they all knew the excuses people use to make it seem less dangerous.

Lisa admitted that she had known all along about the dangers of drinking. In the eighth grade, for example, when she was still getting good grades, she learned that abusing alcohol could damage her liver and heart. But that did not stop her from trying a "drink or two." Soon trying became *needing* a drink.

Two years later, Lisa's parents found a bottle hidden in her room. But she managed to talk them out of any punishment. They, too, wanted to believe her drinking was not a problem. Then, one day Lisa was in a serious car accident. She had been drinking. The judge took away her license and warned her that next time she would be sentenced to jail. But that did not scare her. What did, however, was the fact that she had come very close to killing a young baby and his mother. At that moment, Lisa felt ready to face the fact that she had a serious problem.

When Lisa finished speaking, the people at Alcoholics Anonymous did not judge her or accuse her of being foolish. They knew that alcohol abuse is a disease. But they also knew that Lisa's road to recovery from this disease would be difficult—and would last her entire life. With help, it was a life Lisa could save.

A poster showing the types of people who abuse drugs

19–1 Drug Abuse

Drugs! You probably hear the word a lot. On television, for example, one commercial after another seems to be advertising medicines such as aspirin. "Take aspirin for headaches . . . , take aspirin to relieve muscle aches," the announcers shout. Aspirin is but one of the many **over-the-counter drugs** available in this country. Over-the-counter drugs can be bought in drugstores and many supermarkets.

In the newspapers, you may read about a new product that helps people who have heart disease. Or you may read about a medicine for people with diabetes mellitus. Such drugs cannot be bought over the counter. They require a doctor's prescription and are called **prescription drugs.**

Perhaps when you hear the word *drug,* you think of the problems caused by drugs whose use is normally illegal. Heroin and cocaine are examples of such drugs. Or you may think of the drug alcohol. The use of alcohol may also be illegal, depending on your age and the community in which you live.

The term *drug* means different things to different people. To a scientist, the word *drug* has a very broad meaning. A **drug** is any substance other than

Figure 19–1 *This clay tablet contains the world's oldest known prescriptions dating back to about 2000 B.C. The prescriptions show the use of plants in medicine.*

food that has an effect on the body. Within this definition are many substances you do not normally consider drugs. For example, coffee and tea contain a substance called caffeine. It stimulates the activity of the heart and brain. Clearly, then, caffeine is considered a drug.

Drug Abuse

Like the word *drug*, the term **drug abuse** means different things to different people. A person who injects the illegal drug heroin is obviously a victim of drug abuse. But what about the person who drinks ten cups of coffee each morning? Is that person also abusing drugs?

A drug is abused if too much of the drug is used. A drug is abused if it is used in a way that most doctors would not approve. Under this definition, drinking 20 cups of coffee each morning might well be drug abuse. In fact, that much caffeine can cause a serious reaction in some people, including extreme nervousness and depression.

In this chapter, you will read about some of the drugs that are commonly abused and the problems that may result. But it is important to remember the definition of drug abuse and the fact that any drug can be abused.

The Dangers of Drug Abuse

Drug abuse is dangerous for a number of reasons. Figure 19–4 on page 428 lists some commonly abused drugs. You will examine these drugs in more detail later. But before you do, take a look at the last four columns in the table. They detail some of the most serious side effects of abusing drugs. So, let's look into what these columns mean.

DEPENDENCE People who abuse drugs do so for a variety of reasons. Some seek to "escape" life's problems, others to "increase" life's pleasures. Some take drugs because their friends do, others because their friends do not. Some people abuse drugs to feel grown up, others to feel young again. In short, the list is as endless as the range of human emotions. For many people, the pleasure or relief they feel

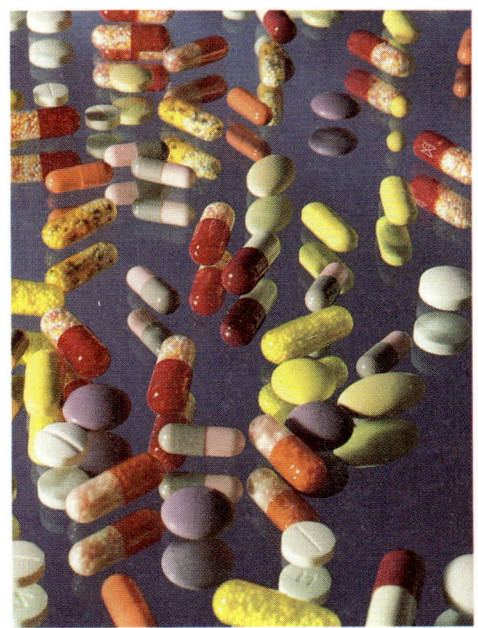

Figure 19–2 You can see in this photograph that drugs come in many sizes and shapes. They also have many different effects on the body. Some speed up body processes, while others slow these processes down. What type of drugs can be bought without a prescription in drugstores?

Figure 19–3 Notice the expiration date on this container of pills. Using medication after the prescription has expired is an example of drug abuse. What is a prescription drug?

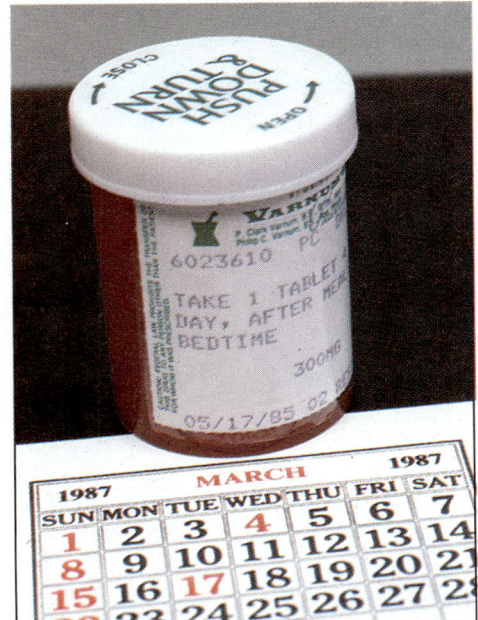

COMMONLY ABUSED DRUGS

Type of Drug	Examples	Basic Action on Central Nervous System	Psychological Dependence	Physical Dependence	Withdrawal Symptoms	Development of Tolerance
Opiates and related drugs	Heroin Demerol Methadone	Depressant	Yes, strong	Yes, very fast development	Severe, but rarely life-threatening	Yes
Barbiturates	Phenobarbital Nembutal Seconal	Depressant	Yes	Yes	Severe, life-threatening	Yes
Tranquilizers (Minor)	Valium Miltown Librium	Depressant	Yes	Yes	Yes	Yes
Alcohol	Beer Wines Liquors Whiskey	Depressant	Yes	Yes	Severe, life-threatening	Yes, more in some people than in others
Cocaine	As a powder Crack	Stimulant	Yes, strong	Yes	Yes	Possible
Cannabis	Marijuana Hashish	Ordinarily a depressant	Yes, moderate	Probably not	Probably none	Possible
Amphetamines	Benzedrine Dexedrine Methedrine	Stimulant	Yes	Possible	Possible	Yes, strong
Hallucinogens	LSD Mescaline Psilocybin	Stimulant	Yes	No	None	Yes, fast
Nicotine	In tobacco of cigarettes, cigars; also in pipe tobacco, chewing tobacco	Stimulant or depressant	Yes, strong	Yes	Yes	Yes
Caffeine	Coffee, tea, some "pep" pills, many cola drinks	Stimulant	Yes, mild	Possible	Possible but very few	Yes, but very little

Figure 19–4 *According to this chart of commonly abused drugs, what is the action of barbiturates on the central nervous system?*

when they take a drug comes to be linked to that drug. In time, they can no longer feel good without the drug. Such people have developed a **psychological dependence.** Psychological dependence is an emotional need for a drug. Psychological dependence is strongest with nicotine, cocaine, and opiates, such as heroin.

Any abused drug will lead to psychological dependence. However, some abused drugs can cause **physical dependence** as well. Physical dependence is when the body cannot function properly without the drug. In this case, drug abuse has become a serious physical disease. And because physical dependence is almost always combined with psychological dependence, it is a disease most people cannot cope with unless they receive medical help.

WITHDRAWAL   A person who is dependent on heroin needs to take a dose of the drug at least three to four times a day. Miss a dose and the body begins to react. The nose runs, the eyes tear. Miss several doses and the body reacts more violently. A cold sweat forms. Vomiting may occur. Miss a few more doses and the heroin abuser begins heroin **withdrawal.** Withdrawal occurs when a person who is physically dependent on a drug is suddenly taken off that drug. With heroin withdrawal, some of the more obvious reactions include uncontrolled vomiting, fevers followed by chills, and muscle aches that cannot be relieved. In time the muscles begin to jerk wildly, kicking out of control.

It takes a few days, but the most painful effects of heroin withdrawal pass. However, heroin abusers are never entirely free of the need for the drug. In fact, far too many heroin abusers who have "kicked the habit" return to the drug again unless they receive medical, psychological, and social help.

TOLERANCE   People who abuse some drugs soon discover that they must take more and more of the drug each time they use it to get the same effect. Such people have developed a **tolerance** to the drug. In time, tolerance can become so high that the daily intake of an abuser would easily kill another person.

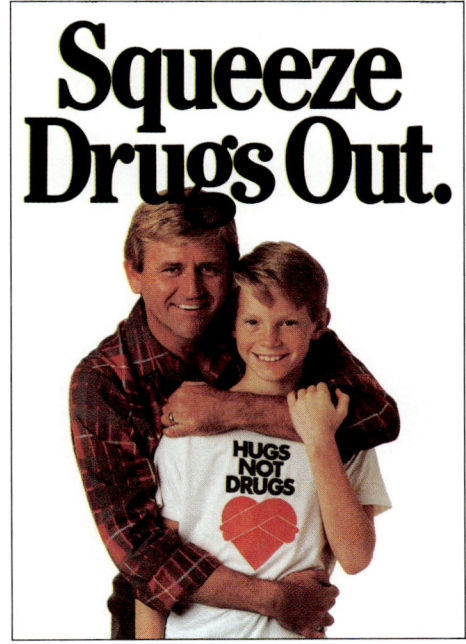

Figure 19–5   *Posters, such as this, alert people to the dangers of drug abuse. When is a drug abused?*

Figure 19–6   *In some cases of drug abuse, counselling is helpful. These people are attending a group session where they discuss some of their problems.*

SECTION REVIEW

1. Define drug abuse.
2. Compare psychological dependence to physical dependence.
3. Why do people develop a tolerance to a drug?
4. Why do heroin abusers often need medical help while they are going through withdrawal?

19–2 Alcohol

Alcohol is a drug that is used widely throughout the world. For most people, the moderate use of alcohol does not result in health problems. But drinking large amounts of alcohol on a daily basis can be a serious health problem. Such alcohol abuse is a disease called **alcoholism.** Alcoholism can damage the body in many ways. **Excess alcohol destroys liver and brain cells and causes both psychological and physical dependence.**

Because alcohol is not a food, it does not have to be digested. It leaves the digestive system and enters the blood almost immediately after it is drunk. Alcohol in the blood passes through the liver. There the alcohol is changed into carbon dioxide and water. However, the liver can process only a few drops of alcohol at a time. The remaining alcohol stays in the blood until it reaches the brain.

In the brain, alcohol acts as a **depressant.** A depressant slows down the actions of the nervous system. But, you may say, people who drink do not seem depressed at all. They seem quite energetic. For a while, that may be true. But it is a false sense of energy. Alcohol first acts on the parts of the brain that control certain human emotions and limits certain actions. When these areas are dulled, people do things that they might not ordinarily do. At this time they may seem quite stimulated. But if

ALCOHOL'S EFFECTS ON THE BRAIN

BAC	Part of Brain Affected	Behavior
.05%		Lack of judgment, lack of inhibition
.1%		Reduced reaction time, difficulty walking and driving
.2%		Saddened, weeping, abnormal behavior
.3%		Double vision, inadequate hearing
.45%		Unconscious
.65%		Death

Figure 19–7 *Blood alcohol concentration, or BAC, is a measure of the amount of alcohol in the bloodstream per 100 mL of blood. The BAC is given as a percent. The higher the BAC, the more powerful is the effect of the alcohol on the brain. What type of behavior occurs if the BAC is 0.3%?*

Figure 19–8 *Alcohol can interfere with reaction time when driving and thereby cause serious car accidents. What effect does alcohol have on the central nervous system?*

they continue to drink, the alcohol in their blood begins to affect other brain areas. See Figure 19–7.

If alcohol abuse goes on for a long time, a condition called **cirrhosis** (suh-ROH-sihs) may develop. In cirrhosis, healthy liver cells are replaced by scar tissue and the liver slowly stops functioning. In addition, alcohol abuse over many years destroys brain cells. Mental disturbances such as blackouts and hallucinations may result. Alcohol also damages the linings of the stomach and small intestine.

SECTION REVIEW

1. Describe the effects of alcohol abuse on the body.
2. What is cirrhosis?
3. Why is it dangerous for a person to drive after they have been drinking?

19–3 Tobacco

"Cigarette smoking is hazardous to your health." You have probably heard that warning hundreds of times. **Cigarette smoking affects both the respiratory and the circulatory systems.** But what makes tobacco smoking so hazardous?

When a cigarette burns, over 4000 substances are produced, many of which are harmful. Cigarette smoke, for example, contains the poisonous gas, carbon monoxide. Carbon monoxide may be picked up

HELP WANTED: EMERGENCY MEDICAL TECHNICIAN Registered Emergency Medical Technician—Ambulance. To maintain and run a well-equipped ambulance. Will be on call for certain time periods. About 40 hours per week. Apply in person to City Police Department.

A call reporting a traffic accident caused by a drunk driver comes into the dispatcher. A team of four people jumps into an ambulance and heads out. The team consists of a driver, a "navigator," who directs and guides the driver to the emergency quickly and safely, and two **emergency medical technicians,** or EMTs. Emergency medical technicians are specially trained to provide on-the-scene care and treatment to emergency victims.

To become an emergency medical technician, you must be 18 years old and have a high school diploma. You must also complete a program that includes 81 hours of EMT instruction, followed by an examination and six months of emergency ambulance or rescue service. EMT's are often specially trained to recognize the symptoms of drug abuse and overdose. In high school, you should take health and science courses as well as driver education.

EMTs should have good judgment and eyesight. They should be physically coordinated and able to lift up to 50 kilograms. Many go on to become paramedics, nurses, or doctors. To learn more about this exciting career, write to the National Association of Emergency Medical Technicians, P.O. Box 334, Newton Highlands, MA 02161.

Figure 19–9 *This anti-smoking poster shows that smoking is not only a hazard to your health but it is unattractive.*

by red blood cells. But these cells cannot easily release the carbon monoxide. So they are then unable to pick up and transport oxygen to body cells.

Another group of dangerous substances in cigarette smoke is tars. Tars coat the lining of the lungs and can lead to certain cancers. Nicotine is another substance in tobacco. Look at Figure 19–4 on page 428. How does nicotine affect the body?

As smoke travels to the lungs, the tars and poisons irritate and damage the entire respiratory system. This damage is particularly serious in the developing lungs of teenagers. Long-term smoking can lead to lung diseases such as bronchitis (brahng-KIGHT-ihs) and emphysema (ehm-fuh-SEE-muh). These diseases are often life threatening. In addition, a heavy smoker has a 20 times greater chance of getting lung cancer than a nonsmoker has.

Aside from lung irritation, smoking increases heartbeat, lowers skin temperature, and causes blood vessels to constrict, or narrow. Constriction of the blood vessels increases blood pressure and makes the heart work harder. Heavy smokers are twice as likely to develop some forms of heart disease than are nonsmokers.

In 1982, the U.S. Surgeon General's report presented a long list of cancers associated with smoking. Besides lung cancer, the report named cancer of the bladder, mouth, esophagus, pancreas, and larynx. In fact, at least a third of all cancer deaths may be caused by smoking.

Smoking is especially dangerous for pregnant women. When they smoke, they may be damaging the health of their unborn children.

Increasing information about cigarette smoke has begun to worry nonsmokers also. Studies have shown that the children of smokers are twice as likely to develop respiratory problems as the children of nonsmokers. Worry about the possible dangers to nonsmokers of smoke-filled air has led to laws that forbid smoking in elevators. Some laws also separate smokers from nonsmokers on trains and planes, in restaurants, and in offices.

Tobacco is dangerous even when it is not smoked. Chewing tobacco, for example, can lead to cancer of the mouth.

Figure 19–10 *In addition to these harmful substances, there are over 4000 substances that are given off as a cigarette burns. Which of these substances coat the lining of the lungs and can lead to certain cancers?*

SECTION REVIEW

1. Which of the body systems are most greatly affected by tobacco?
2. Why is tobacco considered a drug?
3. Almost 30 percent of the people in the United States smoke. Explain why people smoke even when they know the dangers.

19–4 Commonly Abused Drugs

Section Objective

To classify some of the commonly abused drugs

Alcohol and tobacco are two drugs that are often abused by people. But there are many others. **Marijuana, depressants, stimulants, hallucinogens, and opiates are the most commonly abused drugs.**

Marijuana

Marijuana is a drug that is usually taken into the body through smoking. It is found in the leaf of the marijuana plant, *Cannabis sativa*. The effects of marijuana are due mainly to a chemical in the plant called THC.

The THC in marijuana affects people differently. The effect is often hard to describe. Some users report a sense of well-being, a feeling of being able to think clearly. Others get suspicious of people and cannot keep their thoughts from racing. For many, marijuana distorts the sense of time.

Research findings point to a variety of possible health problems caused by marijuana. It slows reaction time and, like alcohol, is the direct cause of many highway accidents. Marijuana also seems to have an effect on short-term memory. Heavy users often have trouble concentrating. And there can be no doubt that marijuana irritates the lungs and leads to some respiratory damage.

Depressants

Earlier you read that alcohol is a depressant. Depressants, you may recall, slow down or decrease the actions of the nervous system. Two other commonly abused depressants are the powerful barbiturates (bahr-BIHCH-uhr-ihts) and the weaker tranquilizers (TRANG-kwuh-ligh-zuhrs). Usually these drugs are taken in pill form. In the past, doctors prescribed these depressants for people who had sleeping problems or who suffered from nervousness. That is because these depressants calm the body and can cause sleep. However, these depressants, partic-

Figure 19–11 *Under the influence of certain drugs, a cross spider (left) will not be able to complete its web (right).*

ularly the barbiturates, cause both physical and psychological dependence. Withdrawal from barbiturate abuse is especially severe and without medical care can result in death. Also, because tolerance to depressants builds up quickly, a person needs to take more and more pills. This can lead to an overdose, or too much of the drug at one time. An overdose of depressants can lead to death, particularly when pills are taken with alcohol.

Stimulants

Drugs that are classified as **stimulants** increase the activities of the nervous system. Caffeine, the drug in coffee, is a stimulant. However, the caffeine in coffee is a mild stimulant. Nicotine, a drug in tobacco, is also a stimulant. Far more powerful stimulants make up a class of drugs called the amphetamines (am-FEHT-uh-meenz).

Today, except in certain cases, the legal use of amphetamines is limited. Yet many people abuse amphetamines illegally. They seek the extra pep an amphetamine pill may bring. Long-term amphetamine abuse can lead to serious mental and physical problems. Perhaps no side effect of amphetamine abuse is more dramatic than the feelings of dread and suspicion that go hand in hand with amphetamine abuse. People who use amphetamines come to believe everyone is against them. They regard everyone as a danger. In turn, they may become a threat to people around them.

A stimulant that has been increasingly abused in recent years is cocaine. Cocaine comes from the leaves of coca plants that grow in South America. Cocaine may be injected by needle, but is usually inhaled as a powder through the nose. In time, the linings of the nose become irritated from the powder. If abuse continues, it is not uncommon for a hole to be eaten through the walls of the nose. Psychological dependence from cocaine use is very powerful. For that reason, it is difficult to quit. But quitting is vital. For long-term cocaine abuse can lead to the same problems as amphetamine abuse.

Around 1981, an extremely dangerous form of cocaine commonly known as "crack" began to be used. Crack is smoked rather than injected or inhaled. It is absorbed rapidly through the lungs and

Figure 19–12 *When the taking of drugs, such as barbiturates, is combined with alcohol, the results are often deadly. Why is this so?*

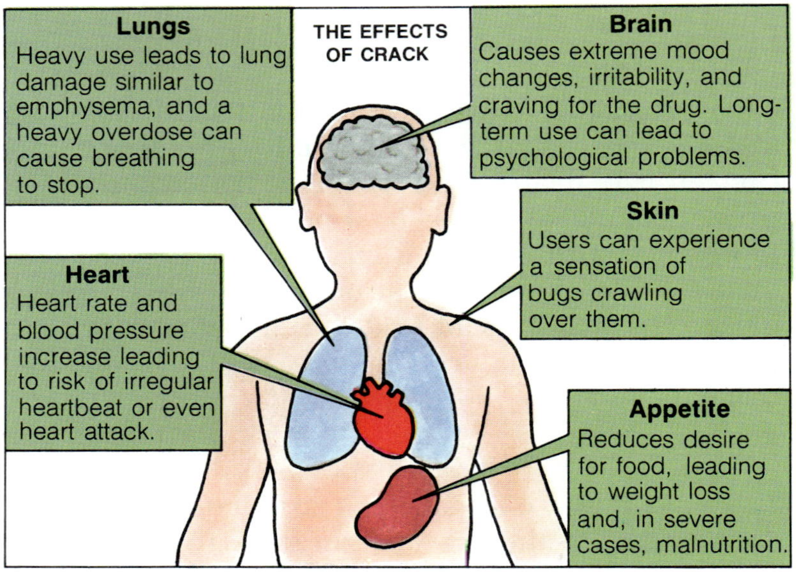
THE EFFECTS OF CRACK
Lungs
Heavy use leads to lung damage similar to emphysema, and a heavy overdose can cause breathing to stop.
Brain
Causes extreme mood changes, irritability, and craving for the drug. Long-term use can lead to psychological problems.
Skin
Users can experience a sensation of bugs crawling over them.
Heart
Heart rate and blood pressure increase leading to risk of irregular heartbeat or even heart attack.
Appetite
Reduces desire for food, leading to weight loss and, in severe cases, malnutrition.

Figure 19–13 *Crack is an extremely dangerous form of cocaine. What effect does crack have on the heart?*

reaches the brain in just seconds. The effects of crack on the body are much more intense than those of regular cocaine. And, dependence on crack develops even more quickly than dependence on regular cocaine. Figure 19–13 shows some of the effects that crack has on the body.

Hallucinogens

Hallucinogens (huh-LOO-suh-nuh-jehns), such as LSD or mescaline (MEHS-kuh-lihn), are a group of drugs that produce powerful hallucinations. Hallucinations occur when a person sees or hears things that are not really there, or when a person sees or hears things that are there in a distorted way. Many hallucinations are extremely unpleasant. Some people have nightmarish "trips" in which all sense of reality is lost. Fears are heightened and a feeling of dread overcomes the abuser. At this time, accidental deaths and even suicide are not uncommon. A small percentage of abusers do not recover from an LSD experience for months and have to be hospitalized.

The most deadly hallucinogen is PCP, or "angel dust." Sometimes marijuana is dusted with PCP to heighten the effects of the marijuana. Many people who buy marijuana smoke PCP without knowing it. This is unfortunate because PCP is much more harmful than marijuana. Smoking "angel dust" sometimes makes the user feel calm and detached. But it can also distort vision and cause panic.

Sharpen Your Skills

Generic Drugs

Today there is a great deal of talk about generic drugs. Using reference books and materials in the library, find out what the term *generic drug* means. What are the advantages of generic drugs? What are the possible disadvantages?

Figure 19–14 *A close-up of a ripe poppy pod (left) shows some of the opium juice oozing out. The round yellowish structure in the middle of this poppy flower (right) is the pod. Which group of drugs is produced from the sap of the poppy plant?*

Opiates

The **opiates** are pain-killing drugs. They can provide great relief to a person suffering from severe pain. Opiates are produced from the sap of the opium poppy plant. The major drugs obtained from the opium poppy include opium, morphine, codeine, paregoric (par-uh-GOR-ihk), and heroin.

Unfortunately, *any* drug obtained from the opium poppy causes strong physical and psychological dependence. In this country, at least 500,000 people are dependent on opiates such as heroin. Heroin abuse, in fact, is a major drug problem.

The dangers of opiate abuse are many. The most obvious to nonabusers is the antisocial behavior of people who have a strong need for the drug and little opportunity to earn the money to pay for it. Such abusers become trapped in a world they cannot escape without help. And the longer they use opiates, the longer they are likely to suffer the dangers of abuse. Most abusers, for example, risk an overdose. And overdoses all too often lead to death. Also, life-threatening diseases, such as hepatitis and AIDS, can be caught from shared needles. In their weakened condition, opiate abusers are no match for such serious diseases.

Figure 19–15 *The bright spots in this cross section of the spinal cord are opiate receptors. The receptors are those areas where opiates attack the nerve cells. What effect do opiates have on the body?*

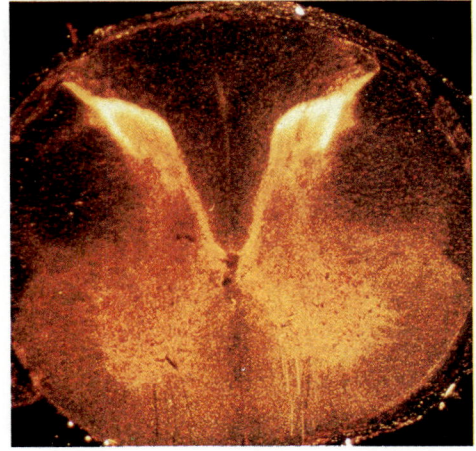

SECTION REVIEW

1. Name and briefly describe five types of drugs.
2. What is an overdose?
3. Compare the effects of crack to regular cocaine.

Analyzing Smoking Advertisements

Problem

How are advertisements used to convince people to smoke or not to smoke?

Materials *(per group)*

magazines
paper

Procedure

1. Choose two or three different types of magazines. Glance through the magazines to find advertisements for and against smoking.
2. On a sheet of paper, make a chart like the one shown. Then, for each advertisement you found, fill in the information in the chart. In the last column, record the technique that the advertisement uses to attract the public to smoke or not to smoke. Examples of themes used to attract people to smoke are "Beautiful women smoke Brand X"; Successful people smoke Brand Y''; "Brand Z tastes better." Examples of themes used to stop people from smoking are "Smoking is dangerous to your health"; "Smart people do not smoke"; "If you cared about yourself or your family, you would not smoke."

Observations

1. Were there more advertisements for or against smoking?
2. Which advertising themes were used most often? Least often?

Conclusions

1. Which advertisements appeal to you personally? Why?
2. In general, how are the advertising themes used related to the type of magazine in which the advertisements appear?

Magazine	Advertisements for Smoking (specify brand)	Advertisements Against Smoking (specify advertisement)	Theme

CHAPTER REVIEW

SUMMARY

19–1 Drug Abuse

❑ A drug is any substance that has an effect on the body.

❑ Illegal or incorrect use of drugs is known as drug abuse. Dependence, withdrawal, and tolerance are serious dangers of drug abuse.

❑ Drug abusers can develop a psychological or a physical dependence or both.

19–2 Alcohol

❑ Drinking large amounts of alcohol on a daily basis can cause alcoholism, a serious health problem.

❑ In the brain, alcohol acts as a depressant by slowing down the actions of the nervous system.

❑ Long-term alcoholism can cause cirrhosis, which is a loss of liver function due to cells destroyed by alcohol.

❑ Treatment for alcoholism includes medical and psychological help.

19–3 Tobacco

❑ When tobacco burns, carbon monoxide, a poisonous gas, is given off. Carbon monoxide interferes with the transport of oxygen through the circulatory system.

❑ Tars, another substance in tobacco smoke, irritate and damage the respiratory system.

❑ Nicotine is a poison found in the leaves of the tobacco plant.

❑ Cigarette smoking is the most important cause of lung cancer. Cigarette smoking irritates the lining of the nose, throat, and mouth, increases heartbeat, lowers skin temperature, and constricts blood vessels.

19–4 Commonly Abused Drugs

❑ A chemical called THC in marijuana affects people in different ways. Some report a sense of well-being; others become suspicious of people and cannot keep their thoughts from racing.

❑ Like alcohol, marijuana slows down reaction time.

❑ Depressants, such as barbiturates and tranquilizers, slow down the actions of the nervous system.

❑ Stimulants, such as amphetamines and cocaine, speed up the actions of the nervous system.

❑ Crack is an extremely dangerous form of cocaine. The effects of crack are much more dangerous than those of regular cocaine. Dependence on crack develops more quickly than dependence on regular cocaine.

❑ Hallucinogens, such as LSD, mescaline, and PCP, are drugs that produce hallucinations.

❑ Opiates are used as pain killers and are produced from the opium poppy. Examples of opiates are opium, morphine, codeine, paregoric, and heroin.

VOCABULARY

Define each term in a complete sentence.

alcoholism

cirrhosis

depressant

drug

drug abuse

hallucinogen

opiate

over-the-counter drug

physical dependence

prescription drug

psychological dependence

stimulant

tolerance

withdrawal

CONTENT REVIEW: MULTIPLE CHOICE

On a separate sheet of paper, write the letter of the answer that best completes each statement.

1. Which drug can be bought over the counter?
   a. marijuana   b. aspirin   c. PCP   d. morphine
2. Which is a drug that can be abused?
   a. alcohol   b. cold medicine   c. stimulant   d. all of these
3. When a person who is physically dependent on a drug is suddenly taken off that drug, they go through
   a. tolerance.   b. denial.
   c. withdrawal.   d. psychological dependence.
4. Alcoholism is a
   a. drug.   b. disease.   c. symptom.   d. depressant.
5. Excess alcohol causes
   a. psychological dependence.
   b. physical dependence.
   c. both psychological and physical dependence.
   d. neither psychological nor physical dependence.
6. An amphetamine is a (an)
   a. stimulant.   b. depressant.   c. hallucinogen.   d. opiate.
7. Which is *not* an effect of cigarette smoking?
   a. lung irritation   b. mood changes
   c. lower skin temperature   d. increased heartbeat
8. A chemical in marijuana is
   a. PCP.   b. LSD.   c. THC.   d. nicotine.
9. Cocaine is a (an)
   a. hallucinogen.   b. opiate.   c. depressant.   d. stimulant.
10. Which is produced from the sap of the opium poppy plant?
    a. codeine   b. cocaine   c. tranquilizers   d. angel dust

CONTENT REVIEW: COMPLETION

On a separate sheet of paper, write the word or words that best complete each statement.

1. Any substance that has an effect on the body is a (an) _____.
2. _____ dependence is an emotional need for a drug.
3. The replacement of liver cells with scar tissue is called _____.
4. _____ increase the activities of the nervous system.
5. Cigarette smoke contains the poisonous gas _____.
6. Bronchitis is a disease that affects the _____ system.
7. Barbiturates belong to the group of drugs called _____.
8. _____ is a form of cocaine that is smoked.
9. A (An) _____ occurs when a person sees something that is not really there.
10. Heroin belongs to the group of drugs called _____.

CONTENT REVIEW: TRUE OR FALSE

Determine whether each statement is true or false. Then on a separate sheet of paper, write "true" if it is true. If it is false, change the underlined word or words to make the statement true.

1. <u>Physical</u> dependence occurs when a drug abuser's body cells cannot function without the drug.
2. Alcohol acts as a <u>stimulant</u>.
3. Nicotine is a <u>depressant</u> found in cigarette smoke.
4. Smoking can cause blood vessels to <u>narrow</u>.
5. A heavy smoker has a <u>20</u> times greater chance of getting lung cancer than a nonsmoker has.
6. The <u>opium poppy</u> plant is also called *Cannabis sativa*.
7. <u>Depressants</u> slow down the activities of the nervous system.
8. The effects of crack on the body are <u>less</u> intense than the effects of regular cocaine on the body.
9. <u>Hallucinogens</u> provide great relief to a person suffering from severe pain.
10. LSD is a powerful <u>opiate</u>.

CONCEPT REVIEW: SKILL BUILDING

Use the skills you have developed in the chapter to complete each activity.

1. **Relating cause and effect** How is cigarette smoking related to heart disease?
2. **Making inferences** How can the use of over-the-counter drugs be dangerous to human health?
3. **Applying definitions** Why is alcohol considered a legal drug?
4. **Making inferences** Doctors might prescribe a nicotine-containing chewing gum for people who want to quit smoking. What is the benefit of this gum?
5. **Relating facts** Why is cigarette smoke harmful to nonsmokers as well as to smokers?
6. **Expressing an opinion** What methods can be used to discourage people from abusing drugs?
7. **Designing an experiment** A scientist developed a new drug, Drug X, which she hopes will cure cancer. Before giving the drug to human beings, however, she must determine the effects of the drug on rats. Design an experiment that she might use to do this. Remember to include a hypothesis, a control, and a variable in your setup.

CONCEPT REVIEW: ESSAY

Discuss each of the following in a brief paragraph.

1. Why is alcohol considered a depressant?
2. Explain what is meant by dependence, withdrawal, and tolerance.
3. Name and describe seven types of drugs.
4. Describe at least four diseases that can result from smoking cigarettes.
5. Explain why any drug can be abused.
6. What is a depressant, a stimulant, and a hallucinogen? Give three examples of each.
7. Describe the effects of marijuana and PCP on the body.

Adventures in Science

Doctor Robert Gale on the Scene of a NUCLEAR DISASTER

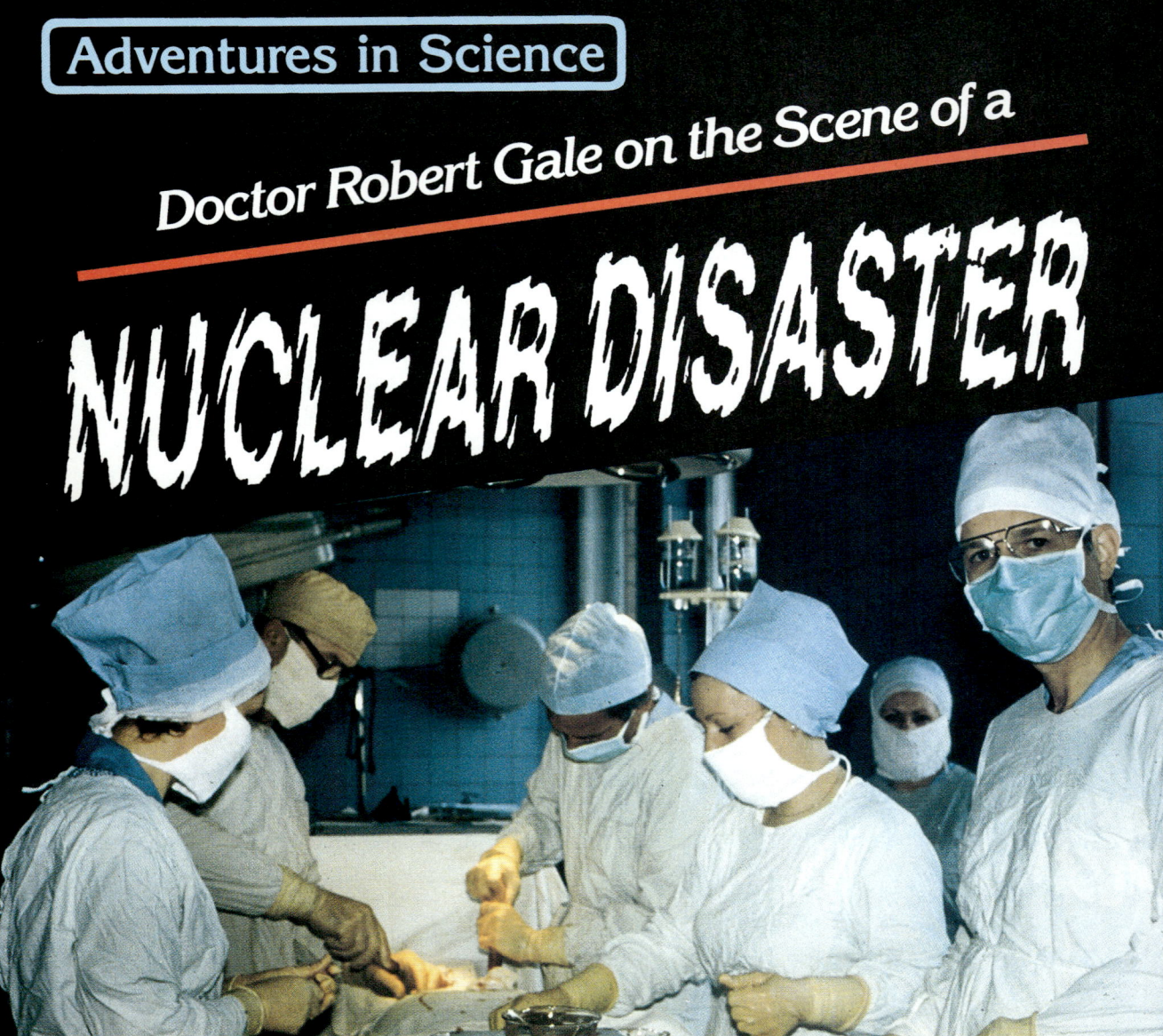

On April 23, 1986, the world's worst nuclear power plant accident occurred. The accident took place at the Chernobyl power plant near Kiev, in the Soviet Union. During an experiment, the cooling system in the reactor was shut off and the uranium fuel in the nuclear reactor overheated. Temperatures inside the reactor rose to nearly 5000°C. Soon an explosion tore off the roof of the reactor. Fire roared out of control for days. A deadly cloud of radiation leaked from the reactor and spread throughout the vicinity. Eventually, the cloud would spread as winds carried it over Western Europe.

News broadcasts spread the word that a Soviet nuclear power plant was burning out of control, and other nations rallied to offer support and help. The Soviet Union politely declined some offers. But they quickly accepted one offer of help from an American physician, Robert Peter Gale.

Dr. Gale is a specialist in transplanting bone marrow. Bone marrow, found inside some bones of the body, makes blood cells. But, bone marrow is quickly destroyed when people are exposed to radiation. And the body cannot produce new bone marrow once it is destroyed. Dr. Gale knew that the bone marrow in many of the people near the Chernobyl disaster would be damaged by the radiation that had been released. He also knew these people would surely die without help.

Dr. Gale accepted the offer from the Soviet Union to come to the site of the disaster and help in any way he could. He knew that the only hope of survival for the affected people was for bone marrow transplants to be performed as soon as possible.

The bone marrow transplant operation is relatively simple. Bone marrow is removed from healthy people and inserted into a person suffering from radiation damage. However, before the operation can begin, donors of bone marrow must be matched accurately with the people who are to get their bone marrow. If this matching process is not done, the body of the person receiving the bone marrow may not accept it. The ideal donor, Dr. Gale knew, is an identical twin. But, as you might expect, most people are not likely to have identical twins. So the first step in the transplant process for Dr. Gale was to find

Dr. Gale (above) played an important role treating victims of the Chernobyl disaster. He performed a series of bone marrow transplants in a Moscow hospital (opposite page).

and interview relatives of the Chernobyl victims. Next, Dr. Gale ran tests to determine which relatives, if any, could donate bone marrow.

Once the matching process was over, the next problem for Dr. Gale was trying to perform a great many bone marrow transplant operations in a short period of time. Dr. Gale knew that the victims at Chernobyl had to have their transplants as soon as possible. Two other American physicians, Dr. Richard Champlin and Dr. Paul Teransaki, as well as Soviet physicians, helped Dr. Gale complete all the bone marrow transplants within weeks.

Dr. Gale does not know how many of the people who received bone marrow transplants will live. But he knows that even if only one patient is helped, then this trip was worthwhile. Dr. Gale also hopes that his work may begin a new spirit of cooperation between American and Soviet physicians.

On a return visit to the Soviet Union in June 1986, Dr. Gale flew over Chernobyl in a helicopter. The reactor looked like an ordinary burned building, but the nearby town of Pripyat was empty.

"I thought, this is it. This is what we've been afraid of all these years: a city devoid of all human life because of radiation." Although the city was abandoned, many of the people from Pripyat have survived. And Dr. Gale continues to help these people through his talents as a doctor and because of his compassion.

Issues in Science

BOXING & BRAIN DAMAGE

Can Boxing Be Made Safer?

The bell sounded and the long fight entered its fourteenth and final round. A left hook caught the challenger on the side of his head. Then a right to the jaw, followed by another left. Finally, a power-house punch to the point of his jaw sent the boxer, Duk Koo Kim, to the mat. The fight was over. The fans went home, but Duk Koo Kim was rushed to a nearby hospital to be treated for severe head injury.

Doctors had little hope for Kim's recovery even after emergency brain surgery. Kim's surgeon, Dr. Lonnie Hammargren, said after the operation that Kim's brain had shifted within his skull and had been bruised by the jarring impact of several punches to his head. During the fight, blood vessels in Kim's head burst. Blood poured into the small space between his brain and his skull. This type of bleeding is the most common cause of death in the boxing ring. Five days after surgery, Duk Koo Kim became the 353rd boxer to die from such injuries since 1945.

Recently, neurologists, doctors who specialize in the body's nerves, have begun to explore the long-term effects of boxing. One effect many fighters experience is a

George Bellows captured some of the flavor and excitement of boxing in this painting.

condition called *dementia pugilistica*. Fighters with this condition are said to be "punch-drunk" because they show symptoms similar to drunkenness. However, a person who is punch-drunk does not recover after a good night's sleep. The slurred speech, shuffling walk, memory loss, and lack of coordination are permanent and become more severe as the boxer ages. Sometimes these symptoms do not appear until many years after a boxer's career in the ring has ended.

Olympic boxers wear protective headgear during their match, which lasts for three rounds. Do you think the headgear and the short fight period protect the fighters from injury?

Doctors agree that the risk of becoming punch-drunk increases with each fight. Every jarring blow to the head bounces and twists a boxer's brain within his skull. Tissues in the brain are torn, brain cells die. And when brain cells die, no new brain cells replace them. Brain cells do not regrow.

In 1984, the American Medical Association called for a ban on all amateur and professional boxing because of the brain damage that may occur. But many fans and boxing professionals claim that boxers know the risk of boxing and insist that boxers are entitled to take these risks. They believe that a complete ban on boxing is not needed and they point to the fact that fewer deaths result from boxing injuries than from other sports, such as mountain climbing, auto racing, and skiing.

Other people suggest that changes in boxing rules could reduce the risks of boxing-related injuries. For example, shorter rounds, two minutes rather than three minutes, and a longer rest period between rounds could reduce the risks of injury to a boxer. Shorter rounds might reduce the most serious injuries, which occur when a boxer becomes too tired to defend himself properly. Other suggestions include the use of protective headgear and heavily padded gloves. The headgear and the padded gloves would lessen the impact of punches to a boxer's head.

In some states, boxers are required to obtain a brain wave scan as well as a complete physical examination before a fight. A clean bill of health is necessary before a boxer is permitted to enter the ring. If an injury is detected during the pre-fight examination, a boxer is prohibited from entering the ring.

Some people argue that boxing provides the only opportunity for some people to achieve success. But after Kim's death in 1982, the Journal of the American Medical Association published this statement: "Some have argued that boxing has redeeming social value in that it allows a few individuals an opportunity to rise to spectacular fame and wealth. This does occur, but at what price?"

Many sports have risks. Boxing is no exception. Should boxing continue as it has, or should the rules be changed to make boxing safer? What do you think?

Heredity and Adaptation

Deep in Georgia's Okefenokee Swamp, a red rat snake slithers up a decaying pine tree. As it nears its prey, the snake runs into an unexpected ooze of thick resin. The resin is poisonous to the snake. When it touches the resin, the snake loses its grip and tumbles 12 meters to the ground below. Nearby, the snake's intended victim, the red-cockaded woodpecker, watches unharmed.

For its nesting hole, the red-cockaded woodpecker chooses an old pine tree that has "red heart." Red heart is a disease caused by a fungus. The disease makes the tree's wood and bark soft. After the woodpecker digs a hole in the soft wood for its nest, a thick sticky resin oozes from the hole. The resin is made of sap and the fungus that causes red heart disease. Chemicals in the resin are poisonous to the snake. The woodpecker's booby-trapped nest provides a unique protection against the red rat snake.

Like the woodpecker, all organisms have adaptations that help them survive in their surroundings. As you read the chapters in this unit, you will learn more about adaptations and how they are passed from generation to generation. You will also discover the important role that heredity plays in the development and survival of living things.

CHAPTERS

Resin oozing from the woodpecker's nesting hole keeps the red rat snake at a safe distance.

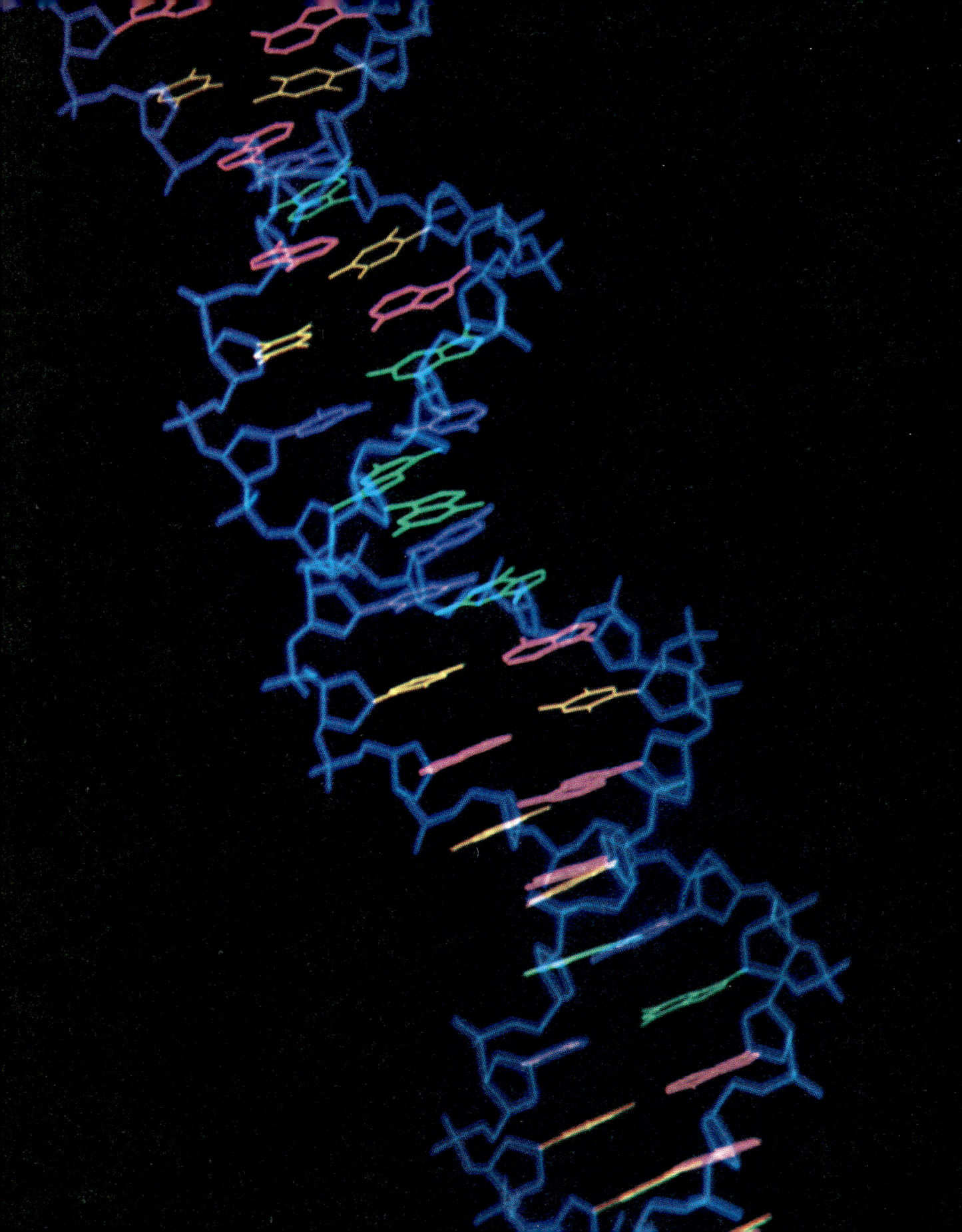

Genetics

20

CHAPTER OBJECTIVES

After completing this chapter, you will be able to

20–1 Identify dominant and recessive traits.

20–2 Apply probability to the study of genetics.

20–2 Compare genotype and phenotype.

20–3 Describe the process of meiosis.

20–3 Describe the functions of genes and chromosomes.

20–4 Describe the structure of DNA and how it replicates.

20–5 Explain how sex-linked traits are inherited.

20–5 Define genetic disease and list several examples.

Suddenly, the courtroom became quiet. The district attorney carefully asked her question: "Are you sure, doctor, that the fingerprints in the hair sample were those of the defendant?" The jury listened for the answer. "Yes, without a doubt. The identification was certain."

"Fingerprints in hair?" you may be saying to yourself. Sounds unusual, but these are not the kind of fingerprints that are found on your fingers. The fingerprints the district attorney was talking about are the DNA fingerprints that were found in a tiny sample of a hair root.

DNA is the basic substance of heredity and is found in every cell of the human body. In this case, the DNA was found in the cells of the hair root. DNA is the substance that determines whether you have brown eyes or blue eyes, brown hair or blonde hair, lots of freckles or none at all.

Although human DNA is very similar from person to person, there are many differences. No two people have the same type of DNA. These differences were found in the recently developed test called DNA fingerprinting. This test was developed by British scientists in 1986. At present, DNA fingerprinting is done in rare cases in the United States. But within the next two to five years, the scene at the beginning of this story could become an everyday occurrence. As you read the chapter, you will discover more about DNA and the study of heredity.

A computer-generated image of DNA is shown in this photograph.

To describe how traits are passed on from one generation to another

Figure 20–1 *More than 100 years ago, Gregor Mendel studied how traits are passed on from parents to offspring. What organisms did Mendel study?*

During the 1860s, an Austrian monk and biologist named Gregor Mendel worked among hundreds of pea plants in the garden of a small monastery in Czechoslovakia. Mendel experimented with pea plants to see if he could find a pattern in the way certain characteristics are handed down from one generation of pea plants to the next generation. Another word for the characteristics of an organism is **trait.** So Mendel actually studied the way certain traits are passed on from one generation of organisms to the next generation. Pea plant traits include how tall the plants grow, the color of their seeds, and the shape of their seeds.

Although Mendel did not realize it at the time, his experiments would come to be considered the beginning of **genetics.** Genetics is the study of heredity, or the passing on of traits from an organism to its offspring. For this reason, Mendel is called the Father of Genetics.

Mendel chose pea plants for his experiments because pea plants grow and reproduce quickly. So he knew that he could study many generations of pea plants in a short amount of time. Mendel also knew that pea plants had a variety of different traits that could be studied at the same time. That is, he could study plant height, plant seed color, plant seed shape, and other traits in the same experiment.

Figure 20–2 *Notice how this young grizzly bear has traits that are similar to those of its mother. What is the passing of traits from parents to offspring called?*

Mendel also chose pea plants because he could easily breed, or cross, them through pollination. Pollination is the transfer of pollen from the male reproductive part of the plant's flower to the female reproductive part. Usually, a pea plant pollinates itself in a process called **self-pollination.** But Mendel found that he could cross the pea plants by transferring pollen from the male part of one flower to the female part of another flower. This process is called **cross-pollination.** In this way, Mendel could cross different plants.

The Work of Gregor Mendel

Mendel discovered that when he transferred pollen from the male part of pea plants with short stems to the female part of short-stemmed pea plants, only short-stemmed plants grew from their seeds. So the next generation of these short-stemmed plants was also short-stemmed. This result was what he, and everyone else at that time, expected. The plants always resembled the parent plants. Mendel called these short-stemmed plants true-breeding plants. By true-breeding plants, Mendel meant those plants that always produce offspring with the same traits as the parents.

Mendel then repeated his experiments by crossing tall-stemmed pea plants with other tall-stemmed pea plants. Naturally, he expected all of the offspring to be tall. But Mendel was in for a surprise. Some of the tall-stemmed pea plants produced only other tall-stemmed pea plants. These plants were true breeders for tall stems. However, some of the tall-stemmed pea plants produced both tall- and short-stemmed offspring. Mendel was not, as yet, able to explain how tall-stemmed plants could produce some short-stemmed offspring. But he realized quickly that not all tall-stemmed plants are true breeders. Why?

Mendel knew that more experiments had to be performed before he could show how pea plants pass on their traits to their offspring. He began to wonder what would happen if he crossed true-breeding tall plants and true-breeding short plants. Mendel called these true-breeding parent plants the parental generation, or P_1. When Mendel crossed

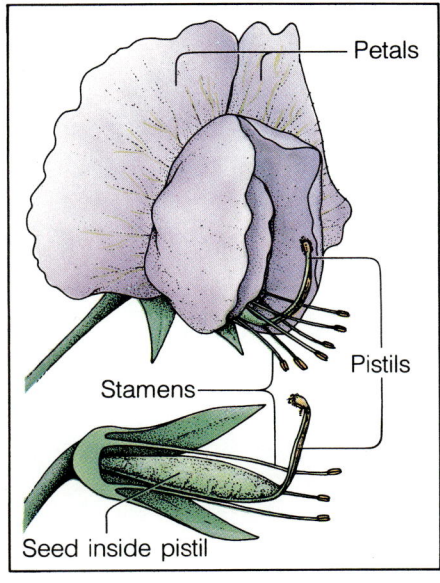

Figure 20–3 *Stamens produce pollen, which contains sperm cells. Flowers also contain a pistil, which produces eggs. Pollination occurs when pollen lands on top of the pistil. Compare self-pollination and cross-pollination.*

Figure 20–4 *Gregor Mendel experimented with pea plants because their life cycle, from flower to new plant, occurs very quickly. Why was this an advantage to Gregor Mendel?*

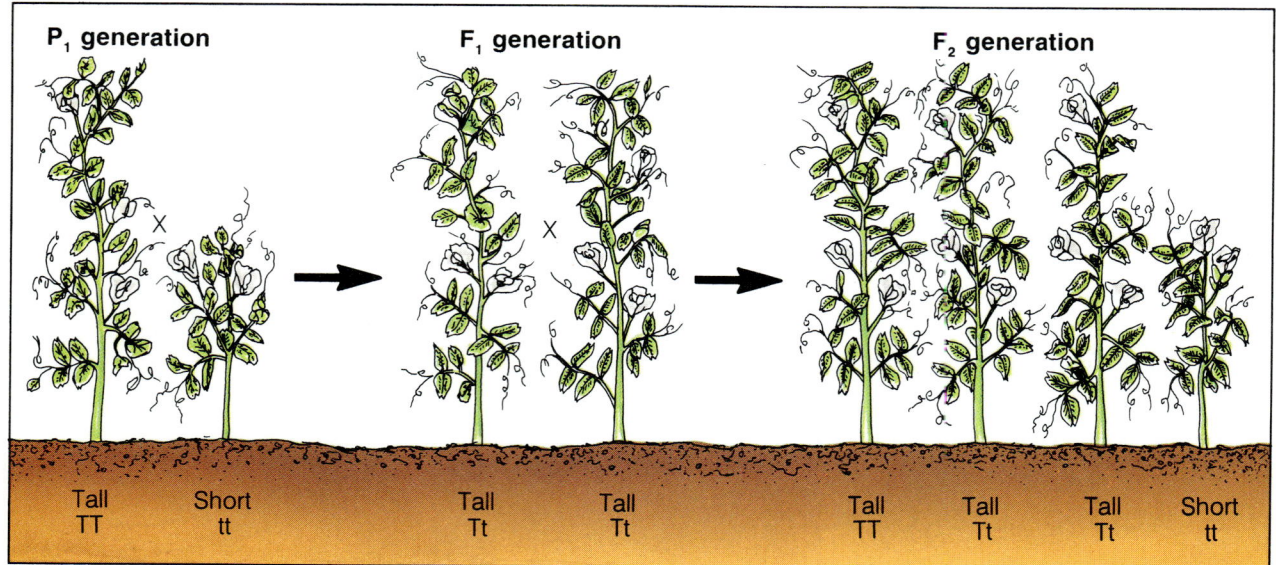

| P₁ generation | | F₁ generation | | F₂ generation | | | |
| Tall TT | Short tt | Tall Tt | Tall Tt | Tall TT | Tall Tt | Tall Tt | Short tt |

Figure 20–5 *Mendel crossed tall and short pea plants. He discovered that the offspring in the first generation were all tall. What kind of plants were produced in the second generation?*

the P₁ generation, he discovered that all of the plants in the next generation were tall. It was as if the traits for shortness that had existed in some of the parent plants had simply disappeared in the next generation! He called the second generation of plants the filial (FIHL-ee-uhl) generation, or the F₁ plants. See Figure 20–5.

What happened next was even more of a mystery. When Mendel crossed the tall plants of the F₁ generation, he expected that they would produce only tall plants. Instead, some of the plants of the second filial generation, or F₂, were tall and some were short. The trait for shortness seemed to have reappeared!

Dominant and Recessive Genes

By keeping careful records of his work, Mendel realized that the tall plants of the F₁ generation were not true breeders. This is because some of the tall F₁ plants produced both short and tall plants in the F₂ generation. But how could the shortness trait disappear in the F₁ generation and then mysteriously reappear in the F₂ generation? Obviously, the trait did not disappear—it was merely hidden, or masked, in some of the plants.

Mendel realized that each plant must contain two factors for a particular trait. In some plants, the two factors were both for the tallness trait. Plants with two tallness factors were true breeders and only produced other tall plants. In the same way, short plants contained two shortness factors and were true-breeding short plants. But some plants, Mendel reasoned, contained a tallness factor and a shortness factor. If a plant contained both tallness and shortness factors, the plant would grow tall. But it might still pass on its shortness factor to the next generation of plants. Today, Mendel's factors, or units of heredity, are called **genes.** And because each plant received two genes for each trait, each plant is said to have a gene pair for each trait.

From his observations, Mendel knew that if a plant contained any genes for tallness, it would be tall. Even if the plant had one tallness gene and one shortness gene, the plant would be tall. So Mendel reasoned that the tallness gene masked the shortness gene. Today, genes that mask other genes are called **dominant** genes. The "weaker" genes, or the ones that seem to be masked, are called **recessive** genes.

Scientists use symbols to represent different forms of a trait. Capital letters represent dominant genes. For example, tallness in pea plants is written as "T." Recessive genes are represented by small letters. Shortness in pea plants is written as "t."

Mendel studied not only the height of pea plants, he studied other traits. These traits were seed shape, seed color, seed coat color, pod shape, pod color, and flower position. In every case, crossing two true-breeding plants with opposite traits did

Figure 20–6 *The chart shows the seven characteristics that Mendel studied in pea plants. Each characteristic has a dominant and a recessive gene. In pea plants, which seed shape is dominant?*

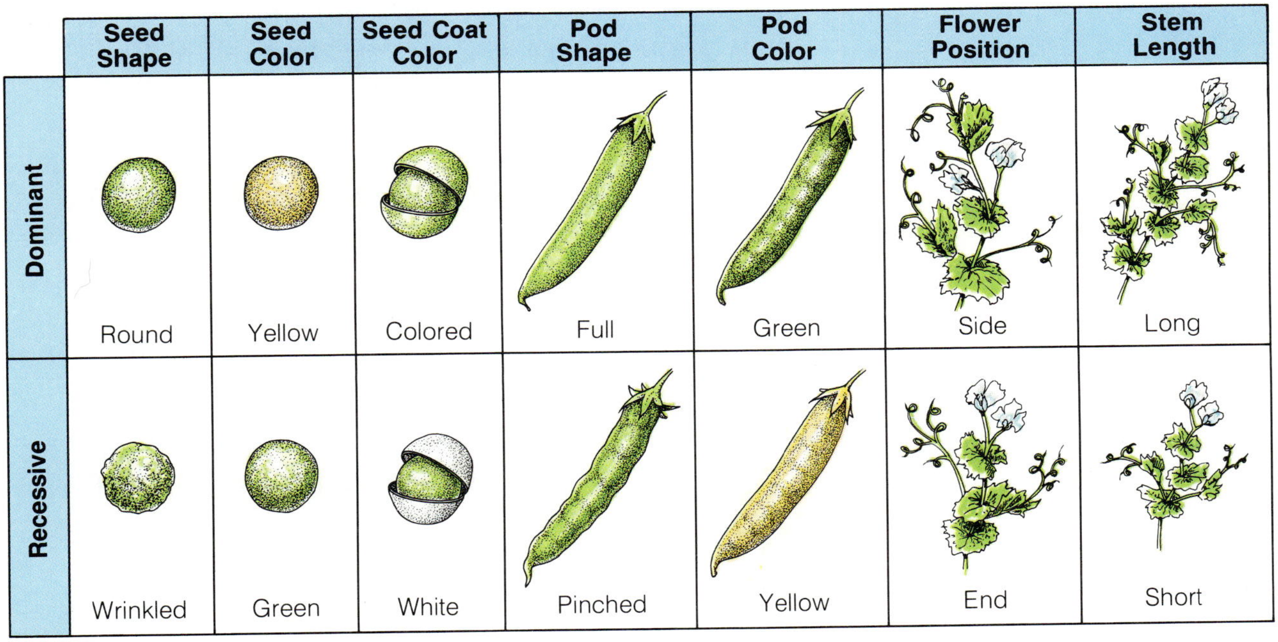

	Seed Shape	Seed Color	Seed Coat Color	Pod Shape	Pod Color	Flower Position	Stem Length
Dominant	Round	Yellow	Colored	Full	Green	Side	Long
Recessive	Wrinkled	Green	White	Pinched	Yellow	End	Short

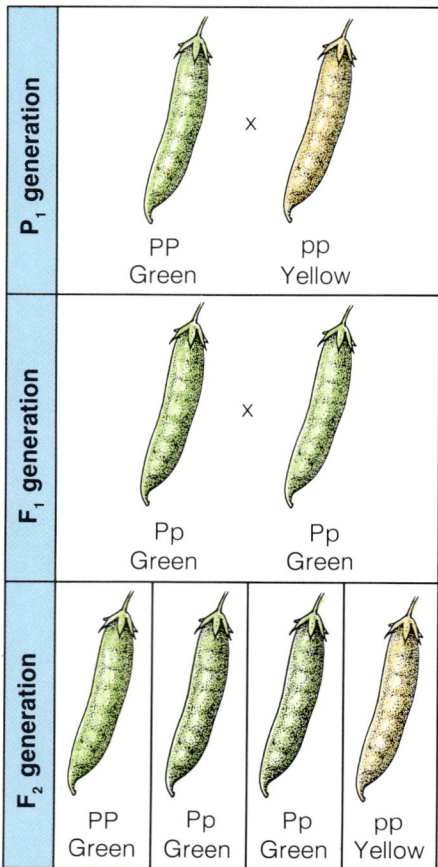

Figure 20-7 *This illustration shows a cross between two pea plants. Examine the first and second generations. Based on these results, predict which color pod is recessive.*

not result in mixtures of the trait. For example, Figure 20-7. shows a cross between a plant with green pods (PP) and a plant with yellow pods (pp). The type of pods that are produced in the F₁ generation are all green. Why? The gene for green pods, P, is dominant. It masks, or hides, the gene for yellow pods, which is recessive. Therefore, all the pods are green. What happens when the F₁ generation pollinates itself? In this case, most of the pods are green and some are yellow. The recessive trait for yellow pods reappears in the F₂ generation.

Scientists call organisms that have genes that are alike for a particular trait, such as PP or pp, pure-bred. An organism that has genes that are different for a trait (Pp) is called a **hybrid** (HIGH-brihd). Is a short-stemmed pea plant a purebred or a hybrid?

Inheriting Traits

By examining his observations and results, Mendel formed a **hypothesis** about how traits are inherited. A hypothesis is a suggested explanation for a scientific problem. Mendel's hypothesis was that each pea plant parent has a pair of factors, or a gene pair. Mendel further reasoned that each parent could contribute only one of these factors to each pea plant in the next generation. In that way, the next generation also had a gene pair for each trait, one gene from each parent.

Now Mendel was finally able to account for the results of his pea plant experiments. You may recall that when he crossed true-breeding tall plants, all the generations that followed were tall. The reason for this was that all the tall plants were purebreds. That is, the gene pair in each of the tall plants consisted of only tallness genes. So the gene pair for these plants was TT. Each of these plants could contribute one tallness gene (T) to its offspring. So, each offspring received two tallness genes (TT), one from each parent. And all the offspring were tall.

The same was true for Mendel's cross between two short plants. The gene pair for each parent plant was tt. So each plant in the next generation, the F₁ generation, received a shortness gene (t) from each parent. The offspring all had two shortness genes (tt) and all were short.

What happened when Mendel then crossed a purebred tall plant (TT) with a purebred short plant (tt) to produce the F_1 generation? Each parent could contribute one gene to its offspring. As a result, the offspring were hybrids (Tt) because they received a tallness gene (T) and a shortness gene (t). Because the dominant tallness gene masked the recessive shortness gene, all of the offspring were tall. Although the offspring had a shortness gene, it was masked by the tallness gene. The shortness trait seemed to disappear in the F_1 generation.

When Mendel crossed the hybrid plants of the F_1 generation to produce the F_2 generation, some of the plants were tall and some of them were short. The reason for this was that each parent in the F_1 generation had one tallness gene and one shortness gene (Tt). So each parent could contribute a tallness gene or a shortness gene to the next generation. Most of the time, one of the parents contributed a tallness gene to the F_2 generation. So most of the F_2 generation had at least one tallness gene. Therefore, most of the plants were tall. But some of the time

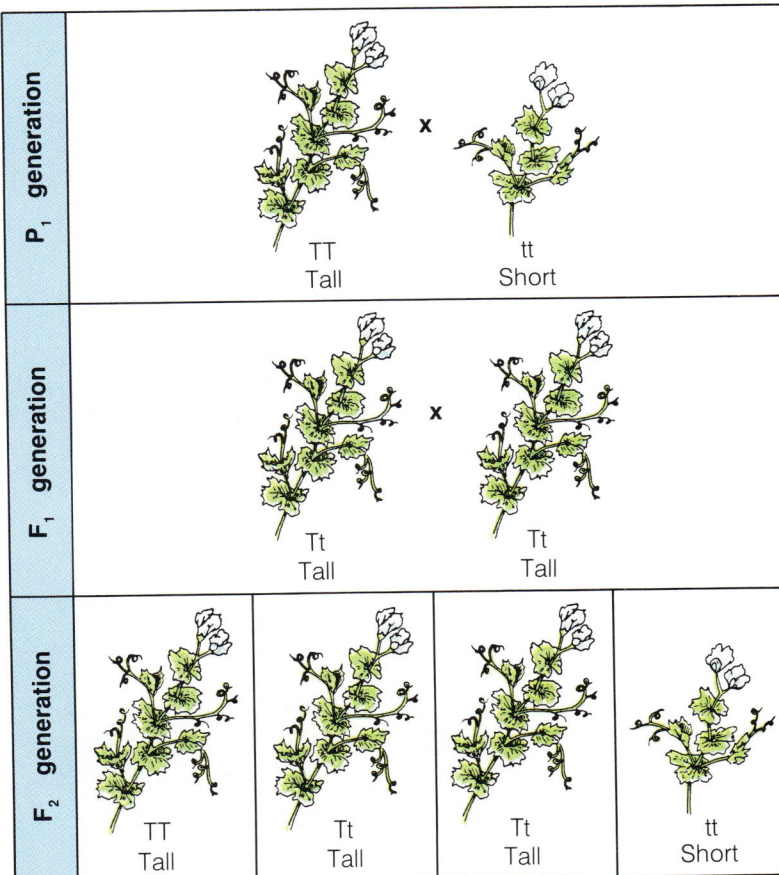

Figure 20–8 *Mendel discovered that the shortness trait in the parent generation disappeared in the first generation. Explain why the shortness trait then reappeared in the second generation.*

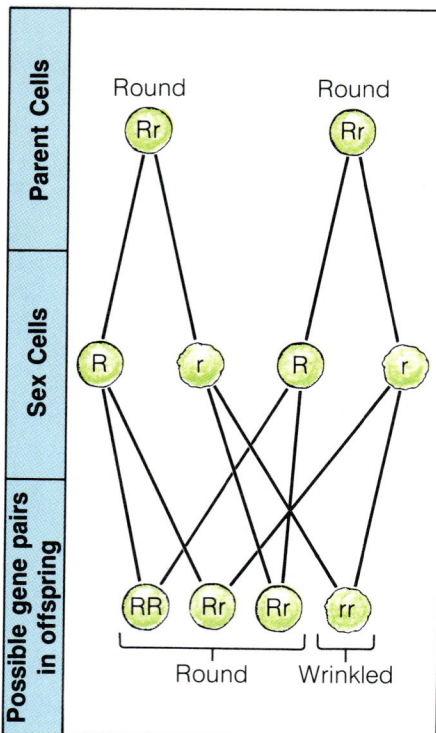

Parent Cells Round Rr | Round Rr

Sex Cells R | r | R | r

Possible gene pairs in offspring RR | Rr | Rr | rr

Round | Wrinkled

Figure 20–9 *Mendel discovered that a pea plant with wrinkled seeds can develop from a cross between parents with round seeds. How does this happen?*

Sharpen Your Skills

Incomplete Dominance

1. Obtain two coins.

2. Cut four squares of masking tape that are equal in size. The squares should not hang over the edges of the coins.

3. Place the letter R on one side of a coin and the letter W on the other side.

4. Repeat this procedure for the second coin.

5. Toss both coins 100 times, recording the genotype for each toss.

What is the percentage of occurrence for each genotype? Construct a Punnett square showing this cross.

both parents contributed a shortness gene to the F_2 generation. Those plants in the F_2 generation that received a shortness gene from both parents had the gene pair tt. And all of these plants were short. The shortness trait that had mysteriously disappeared in the F_1 generation had suddenly reappeared in the F_2 generation. But now you know that this was no mystery at all. Most of the F_2 plants had at least one tallness gene, so they grew tall. However, the F_2 generation contained some plants that had only shortness genes, so they were short. Were the tall plants in the F_2 generation purebreds or hybrids? What about the short plants?

Mendel applied his experimental results to the other traits he studied. For example, he could explain why a cross between parents with round seeds could produce some plants with wrinkled seeds. Why? The round seed gene is dominant. But the plants were not purebred. That is, the plants with round seed genes also contained a recessive wrinkled seed gene. The gene for round seed can be written as R. The gene for wrinkled seed can be written as r. A parent that has Rr genes would have round seeds. But if plants with Rr genes were crossed, some of the offspring would receive a recessive wrinkled seed gene (r) from both parents. Those offspring would have two wrinkled seed genes (rr) and would have wrinkled seeds—even though both their parents had round seeds! See Figure 20–9.

Incomplete Dominance

Mendel's ideas that genes are always dominant or recessive often holds true—but not always. In 1900, Karl Correns, a German botanist, made an important discovery. Correns discovered that in some gene pairs the genes are neither dominant nor recessive. These genes show **incomplete dominance.** In incomplete dominance, neither gene in a gene pair masks the other. Put another way, each gene blends with the other.

Working with four-o'clock flowers, Correns discovered that when he crossed purebred red four-o'clocks (RR) with purebred white four-o'clocks (WW), the result was all pink four-o'clock flowers (RW). See Figure 20–10. Notice that the symbols for the gene pairs for red, white, and pink are all capital letters. No one gene is dominant over the others.

Incomplete dominance also occurs in animals. In Andalusian (an-duh-LOO-zhuhn) fowl, a kind of chicken, neither black nor white feathers are dominant. When a gene for black feathers and a gene for white feathers are present, the fowl appears to have black feathers with tiny white dots. This gives the fowl's feathers a blue-gray color.

Principles of Genetics

Through the work of scientists such as Mendel and Correns, certain basic principles of genetics have been established. Following is a review of the basic principles of genetics:

- **Traits, or characteristics, are handed down from one generation of organisms to the next generation.**

- **The traits of an organism are controlled by genes.**

- **Organisms inherit genes in pairs—one gene from each parent.**

- **Some genes are dominant, while other genes are recessive.**

- **Dominant genes mask recessive genes when both are inherited by an organism.**

- **Some genes are neither dominant nor recessive. These genes blend together when they are inherited.**

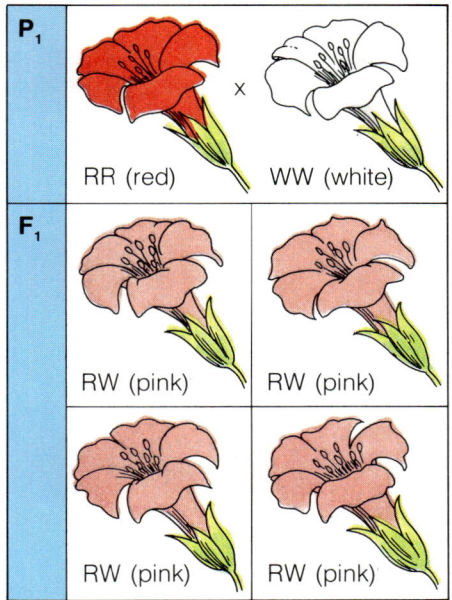

Figure 20–10 *In four o'clock flowers neither the red nor the white gene for flower color is dominant. When these two genes are present in the same plant, a pink flower results. What kind of inheritance does this show?*

SECTION REVIEW

1. How are traits passed on from one generation to another?
2. Define genetics.
3. Compare self-pollination and cross-pollination.
4. What is a dominant gene? A recessive gene?
5. What is incomplete dominance?
6. Can a short-stemmed pea plant ever be a hybrid? Explain.

20–2 Genetics and Probability

In one of Mendel's experiments, he crossed two plants that were hybrid for yellow seeds (Yy). When he examined the plants that resulted from this cross, he discovered that about one of every four seeds was green. Put another way, you would say that the possibility of such a cross producing green seeds is ¼, or 25 percent. In mathematics, the term **probability** is used to describe the possibility that an event may or may not take place. **Scientists use probabilities to predict the results of genetic crosses.**

Probability

Suppose you flipped a coin. What are the chances that it will land heads up? If you said a 50 percent chance, you are correct! You, like Gregor Mendel, used probabilities. Probabilities are usually written as fractions or percentages. For example, the chance that a sperm or an egg, or a sex cell, will receive a Y gene from a Yy parent is ½, or 50 percent.

In probability, the results of one chance event do not affect the results of the next. Each event happens independently. For example, if you toss a coin 10 times and it lands heads up each time, the probability that it will land heads up on the next flip is still ½, or 50 percent. The first 10 flips do not influence the result of the eleventh flip. What is the probability that a coin will land heads or tails?

Punnett Squares

In addition to probability, scientists use a special chart to show possible combinations of the cross between two organisms. This chart is called a Punnett square. It was developed by Reginald C. Punnett, an English geneticist.

To see how a Punnett square works, look at Figure 20–12. It shows a cross between two dogs. Each of the genes in the female sex cells is listed along the top. The genes in the male sex cells are listed along the left side. Remember that when male and female sex cells join, a zygote forms. A zygote is a

Figure 20–11 *According to the law of probability, a coin will land heads up 50 percent of the time and tails up 50 percent of the time. If a coin is tossed nine times and come up heads each time, what is the probability that it will come up heads on the tenth toss?*

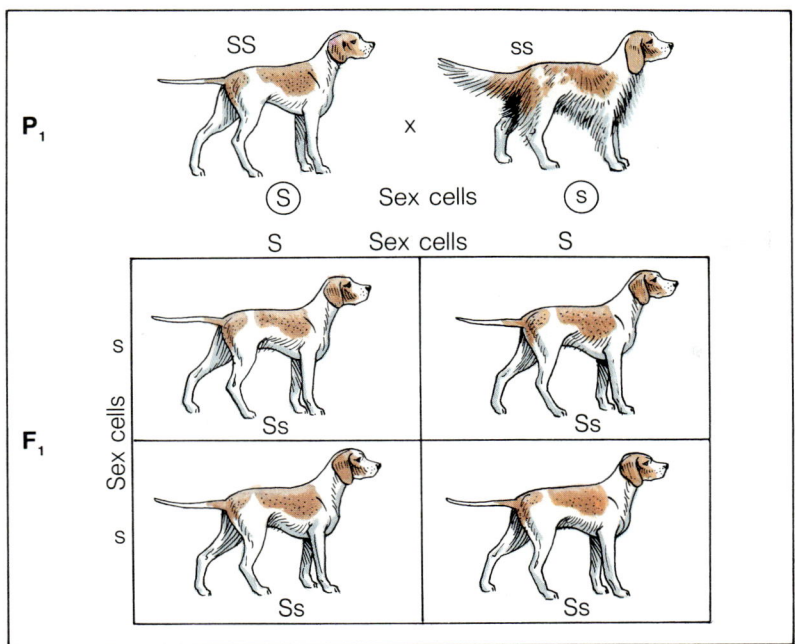

Figure 20-12 *This Punnett square shows a cross between a long-haired dog and a short-haired dog. What is the phenotype of the offspring?*

fertilized egg that develops into an organism. Each box in the chart represents a possible zygote.

Notice in Figure 20-12 that in the P₁ generation, the female has both genes for short hair (SS). The male has both genes for long hair (ss). Short hair is the dominant gene. As a result, all of the F₁ generation are hybrid short-haired dogs (Ss). If you were to look at hybrid short-haired dogs, you would not be able to tell the difference between them and purebred short-haired dogs. The **phenotypes** (FEE-nuh-tighps), or visible characteristics of both are the same. The **genotypes** (JEE-nuh-tighps), however, are

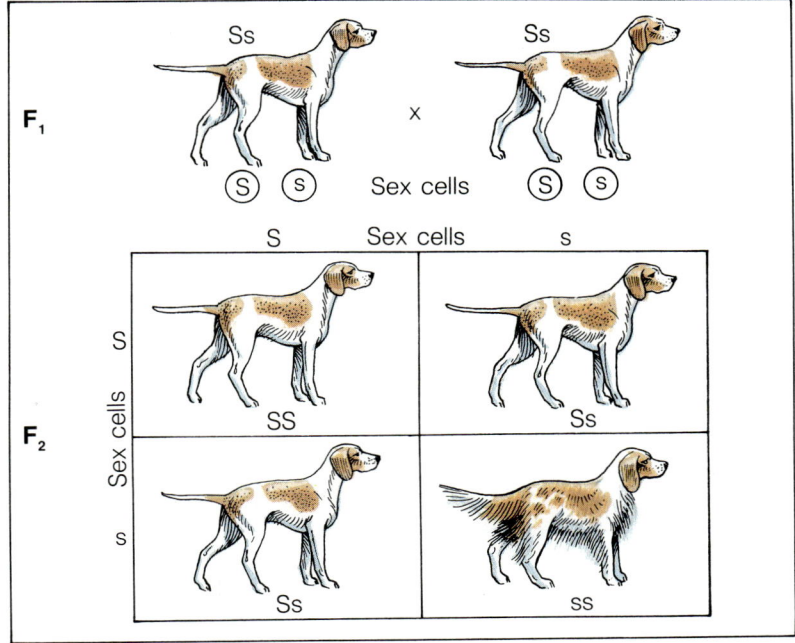

Figure 20-13 *This Punnett square shows a cross between two hybrid short-haired dogs. What phenotypes are present in the offspring?*

A Visit to Solaron

On an imaginary planet called Solaron, green skin is dominant over red skin. If a hybrid green-skinned male married a hybrid green-skinned female, what is the probability that they will have children with green skin? With red skin? Draw a Punnett square to show this cross.

different. A genotype is the actual makeup of the genes. For example, a purebred short-haired dog has the genotype SS, while a hybrid short-haired dog has the genotype Ss, even though they both have short hair.

Figure 20–13 on page 459 shows a cross between two of the hybrid short-haired dogs from the F_1 generation. The probability that a purebred long-haired dog may result is ¼, or 25 percent. The probability that a hybrid short-haired dog may result is ½, or 50 percent. What is the probability that a purebred short-haired dog may result?

SECTION REVIEW

1. What is probability?
2. If yellow seed (Y) is dominant over green seed (y) in pea plants, what is (are) the possible genotype(s) of a pea plant that has yellow seeds?
3. Explain why a plant with the genotype Tt is tall instead of medium-sized?
4. What is the probability that two plants with the genotype Rr will produce an offspring that has wrinkled seeds? (R = smooth; r = wrinkled)
5. If the gene for freckles (F) is dominant over the gene for no freckles (f), what are the possible genotypes of the parents of a child who does not have any freckles?

20–3 Chromosomes

At about the same time that Karl Correns was doing his work on incomplete dominance, Walter Sutton, a 25-year-old American graduate student, began to develop a new theory of heredity. Sutton knew that the factors Mendel had discovered were actually genes. And he knew that an organism usually contains a pair of genes—one from each parent—for every trait.

Sutton believed that the genes in an organism are found on **chromosomes.** The chromosomes are rod-shaped structures that are located in the nucleus of every cell in an organism. Sutton's belief that genes are found on chromosomes came to be known as the **chromosome theory** of genetics. According to the chromosome theory, hereditary factors, or genes, are carried from the parental generation to the next generation on chromosomes. Actually, it is really quite amazing that Mendel was able to do his work and arrive at so many correct conclusions without knowing anything about chromosomes!

How Chromosomes Control Traits

If you look at Figure 20–14, you can see a number of chromosomes in the nucleus of a cell. Chromosomes control all the traits of an organism. How do they perform this complex task?

The main function of chromosomes is to control the production of substances called proteins. Proteins determine the traits of organisms. All organisms are made mainly of proteins. They determine the size, shape, and characteristics of all living things. The kind and number of proteins in each organism determine the traits of that organism. So by controlling the kind and number of proteins produced in a living organism, chromosomes are able to control the traits of that organism. Keep in mind that chromosomes are passed on—along with the genes on the chromosomes—from parents to offspring. So chromosomes control the way traits are passed on from parents to offspring.

If you have not guessed by now, it is actually the genes on the chromosomes that control the production of proteins. In fact, each gene's major role is to control the production of a specific protein. So a chromosome, which contains many genes, actually controls the production of a wide variety of proteins. Now let's see how the chromosomes are passed on from parents to offspring.

Meiosis

You have read that organisms usually have a gene pair for every trait. The chromosomes within the nucleus of an organism's cells are also found in pairs. In general, for any particular trait, one gene contributed by a parent is on one of the paired chromosomes. The other gene for that trait from the other parent is on the opposite chromosome. Of

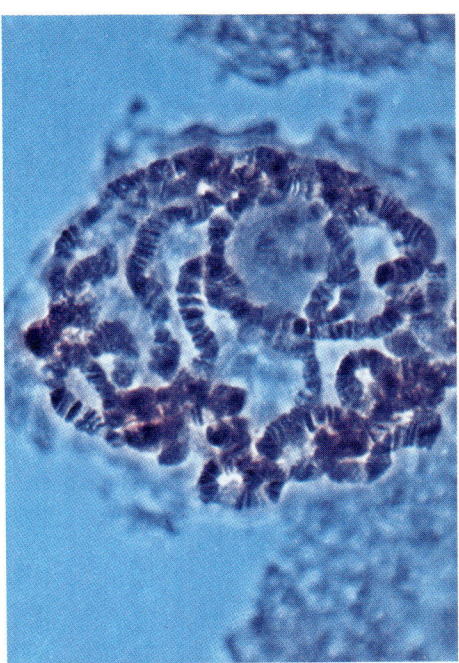

Figure 20–14 *This photograph shows the chromosomes of a fruit fly magnified 400 times. What is the main function of chromosomes?*

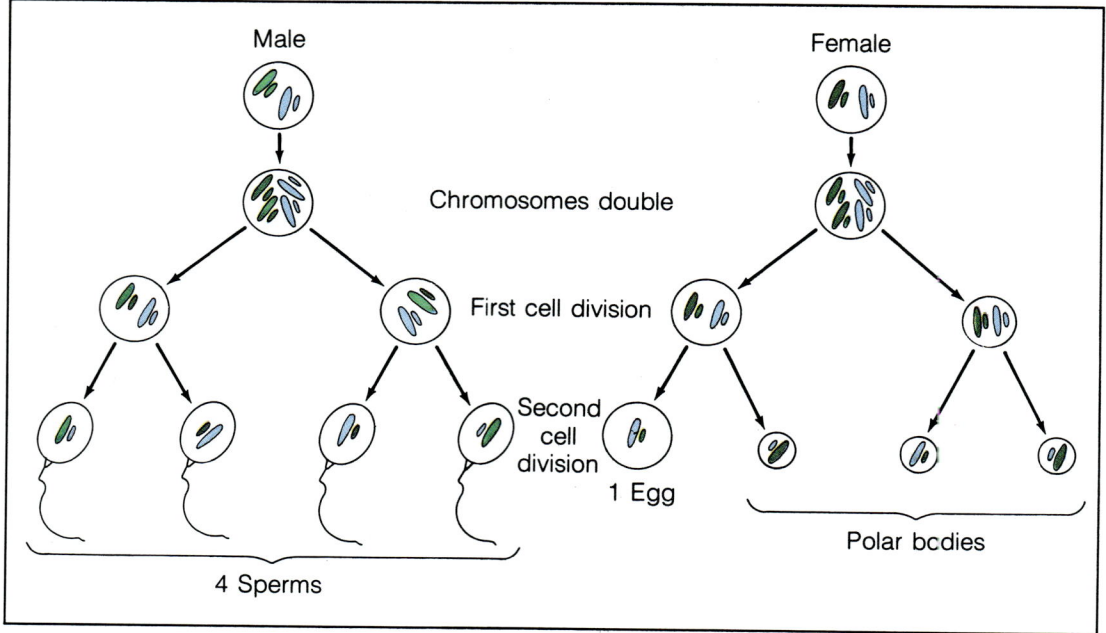

Figure 20–15 *During meiosis, a cell undergoes two divisions resulting in cells that have half the normal number of chromosomes. The sex cells, egg and sperm, are formed by meiosis. The polar bodies disintegrate, leaving only one egg cell.*

Figure 20–16 *During meiosis, chromosomes double and then separate. Different gene pairs on different chromosomes separate independently.*

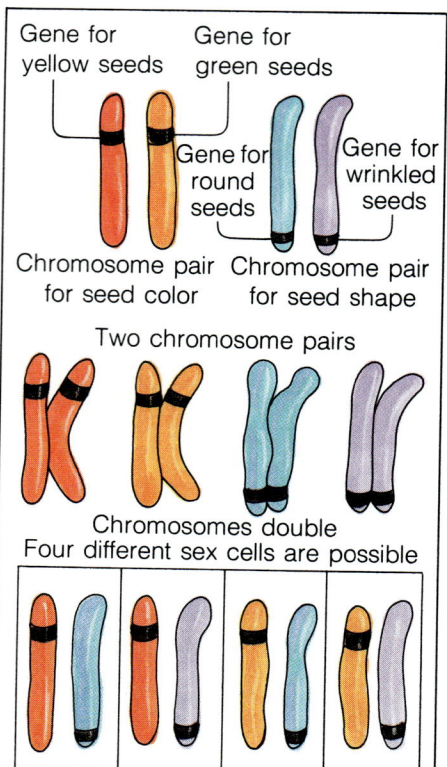

course, if each parent contributed all its chromosomes and genes to an offspring, then the offspring would have twice the number of chromosomes as its parents. This does not happen because of a process called **meiosis** (migh-OH-sihs).

The process of meiosis produces the sex cells—either sperm or egg cells. As a result of meiosis, the number of chromosomes carried by each sex cell is exactly half the normal number of chromosomes found in the parent. When sex cells combine to form the offspring, each sex cell contributes half the normal number of chromosomes. In the end, the offspring gets the normal number of chromosomes—half from each parent.

You can see in Figure 20–15 how meiosis works. In this example, each parent cell has four chromosomes, which come in two pairs. First the chromosomes in the cell double, producing eight chromosomes. Then the cell divides. During this cell division, the chromosomes separate and are equally distributed. So each of the two cells formed by this cell division has four chromosomes. Next, these cells divide again. So each of the resulting cells now has two chromosomes. That is, each of the last group of cells produced by meiosis has half the number of chromosomes as the original parent cell.

Chromosomes and Sex Determination

In 1907, American zoologist Thomas Hunt Morgan began his own study of chromosomes and genetics. Morgan experimented with tiny insects called fruit flies. He chose the fruit fly for three reasons: (1) they are easy to breed, (2) they produce new generations very quickly, and (3) their chromosomes are easy to study.

Morgan rapidly discovered something peculiar about the four pairs of chromosomes that each fruit fly had. In females, the chromosomes of each pair were the same shape. But in males, the chromosomes of one pair were not the same shape. One chromosome, which Morgan called the X chromosome, was rod-shaped. The other chromosome in the pair, which Morgan called the Y chromosome, was hook-shaped.

Through a number of experiments, Morgan discovered that the X and Y chromosomes determine the sex of an organism. In general, any organism that has two X chromosomes is a female. Any organism that has one X and one Y chromosome is a male. These X and Y chromosomes are called **sex chromosomes.**

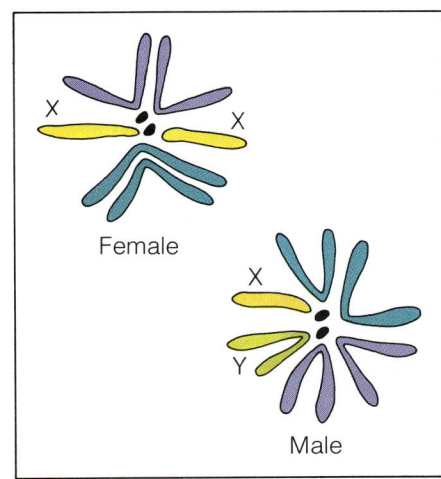

Figure 20–17 *A fruit fly has four pairs of chromosomes. A female fruit fly has two X chromosomes. A male fruit fly has an X and a Y chromosome. What are the X and Y chromosomes called?*

Mutations

In 1886, while out on a walk, Hugo De Vries (duh-VREES), a Dutch botanist, made an accidental discovery that would go beyond Mendel's work. De Vries came across a group of American evening primroses. As with pea plants, some primroses were

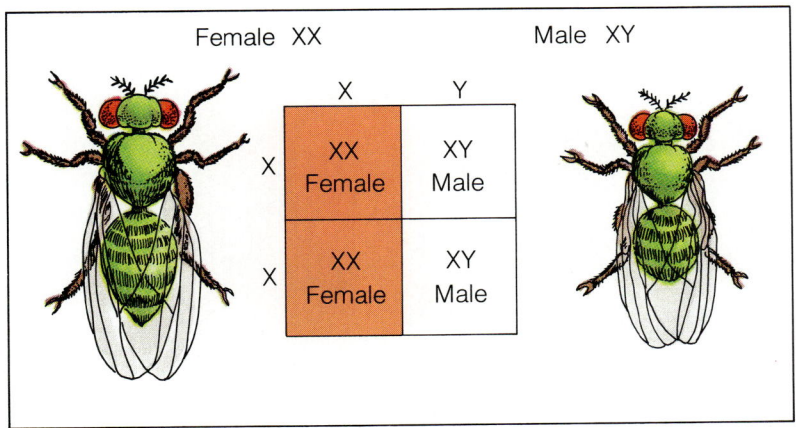

Figure 20–18 *You can see from the drawings of the male and female fruit flies that they have some physical differences. The Punnett square shows the probability of the sex of their offspring.*

Figure 20–19 *The albino koala (left) and seedless navel oranges (right) are the result of mutations. What is a mutation?*

very different from others. De Vries bred the primroses and got results similar to the results of Mendel's work with pea plants. But he also found that every once in a while, a new variety of primrose would grow—a variety that could not be accounted for by genetic laws. De Vries called these sudden changes in characteristics **mutations.** A mutation is a change in genes or chromosomes, which causes a new trait to be inherited.

If a mutation occurs in a body cell such as a skin cell, the mutation affects only the organism that carries it. But if a mutation occurs in a sex cell, then that mutation can be passed on to the next generation. The mutation may then cause a change in the characteristics of the next generation.

Most mutations are harmful, that is, they reduce an organism's chances for survival or reproduction. For example, sickle-cell anemia is a serious blood disease caused by a mutation in a gene. Sickle-cell anemia results in blood cells that are shaped like a sickle or half circle. People who have sickle-cell anemia have difficulty obtaining enough oxygen. This happens because the sickle-shaped cell cannot carry enough oxygen to all the cells in the body. The sickle-shaped cells may clump and clog tiny blood vessels.

Some mutations are helpful and cause desirable traits in living things. For example, when mutations occur in crop plants, the crops may become more useful to people. A gene mutation in potatoes has produced a new variety of potato called the Katahdin potato. This potato is resistant to diseases that attack other potatoes. Also, the new potato looks and tastes better than other types of potatoes.

It may seem as if mutations produce only helpful or harmful traits. This is not so. Many mutations are neutral and do not produce any obvious changes. Still other mutations are lethal, or deadly, and result in the death of an organism.

SECTION REVIEW

1. What is the role of chromosomes in heredity?
2. Explain the chromosome theory of heredity.
3. What is meiosis?
4. What is a mutation?
5. Why are mutations that do not occur in sex cells not passed on to future generations?

20–4 DNA

"We wish to suggest a structure for the salt of deoxyribose nucleic acid." So began a letter from two scientists in 1953 to a scientific journal. What followed in this letter was a description of the structure that would help unlock the deepest secrets of genetics.

The structure written about in the letter was deoxyribonucleic (dee-AHK-sih-righ-boh-noo-KLAY-ihk) acid, or **DNA.** DNA is the basic substance of heredity. Sometimes DNA is called the code of life because it contains all the information needed to make and control every cell of an organism.

The two scientists who wrote the letter about DNA were James Watson, an American biologist, and Francis Crick, a British biologist. In 1962, they, along with Maurice Wilkins, were awarded the Nobel Prize for physiology or medicine for their work on the structure of DNA. Many scientists believe that the discovery of the structure of DNA was the most important biological event of this century.

Watson and Crick's discovery showed that chromosomes are made of molecules of DNA. And the DNA molecules in chromosomes contain the genes. So DNA is the hereditary material that carries the genes that control all traits passed on from parents to offspring.

Figure 20–20 *This computer image shows the structure of a DNA molecule. What do the letters DNA stand for?*

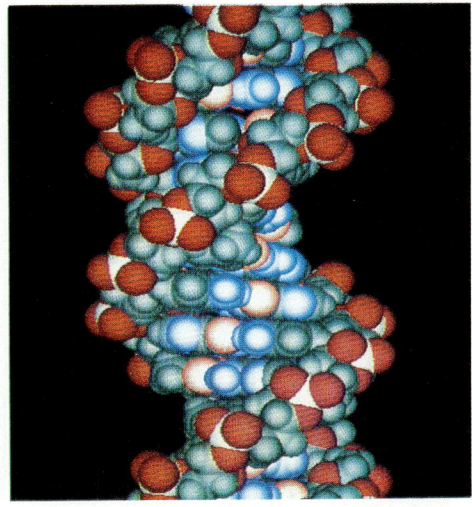

465

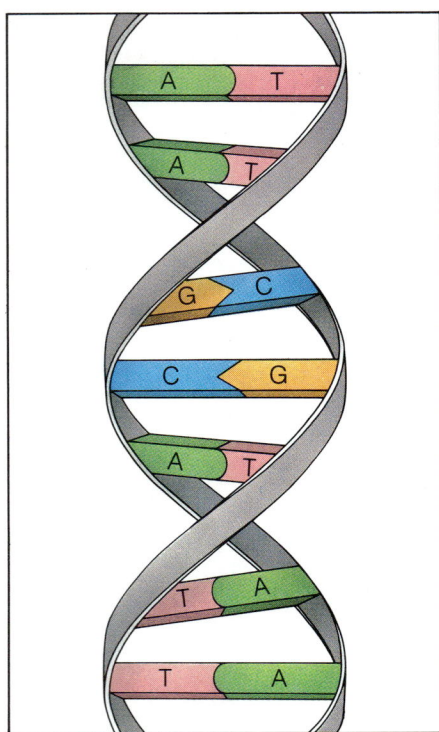

Figure 20–21 *This illustration shows the ladderlike structure of a DNA molecule. What substances make up the rungs of the DNA molecule?*

The Structure of DNA

The structure of DNA looks like a twisted ladder, or spiral staircase, with steps made of nitrogen bases. See Figure 20–21. Notice that the steps, or rungs, of this twisted ladder are formed by pairs of substances called **nitrogen bases.** Nitrogen bases are substances that contain the element nitrogen. There are four different nitrogen bases in DNA: adenine (AD-uhn-een), guanine (GWAH-neen), cytosine (SIGHT-uh-seen), and thymine (THIGH-meen).

You can see in Figure 20–22 that the two nitrogen bases that make up each step of the DNA ladder combine in very specific ways. Scientists use the capital letters A, G, C, and T to represent each of the four nitrogen bases. In the DNA ladder, adenine (A) always pairs with thymine (T). Guanine (G) always pairs with cytosine (C).

The DNA molecule that makes up a single chromosome may have hundreds or even thousands of pairs of nitrogen bases. So a DNA molecule may contain hundreds or thousands of rungs. In addition to discovering DNA's structure, Watson and Crick found out that the order of the nitrogen bases determines the particular genes of that DNA molecule. That is why DNA is said to contain the genetic code. The genetic code is actually the order of nitrogen bases on the DNA molecule.

Because a DNA molecule can have thousands of rungs and they can be in any order, the number of different genes that can form is almost limitless. This is why different organisms on the earth can

Figure 20–22 *In a DNA molecule, the nitrogen bases that make up the rungs of the DNA ladder always combine in a specific way. Which nitrogen base always combines with thymine?*

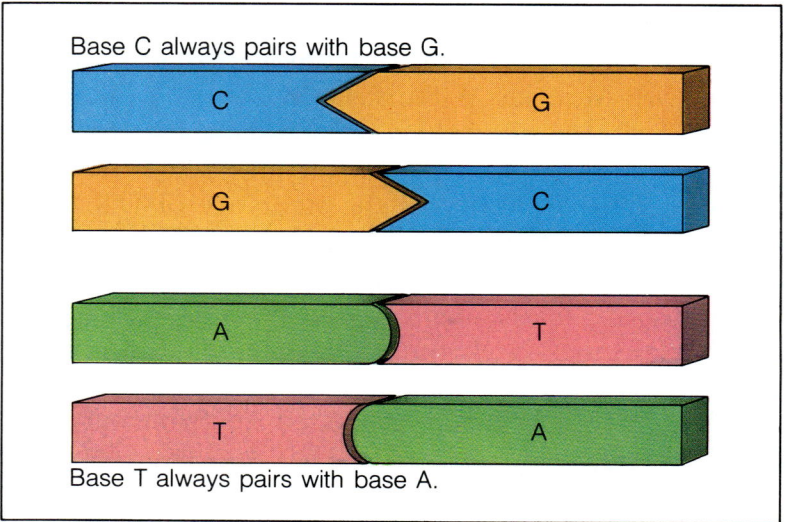

Base C always pairs with base G.

Base T always pairs with base A.

have so many different traits. By changing the order of one pair of nitrogen bases in a DNA molecule that has a thousand pairs, a new gene can form.

DNA Replication

In Chapter 3, you read how body cells divide into two similar cells in a process called **mitosis.** Before a body cell can divide into two cells during mitosis, the DNA in the nucleus must be duplicated, or copied, so that each new cell gets the same amount and kind of DNA as the original parent cell. During meiosis, DNA must also be duplicated. But just how does this happen?

The process in which DNA molecules form exact duplicates is called **replication** (rehp-luh-KAY-shuhn). During replication, the DNA ladder separates, or unzips. See Figure 20–24. The separation, or unzipping, occurs between the two nitrogen bases that form each rung of the DNA molecule. So after the first step in replication, the DNA has split into two "halves of a ladder." Next, free nitrogen bases that are "floating" in the nucleus begin to pair up with the nitrogen bases on each half of DNA. Remember that adenine (A) attaches to thymine (T) and guanine (G) attaches to cytosine (C).

Once each new base has attached itself to a base on each half of the DNA molecule, two new DNA molecules form. And each molecule is an exact

Figure 20–23 *This scientist is holding a flask that contains the nitrogen base cytosine. The cytosine is synthetic and was produced in the laboratory.*

Figure 20–24 *In this illustration, you can see how a DNA molecule duplicates itself in the process of replication.*

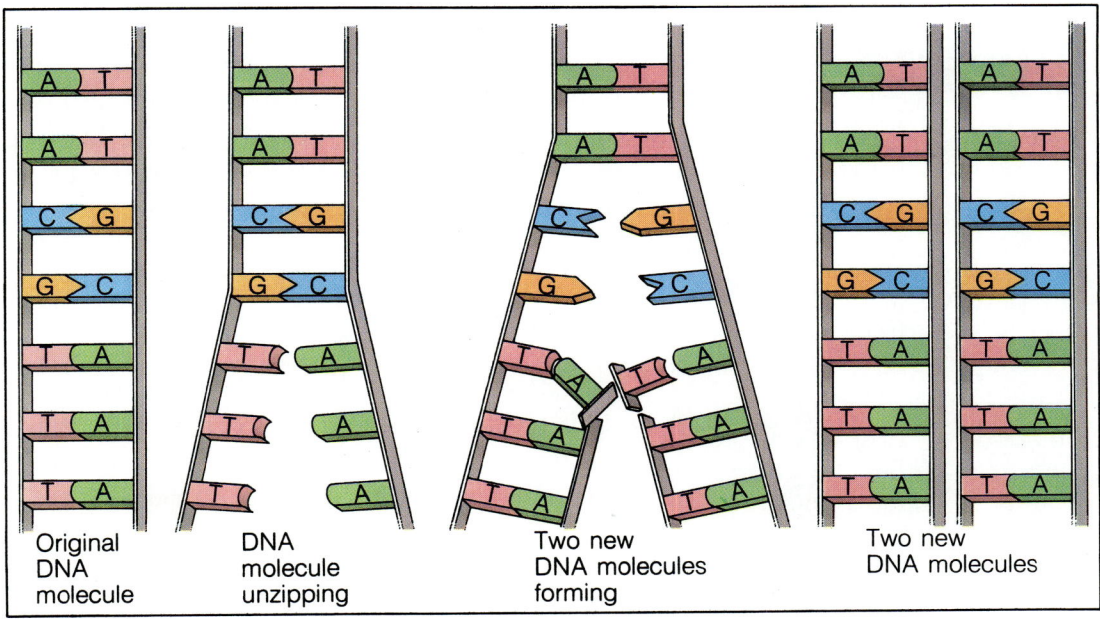

Original DNA molecule

DNA molecule unzipping

Two new DNA molecules forming

Two new DNA molecules

duplicate of the original DNA molecule. Look at Figure 20–24 on page 467 again.

SECTION REVIEW

1. Describe the structure of DNA.
2. List the four nitrogen bases.
3. What is replication?
4. On one-half of a molecule of DNA, the nitrogen bases are arranged in the following order: AGTTCT. What is the order of bases on the other half of the DNA molecule?

Section Objective

To relate the principles of genetics to human heredity

Figure 20–25 *In this photograph, you can see some of the 23 chromosome pairs that are found in human cells. How many chromosomes does each human body cell contain?*

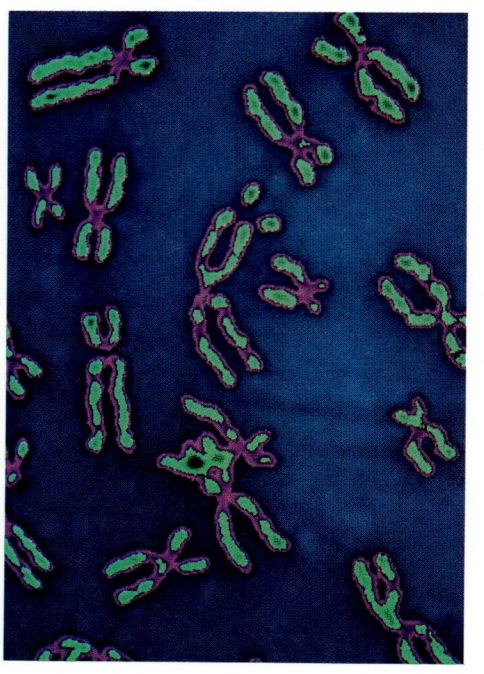

20–5 Human Genetics

Human beings, like all living things, are what they are because of the genes they inherit from their parents. These genes—and there are about 100,000 of them—are located on the 46 chromosomes in the nucleus of almost every body cell. An exception is the sex cells, which have half the number of chromosomes. The 46 chromosomes consist of 23 pairs. Each pair has matching genes for a particular trait such as eye color, hair color, and ear lobe shape.

Because a person gets one matching chromosome from each parent, the person gets matching genes from each parent. For example, you received genes for eye color from each of your parents. The way these genes combined determined your eye color.

As you can see, human genes seem to follow the same pattern of inheritance as the genes in the pea plants that Mendel studied more than a century ago. **Scientists now know that the basic principles of heredity discovered by Mendel can be applied to human heredity.** Now read on to discover more about human traits.

Multiple Alleles

Earlier in this chapter, you learned that a trait, such as the color of a flower, is determined by how a pair of genes act. You learned that one gene in the pair may be dominant or recessive. And the same can be said of the other gene in the pair.

In human beings, some traits are not so easily determined. Skin color, for example, is determined by a blending of four pairs of genes. Various combinations of these eight individual genes can produce all the skin colors of all people. Each member of a set of genes for a particular trait is called an **allele** (uh-LEEL). In many flowers, there are only two alleles for flower color. In human skin, there are eight alleles for color. So scientists say that the trait of human skin color is determined by **multiple alleles,** or more than two alleles.

The four major human blood groups are also determined by multiple alleles. These groups are called A, B, AB, and O. Scientists know blood groups are determined by multiple alleles because there is no way a single pair of alleles can produce four different characteristics.

Both the allele for group A blood and that for group B blood are dominant. When this combination is inherited—an A from one parent and a B from the other—the child will have AB blood. The O allele, however, is recessive. So a person who inherits an O and an A allele will have group A blood. A person who inherits an O and a B will have group B blood. What two alleles must a person inherit to have group O blood?

Inherited Diseases

Sometimes the structure of an inherited gene contains an error. If the gene controls the production of an important protein such as hemoglobin, the hemoglobin will also have an error in its structure. Hemoglobin is the red pigment in blood. In such a case, the hemoglobin may not do its job well. This is an example of what scientists call an inherited disease.

Although some inherited diseases can be treated, until recently there was no hope of curing them. To cure an inherited disease meant correcting the error in a gene. And no one knew how to do this.

Then in the early 1980s, scientists discovered a chemical that could slightly change the structure of genes. It was this structure that needed to be changed to get people with **sickle-cell anemia** to produce greater amounts of hemoglobin. A person

BLOOD GROUP ALLELES

Blood Groups	Combination of Alleles
A	AA or AO
B	BB or BO
AB	AB
O	OO

Figure 20–26 *Blood group in humans is determined by multiple alleles. What two possible allele combinations might a person with group A blood have?*

Sharpen Your Skills

Human Genetic Disorders

Using reference materials in the library, find out about Klinefelter's syndrome, Turner's syndrome, phenylketonuria, and Tay–Sachs disease. Write a report on each one of these disorders.

In your report, include how these genes are inherited, the symptoms of the disorder, and the treatment that is available for the disorder. Include a drawing on posterboard to show how each disorder is inherited.

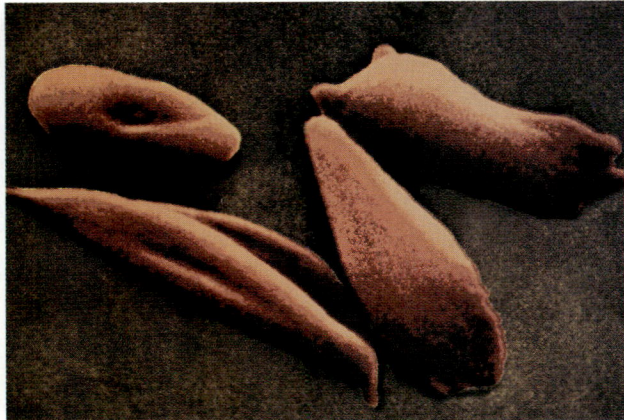

Figure 20–27 *The photograph on the left shows the normal shape of red blood cells. The photograph on the right shows the shape of red blood cells in people who have sickle-cell anemia. Can a person inherit sickle-cell anemia from only one parent?*

who has sickle-cell anemia inherits a damaged gene for the manufacture of hemoglobin from each parent. These genes are recessive and, among other things, cause red blood cells to become sickle shaped. See Figure 20–27.

In December 1982, scientists at three different hospitals in the United States reported that they had given the new chemical to three patients with sickle-cell anemia. Every patient's condition improved. But again, the scientists cautioned that more tests would have to be performed before the chemi-

---

CAREER

Genetic Counselor

HELP WANTED: GENETIC COUN-SELOR. Master's degree in genetic counseling, social work, or nursing required. Experience in medical genetics and counseling desired. Certification as genetic counselor a plus.

Human genetics is one of the fastest changing fields in medical science today. Although there are doctors who specialize in genetic disorders, they often do not have enough time to explain to patients and family members about a genetic disorder. Instead, **genetic counselors** provide this information. Genetic counselors talk with parents who are concerned that they might be carrying genes for a disease or trait that they could pass on to their children.

Most jobs for genetic counselors are found in medical centers and teaching hospitals. In addi-

tion to providing support services to families, counselors also become resource people for the public or do research in the field of medical genetics. Those interested in this fascinating and rewarding career can write to the March of Dimes, Birth Defects Foundation, 1275 Mamaroneck Avenue, White Plains, NY 10605.

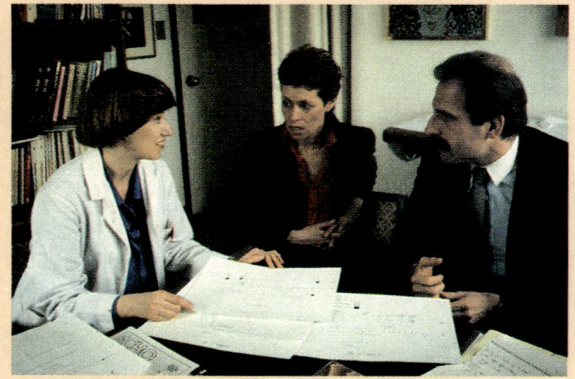

cal could be tried with large groups of patients. Then if the tests proved successful, this genetic disease might be curable.

Sex-Linked Traits

You will remember from earlier in the chapter that X and Y chromosomes are sex chromosomes. The X and Y chromosomes are the only chromosome pairs that do not always match each other. All body cells of normal human males carry one X chromosome and one Y chromosome. Females have two matching X chromosomes, or XX.

X chromosomes also carry genes for traits other than sex. However, Y chromosomes carry few, if any, genes other than those for maleness. Therefore, any gene—even a recessive gene—carried on an X chromosome will produce a trait in a male who inherits the gene. This is because there is no matching gene on the Y chromosome. Such traits are called **sex-linked traits** because they are passed from parent to child on a sex chromosome, the X chromosome. Because a female has two X chromosomes, a recessive gene on one X chromosome can be masked, or hidden, by a dominant gene on the other X chromosome.

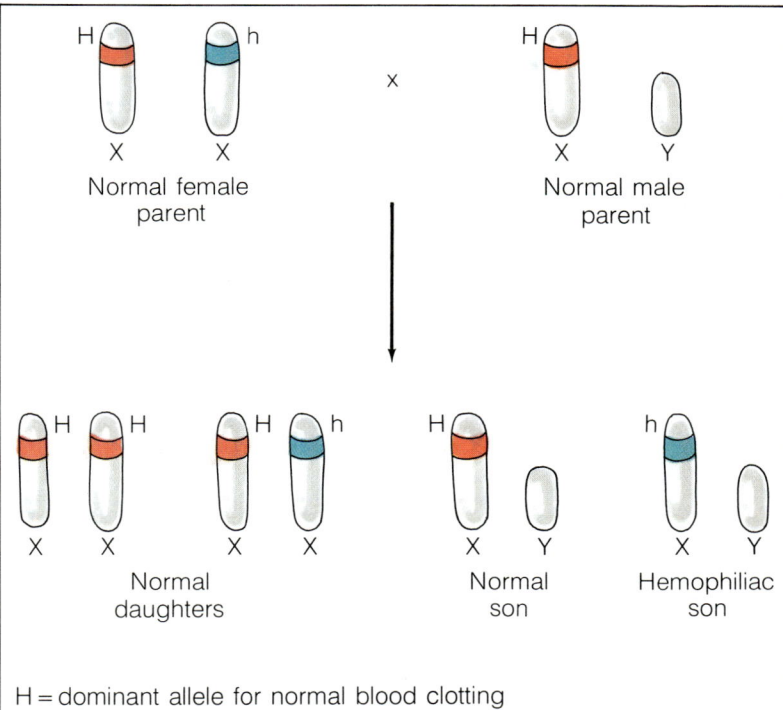

H = dominant allele for normal blood clotting
h = recessive allele for hemophilia

Figure 20–28 *You can see in this illustration that there is a 25 percent chance that a female who carries a gene for hemophilia and a normal male will have a male child with hemophilia. Hemophilia is a sex-linked trait.*

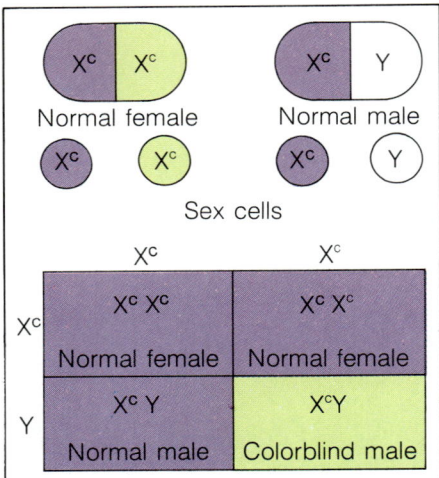

Figure 20–29 *Colorblindness is a sex-linked trait. Very few females are colorblind because a female has two X chromosomes. So a recessive gene on one X chromosome can be hidden by a dominant gene on the other X chromosome.*

Figure 20–30 *Down syndrome is a condition in which all the body cells have an extra twenty-first chromosome. What is the total number of chromosomes in the body cells of a person with Down syndrome?*

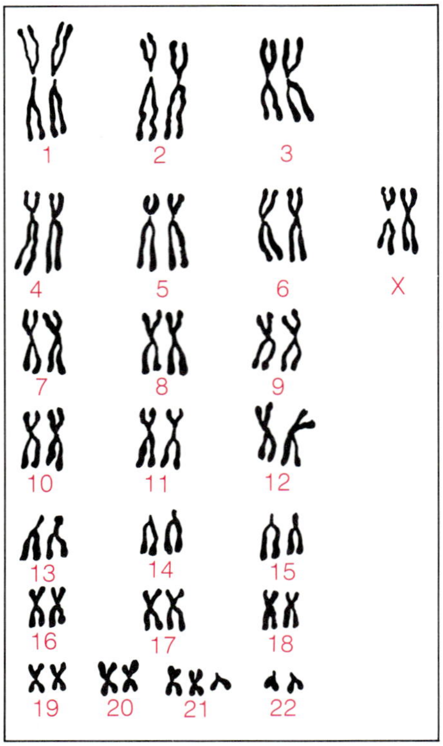

An example of a sex-linked recessive trait is **hemophilia** (hee-muh-FIHL-ee-uh). Hemophilia is an inherited disease that causes the blood to clot slowly or not at all. See Figure 20–28 on page 471. This disease was very common in the royal families of Europe. During the nineteenth century, Queen Victoria of England had a son and three grandsons with hemophilia. At least two of her daughters and four of her granddaughters carried the gene for hemophilia on one X chromosome. But they did not have the disease because they carried a gene for normal blood clotting on their other X chromosome. Hemophilia spread through the royal families in Europe as Victoria's descendants married other royalty and passed the hemophilia gene on.

Another sex-linked recessive trait is **colorblindness.** For this reason, there are more males who are colorblind than there are females. A female has to inherit two recessive genes for colorblindness. A male has to inherit one such gene. The most common type of colorblindness involves difficulty in distinguishing between the color red and the color green.

Nondisjunction

During meiosis, or the process through which sex cells are formed, chromosome pairs usually separate. But in rare cases a pair may remain joined. This failure of chromosomes to separate is known as **nondisjunction** (nahn-dihs-JUHNGK-shuhn). When this happens, body cells inherit either extra or fewer chromosomes than normal.

Look carefully at Figure 20–30. This is an example of a human karyotype (KAR-ee-uh-tighp). A karyotype shows the size, number, and shape of chromosomes in an organism. Usually, people have 46 chromosomes, or 23 chromosome pairs. But now look again at Figure 20–30. How many chromosomes are there? If you said 47, you are correct. The extra chromosome is found in what would normally be the twenty-first pair. When a person has an extra chromosome on the twenty-first pair, a condition called **Down syndrome** results. People with Down syndrome may have various physical problems and some degree of mental retardation. How-

Figure 20–31 *Although people with Down syndrome are physically challenged, many lead full, active, and productive lives.*

ever, many people with Down syndrome hold jobs and provide important contributions to society.

Is there a way of determining before a child is born whether he or she will have Down syndrome or another inherited problem? There are a number of ways. One way to determine before a child is born whether the child will have an inherited disease is called **amniocentesis** (am-nee-oh-sehn-TEE-sihs). Amniocentesis involves taking a little bit of fluid out of the sac that surrounds a baby while it is still in its mother. This fluid contains some of the baby's cells. Using special techniques, doctors and scientists can examine the chromosomes of the cells. By doing this they can, for example, discover whether an unborn child does or does not have Down syndrome. Various other tests can reveal the presence of a variety of inherited disorders. Scientists hope that such tests will eventually lead to the treatment of some disorders before babies are born.

SECTION REVIEW

1. What are multiple alleles?
2. If a woman with blood group O marries a man with blood group B, can they have a child with blood group A? Explain.
3. What are sex-linked traits?
4. A man and a woman each have a gene for sickle-cell anemia. What is the probability that their child will inherit one gene or both genes for sickle-cell anemia?

LABORATORY INVESTIGATION

Observing Dominant and Recessive Traits

Problem

What are the phenotypes of dominant and recessive human traits?

Materials *(per student)*

paper
pencil

Procedure

1. On a sheet of paper, copy the data table. Tongue-rolling is the ability to roll the tongue into a U shape. Free ear lobes are those that hang below the point of attachment to the head. Attached ear lobes are those that are attached directly to the side of the head. A widow's peak is a hairline that forms a distinct point in the center of the forehead.
2. Count the number of your classmates who have each of the traits listed in the table. Fill in the numbers for each trait in your data table.
3. To determine the percentage of students demonstrating each trait, divide the number who have the trait by the total number of students in your class and multiply by 100.

Observations

1. Which trait is the most common in your class? The least common?
2. Are there any students who have traits that are intermediate between the dominant and recessive trait? How many and what kind?

Conclusions

1. Do dominant traits occur more often than recessive traits? Explain.
2. Predict what would happen to your results if you were to observe four other classes of students.
3. How do you account for those students whose traits are intermediate between dominant and recessive?

Traits		Number of Students Demonstrating Dominant Trait	Number of Students Demonstrating Recessive Trait	Percentage Demonstrating Dominant Trait	Percentage Demonstrating Recessive Trait
Dominant	*Recessive*				
Tongue-roller	**Non-roller**				
Free ear lobes	**Attached ear lobes**				
Dark hair	**Light hair**				
Widow's peak	**Straight hair line**				
Non-red hair	**Red hair**				
Total					

CHAPTER REVIEW

SUMMARY

20–1 Inheritance of Traits

❏ Gregor Mendel found patterns in the way traits are inherited.

❏ Genetics is the study of the passing on of traits from an organism to its offspring.

❏ A purebred trait has genes that are alike. A hybrid has genes that are different.

❏ "Stronger" traits are called dominant. "Weaker" traits are called recessive.

❏ In some gene pairs, neither gene is dominant nor recessive. This is known as incomplete dominance.

20–2 Genetics and Probability

❏ Scientist use probability in predicting the results of genetic crosses.

❏ Punnett squares show possible results of a cross between two organisms.

❏ A phenotype is a visible characteristic, while a genotype is the gene makeup.

20–3 Chromosomes

❏ The chromosome theory states that chromosomes are the carriers of genes.

❏ The process of meiosis produces sex cells.

❏ A mutation is a change in the genes that cause the inheritance of a new trait.

20–4 DNA

❏ DNA, the basic substance of heredity, resembles a twisted ladder, with rungs made of nitrogen bases.

❏ In DNA replication, two identical copies of the original DNA molecule are formed.

20–5 Human Genetics

❏ Many human traits are controlled by multiple alleles.

❏ In sickle-cell anemia, a damaged gene is responsible for making abnormal hemoglobin.

❏ Sex-linked traits are passed from parent to child on the X chromosome.

VOCABULARY

Define each term in a complete sentence.

allele	dominant	hypothesis	nondisjunction	sex-linked trait
amniocentesis	Down syndrome	incomplete dominance	phenotype	sickle-cell anemia
chromosome	gene	meiosis	probability	trait
chromosome theory	genetics	mitosis	recessive	
colorblindness	genotype	multiple allele	replication	
cross-pollination	hemophilia	mutation	self-pollination	
DNA	hybrid	nitrogen base	sex chromosome	

CONTENT REVIEW: MULTIPLE CHOICE

On a separate sheet of paper, write the letter of the answer that best completes each statement.

1. The study of the passing on of characteristics from one organism to its offspring is
   a. microbiology.   b. ecology.   c. genetics.   d. botany.

2. The factors, or units of heredity, that produce traits are
   a. genes.   b. mutations.   c. hybrids.   d. triplets.

3. The process by which sex cells are formed is known as
a. mitosis.　　b. incomplete dominance.　　c. replication.　　d. meiosis.

4. An organism that has genes that are different for a trait is known as a
a. multiple allele.　　b. dominant.　　c. recessive.　　d. hybrid.

5. When the genes in a gene pair are identical, the organism is said to be
a. dominant.　　b. hybrid.　　c. purebred.　　d. hemophilic.

6. The gene makeup of an organism is called its
a. genotype.　　b. multiple allele.　　c. mutation.　　d. phenotype.

7. Which states that chromosomes are the carriers of genes?
a. replication　　b. nondisjunction　　c. chromosome theory　　d. incomplete dominance

8. The basic substance of heredity is
a. guanine.　　b. DNA.　　c. RNA.　　d. cytosine.

9. If a girl inherits the allele for group B blood from her mother and the allele for group O blood from her father, she will have blood group
a. A.　　b. B.　　c. AB.　　d. O.

10. A condition that is caused by nondisjunction of the twenty-first chromosome pair is
a. hemophilia.　　b. colorblindness.　　c. sickle-cell anemia.　　d. Down syndrome.

CONTENT REVIEW: COMPLETION

On a separate sheet of paper, write the word or words that best complete each statement.

1. A (An) _____ is a characteristic of an organism.

2. Organisms that have genes that are different for a particular trait, such as Tt, are called _____.

3. If a purebred red (RR) and a purebred white (WW) four-o'clock plant were crossed, the phenotype of the offspring would be _____.

4. The possibility that an event may or may not take place is known as _____.

5. Genes are located along rod-shaped structures called _____.

6. The X or Y chromosome is called a (an) _____.

7. A sudden change in the chromosomes or genes of an organism is known as a (an) _____.

8. In DNA, the nitrogen base adenine bonds with the nitrogen base _____.

9. The process by which new molecules of DNA are produced is known as _____.

10. Colorblindness is an example of a (an) _____.

CONTENT REVIEW: TRUE OR FALSE

Determine whether each statement is true or false. Then on a separate sheet of paper, write "true" if it is true. If it is false, change the underlined word or words to make the statement true.

1. Thomas Hunt Morgan is known as the Father of Genetics.

2. In the genotype Yy, y represents the recessive gene.

3. An organism that has genes that are different for a trait is called a hybrid.

4. The process of meiosis produces sex cells.

5. X and Y chromosomes are <u>body</u> chromosomes.
6. In DNA, cytosine is attached to <u>thymine</u>.
7. The process of making more DNA is called <u>meiosis</u>.
8. Human blood groups are determined by <u>multiple alleles</u>.
9. Colorblindness is a <u>sex-linked</u> trait.
10. The failure of chromosomes to separate is known as <u>nondisjunction</u>.

CONCEPT REVIEW: SKILL BUILDING

Use the skills you have developed in the chapter to complete each activity.

1. **Sequencing events** Make a time line of the major events in the history of genetics described in this chapter. Use the events for which dates are given. Begin with Mendel's work.
2. **Making predictions** In a family, there are four daughters. What is the probability that a fifth child will be female? Does the fact that the family has four daughters increase their probability of having another daughter? Explain.
3. **Applying concepts** Using probability, how many times is "heads" likely to occur when a coin is tossed 50 times?
4. **Applying concepts** If a person developed a mutation in the skin cells, could this mutation be passed on to his or her children? Explain.
5. **Relating concepts** A mouse's skin cell has 40 chromosomes. How many chromosomes did the mouse receive from each of its parents? How many chromosomes will be present in a sex cell of this mouse?
6. **Applying concepts** In horses, black color (B) is dominant over chestnut color (b). If a hybrid black horse is mated to a chestnut horse, what is the probability of their producing a colt that is black?
7. **Interpreting diagrams** In one family, a male affected with an inherited disease married a woman who did not have the disease. They had four children. Two of these children married and had their own children. Using the diagram, describe how the disease is inherited.

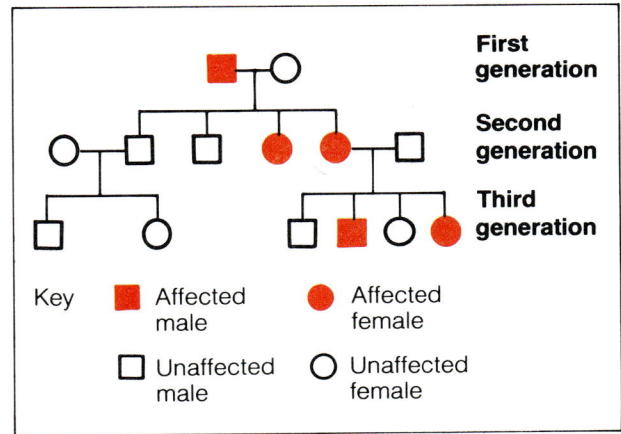

Key
- ■ Affected male
- ● Affected female
- □ Unaffected male
- ○ Unaffected female

First generation
Second generation
Third generation

CONCEPT REVIEW: ESSAY

Discuss each of the following in a brief paragraph.

1. What is meant by a dominant trait? A recessive trait?
2. Describe the difference between a purebred and a hybrid organism.
3. In a Punnett square, show the cross between a hybrid black guinea pig and a purebred white guinea pig. Which of these traits is dominant?
4. How are mutations harmful? How are they helpful? Give an example of each.
5. Explain the contribution of Thomas Hunt Morgan to genetics.
6. Describe DNA replication.
7. Explain how human blood groups are inherited.

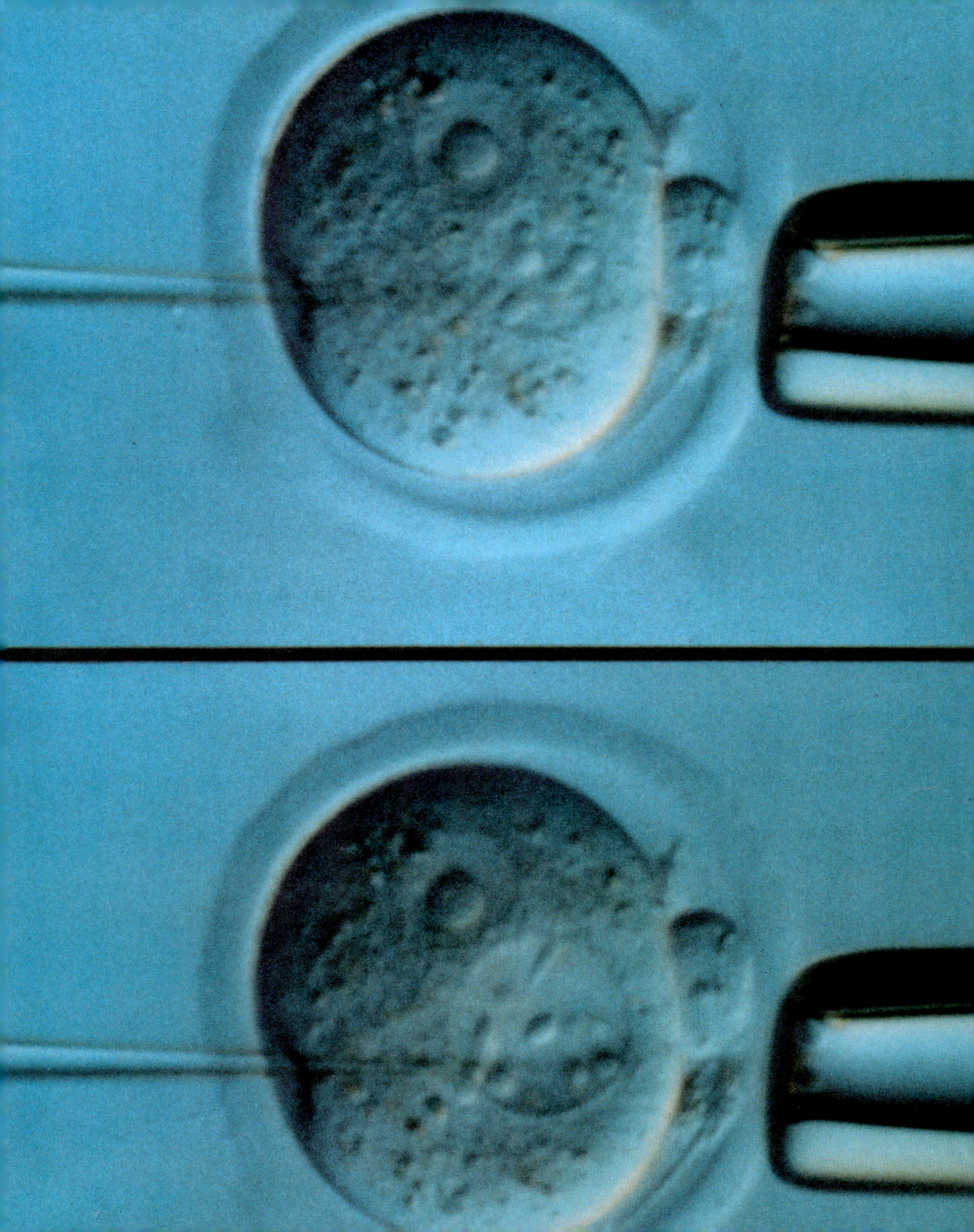

Applied Genetics 21

CHAPTER OBJECTIVES

After completing this chapter, you will be able to

21–1 Define biotechnology.

21–1 Describe a method for producing recombinant DNA.

21–1 List five products of recombinant DNA.

21–2 Explain methods used by breeders to produce organisms with desired characteristics.

21–2 Compare hybridization and inbreeding.

Do you know how to make a supermouse? Myrna Trumbauer does. Trumbauer is a biologist. She works with fertilized mouse eggs much like the ones shown in the photograph on the opposite page.

Trumbauer's "recipe" for a supermouse involves placing a fertilized egg from a normal-sized mouse under a special microscope that magnifies the egg about 400 times. Then, she very carefully injects a clear liquid containing certain human genes into the egg. How does this "recipe" produce a supermouse?

The answer is hidden in the human genes. Remember that human genes carry information that directs the production of certain substances. The genes that Trumbauer injects into a mouse egg produce human growth hormone. This hormone, a chemical messenger that controls growth is normally produced by a gland near the brain. Inside the fertilized mouse egg, the "transplanted" genes cause the mouse to produce the human growth hormone. As the mouse develops, the hormone causes it to grow to twice its normal size. In other words, it grows into a supermouse!

You may wonder of what use is a supermouse. To Trumbauer it is not just a bigger mouse. She and others at the University of Pennsylvania are interested in developing methods of moving genes from one organism to another. Soon, scientists hope to transplant genes into farm animals to make them resistant to certain diseases. This new technique is just one example of applied genetics, genetics used in everyday life. As you read on, you will discover more about other new developments in applied genetics.

The nuclei of these fertilized mouse eggs are being injected with a liquid containing human genes.

21–1 Genetic Engineering

Unlike Gregor Mendel who had to wait weeks for the results of his experiments, today's scientists can pluck a gene for a specific trait from one organism and transplant it into another. This is possible because of great advances in **biotechnology** (bigh-oh-tehk-NAHL-uh-jee). Biotechnology is the application of technology to the study and solution of problems involving living things. For example, scientists have placed genes, or hereditary factors, into one-celled microorganisms, such as yeasts and bacteria, to make copies of human proteins. Some of these proteins include a hormone for healing wounds, a substance for treating cancer, and an enzyme for reducing inflammation.

To date, more than 150 products produced by biotechnology are being used in medicine and in agriculture. Another hundred or more new products are in various stages of development. So you can see that biotechnology may soon influence your life more than any other technology will. As you read on, you will discover some of the other contributions of biotechnology.

Biotechnology is not a new field in science. For thousands of years, microorganisms have been used for various reasons, such as making cheese and causing bread to rise. And, in this century, other microorganisms have been put to work making antibiotics, such as penicillin.

Figure 21–1 *The tiny transparent objects in the dish are synthetic celery seeds. The genetic information inserted into these seeds was removed from celery plants such as the one in this photograph.*

But in the past 10 years or so, a new discovery in biotechnology called **genetic engineering** has produced all sorts of products never before possible or even imagined. **Genetic engineering is the process in which genes, or parts of DNA, from one organism are transferred into another organism.** The transferred parts of the donating organism's DNA are joined to the DNA of the receiving organism. The new pieces of combined DNA are called **recombinant DNA.**

In genetic engineering, DNA usually is transferred from a complex organism, such as a human being, into a simpler one, such as a bacterium or a yeast cell. As the bacterium or yeast cell reproduces, copies of the recombinant DNA are passed from one generation to the next. Bacteria and yeast cells are used because they reproduce quickly.

Making Recombinant DNA

Scientists use special processes to make recombinant DNA. Let's use human and bacterial DNA as examples to illustrate this process. Some bacterial DNA comes in the form of a ring called a **plasmid.** You might think of a plasmid as a circle of rope. Using special techniques, a scientist removes the

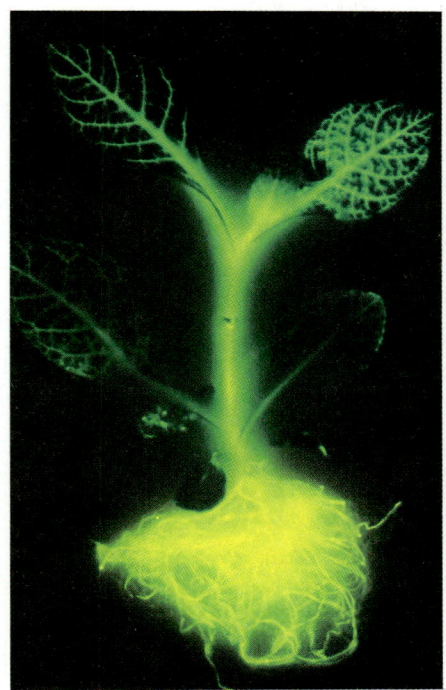

Figure 21–2 *Scientists have changed the genetic makeup of this tobacco plant by placing a gene from a firefly into the plant's DNA. Now the tobacco plant can glow in the dark just like a firefly can.*

Figure 21–3 *During the formation of recombinant DNA, a plasmid from a bacterium, such as* E. coli, *is snipped open. A short piece is then removed from the DNA of a human cell. This human DNA is inserted into the snipped bacterium plasmid. Then the plasmid is placed back into the bacterium. What is this technique called?*

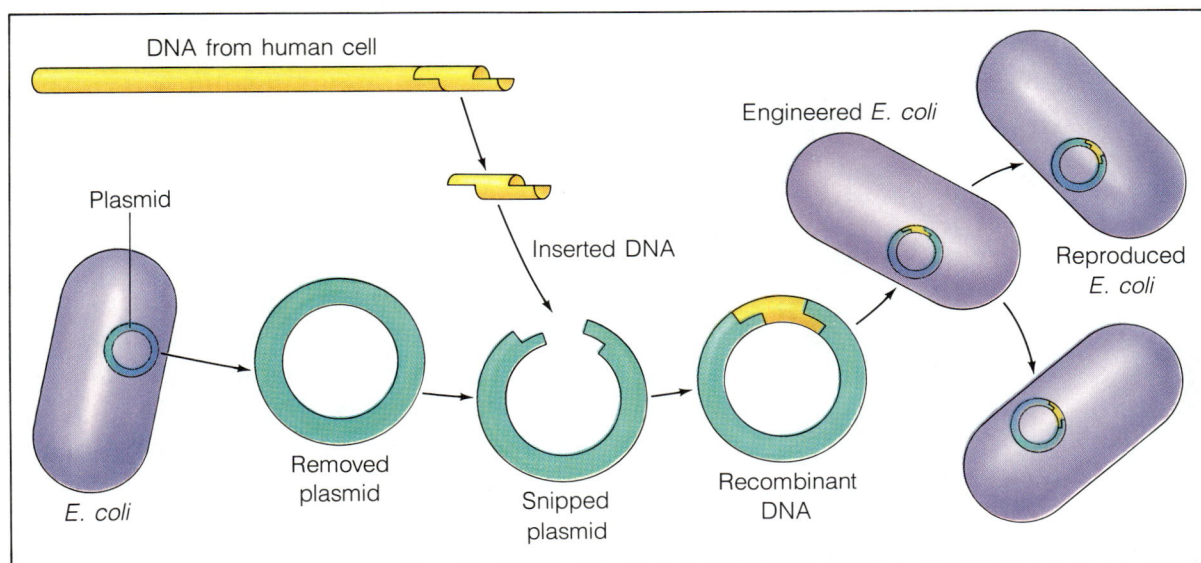

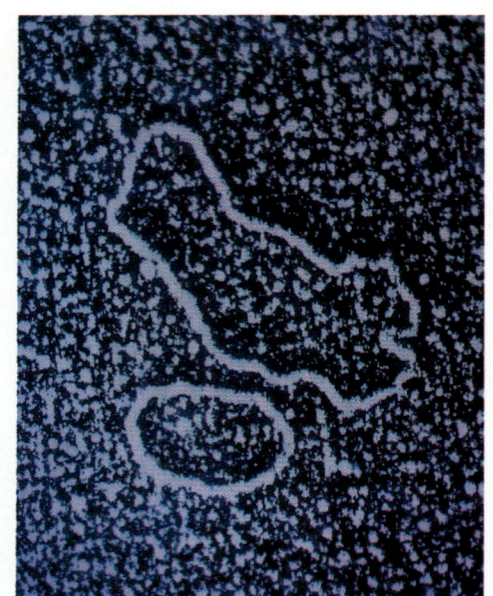

Figure 21–4 *This photograph shows two plasmids from the bacterium* E. coli. *A single plasmid contains more bits of information than a home computer. What is a plasmid?*

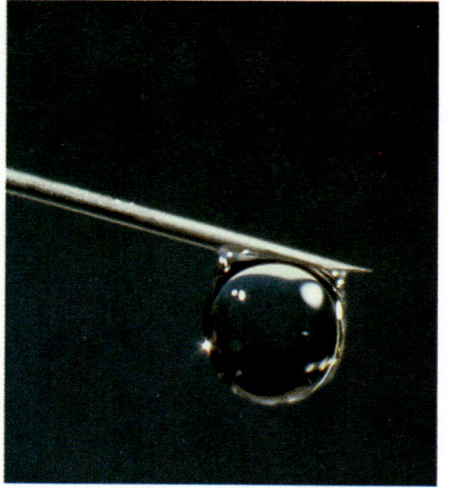

Figure 21–5 *This drop of human insulin was the first drop to be made by genetically engineered bacteria. What disorder occurs when the body does not produce insulin?*

plasmid from a bacterium and "snips" it open. The scientist then removes a short piece of DNA from a human cell. You might think of this as a short length of rope. The scientist "ties" this length of "rope" to the cut ends of the bacterial DNA. Now the bacterial DNA forms a ring again. But this DNA ring has something new in it: genes that direct the production of a human protein!

Finally, the scientist puts this combined DNA back into the bacterial cell. What happens now? The bacterial cell *and all its offspring* will produce the human protein coded for by the piece of human DNA.

Products of Recombinant DNA

Scientists use recombinant DNA to turn certain bacteria and yeast cells into protein factories. Grown in huge containers, billions of genetically engineered bacteria produce large quantities of proteins. These proteins are used in testing for diseases such as AIDS and strep throat. Other proteins are used to treat human disorders. Still others are human and animal vaccines that help organisms fight disease. Today there are more than 150 such protein "products" available!

IN MEDICINE One important product of genetic engineering is human insulin. Without this hormone, the level of sugar in the blood rises and causes a disorder called diabetes mellitus. The treatment for this disorder can involve one or more injections of insulin daily. The insulin used in the past came from animals such as pigs and cattle. However, many people with diabetes were allergic to this insulin. Moreover, the animal insulin was in short supply. Then, a few years ago, human insulin was produced by genetically engineered bacteria. Scientists expect that this insulin will be plentiful, will not cause allergies, and will be inexpensive.

Another substance that bacteria make through genetic engineering is human growth hormone. This chemical, produced by a gland near the brain, controls growth. A lack of this hormone prevents people from growing normally. Until 1981, children who did not have enough human growth hormone were given injections of the hormone. However, there was only a limited supply of human growth hormone and many children could not be treated. Then, in 1982, bacteria were genetically engineered to produce human growth hormone. Now an almost unlimited supply is available. Scientists think that this hormone may also be useful in treating burns and broken bones.

Vaccines, too, can be produced through recombinant DNA. When introduced into a person's body, a vaccine triggers the production of antibodies, or "combat" chemicals. These antibodies protect a person from disease. Vaccines are made from the disease-causing viruses or bacteria. Making vaccines for certain diseases, such as hepatitis B, a serious liver disease, is expensive. Now, scientists can remove a gene from the hepatitis B virus and insert it into a yeast cell. The yeast cell multiplies rapidly and makes large amounts of a protein from the virus. This protein helps to form the hepatitis B vaccine. Now making the hepatitis B vaccine is less expensive than it was before.

Another product produced by genetic engineering is interferon. Interferon is normally produced by human body cells and helps cells fight viruses that enter the body. How does interferon fight viruses? When a virus enters a cell, the cell produces interferon. Interferon then leaves the infected cell and "alerts" the surrounding cells to produce their own antiviral chemicals. For a long time, scientists believed that interferon could be used to treat many viral infections of human beings and, perhaps, even cancer. However, there was not enough interferon

Figure 21–6 *This 800-liter container is used to manufacture large amounts of recombinant DNA. The DNA is used to make the hepatitis B vaccine. What type of microorganism is used to make this vaccine?*

Figure 21–7 *These flasks contain interferon, which was produced by recombinant DNA. What is interferon?*

483

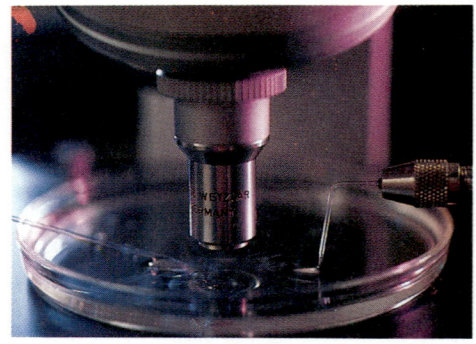

Figure 21–8 *Using a microscope and tiny dissecting instruments, scientists can change the DNA in cells in order to make the DNA produce new products.*

available to test this idea. Today, scientists are able to produce interferon from genetically engineered bacteria. And many experiments are underway to discover whether interferon can be used as a new weapon against some human diseases.

IN AGRICULTURE Imagine being able to engineer plants to resist disease. Or being able to develop a type of wheat that can withstand drought or cold. With genetic engineering, scientists think that plants with new and useful characteristics may be possible by the year 2000.

Recently, scientists have found a way to protect a plant from a disease caused by the tobacco mosaic virus. This virus attacks and damages tobacco and tomato plants. Genetic engineers have inserted a gene from the tobacco mosaic virus into plant cells. For some unknown reason, the gene makes the plant resistant to tobacco mosaic disease.

Another interesting use of genetic engineering in agriculture is the development of the "ice-minus" bacteria. These genetically engineered bacteria help to slow down the formation of frost on plants. Scientists have discovered that the bacteria that normally live on the leaves of many plants have an ice-forming gene. This gene produces a protein that triggers the formation of ice. As the temperature drops below 0°C, the temperature at which water freezes, the moisture on the plant becomes ice. This causes the fluid in the plant to freeze and the cells

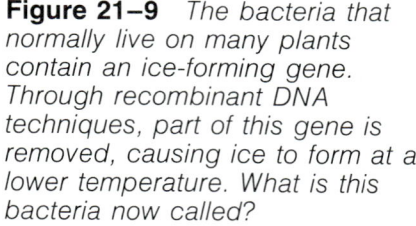

Figure 21–9 *The bacteria that normally live on many plants contain an ice-forming gene. Through recombinant DNA techniques, part of this gene is removed, causing ice to form at a lower temperature. What is this bacteria now called?*

to die. When the ice-forming genes are removed from the bacteria, the protein is not produced. As a result, ice forms, but at a lower temperature, −4°C. These bacteria are now called "ice-minus" bacteria. Scientists hope that one day they can spray the "ice-minus" bacteria onto plants such as strawberries and oranges to protect them from frost damage.

SECTION REVIEW

1. What is the technique of transferring DNA from one cell to another called?
2. What is a plasmid?
3. Why are one-celled microorganisms, such as bacteria and yeast cells, able to produce large amounts of recombinant DNA products?

21–2 Plant and Animal Breeding

Section Objective

To describe the uses of selective breeding

Have you ever seen a square tomato or a lean cow? These unusual organisms were developed by **selective breeding.** Selective breeding is the crossing of animals or plants that have desirable characteristics to produce offspring with desirable characteristics.

Using special methods, animal and plant breeders develop organisms through selective breeding that are larger in size, produce more food, or are resistant to certain diseases. Why is an increase in production important to world population?

Other traits, such as "squareness" in tomatoes, can be developed, too. Why would people want

Figure 21–10 *Using special methods, animal and plant breeders develop organisms that have desirable traits. Stronger work horses (left) and larger eggs (right) are examples of ways in which we benefit from selective breeding. What is selective breeding?*

Figure 21–11 *Hybridization is the crossing of two genetically different but related species of organisms. The mule is a cross between a female horse and a male donkey. What is the purpose of hybridization?*

square tomatoes or lean cows? Square tomatoes can fit more easily into crates than round tomatoes can. And the beef from leaner cows is more healthful than the beef from fatter cows. Can you give any other examples of selective breeding?

Hybridization

How do breeders develop square tomatoes? Sometimes they combine two or more different traits from the parents in one offspring. To do this, breeders use a technique called **hybridization** (high-bruh-duh-ZAY-shuhn). Hybridization is the crossing of two genetically different but related species of organisms. When the organisms are crossed, a **hybrid** is produced. A hybrid is an organism that has the traits of both parents. In some ways, hybrid offspring are often better than either parent. For example, the hybrid may grow faster or larger. These offspring are said to have **hybrid vigor.**

Farmers often use hybridization to produce plants that resist common pests or that produce food. Today, most hybrids are developed through breeding techniques. Scientists estimate that within 10 years, however, genetic engineering may be a primary method of producing hybrids.

Figure 21–12 *A breeding technique that is opposite to hybridization is inbreeding. Inbreeding is used to keep various types of animals, such as cocker spaniels (left) and cheetahs (right), pure so that they keep their basic characteristics. What is inbreeding?*

Inbreeding

A breeding technique that is opposite to hybridization is **inbreeding.** Inbreeding is a type of breeding that involves crossing plants or animals that have the same or very similar sets of genes, rather than different ones. Inbreeding is used to keep various breeds of dogs, such as cocker spaniels, pure.

Unfortunately, inbreeding reduces an offspring's chances of inheriting new gene combinations. Put another way, inbreeding produces organisms with genetic similarity. But this similarity, or rather the lack of genetic difference, concerns many scientists. They are worried that the lack of genetic difference in plants and animals may cause the organisms to be susceptible to certain diseases or changing environmental conditions. If all the organisms have the same genes, then they are all susceptible to the same diseases. As a result, these organisms might eventually become extinct, or die off.

SECTION REVIEW

1. What are some uses of selective breeding?
2. What is hybridization?
3. Define inbreeding.
4. Explain why animals that are produced through inbreeding look so much alike.

Observing the Growth of Mutant Corn Seeds

Problem

What is the effect of a mutation on the growth of corn plants?

Materials (per group)

10 albino corn seeds   string
10 normal corn seeds   tape
flower box   water
potting soil

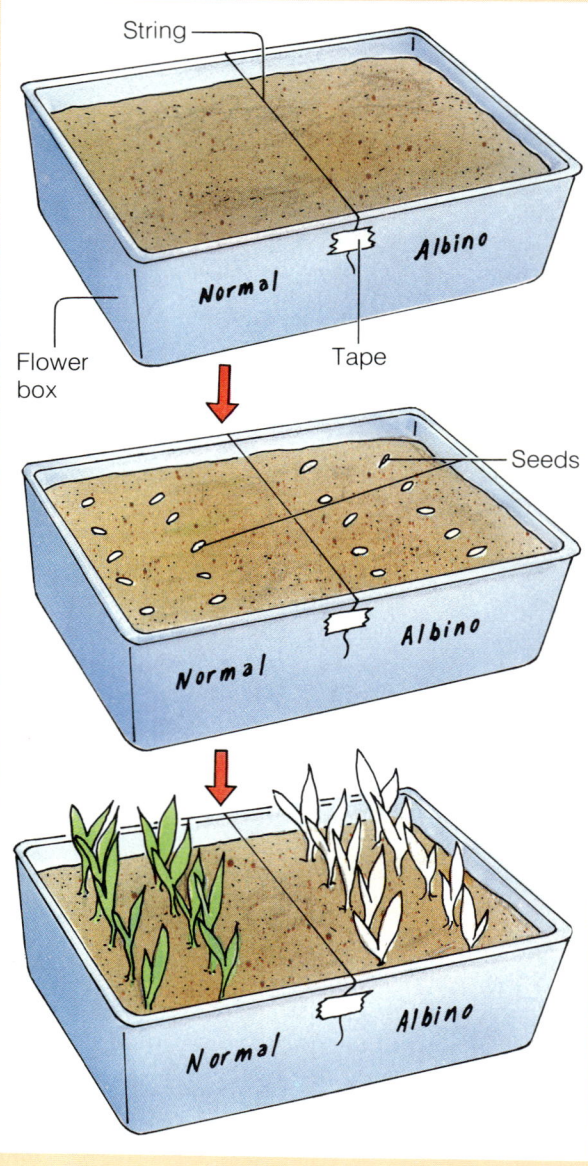

String

Normal   Albino

Flower box   Tape

Seeds

Normal   Albino

Normal   Albino

Procedure

1. Fill a flower box about three-fourths full with potting soil.
2. Place the string widthwise across the top of the flower box so that the string divides the box into two halves.
3. Label the right side of the box "albino." Label the left side of the box "normal."
4. On the side labeled albino, plant each of the albino seeds about 1 cm below the soil's surface. The seeds should be spaced at least 1 cm from each other. Moisten the soil.
5. On the side of the box labeled normal, plant the normal seeds as you did the albino seeds in step 4.
6. Place the flower box on a table near a window or on a windowsill where it will receive sunlight. Keep the soil moist. Observe daily for three weeks.

Observations

1. What was the total number of seeds that sprouted?
2. How many albino seeds sprouted? How many normal seeds sprouted?
3. Describe the difference between the albino plants and the normal plants.

Conclusions

1. Do the albino seeds sprout as well as the normal seeds? Explain.
2. What happened to the plants a week after they sprouted?
3. Which seeds, albino or normal, showed the mutation?
4. What effect did the mutation have on the growth of corn seeds?
5. If you were a farmer, which group of plants, albino or normal, would have the desired characteristic? Explain.

CHAPTER REVIEW

SUMMARY

21–1 Genetic Engineering

❑ Biotechnology is the application of technology to the study and solution of problems of living things.

❑ When DNA from one organism is removed and inserted into another organism, the new piece of DNA is known as recombinant DNA. The technique by which recombinant DNA is made is called genetic engineering.

❑ In genetic engineering, scientists remove a plasmid from a bacterium. Next, a short piece of DNA is removed from another cell such as a human cell. The short piece of DNA is joined to the plasmid after it is opened. Then, the plasmid is returned to the bacterium.

❑ The products of recombinant DNA are used to produce vaccines, insulin, interferon, and human growth hormone.

❑ Genetic engineering is also used to protect plants from the tobacco mosaic virus and to prevent the development of frost on plants.

21–2 Plant and Animal Breeding

❑ Selective breeding is the crossing of animals or plants that have desired traits to produce offspring with desired traits.

❑ A breeding technique in which two genetically different but related species of organisms are crossed is called hybridization.

❑ A breeding technique in which two organisms with the same or very similar sets of genes are crossed is called inbreeding.

VOCABULARY

Define each term in a complete sentence.

biotechnology	hybrid	inbreeding	selective breeding
genetic engineering	hybridization	plasmid	
	hybrid vigor	recombinant DNA	

CONTENT REVIEW: MULTIPLE CHOICE

On a separate sheet of paper, write the letter of the answer that best completes each statement.

1. The application of technology to the study and solution of problems involving living things is known as
   a. hybridization.   b. biotechnology.
   c. inbreeding.   d. genetic engineering.
2. In genetic engineering, the new pieces of combined DNA are called
   a. hybridization.   b. inbreeding.
   c. selective breeding.   d. recombinant DNA.
3. A plasmid is a (an)
   a. enzyme.   b. nitrogen base in DNA.
   c. ringlike form of DNA.   d. growth hormone.
4. Which has been genetically engineered to produce proteins made by other organisms?
   a. plants   b. mice   c. bacteria   d. human beings
5. Which of the following is *not* produced as a result of genetic engineering?
   a. square tomatoes   b. human growth hormone
   c. insulin   d. "ice-minus" bacteria

6. An increase of sugar in blood can lead to a disorder called
a. beta thalassemia.　　b. tobacco mosaic disease.
c. hepatitis.　　d. diabetes mellitus.

7. A substance produced by human body cells that have been attacked by a virus is called
a. interferon.　　b. hemoglobin.
c. insulin.　　d. human growth hormone.

8. The "ice-minus" bacteria permits ice crystals to form on plants
a. above 0°C.　　b. at 0°C.　　c. below 0°C.　　d. at 10°C.

9. The crossing of plants or animals that have desirable characteristics is called
a. hybrid vigor.　　b. recombinant DNA.
c. selective breeding.　　d. interferon.

10. The crossing of two genetically different but related species of organisms is known as
a. genetic engineering.　　b. inbreeding.
c. gene therapy.　　d. hybridization.

CONTENT REVIEW: COMPLETION

On a separate sheet of paper, write the word or words that best complete each statement.

1. When DNA from one organism is removed and inserted into another organism, the new piece of DNA is called _____.

2. Yeast and bacteria are examples of one-celled _____.

3. In genetic engineering, DNA is transferred from complex organisms into simpler ones such as _____ and _____.

4. An important hormone produced by genetic engineering that helps the body use sugar is _____.

5. Bacteria that produce human growth hormone were developed by the technique of _____.

6. _____ is a substance that is produced by human body cells and helps cells fight viruses.

7. Recently scientists have found a way to protect plants from a disease caused by the _____ virus.

8. "Ice-minus" bacteria have been genetically engineered to protect plants against damage from _____.

9. If a hybrid offspring has traits that are better than those of its parents, it is said to have _____.

10. A type of breeding that involves crossing organisms that have similar sets of genes is called _____.

CONTENT REVIEW: TRUE OR FALSE

Determine whether each statement is true or false. Then on a separate sheet of paper, write "true" if it is true. If it is false, change the underlined word or words to make the statement true.

1. Hybridization is the application of technology to the study and solution of problems involving living things.

2. A ringlike form of DNA is a plasmid.

3. Offspring that are often better than either parent are said to have hybrid vigor.

490

4. Selective breeding is the process in which genes from one organism are transferred into another organism.
5. Human growth hormone, interferon, and insulin are human substances that can be genetically engineered.
6. When introduced into a person's body, a vaccine triggers the production of antibodies.

7. The "ice-minus" bacteria help to stop the formation of frost on plants.
8. Genetic engineering is the crossing of two genetically different but related species of organisms.
9. A breeding technique that is opposite to hybridization is inbreeding.
10. A hybrid is an organism that has the traits of both parents.

CONCEPT REVIEW: SKILL BUILDING

Use the skills you have developed in the chapter to complete each activity.

1. **Sequencing events** The diagrams below represent steps in the process of forming recombinant DNA, but the sequence is incorrect. Place the steps in the proper sequence. Describe what is happening in each step.

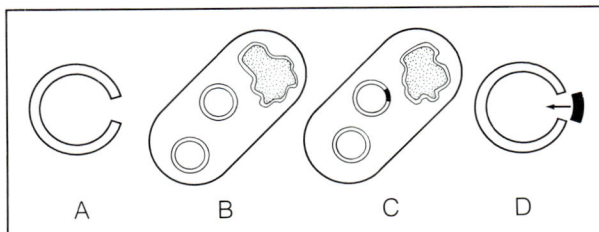

2. **Applying concepts** Texas shorthorn cattle have meat that is of good quality. Brahman cattle have immunity to Texas fever. When these two animals are crossed, the offspring have good quality meat and are immune to Texas fever. Is this cross an example of hybridization or inbreeding? Explain.

3. **Relating concepts** Why is it not possible to predict the traits of offspring from a cross between two dogs that are mixed breeds?
4. **Applying concepts** How would a plant breeder go about producing a wheat plant that is resistant to cold and disease? Why are these characteristics important?
5. **Relating concepts** Many people are concerned about the introduction of new organisms into the environment. Some of these organisms, they think, may prove harmful to other organisms. What guidelines would you develop concerning the introduction of genetically engineered organisms into the environment?
6. **Applying concepts** Suppose you were asked to develop a plant that could grow in a very cold and dry environment. What characteristics might you want your plant to have?

CONCEPT REVIEW: ESSAY

Discuss each of the following in a brief paragraph.

1. Describe the steps involved in the production of recombinant DNA.
2. Name and describe five applications of recombinant DNA technology.
3. What is the advantage of developing "ice-minus" bacteria? Are there any disadvantages? Explain.

4. How do plant and animal breeders produce organisms with desired characteristics?
5. How do hybridization and inbreeding differ? Give an example of each.
6. What is biotechnology? How has biotechnology influenced your life? Give some examples.

Changes in Living Things

22

CHAPTER OBJECTIVES

After completing this chapter, you will be able to

22–1 Define evolution.

22–2 Describe chemical, anatomical, and fossil evidence of evolution.

22–2 Compare radioactive dating to relative dating.

22–2 Describe some of the first primates to be called human.

22–3 Explain how natural selection leads to new and varied species.

22–4 Compare gradualism to punctuated equilibrium.

During the summer of 1984, a team of scientists worked in a quarry near the small town of Post, Texas. A quarry is a place where building stone is dug out of the earth. The scientists worked carefully to uncover what they thought were the ancient bones of a baby dinosaur. Nearly a year and a half later they discovered they were wrong. But the bones turned out to be much more important than they had ever dreamed!

When in 1985, the leader of the discovery team, Dr. Sankar Chatterjee, began to examine the bones, he noticed that they had many birdlike characteristics. For one thing, the bones were hollow. And, there was a well-developed wishbone. In addition, the breastbone had a keel, a piece of bone that would have anchored muscles used in flight. Finally, using a microscope, Chatterjee discovered tiny bumps where feathers had probably been attached. From this evidence, he concluded that the bones must have belonged to a bird, not to a baby dinosaur.

Chatterjee named his find *Protoavis*, or "first bird." Fossils show that *Protoavis* had teeth, a long bony tail, and powerful hind legs that helped it to run. According to scientists, these characteristics suggest that dinosaurs were ancestors of birds.

However, the most interesting thing about *Protoavis* is that it was dug out of 225-million-year-old rocks. Until now, *Archaeopteryx* (ahr-kee-AHP-tuhr-ihks), a 150-million-year-old bird, had been considered the first bird. But, if *Protoavis* is indeed a bird, it is 75 million years older than *Archaeopteryx*. So, in the near future, *Archaeopteryx* may be knocked from its perch.

In the following pages, you will read more about fossils, such as those of *Protoavis*. And you will see how such fossils help us learn about the changes that occur over time in living things.

This fossil found in Texas is believed to be that of one of the world's oldest birds, Protoavis.

22–1 What Is Evolution?

The quagga vanished from its home in South Africa about 100 years ago. Descriptions of the animal create a strange picture. The quagga had stripes like a zebra. But the stripes covered only its head, neck, and the front part of its body. Was this animal closely related to the zebra? Or was it a completely different kind of creature?

Scientists often asked these similar questions about other animals and plants that appeared to be closely related. The scientists knew that in one way or another all living things are related to each other. But which are closely related and which are only distantly related?

Today scientists believe that the process of **evolution** holds the key to these answers. **Evolution is a change in a species over time.** But how do living things change? And why do some organisms survive while others die off? During the history of life on this planet, chance mutations of genes produce new or slightly modified living things. Most of these new living things cannot compete with other organisms and soon die off. But some new life forms do survive. They survive and reproduce because they just happen to meet the demands of their environment better than other organisms do. Because of this process, many living things that inhabit our planet today did not exist millions of years

Figure 22–1 *The quagga is a South African mammal that became extinct about 100 years ago. Notice the stripes on its head, neck, and front part of its body. What animal does it remind you of?*

HISTORY OF LIFE ON EARTH

Era	Years Before Present Time	Changes in Plant and Animal Life	
Precambrian	4,600,000,000	First life forms develop — bacteria, algae, jellyfish, corals, and clams. 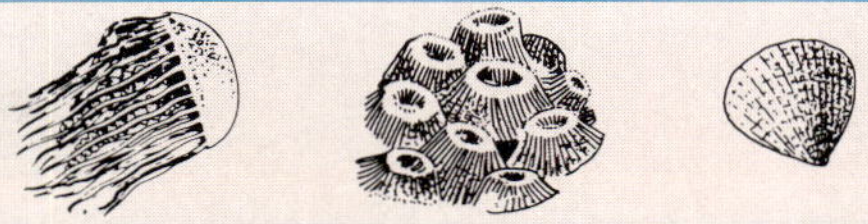	
Paleozoic	570,000,000	First land plants, fishes, insects, amphibians, and first reptiles develop. 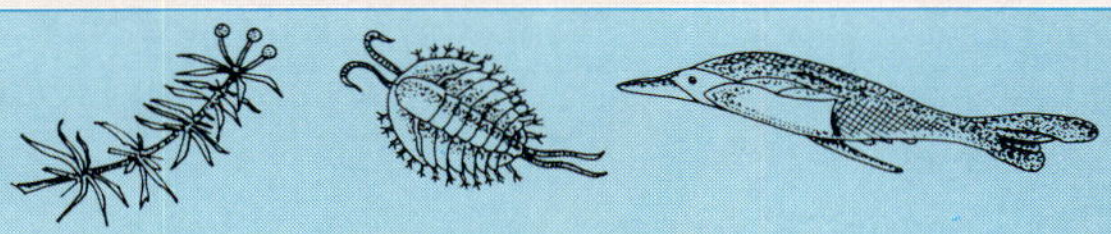	
Mesozoic	225,000,000	Cone-bearing and flowering plants, dinosaurs, and first birds and mammals appear. 	
Cenozoic	65,000,000	Modern plants and mammals develop; human beings appear. 	

Figure 22–2 *This chart shows a history of the development of life on earth. During which era did flowering plants first appear?*

ago. Figure 22–2 shows the history of life on earth. During which era did human beings appear?

SECTION REVIEW

1. What is evolution?
2. During which era did the first amphibians appear?

22–2 Evidence of Evolution

Section Objective

To describe evidence for the theory of evolution

There is a great deal of scientific evidence to support the theory of evolution. You will be able to examine this evidence for yourself in this chapter. This evidence has allowed scientists to develop the theory of evolution. A scientific theory is a very powerful and useful idea. It is an explanation of facts and observations. Moreover, it is an explanation that has been tested many times by many scientists. If a scientific theory fails a test, the theory is modified. Scientific theories also let scientists make predictions of future events. If these predictions come true, the theory passes a key test. **There are many different types of evidence that support the theory of evolution.**

Figure 22–3 *Notice how the arrangement and number of bones in the front limbs of the lion (left), bat (center), and dolphin (right) are similar. What is the name for such similar structures?*

Anatomical Evidence

In the early 1800s, a French biologist named Jean-Baptiste de Lamarck began to develop some theories about evolution based on **anatomy.** Anatomy is the study of the structure of living things.

In his book *Philosophie zoologique,* Lamarck suggested that all forms of life could be organized into one vast "family tree." Lamarck looked to anatomy for evidence of evolution. Look at the bones of a lion's foreleg, a bat's wing, and a dolphin's flipper in Figure 22–3. Can you see any similarities in the shape and arrangement of the bones of these animals? When body parts and organs are similar in structure they are said to be **homologous** (hoh-MAHL-uh-guhs). A lion's foreleg, a bat's wing, and a dolphin's flipper are homologous structures. It was Lamarck who first suggested that organisms with such homologous structures are closely related.

Lamarck's theory that organisms with homologous structures are related still holds true. However, Lamarck is also known for another theory that proved to be wrong. This theory is known as the theory of evolution by inheritance of acquired characteristics. An acquired characteristic is one that an organism gets, or acquires, during its lifetime.

According to Lamarck's theory, an organism acquired a characteristic through the use or disuse of body parts. For example, giraffes acquired their long necks by stretching to reach the leaves of tall trees. The giraffes then passed the characteristics on to their offspring. See Figure 22–4. Scientists now know that acquired traits cannot be passed on to offspring. Why? Acquired traits affect body cells not sex cells. And it is only the genes in sex cells that are inherited.

Fossil Evidence

Believe it or not, whales or their whalelike ancestors once walked! The evidence that these huge mammals once moved across land lies in the fact that they possess the bone structure of long-lost hind legs. However, such anatomical evidence alone could not have convinced the majority of scientists that whales once walked were it not for the added evidence of **fossils.**

Fossils are the imprints or remains of plants or animals that existed in the past. In 1983, fossil hunters found a buried skull belonging to an animal that had lived more than 50 million years ago. The skull was very similar to that of a whale. But the body structure that allowed the animal to hear could not have worked underwater. Scientists believe that this whalelike skull belonged to a land-dwelling ancestor of today's whales. Similar kinds of fossil evidence allow scientists to trace the evolution of many of the world's living things.

Figure 22–4 *According to the theory of evolution by inheritance of acquired characteristics, giraffes acquired their long necks by stretching to reach the leaves of tall trees. The giraffes then passed the characteristics on to their offspring. Why can this theory not be true?*

Figure 22–5 *Fossils help scientists trace the evolution of many organisms. This fossil (right) is of a trilobite, a now extinct marine animal. You can see from the fossil of this starfish (left) that starfish have changed very little during the course of earth's history. What is a fossil?*

Sharpen Your Skills

Dating Fossils

If seven-eighths of the carbon-14 in a fossil has decayed to nitrogen, approximately how old is the fossil?

Radioactive Dating

A method used by scientists to measure the age of fossils or the age of the rocks in which fossils are found is called **radioactive dating.** Radioactive dating involves the use of radioactive elements. A radioactive element is one whose atoms give off radiation as it decays, or breaks down, into the atoms of a different element.

Some radioactive elements decay in a few seconds. Some take thousands, millions, or even billions of years to decay. But the rate of decay for each element is steady. Scientists measure the decay rate of a radioactive element by a unit called a **half-life.** The half-life is the time it takes for half of a radioactive element to decay. For example, carbon-14 is the radioactive form of carbon. The half-life of carbon-14 is 5770 years. Carbon-14 is present in all living things. When a living thing dies, the carbon-14 begins to decay.

By measuring the amount of carbon-14 in a fossil, scientists can tell how old the fossil is. Suppose, for example, the fossil of an organism contained only one-fourth of the amount of carbon-14 found in a similar living organism. The fossil would be about 11,540 years old. See Figure 22–6.

Carbon-14 can be used to date or estimate the age of fossils that were formed within the last 50,000 years. Other radioactive elements with longer half-lives are used to date other fossils. By using radioactive dating to determine the age of different fossils, scientists can learn how organisms have changed over time.

Figure 22–6 *This diagram illustrates how carbon-14 radioactive dating can be used to determine the age of a fossil. How old would a fossil be if it contained only ⅛ of the amount of carbon-14 of a similar living organism?*

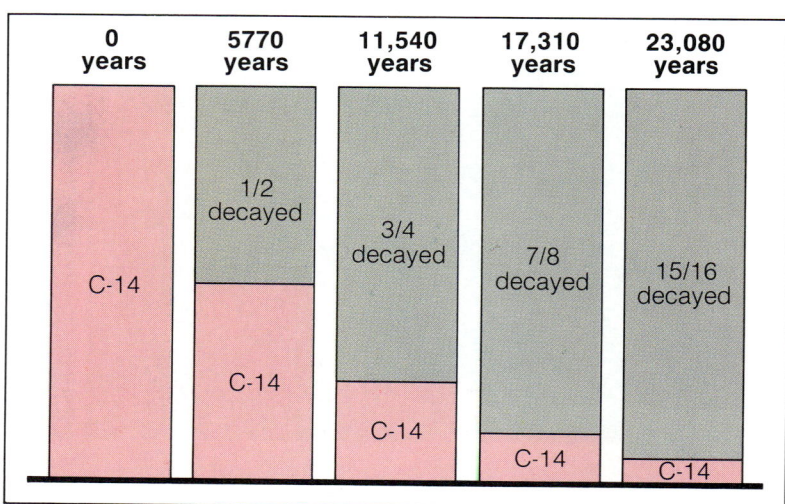

0 years	5770 years	11,540 years	17,310 years	23,080 years
C-14	1/2 decayed / C-14	3/4 decayed / C-14	7/8 decayed / C-14	15/16 decayed / C-14

Relative Dating

Another way to date fossils is to compare the rock layers in which they are formed. This type of dating is called **relative dating.**

To understand relative dating, think of the Grand Canyon, the deepest visible crack in the earth's land surface. A ride on muleback into the Grand Canyon is like a descent to a time 2 billion years ago. Its stacked layers of **sedimentary rock** tell the history of life on earth. Sedimentary rock is a type of rock formed from layers of mud and sand that harden slowly over time. Even the canyon's topmost sedimentary rock layer is 200 million years old. By the time you reach the layer at the bottom, you are looking at rocks that were formed 2 billion years ago. Many of the layers of these rocks hold the fossils of plants and animals that lived when the layers first formed.

Clearly, the lower layers are older than the upper layers. So the fossils found in the upper layers are of plants and animals that lived more recently. Using this method of relative dating, scientists have determined the age of fossils in various rock layers all over the world. Once scientists know the age of fossils, they can use the fossils as evidence of how organisms have evolved.

The Fossil Record

All the fossil evidence scientists have collected forms what is known as the **fossil record.** The fossil record is the most complete biological record of life on earth.

Here is an example of how scientists read the fossil record. Some years ago in the Mississippi River valley scientists found the fossil bones of a leg and a foot of an unknown animal. The fossils were in a deposit of sedimentary rock. From the bones, scientists discovered that the animal had four toes on each of its front feet. Moreover, the toes were spread apart.

The discovery of other fossils in the same sedimentary layer helped scientists piece together what the entire animal probably looked like. It was about the size of a modern-day cat. But except for its

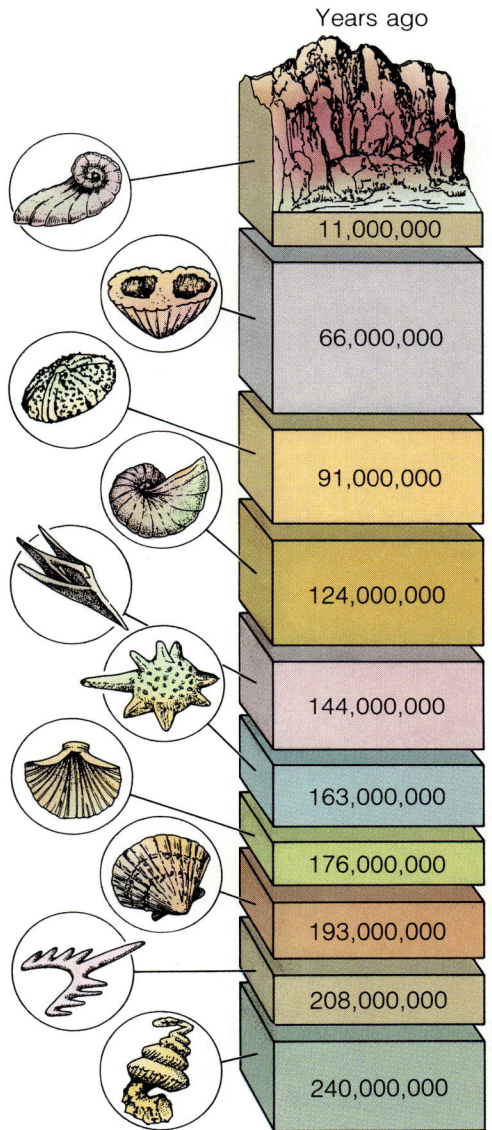

Years ago

Figure 22–7 *Fossils are found in sedimentary rock, which is formed from layers of mud and sand that slowly harden over time. Because the lower layers of rock are older than the upper layers, scientists can trace the history of life on earth.*

Figure 22–8 *Scientist have determined that* Eohippus, *or "early horse," lived about 50 million years ago. How do they know this?*

Figure 22–9 *This diagram shows how the horse has evolved over the last 50 million years. How is the forefoot of the modern horse different from that of* Eohippus?

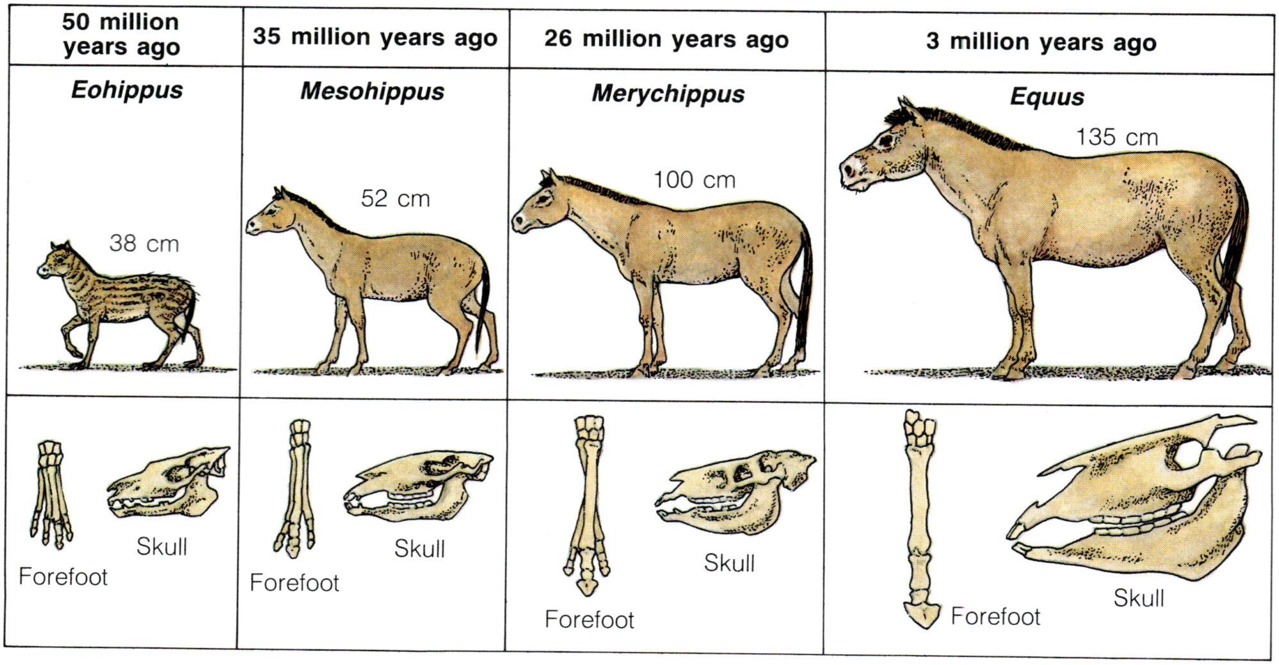

small size, short teeth, and four toes, it looked a lot like a horse. Scientists named it *Eohippus,* meaning "dawn horse" or "early horse."

The scientists used radioactive dating to measure the age of the rock layer in which *Eohippus* was found. The rock layer was about 50 million years old and, therefore, so was *Eohippus.*

From the fossils of plants and animals found in the same layer of sedimentary rock as *Eohippus,* scientists also were able to tell what the Mississippi River valley may have been like when *Eohippus* lived. Most of the other fossils found in the *Eohippus* layer resemble plants and animals that live in warm, wet climates today. This is evidence that *Eohippus* probably lived in a tropical climate, surrounded by swamps and mud.

Toes that are spread apart are very useful for walking on soft mud. They spread an animal's mass over a large area, almost like snowshoes or skis. The structure of *Eohippus's* foot was well suited to the kind of surroundings in which it lived.

Today's horse spends most of its life on dry, hard ground. Its single hard hoof lessens the shock of walking more than four spread toes could ever do. The modern horse also has larger teeth than *Eohippus.* As a result, it can chew the tough, dry grasses of the prairie.

50 million years ago	35 million years ago	26 million years ago	3 million years ago
Eohippus	**Mesohippus**	**Merychippus**	**Equus**
38 cm	52 cm	100 cm	135 cm
Forefoot Skull	Forefoot Skull	Forefoot Skull	Forefoot Skull

Scientists conclude that the fossils of *Eohippus* and other plants and animals are evidence that generations of species go through changes. A plant or animal with a new kind of trait may first come into being, perhaps as a result of a mutation. If the change makes it easier for the plant or animal to survive in its surroundings, then it will have a chance to pass this new trait on to its offspring. How does this idea of how species change compare to Lamarck's theory of acquired characteristics?

A change that increases an organism's chances of survival is known as an **adaptation.** The feet of *Eohippus* were well adapted to walking on soft, wet ground but not to walking on hard earth. Suppose a change in the climate of the Mississippi River valley occurred millions of years ago and dried up the land. If so, animals with hooves would have been better adapted to the land than was *Eohippus*. *Eohippus* may have become **extinct,** or died out, because of this. Over long spans of time, so many small adaptations may occur that a species will no longer resemble its ancient ancestors. Such major adaptations probably require many generations. Gradually, a new species forms.

CAREER   *Museum Curator*

HELP WANTED: MUSEUM CURATOR, GEOLOGY. Master's degree in geology with a specialization in paleontology required. Three years of work experience in a museum, educational institution, or research organization.

When the museum decided to expand its geology department, a new **museum curator** was hired. Museum curators are specialists in areas that relate to their museum's collection. They are responsible for the care, interpretation, and identification of all items belonging to the museum.

Other tasks of curators include suggesting which items the museum should obtain, classifying the items already in the collection, and determining whether an item is genuine. They must also research the museum holdings and publish the results of that research. Curators have managerial duties and are responsible for seeing that museum holdings are displayed in a creative, educational way.

Anyone interested in a career as a museum curator should develop good communication skills and a sound knowledge of the type of collection he or she wishes to oversee. For more information, write to the Museum Reference Center, Smithsonian Institution Libraries, Washington, DC 20560.

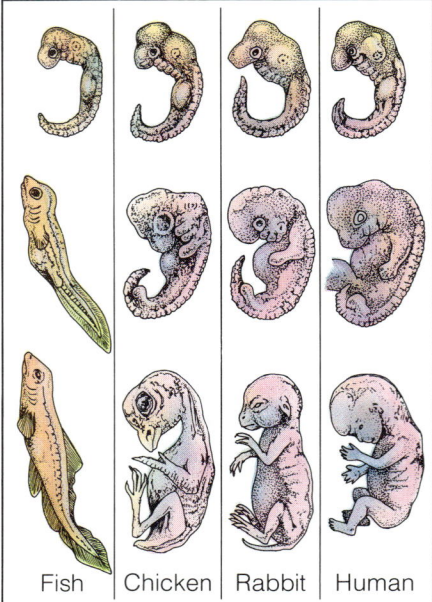

Figure 22--10 *The more similar the structure of the embryos of different organisms, the more closely related those organisms are. What is the study of developing organisms called?*

Figure 22–11 *This modern-day plains zebra shared a common ancestor with the quagga about 3 million years ago. What evidence have scientists used to determine this?*

Embryological Evidence

Another type of evidence for evolution is based on **embryology,** or the study of developing organisms. Scientists can compare the embryos, or developing organisms, of different species to see how closely related they are.

The embryos of vertebrates, or animals with backbones, are very similar in the early stages of development. The more similar the structure of the embryos of different organisms, the more closely related those organisms are. Scientists think that the development of different organisms is similar because they inherit their traits from a common ancestor. See Figure 22–10.

Chemical Evidence

Think back for a moment to the quagga—the animal that resembles a zebra. One type of evidence that might show if a zebra evolved from the quagga is chemical similarities in their DNA. In Chapter 20, you learned that DNA is the basic substance of heredity.

Chemical similarities in DNA molecules are one kind of evidence of evolution. In fact, this kind of evidence was used to show the relationship between quaggas and zebras.

Among other things, the theory of evolution predicts that the more closely related two living things are, the more similar the structure of their DNA molecules will be. In 1985, Dr. Allen Wilson of the University of California at Berkeley analyzed DNA taken from the muscle tissue attached to the preserved skin of a quagga. The skin had been stored for more than a hundred years in a German museum. Dr. Wilson also analyzed DNA taken from a modern-day plains zebra. The structure of the DNA in the two samples was 95 percent identical!

Dr. Wilson concluded that the quagga and the plains zebra were, indeed, close relatives who shared a common ancestor about 3 million years ago. In other words, both the quagga and the plains zebra evolved from an animal that lived earlier. Scientists use this method of DNA comparison to show the relationship between many other types of organisms as well.

Molecular Clock

You have just seen how similarities in DNA structure can be used to show relationships between organisms. But similarities in protein structures can show relationships between organisms as well. Proteins, as you recall from Chapter 13, are made up of amino acids and are used to build and repair body parts. Scientists believe that the more similar the structure of the protein molecules of different organisms, the more recent their common ancestor.

Figure 22–12 *The process by which many different species evolve from a common ancestor, such as the cotylosaur, is called adaptive radiation.*

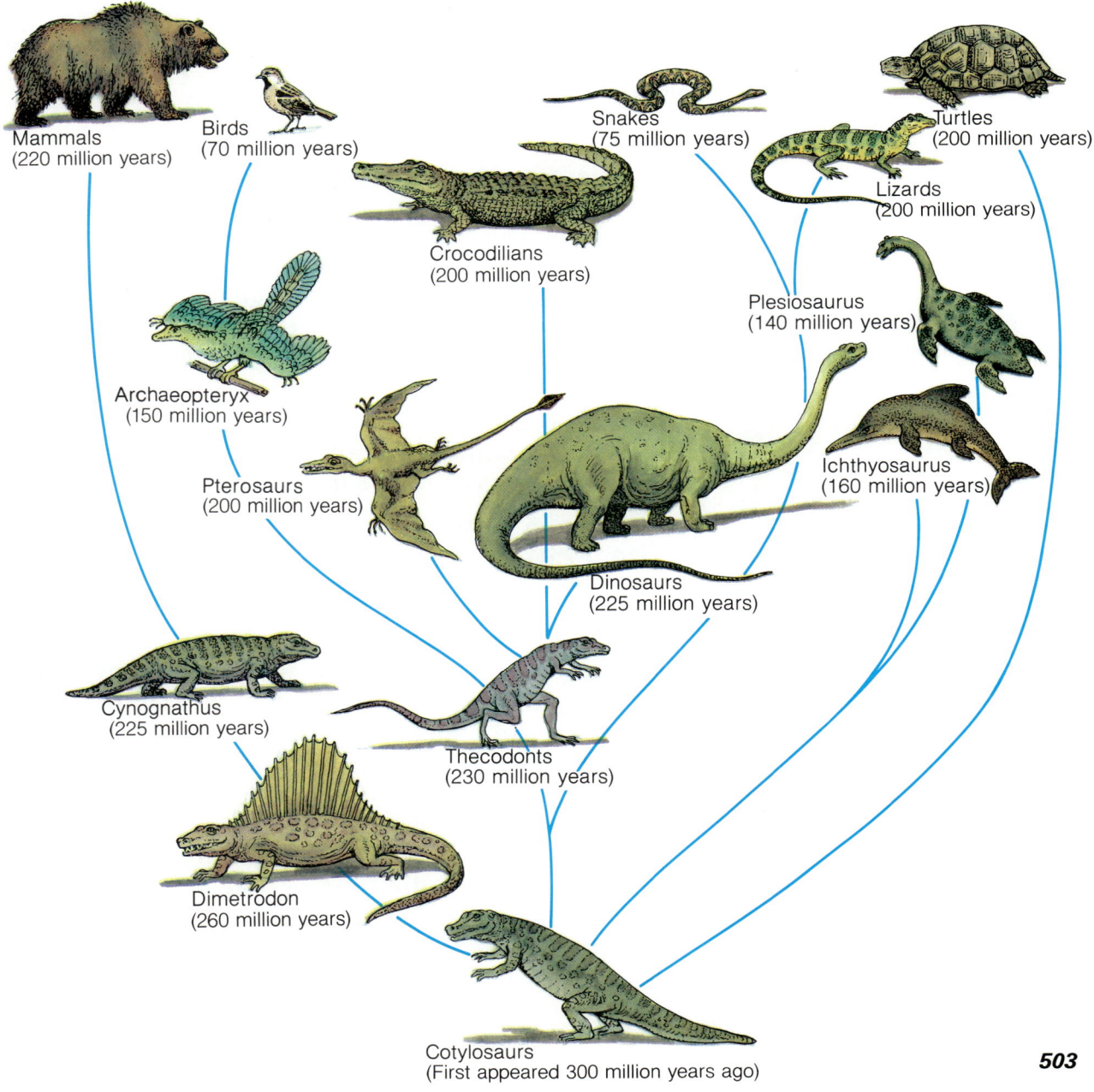

Mammals
(220 million years)

Birds
(70 million years)

Snakes
(75 million years)

Turtles
(200 million years)

Crocodilians
(200 million years)

Lizards
(200 million years)

Archaeopteryx
(150 million years)

Plesiosaurus
(140 million years)

Ichthyosaurus
(160 million years)

Pterosaurs
(200 million years)

Dinosaurs
(225 million years)

Cynognathus
(225 million years)

Thecodonts
(230 million years)

Dimetrodon
(260 million years)

Cotylosaurs
(First appeared 300 million years ago)

Figure 22–13 *Both the lemur staring out at you from the tree (left) and the chimpanzee searching for termites to eat (right) belong to the same order of mammals as you. What is the name of that order?*

Figure 22–14 *This artist's conception shows what Lucy and others like her might have looked like. How old do scientists believe Lucy to be?*

Scientists have developed a method of measuring the difference between the proteins of different species. They have also developed a scale that can be used to estimate the rate of change in proteins over time. This scale of protein change is called a **molecular clock.**

By comparing the similarities of protein structures between organisms, scientists can determine that the organisms had a common ancestor. But more importantly they can determine how long ago the organisms separated from that ancestor.

The Path To Modern Humans

Various types of evidence for evolution support the theory that both humans and apes evolved from a common **primate** ancestor. Primates are an order of mammals that includes humans as well as apes, monkeys, and about 200 other species of living things. Scientists believe that humans developed from an apelike animal over millions of years. During that time the posture of humans became more upright, their jaws became shorter, their teeth became smaller, and their brains became larger.

The oldest humanlike fossil found to date was found in Africa in 1977. The fossil was named *Australopithecus afarensis* (aw-stray-loh-PIHTH-uh-kuhs af-uh-REHN-sihs) and nicknamed "Lucy." Lucy is believed to be about 3.5 million years old.

A more recent species of primate that was actually called "human" was *Homo erectus* (HOH-moh ih-REHKT-uhs), or "upright man." Fossils of this species, which lived 1.6 to 0.5 million years ago, were first found on the island of Java in Southeast Asia. *H. erectus* had thicker bones than modern humans, a sloped forehead, and a very large jaw. *H. erectus* was able to use fire and make tools.

About 300,000 years ago, another species of human appeared. Because this species had a larger brain capacity than other humans, scientists named the species *Homo sapiens* (HOH-moh SAY-pee-uhnz), or "wise man." Two early types of *Homo sapiens* were the Neanderthals (nee-AN-duhr-thahlz) and the Cro-Magnons (kroh MAG-nuhnz).

Neanderthals lived in Europe and Asia from about 100,000 to 35,000 years ago. Neanderthals were shorter than modern humans. Their average height was only about 150 centimeters. They were able to hunt, fish, cook, and make tools. Neanderthals buried their dead with tools or animal bones. Sometimes they arranged the bones in patterns that suggest religious rituals.

Eventually the Neanderthals died out and were replaced by Cro-Magnons. The Cro-Magnons were taller—over 180 centimeters. And Cro-Magnons looked more like modern humans with long faces, straight high foreheads, and small teeth. The toolmaking skills of these humans far surpassed those of earlier humans. Cro-Magnons are believed to have worked together to make tools, build shelters, and hunt. To do so they probably used language.

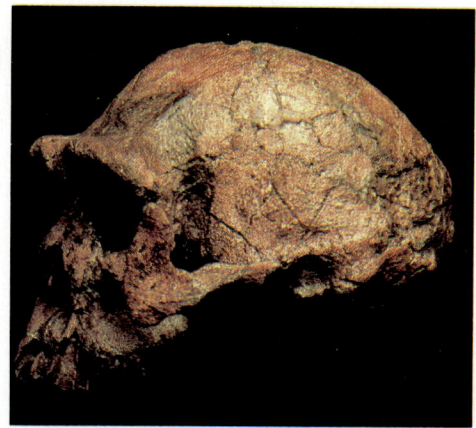

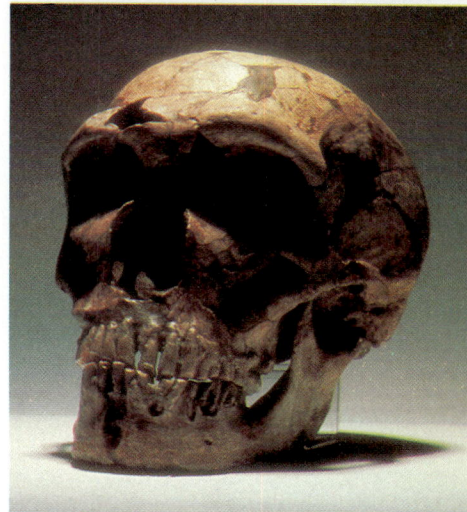

Figure 22–15 *Notice the large brain case on the skull of* Homo erectus *(top). The Neanderthal skull (bottom) is between 35,000 and 53,000 years old. The pattern of wear on the teeth shows that this person probably used his or her teeth for more than eating, perhaps for softening hides.*

SECTION REVIEW

1. What are six types of evidence used to develop the theory of evolution?
2. Give an example of homologous structures.
3. What is a fossil?
4. Compare radioactive dating to relative dating.
5. How can scientists learn about evolution by studying embryology?
6. What are the differences and similarities between Neanderthals and Cro-Magnons?
7. Would you expect there to be more similarities in the DNA molecules of a cat and a lion or in those of a cat and a dog? Explain.

22–3 Natural Selection

The Galápagos Islands rise out of the Pacific Ocean about 1000 kilometers from the west coast of South America. The islands received their name from the giant Galápagos tortoises that live there. The tortoises' long necks, wrinkled skin, and mud-caked shells make them look like prehistoric creatures. Sharing the islands with the tortoises are many other animals, including penguins, long-necked diving birds called cormorants, and large, crested lizards called iguanas (ih-GWAH-nuhz).

The most striking thing about the animals of the Galápagos is how they differ from related species on the mainland of South America. For example, the iguanas on the Galápagos have extra large claws that allow them to keep their grip on slippery rocks, where they feed on seaweed. On the mainland, the same animals have smaller claws. Smaller claws allow the mainland iguanas to climb trees, where they feed on leaves.

In 1831, a young British student named Charles Darwin set sail for a five-year sea voyage on a ship called the *Beagle*. Serving as the ship's naturalist, Darwin studied animals and plants at every stop the ship made. When Darwin arrived at the Galápagos, he soon noticed many differences between island and mainland creatures. As he compared the animals on the mainland to those on the islands, he realized something very special. It appeared that each animal was perfectly adapted to survival in its particular environment. Darwin took many notes on

Figure 22–16 The giant Galápagos tortoise is among the strange animals that live on the Galápagos Islands. The tortoises can have a mass of more than 230 kilograms. In fact, the Spanish word for tortoise, "galápagos," gave the islands their name.

his observations. For the next 20 years, he tried to find an underlying theory that could explain his observations. Then, in 1858, he and another British biologist named Alfred Wallace presented a new and exciting concept—the theory of evolution.

This theory was further enlarged by Darwin in a book he published a year later entitled *The Origin of Species*. Some of the ideas in this book had already been suggested by others. But one idea was entirely new—the concept of **natural selection.**

Darwin used the concept of natural selection to explain the theory of evolution. **Natural selection is the survival and reproduction of those organisms best adapted to their surroundings.** Before you can grasp what Darwin meant by natural selection, you must understand the role of overproduction in nature.

Figure 22–17 *The interesting thing about the animals of the Galápagos Islands is the way they evolved differently from related species on the mainland of South America. What differences can you see between the Galápagos Island iguana (left) and the iguana of the South American mainland (right)?*

Figure 22–18 *Through overproduction, nature assures that at least some of the seeds of this thistle plant will survive to continue the species. The seeds best adapted to their surroundings will survive while the others die. What is this process sometimes called?*

Overproduction and Natural Selection

Anyone who carefully observes nature will notice that many species seem to produce more offspring than can be supported by the environment. Every year, for example, dandelions grow seeds with sails that form into a white puff on the stem. The wind blows the seeds from the plant and carries them through the air. Most land in places where conditions are wrong for new dandelion growth. Only a few seeds land in a place with the right soil, light, and water conditions. These seeds grow into new dandelion plants. Through overproduction, nature assures that at least some seeds will survive to continue the species.

1. Using a metric ruler, measure the length of one hand of each of your classmates. Measure from the tip of the longest finger to the wrist. Record your results.

2. Construct a bar graph or a line graph that shows the relationship between the length of the hand of each student and the number of students.

Based on your graph, what conclusions can you make about variations in hand length? In what situations might large hands have been an advantage for prehistoric people? A disadvantage?

Figure 22–19 *This photograph shows variations among humans. What variations can you see?*

Sometimes overproduction of offspring results in competition for food or shelter among the different members of a species. In the case of tadpoles, which hatch from frog eggs, competition can be fierce. The food supply in a pond often is not large enough for every tadpole to survive. Only those strong enough to obtain food and fast enough to avoid enemies will live. These animals will reproduce. The others die.

The process in which only the best-adapted members of a species survive is sometimes called "survival of the fittest." In a sense, the fittest animals are selected, or chosen, by their surroundings to survive. This is what Darwin meant by natural selection—nature "selects" the fittest.

Variation and Natural Selection

Although all members of a species are enough alike to mate, normally no two are exactly the same. Put another way, even members of the same species have small **variations,** or differences. For example, some polar bears have thicker coats of fur than others. This thicker fur gives them more protection against the cold. Such polar bears are fitter and more likely to survive and pass on this characteristic.

In the same way, members of the same plant species may show minor variations in the length and thickness of roots. A plant with a deep root system can reach deep underground water more easily and will have a better chance for survival than a plant with shorter roots. So those plants with deeper roots are likely to be "selected" by nature and pass on their traits to the generations that follow. On the other hand, a plant with a shallow root system would have a better chance for survival in an area in which most water was close to the surface. As you can see, variations among the members of a species are another reason natural selection can lead to new species over long periods of time.

Minor variations in a species are common. Sometimes, however, a mutation can cause a change in an organism's characteristics that is far from minor. Most scientists believe, for example, that through a series of mutations certain small carnivorous, or meat-eating, dinosaurs evolved into the ancient fore-

Figure 22–20 *These puppies are members of the same species. However, as you can see, there are variations among them.*

runners of modern birds. Indeed, *Protoavis,* the bird you read about on page 493, may have been one of those birds.

In Chapter 20, you learned that helpful mutations better enable an organism to survive and reproduce. For example, in White Sands National Monument in New Mexico, the sand dunes are white. White mice live on these dunes. The light color of the mice was the result of a helpful mutation. Because the mice blend in better with their environment, they are less likely to be eaten by predators. If a mutation occurred that darkened some of the mice, they would not be able to blend in with their surroundings. The darker color would be the result of a harmful mutation. In this case, natural selection would "weed out" the mice with the harmful mutation. As generations of mice reproduced, the darker mice, along with their harmful mutation, would be eliminated from the species. The light-colored mice with the helpful mutation would survive and multiply.

Environment and Evolution

In some ways, living things become a mirror of the changes in their surroundings. The British peppered moth is a recent example of this process. In the 1850s, most of the peppered moths near Manchester, England, were gray in color. Only a few black moths existed. Because the gray moths were almost the same color as the tree trunks they lived on, they were nearly invisible to the birds that

Sharpen Your Skills

Survival of the Fittest

Scatter a box of red and green toothpicks in an area of your lawn or schoolyard. Then have a friend pick up the toothpicks one at a time. Which color toothpicks did your friend spot first? Why? How does this relate to the idea of natural selection?

Figure 22–21 *Changes in the environment often play a role in natural selection. As trees in Manchester, England, were blackened by factory soot, the number of black moths in the area increased. What advantage does a black moth have over a white moth?*

hunted them for food. Most of the black moths were spotted by the birds and were eaten. The species as a whole survived because of the gray moths. Then environmental conditions changed, and these changes had a drastic effect on the moths that lived in the area.

As more factories were built in the area, soot from the chimneys blackened the tree trunks. The gray moths could then be seen against the tree trunks. More and more gray moths were eaten by the birds. The few surviving black moths, however, now blended in with the tree trunks and survived. These moths produced more black offspring. In time, practically all peppered moths were black. Again the species as a whole survived. The tale of the peppered moths shows how natural selection was able to turn an unusual trait into a common one in a relatively short amount of time. It also shows how the environment in which a species lives determines which traits are desirable for that species.

SECTION REVIEW

1. What is natural selection?
2. What is the role of overproduction in nature?
3. Recently, there has been an increase in the number of white peppered moths near Manchester, England. Suggest a possible reason for this change.

22–4 Change: Slow or Rapid?

Most scientists accept the theory of evolution by natural selection. However, not all scientists agree on the rate at which evolution occurs. **Some scientists believe that evolution is a gradual process while others think evolution occurs in a series of rapid changes.**

According to Darwin, evolution is a slow and steady process. This is called **gradualism** (GRAJ-oo-wuhl-ihz-uhm). Scientists, however, have found very few fossil records that show gradual change. But this lack of evidence is not too surprising. Not all organisms become fossils and many fossils are destroyed with the passage of time.

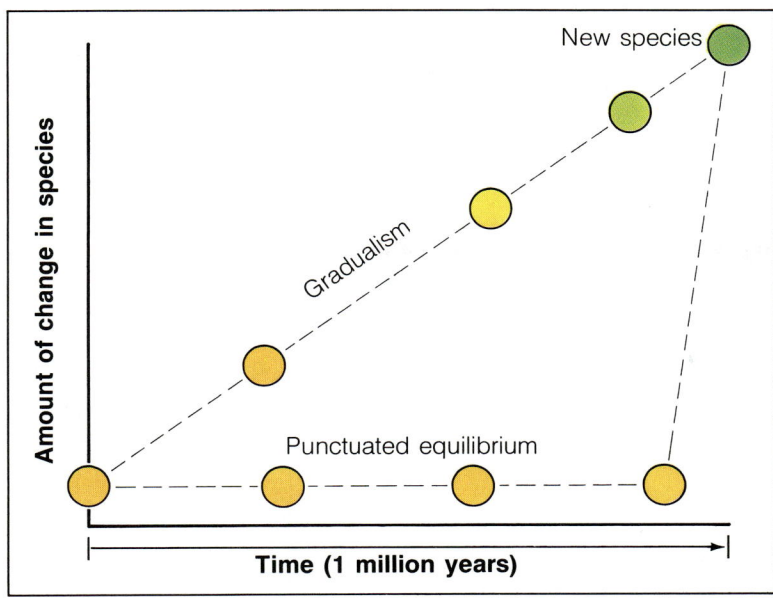

Figure 22–22 *This graph illustrates the difference between the gradualism and punctuated equilibrium theories of evolution. Which of these theories did Darwin develop?*

But in 1972, two scientists, Stephen Jay Gould and Niles Eldridge, came up with a possible explanation for the empty spaces in the fossil record. Their theory is known as **punctuated equilibrium** (PUHNGK-choo-wayt-uhd ee-kwuh-LIHB-ree-uhm). According to punctuated equilibrium, a species may change very little or not at all for a long period of time. Then, this equilibrium, or balance, is punctuated, or interrupted, by a rapid and sudden change. "Rapid," of course, does not mean in a second, minute, hour, or day. In this case, rapid means *thousands of years.* When you consider that the earth is 4.6 billion years old, a few thousand years—or even tens of thousands of years—is really a short time.

Today, many scientists conclude that the gradual evolution Darwin described is partly correct. But they believe gradualism has often been combined with the relatively fast changes in a species described by punctuated equilibrium. This combination has led to all the plants and animals living on the earth at this time.

SECTION REVIEW

1. What is the difference between gradualism and punctuated equilibrium?
2. Who developed the theory of punctuated equilibrium?
3. How could fossils provide more evidence for either the theory of gradualism or the theory of punctuated equilibrium?

Making a Fossil Imprint

Problem

How are fossil imprints formed?

Materials *(per group)*

seashell or bone
petroleum jelly
modeling clay
plastic container
water
plaster of Paris

Procedure

1. Coat the outside of a seashell or bone with petroleum jelly.
2. Gently press the shell or bone into a lump of clay until most of it is surrounded by the clay.
3. Carefully remove the shell or bone so that you have a sharp impression of the object in the clay. You now have a fossil imprint of the object.
4. Coat the impression you made in the clay with an even layer of petroleum jelly.
5. Use the plastic container to mix water with plaster until you have a paste. Add just a little water at a time.
6. Pour the plaster mixture into the impression. Allow it to dry for about 30 minutes.
7. Carefully peel away the clay from the plaster. You now have a cast of the original seashell or bone.
8. Place the cast and the shell or bone you used to make the mold on a table with those of your classmates. Try to match the original objects with their casts.

Observations

1. What are the differences between the molds and the casts? The similarities?
2. What types of objects made the best imprints and casts?

Conclusions

1. Making a cast is one way in which scientists can study the living thing that formed a fossil imprint. What kinds of things can a scientist learn from a cast?
2. Can a cast be made naturally? Explain.
3. Suppose one of the casts looked like nothing you had ever seen before. How would you decide what kind of animal or plant had made the fossil imprint? Would it help to know where the imprint had been found?
4. What type of environment do you think is most suitable for the formation of fossils?
5. Why is it difficult to find fossils of entire organisms?

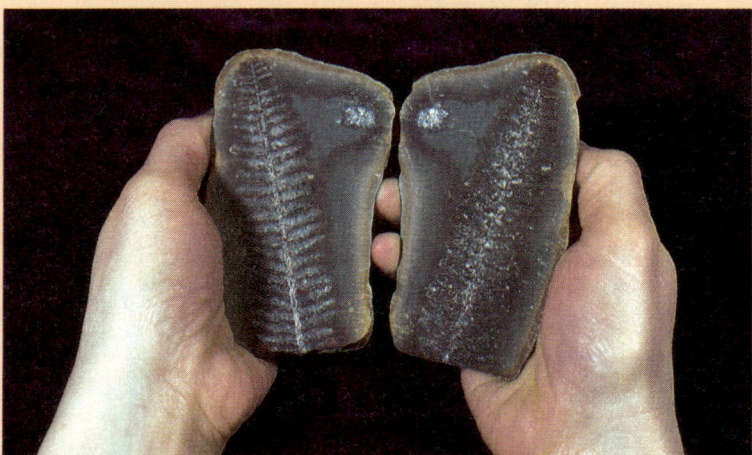

CHAPTER REVIEW

SUMMARY

22–1 What Is Evolution?

❏ Evolution is a change in a species over time.

❏ Some organisms survive and reproduce because they meet the demands of their environment better than other organisms do.

22–2 Evidence of Evolution

❏ Homologous structures provide evidence that various living things may share a common ancestor.

❏ Scientists use a wide variety of fossils to gain information about organisms from the distant past.

❏ By measuring the decay of radioactive substances in fossils or rocks, scientists are able to date fossils.

❏ Sedimentary rock is formed from layers of mud and sand that harden slowly over time. The layer in which a fossil is found is important in dating that fossil.

❏ Through the fossil record, scientists can show that species have evolved.

❏ An adaptation is any change that increases a species' chances of survival.

❏ The more similar the structure of the embryos of different organisms, the more closely related those organisms are.

❏ Chemical similarities in the DNA molecules and proteins of different organisms provide evidence of evolution.

❏ A scale that scientists use to estimate the rate of change in proteins over time is called a molecular clock.

❏ Primates are an order of mammals that includes humans, apes, monkeys, and about 200 other species of living things.

❏ According to the theory of evolution, both humans and apes evolved from a common ancestor.

22–3 Natural Selection

❏ Natural selection refers to the process in which the most able or fit organisms survive while the least fit die off.

❏ Overproduction in nature leads to competition within a species for food, water, and shelter. Only the organisms most able to compete survive.

23–4 Change: Slow or Rapid?

❏ The view that evolution is a slow and steady process is called gradualism.

❏ According to the theory of punctuated equilibrium, changes in species over time were sudden and rapid rather than gradual.

VOCABULARY

Define each term in a complete sentence.

adaptation	half-life	punctuated equilibrium
anatomy	homologous	
embryology	molecular clock	radioactive dating
evolution		
extinct	natural selection	relative dating
fossil		sedimentary rock
fossil record	primate	variation
gradualism		

CONTENT REVIEW: MULTIPLE CHOICE

On a separate sheet of paper, write the letter of the answer that best completes each statement.

1. A term that can be described as descent with modification is
   a. extinction.   b. evolution.   c. natural selection.   d. variation.
2. Which is *not* used as evidence for the theory of evolution?
   a. fossils   b. similarities in DNA
   c. intelligence   d. homologous structures
3. A bat's wing and a dolphin's flipper are examples of
   a. natural selection.   b. homologous structures.   c. variation.   d. fossils.
4. A technique used to measure the age of fossils is
   a. anatomical dating.   b. radioactive dating.
   c. homologous dating.   d. natural selection.
5. The name scientists gave to the early horse was
   a. *Protoavis*.   b. quagga.   c. *Archaeopteryx*.   d. *Eohippus*.
6. A change that increases an organism's chances of survival is known as a (an)
   a. adaptation.   b. variation.   c. mutation.   d. extinction.
7. The oldest humanlike fossil ever found was
   a. *Australopithecus afarensis*.   b. *Homo sapiens*.
   c. *Homo erectus*.   d. Neanderthal.
8. Another way to say "survival of the fittest" is
   a. competition.   b. natural variation.
   c. natural selection.   d. overproduction.
9. Small differences between members of a species are called
   a. variations.   b. homologous structures.
   c. punctuations.   d. modifications.
10. The theory that evolution is a slow and steady process is called
    a. punctuated equilibrium.   b. natural selection.
    c. survival of the fittest.   d. gradualism.

CONTENT REVIEW: COMPLETION

On a separate sheet of paper, write the word or words that best complete each statement.

1. The name given to the fossil bird recently found that may be older than *Archaeopteryx* is _____.
2. _____ is a theory that explains changes in organisms over time.
3. The study of the physical structure of living things is _____.
4. Imprints or remains of once-living organisms are called _____.
5. The decay rate of a radioactive substance is measured by a unit called a (an) _____.
6. _____ rock is formed from layers of mud and sand that harden slowly over time.
7. All the evidence scientists have gathered from fossils is the _____.
8. *Eohippus* is a species that died out, or became _____.
9. The study of developing organisms is known as _____.
10. Sometimes overproduction of offspring results in _____ for food and shelter among members of a species.

CONTENT REVIEW: TRUE OR FALSE

Determine whether each statement is true or false. Then on a separate sheet of paper, write "true" if it is true. If it is false, change the underlined word or words to make the statement true.

1. Natural selection is a change in a species over time.
2. The idea of evolution is a law.
3. Homologous organs are organs that have similar functions.
4. The half-life of carbon-14 is 5770 years.
5. Fossils in lower layers of rock are younger than fossils in upper layers.
6. Chemical similarities in DNA show that the quagga and the zebra are closely related.
7. Primates are a family of mammals.
8. Cro-Magnons probably used language.
9. Lamarck worked with Darwin to develop the theory of evolution.
10. Most species underproduce the number of young that the environment can support.

CONCEPT REVIEW: SKILL BUILDING

Use the skills you have developed in the chapter to complete each activity.

1. **Classifying objects** Which of the following would be considered fossils? Explain.

    a. an oyster shell found on a beach
    b. an imprint of an oyster shell in a rock
    c. a footprint of a person in cement
    d. a footprint of a dinosaur in rock

2. **Making observations** Observe two animals in your home, classroom, or a pet store. Make a list of at least five characteristics you observe such as hair length or color. Then list possible variations for each characteristic. Finally, explain how each variation might make an animal more "fit" for survival in its natural environment.

3. **Making inferences** Certain snails that live in woods and in grasses are eaten by birds. The snails that live on the woodland floor are dark-colored. The snails that live in grasses are yellow. How have the snails become adapted to their environments through natural selection?

4. **Applying definitions** Are the wing of a butterfly and the wing of a bat homologous structures? Why or why not?

5. **Relating concepts** Fossils of organisms that lived in oceans have been found on the tops of mountains. Provide an explanation for this.

6. **Making comparisons** Why is genetic analysis, such as comparison of DNA molecules, often a more accurate way of comparing organisms than is anatomical evidence?

7. **Applying concepts** Do you think evolution has stopped on this planet? Do you think it will ever stop? Explain.

CONCEPT REVIEW: ESSAY

Discuss each of the following in a brief paragraph.

1. Evolution is an ongoing process. It continues as it has for millions of years. Why are people usually unable to see the effects of evolution during their lives?

2. List two differences between *Eohippus* and the modern horse and provide an explanation for each difference.

3. Describe the role of overproduction and variation in natural selection.

4. Discuss the changes that have occurred in the peppered moths near Manchester, England, over the last 150 years. Include in your discussion an explanation for the changes.

Adventures in Science

Barbara McClintock and the

On a chilly day in October, reporters swarmed through a small cornfield in Cold Spring Harbor, New York. They were searching for a small, shy, gray-haired woman. Overnight the woman, Barbara McClintock, and the cornfield she tends had become famous. A discovery that occurred in this cornfield more than 30 years ago has changed the study of modern genetics.

Eighty-four-year-old Dr. Barbara McClintock is a biologist who has been studying the genetic makeup of maize, or Indian corn, for more than 60 years. In the 1920s, when she began her research, she and a few other scientists at Cornell University learned that maize plants could be used to observe the passing of traits from one generation to another. Through many experiments and by examining maize chromosomes with a microscope, Barbara McClintock was able to discover exactly how chromosomes control certain traits in maize.

Later, Dr. McClintock observed strange variations in the color of some of the corn kernels. These kernels, speckled with a variety of hues, seemed to be genetically unrelated to the kernels of previous generations.

Dr. McClintock began a new series of breeding experiments designed to show how the speckled kernels were inherited. With a microscope, she began to study the genes that control the inheritance of color in corn kernels. Dr. McClintock took careful notes of the position of each gene on the corn's chromosomes. You can imagine her surprise when she discovered something that broke one of the established rules of genetics. Instead of staying in the same position on the chromosomes, some genes seemed to "jump" or shift places.

It took a few more years of work to show that the speckling of the corn kernels was actually caused by these jumping genes. Dr. McClintock found that when a jumping gene landed next to a gene that controls

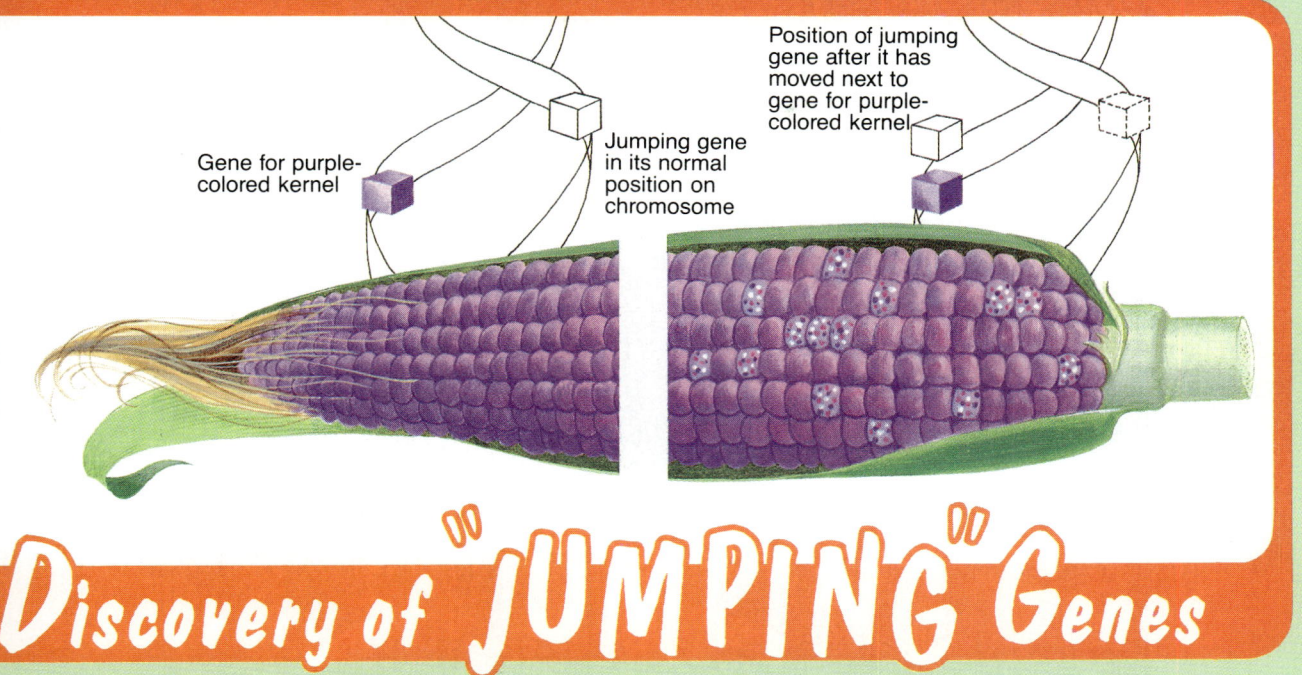

Gene for purple-colored kernel

Jumping gene in its normal position on chromosome

Position of jumping gene after it has moved next to gene for purple-colored kernel

Discovery of "JUMPING" Genes

kernel color, it sometimes blocked the action of that gene. The jumping gene caused color changes in the kernel. For example, if a jumping gene landed near a gene that usually produces a purple color, the action of the purple gene was blocked. The kernel became speckled with purple, pink, and white.

Dr. McClintock wrote a report based on her research and presented it at a meeting of scientists in Cold Spring Harbor in 1951. Nearly everyone reacted with disinterest or disbelief. "They called me crazy," Dr. McClintock recalls with a smile. At that time, no one knew that Dr. McClintock's findings would prove to be one of the most important discoveries in genetics to occur in this century.

Despite the fact that she received no credit for her discovery, Dr. McClintock continued her research. She did not gain recognition for her ideas until the 1970s, when more sophisticated ways of studying chromosomes

were developed. Then, after years of neglect, people began to call her a genius. When she was awarded a Nobel prize on October 10, 1983, she reacted to its lateness without bitterness. "If you know you're right, you don't care," she smiled. "You know that sooner or later it will come out in the wash."

In the meantime, researchers have found jumping genes in bacteria that give the bacteria more resistance to antibiotics. Some researchers suspect that jumping genes may play a role in turning normal cells into cancer cells. Each year, the value of Dr. McClintock's discovery grows, as scientists find new applications for it.

Dr. McClintock's fame has done little to change her habits. On the day her Nobel prize was announced, reporters searched for her in vain. She had quietly slipped into the woods, where she searched for walnuts she needed for the brownies she wanted to bake for friends.

Issues in Science

Genetic Engineering: Promise or Peril?

Four plots of land in the Salinas Valley in California are the subject of lawsuits, public protests, scientific debate, and a congressional investigation. These four strawberry fields are quite small—together they add up to less than one-half a hectare. You may wonder why these plots are the subject of so much attention. It is because researchers planned to release the first genetically engineered life forms to live outside of a laboratory on these fields. They wanted to spray the strawberry plants in these fields with bacteria "designed" to pro-

tect the strawberries from the damaging effects of freezing temperatures.

The bacteria appear quite ordinary. However, the genetic material inside their cells has been changed. And, it is this changed genetic material that has caused some people to worry.

Genetic engineering allows scientists to change the genes of an organism or to insert genes from one organism into another. One genetic engineering technique uses a tiny needle smaller than a cell. The needle is inserted into the nucleus of a cell. This needle can remove a gene from the cell. This removed gene can then be inserted into another cell. By exchanging genes between the two cells, a new kind of cell may begin to develop.

In another technique, scientists can remove a piece of DNA from one species and "splice," or join, it to the DNA of another species. This is called recombinant DNA technology or "gene splicing." Gene splicing is now being used in plants. For example, a single gene from a yeast cell has been spliced to the chromosome of a tobacco plant. Scientists hope that the yeast gene will help the tobacco plant resist a serious disease called crown gall.

Recombinant DNA technology has many potential uses. Scientists hope to produce crops with special abilities. For example, these crops may be able to produce their own natural fertilizers. Or they might thrive in poor soil. Other crops may be able to grow well in severe weather conditions. The ability to produce fast-growing, super-sized crops is another possibility. Yet, despite this future promise, today genetic engineering is a topic that stirs a great deal of public debate.

One activist, Jeremy Rifkin, has filed lawsuits to block field tests of genetically altered bacteria. Rifkin says that such experiments need to be examined in great detail before "altered" organisms are brought out of the laboratory and are released into the environment. Rifkin is afraid that releasing genetically altered organisms without proper laboratory testing could produce an environmental disaster.

Here is one kind of situation activists such as Rifkin worry about. A researcher proposed adding genetic material from a cell that produces antibodies, or chemicals that fight disease, to a cancer cell. The cell that produces antibodies is an important

part of the body's defense system. The cancer cell divides quickly. The researcher hoped that this new cell would grow rapidly and produce huge amount of antibodies. In effect, this new cell would become a kind of "antibody factory."

But, some people wonder what would happen if an accident occurred in a laboratory and the genetic material from a cell that causes a contagious disease, such as a strep throat, was added to a cancer cell. They suggest that this new cell might cause cancer and be as easily spread as the common cold. Such a mistake might prove disastrous and harmful to human life.

Some protesters say that genetic engineers are crossing the boundaries of human wisdom. They are afraid that a mistake may be irreversible. One well-respected scientist, Erwin Chargaff, has commented, "You cannot recall a new form of life."

As you can see, genetic engineering holds promise for the future. But it may involve certain risks. Should research in genetic engineering be stopped? Or should it proceed with strict guidelines that are designed to protect the environment and yet let basic research continue?

Scientists hope that special kinds of bacteria will protect these strawberry plants from frost damage.

Ecology

The crew aboard the *Polar Duke* knew they had struck gold when the icy water turned a golden brown. The *Polar Duke* is a new research vessel designed to travel through chunks of ice floating in the waters around Antarctica. The crew's mission was to study Antarctic marine organisms during winter—a season when few dare to travel so far south. The "gold" in the water was actually caused by microscopic plants called diatoms. In Antarctica, diatoms grow beneath the ice and produce food through photosynthesis.

After finding the golden diatoms, the crew then drilled a hole through a chunk of ice. When they did so, they found young krill swimming through the water. Krill are tiny shrimplike animals that swim in large groups. The *Polar Duke* researchers already knew that krill are an important food source for organisms living in the waters of Antarctica. Krill, for example, are eaten by small jellyfishlike organisms, fish, birds, seals, and whales. However, the krill's food source had long been a mystery. Now the *Polar Duke* researchers had an answer! Krill must eat the tiny diatoms growing beneath the ice.

It was obvious to the crew of the *Polar Duke* that organisms living in the icy cold water off Antarctica have much in common with living things in other environments on the earth. For just as the lives of the diatoms, krill, and other marine organisms are interrelated, so are the lives of living things in other environments. In this unit, you will learn more about the interrelationships among living things. And you will discover what happens when the natural balance among living things is disrupted.

CHAPTERS

The Polar Duke *is a new ship that is designed to travel through the icy waters around Antarctica.*

Interactions Among Living Things

CHAPTER OBJECTIVES

After completing this chapter, you will be able to

23–1 Define ecology.

23–1 Relate ecosystem, community, population, habitat, and niche to one another.

23–2 Define producer, consumer, and decomposer.

23–2 Describe a food chain, a food web, and an energy pyramid.

23–3 Compare competition, predation, commensalism, mutualism, and parasitism.

23–4 Describe the process of ecological succession.

23–5 Describe how nitrogen, water, oxygen, and carbon dioxide are recycled in the environment.

The giant Loggerhead sea turtle drags herself across the silvery sands of Bethune Beach on the eastern coast of Florida. She has nested on this stretch of beach for 35 years. Using her forward flippers to pull and her back flippers to push, she drags her almost 500-kilogram body to the edge of the dunes. Then she turns to face the ocean and begins to dig a hole.

If sea turtles had long memories, she would remember hatching from an egg not far from this site. She would remember, too, crawling slowly from her sandy nest to the safety of the ocean. There she swam to a place filled with floating grasses where she fed on the grass and on small animals she found. She was able to hide among the grasses from sharks and other large fish that might have eaten her.

The turtle spent her early years growing in this grassy place. Upon reaching maturity, she began to mate in the early summer of each year. And now, as she has for the last 35 years, this Loggerhead turtle has returned to the place she was born to lay her eggs.

She lays them in the hole she dug on Bethune Beach. Then with her hind flippers she fills in the hole and scatters sand along the bottom of the dune to hide the nest. Her task complete, the giant turtle makes her way back toward the ocean.

The ocean is the turtle's environment. There she is part of a complex system in which all of the parts interact with one another. In the following pages, you will learn more about the kinds of interactions that occur in the environment. And you will learn that you too are a part of the environment.

The giant Loggerhead sea turtle lives in the ocean but lays her eggs on land.

23–1 Living Things and Their Environment

Section Objective

To describe the interdependence between living and nonliving things in an environment

The deep dark sea and a tropical rain forest are examples of different **environments** found on the earth. An environment includes all the living and nonliving things with which an organism interacts. Other environments include deserts, grasslands, polar regions, and cities.

All of the living and nonliving things in an environment are interdependent. That is, they depend on one another. You can think of each kind of environment as being like a giant spider web. However, the threads of this web are not spun from silk. The threads of an environmental web are the plants, animals, soil, water, temperature, and light found there.

ENVIRONMENT

Figure 23–1 *An environment includes all the living and nonliving things that interact with one another. What are the living things in an environment called?*

BIOTIC FACTORS

ABIOTIC FACTORS

Consider for a moment what happens when an insect gets caught in a spider's web. As one thread of the web is disturbed, the shaking motion is transferred to all the threads that are part of the web. So a spider resting some distance from the trapped insect suddenly receives a signal that dinner is nearby!

In an environmental web, all the living and nonliving things make up different threads. The living things in an environment, such as plants and animals, are called **biotic factors.** The nonliving things in an environment, such as soil, light, and temperature, are **abiotic factors.** What are some biotic and abiotic factors in your classroom environment?

As in a real spider web, changes in one thread of an environmental web may affect other threads in the environment. In order to understand the changes that occur in an environment and how they can affect the environment, you must study the science called **ecology.** Ecology is the study of the relationships and interactions of living things with one another and with their environment. Scientists who study these interactions are called ecologists.

Ecosystems

Living things inhabit many different environments. From the poles to the equator, life can be found underground, in air, in water, and on land. A group of organisms in an area that interact with each other, together with their nonliving environment, is often called an **ecosystem.** An ecosystem can be as tiny as a drop of pond water or a square meter of a garden. Or it can be as large as an entire forest or an ocean.

If, for example, you look at a drop of pond water through a microscope, you will see a world of living things. Tiny plants, such as diatoms and green algae, float through the water in this microscopic ecosystem. Rotifers and copepods are animals that chase after these plants and eat them.

In the forest ecosystem, other microscopic organisms can be found. Some live in the soil and feed off dead animals and plants. Much larger living organisms, such as foxes, roam through the forest looking for food, while birds fly above. The foxes eat earthworms, birds, fruits, and small animals such

Figure 23–2 *Like the threads of a spider's web, the various parts of an environmental web are interconnected. Which branch of science involves the relationships and interactions of living things with one another and their environment?*

Figure 23–3 Daphnia *are tiny freshwater animals that eat microscopic plants. Both the* Daphnia *and the plants they eat may be part of a pond ecosystem.*

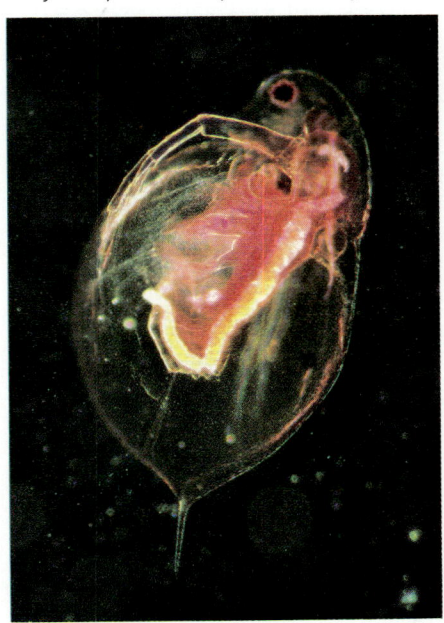

Figure 23–4 *Can you see the young deer in this photograph? To what ecosystem does the deer belong?*

Figure 23–5 *This collared lizard and the cactus plant behind it are part of a desert community. Why is the rock that the lizard rests on not considered part of the community?*

as mice. Fox families live in burrows, which are holes dug into the ground. The trees of the forest supply shelter for birds, which eat insects and earthworms. From this example you can see that the various threads or factors in an ecosystem are interdependent. In other words, the plants, animals, and nonliving parts of the ecosystem all interact with one another.

Communities

The living part of any ecosystem—all the different organisms that live together in that area—is called a **community.** Fish, insects, frogs, and plants are members of a pond community. Insects, birds, small animals, and trees make up a community in a forest ecosystem. Are these living things interdependent? If you have ever observed pond life, you may know the answer to this question. Frogs eat insects, and fish eat tadpoles. Insects eat plants, as do some of the fish.

In a desert ecosystem community, insects and birds feed on plants. Other animals, such as lizards, feed on insects and bird eggs. And larger animals eat lizards. Can you see a pattern to these communities? The plants are a source of food to plant-eating animals. And the rest of the organisms feed on the plant eaters and on other animals. So although not all the animals in these communities eat plants, they could not exist without them! Perhaps you are asking yourself what plants eat. Most plants make their own food, as you will learn in the next section.

Populations

In a community, each kind of living thing makes up a **population.** A population is a group of the *same* type of organism living together in the same area. For example, all the trout living in a lake are a population of trout. All the mesquite bushes surviving in the desert make up another population.

Different ecosystems support different populations of plants and animals. You might expect to see squirrels in a park near your home. The park is the squirrels' natural environment. But, certainly, you would be quite surprised to see a group of zebras grazing there! For the natural environment of zebras is the African plain.

In any ecosystem, the most successful organisms are those that are best adapted to their environment. An example of an animal population that is well adapted to its environment is the tortoise of the Galapagos Islands. Located in the Pacific Ocean off the western coast of South America, the Galapagos Islands are the home of two different kinds of tortoises. On some of the islands, the tortoises have short necks. Their shells hang over the backs of their necks. These shells prevent the tortoises from raising their heads. On other islands, the tortoises have long necks. The part of the shells above their necks is high. These tortoises can raise their heads.

The tortoises with the short necks and low shells live on islands where food is close to the ground. So

Sharpen Your Skills

Human Population Density

Population density is a measurement of the number of individuals of a species in a certain area. Using the information in the table below, construct a bar graph for human population density by continent. What conclusions can you draw from your graph?

Continent	Density
Europe	68 per km^2
Asia	65 per km^2
Africa	19 per km^2
North America	17 per km^2
South America	15 per km^2
Australia	2 per km^2

Figure 23–6 *These zebras and the acacia tree in the background are members of two separate populations of the African plain. How is a population different from a community?*

these tortoises do not have to raise their heads to find food. The tortoises with long necks and high shells live on islands where food is high off the ground. So the tortoises must raise their heads high in order to eat. Each kind of tortoise, then, is adapted to its own environment in ways that help it obtain food.

Habitats and Niches

Each member of a community has a certain place where it lives. The place in which an organism lives is called its **habitat.** A habitat provides food and shelter for an organism. The beautiful orchid, for example, grows from the branches of tall trees in the rain forest. The branches of these trees are the orchid's habitat.

In addition to a habitat, each organism in a community plays a particular role. The role of an organism in its community or environment is called a **niche.** A niche is everything the organism does and everything the organism needs within its habitat. And an organism's niche can affect the lives of other organisms.

Beavers, for example, build dams across streams, creating small lakes. The waters of the stream no longer rush along, wearing down the stream banks. Plants can now grow alongside the lake. Fish, such as trout, can make their home in the calm waters of the lake, and songbirds can nest along the quiet banks. Meanwhile, the beaver lives in a home it makes from sticks and mud. The newly formed lake also makes a habitat in which other animals and plants can thrive.

Organisms in an ecosystem can share the same habitat, but they *cannot* occupy the same niche. This fact seems obvious because sharing the niche would mean competing for the same food and space. Here is an example of how similar organisms in a habitat occupy different niches.

Three types of birds among those found on the Galapagos Islands are the blue-footed booby, the red-footed booby, and the white booby. These three birds all feed on the same kind of fish in the nearby waters. However, the blue-footed booby feeds close to shore, while the red-footed booby flies farther

Figure 23–7 *Each member of a community has a habitat, or a place where it lives. The gray fox (top) lives in a den. Hundreds of tent caterpillars (bottom) make their home in this "tent" among the tree branches. What do you think the caterpillars eat?*

Figure 23–8 *This beaver is building a dam across a stream, thus creating a lake. The newly formed lake will become a habitat for other animals, such as trout.*

out to sea to catch its food. And the white booby dives for its meals in the water in between. Can you think of another example of organisms sharing habitats but not niches?

SECTION REVIEW

1. Compare biotic and abiotic factors.
2. Define ecosystem.
3. What is the difference between a community and a population?
4. Why can organisms share the same habitat but not the same niche?
5. Lions, zebras, and grasses are three populations that live on the plains of Africa. How are these populations interdependent?

23–2 Food and Energy in the Environment

All living things need energy in order to live. They use energy to carry on all the basic life functions. For most living organisms, the immediate source of energy is the food they eat. But this food can be traced all the way back to green plants, which use the sun's energy. Thus, all the energy used by plants and animals comes directly or indirectly from the sun.

Green Plants: Food Factories

Green plants have one very important advantage over other living things. They can make their own food, and most other living things cannot. As you read in Chapter 8, plants make food by the process of **photosynthesis.** During photosynthesis the green parts of plants, especially the leaves, capture the energy of sunlight and use it to make glucose, a type of sugar. The glucose is formed when the plants chemically combine water and carbon dioxide. The reason green plants can carry on photosynthesis is that they contain the green pigment chlorophyll. Chlorophyll captures the energy of sunlight. In addition to glucose, oxygen is also produced during photosynthesis. In fact, photosynthesis is an important source of oxygen on the earth. If it were not for green plants, animals would quickly use up all the oxygen available to them.

Because green plants use the energy of the sun directly to make their own food, they are called **autotrophs** (AWT-uh-trahfs). An autotroph is an organism that is able to make its own food from sim-

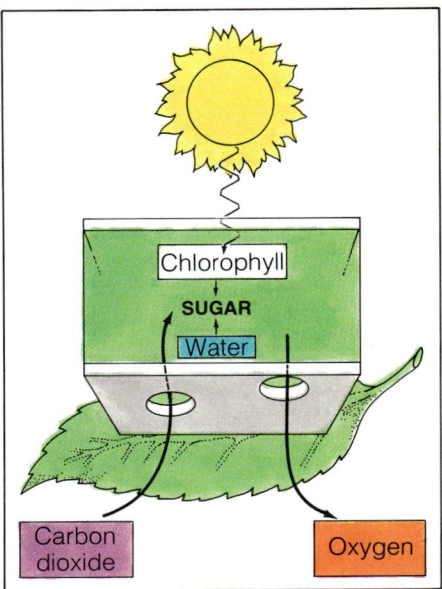

Figure 23–9 *The leaves of green plants are like food factories that capture the energy of sunlight to make glucose. What is this process called?*

ple substances. Plants use glucose to provide energy for carrying out life functions. But plants also combine the glucose with other chemicals to make starches, fats, and proteins. As you can see, photosynthesis is a very important process to green plants. But photosynthesis is also very important to animals. Can you guess why this process is so important?

Interactions Among Organisms

All organisms in an ecosystem are interdependent. **Organisms can be classified into three main groups based on how they obtain energy: producers, consumers, and decomposers.**

PRODUCERS Green plants make their own food and are thus the food **producers** of the ecosystem. Animals cannot make their own food. They must eat either plants or other animals that eat plants.

CONSUMERS An organism that feeds directly or indirectly on producers is called a **consumer.** Consumers are also called **heterotrophs** (HEHT-uhr-uh-trahfs), or "other-feeders."

Figure 23–10 *Indian pipe plants lack chlorophyll and cannot make their own food. But a certain fungus that lives on a nearby green plant forms a bridge to the Indian pipe plants. The fungus absorbs glucose from the green plants, and the Indian pipe plants, in turn, absorb glucose from the fungus.*

PRODUCER

CONSUMER

CONSUMER

DECOMPOSER

Figure 23–11 *Organisms are classified as producers, consumers, or decomposers depending on how they get their food. How do decomposers obtain their food?*

There are many kinds of consumers. Some organisms, such as mice, insects, and rabbits, are **herbivores** (HER-buh-vorz). Herbivores are plant eaters. Snakes, frogs, and wolves are **carnivores** (KAHR-nuh-vorz). Carnivores are flesh-eaters. Other consumers, such as bears, raccoons, and people, are **omnivores** (AHM-nuh-vorz). They eat both plants and animals.

A special type of animal consumer feeds on the bodies of dead animals. These organisms are called **scavengers.** Jackals and vultures are examples of scavengers. So are crayfish and snails, who "clean up" lake waters by eating dead organisms.

DECOMPOSERS After plants and animals die, organisms called **decomposers** use the dead organic matter as a food source. Unlike scavengers, decomposers break down dead plants and animals into simpler substances. In the process, they return important materials to the soil and water. You may be familiar with the term "decay," which is often used to describe this process. Bacteria and mushrooms are examples of decomposers.

Decomposers are essential to the ecosystem for two reasons. They rid the environment of dead plant and animal matter. But even more important, decomposers return substances such as nitrogen, carbon, phosphorus, sulfur, and magnesium to the environment. These substances then are used by

Figure 23–12 *These vultures are feeding on a dead giraffe. But the vultures did not kill the giraffe. What type of consumers are vultures?*

other plants to make food, the cycle continues. If the nutrients were not returned to the environment, organisms within that ecosystem could not survive for long.

Food Chains and Food Webs

Within an ecosystem, there are food and energy links between the different plants and animals living there. A **food chain** shows how groups of organisms within an ecosystem get their food and energy.

Green plants are autotrophs, or producers. So they are the first link in a food chain. Animals that eat the green plants are the second link in the food chain. And animals that eat the animals that eat plants are the third link.

Here are two examples of food chains. In a meadow, a grasshopper eats a plant. A bird then eats the grasshopper. Finally, a snake eats the bird. In a freshwater pond, microscopic green plants are eaten by water fleas. In turn, the water fleas become the food of minnows, or small fish. Larger fish such as perch and bass feed on the minnows. If no animal eats the perch or bass, it will eventually die and become a source of food for a scavenger such as the crayfish. As the perch or bass decays, important substances are returned to the lake water. The microscopic green plants use these substances to make their food. The food chain goes on and on. This food chain could have been extended beyond the perch or bass. Can you think of who the next consumer might be?

Food chains are good descriptions of how food energy is passed from one organism to another. But interactions among organisms often are more complex because most animals eat more than one type of food.

Figure 23–13 *In this ocean food chain, microscopic plants called phytoplankton (top) are the food of tiny water animals called zooplankton (top center). The zooplankton are then eaten by bivalves such as mussels (center). Mussels feed many creatures including crabs (bottom center). The seagull (bottom) feeds on both crabs and mussels, as well as an occasional starfish. Which organisms are the producers in this food chain?*

533

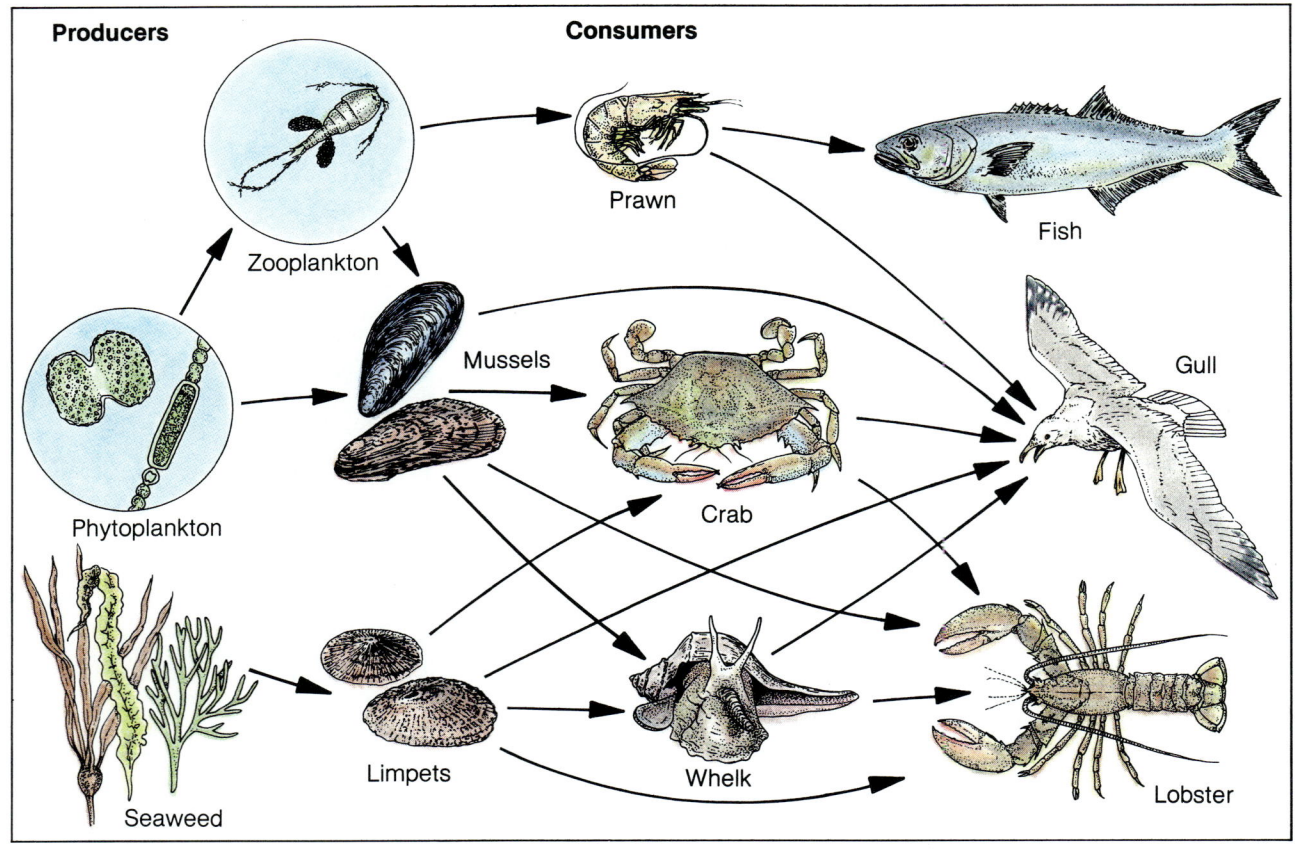

Producers

Zooplankton

Phytoplankton

Seaweed

Limpets

Consumers

Prawn

Fish

Mussels

Crab

Whelk

Gull

Lobster

Figure 23–14 *The ocean food web is a complex overlapping of many individual food chains. Which animals are consumers?*

For example, in a saltwater environment, sea snails eat seaweed. But seaweed is also eaten by other small fish. And these small fish may become food for larger fish, birds, and octopuses. There are several different food chains involved here, and they overlap to form a **food web.** A food web includes all the food chains in an ecosystem that are connected together. See Figure 23–14.

Feeding and Energy Levels

A feeding level is the location of a plant or an animal along a food chain. Because green plants produce their own food, they form the first feeding level. Consumers that eat plants form the second feeding level. And consumers that eat animals that eat plants form a third feeding level.

At each feeding level, however, much of the energy in food is lost. Each organism at a particular feeding level uses up some of the energy locked in food to carry out its life activities. In addition, a great deal of energy is used up as it is transferred from one feeding level to another. What does this mean for living things at higher energy levels? There is less energy available to organisms at each higher level. The loss of energy as it moves through a food chain can be pictured as an energy pyramid. See Figure 23–15. The energy pyramid represents the decreasing amount of energy available at each feeding level.

Because the autotrophs, or producers such as green plants, support all other living organisms on the earth, they form the base of the energy pyramid. This first level is very broad. Each level above the first level is made up of consumers. And each successive level of consumers in the pyramid is narrower than the one beneath it. As energy moves through the pyramid, from the first feeding level up through all the other levels, much of it is lost.

Sharpen Your Skills

Drawing a Food Web

Using the following plants and animals, draw the food web as you think it might exist: grass, mice, deer, mountain lions, vultures, bacteria.

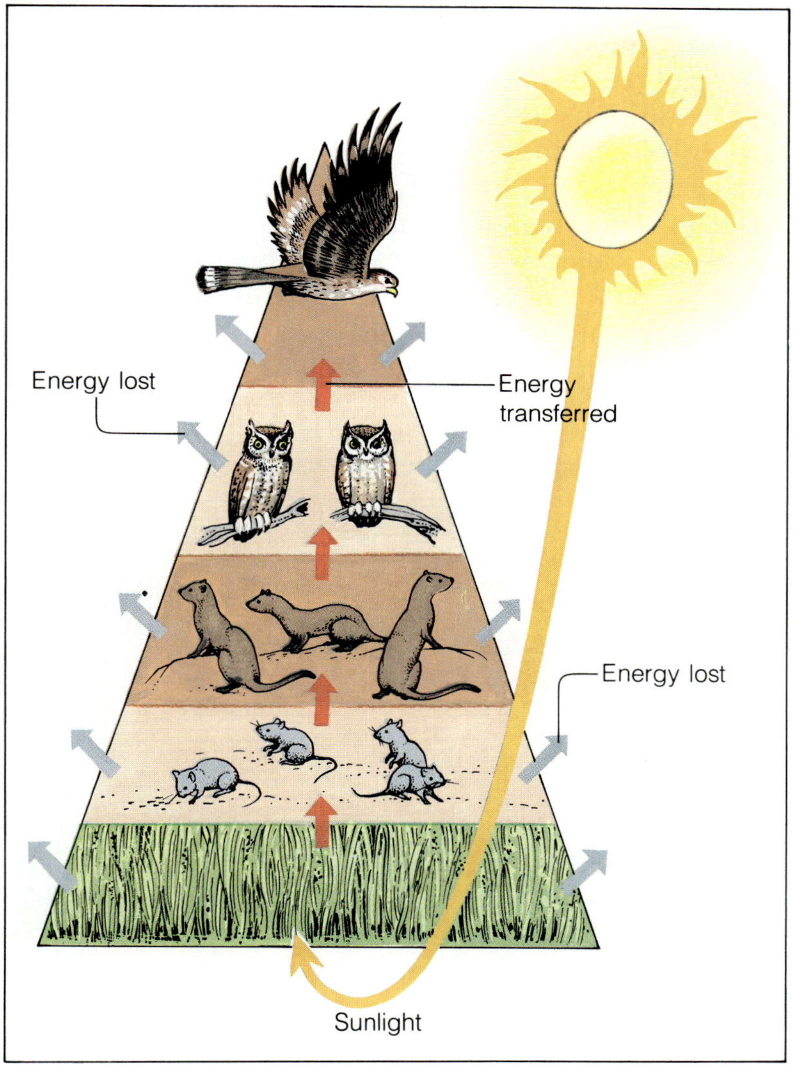

Energy lost
Energy transferred
Energy lost
Sunlight

Figure 23–15 *This energy pyramid shows how energy is lost as it moves through the food chain. What is the grass's energy source?*

SECTION REVIEW

1. Describe the three main groups of organisms in an ecosystem.
2. Why are decomposers important to the ecosystem?
3. What is the difference between a food chain and a food web?
4. Describe a food chain that includes you.

23–3 Relationships in an Ecosystem

Section Objective

To classify relationships between organisms in an ecosystem

Whether an ecosystem is a huge patch of ocean bottom or a dark corner of a forest floor, all the living and nonliving things in it must interact successfully in order for the ecosystem to survive. From their environment, all the living organisms must be able to obtain food, water, and shelter. Given the right conditions, the plant and animal populations can reproduce and increase their numbers. But certain living and nonliving factors in the environment can stop a population from increasing in size. Such factors are called **limiting factors.** Food, space, temperature, and disease are examples of limiting factors. Sometimes the growth of a population is limited as a result of its relationships with other

Figure 23–16 *Living things, such as this adult and young chameleon, must interact with other living and nonliving things in the ecosystem. If conditions are right, organisms will reproduce and increase their populations.*

organisms. **Relationships between organisms include competition, predation, commensalism, mutualism, and parasitism.**

Competition

Ecosystems often cannot satisfy the needs of all the living things in a particular habitat. When there is only a limited amount of food, water, shelter, or even light in an environment, **competition** occurs. Competition is a type of relationship in which organisms struggle with one another and their environment to obtain the materials they need to live.

Most of the relationships among living things in an ecosystem are based on competition. The moose and the snowshoe hare, for example, share the same habitat and compete for the same food. Birch trees are the most important food source for these two animals. However, the small hares can hardly compete with the large, hungry moose. So the hares may face a shortage of food, and possibly death, during the cold winter.

Competition often has a positive effect on an ecosystem. Through natural competition, different populations of animals usually are prevented from growing so large that they disrupt the ecosystem. But sometimes outside forces can unbalance natural competition. For example, people interfered with such a natural plan when they introduced rabbits to Australia in 1788. The rabbits had few natural enemies, so the rabbit population grew very quickly. By the middle of the 1800s, these harmless-looking

Figure 23–17 *The moose (left) and the snowshoe hare (right) share the same habitat and compete for the same food. What is the most important food source for these animals?*

Figure 23–18 *Even though they are members of the same population, these aspen trees in Colorado are in competition with one another. For what do the trees compete?*

animals were stripping the grasslands bare of vegetation. Cattle herds that also grazed on this vegetation were threatened with starvation. Finally, a disease-causing virus was given to many of the rabbits by scientists. These rabbits died, and so the cattle were saved from starvation.

Plants compete with one another for light, carbon dioxide, minerals, and water. In the forest, plants usually compete for light. Tall trees soak up plenty of sunlight. Smaller trees that also need sunlight may not survive in the shade of these big trees. However, shade can be an advantage to some trees. The red maple tree is hardier than the sugar maple tree, and it usually outnumbers the sugar maple tree in a forest. But the sugar maple tree can grow in the shade, and the red maple tree cannot. So if a young red maple tree and a young sugar maple tree are competing for space and food in the shade of a forest, the sugar maple tree will more likely survive.

Predation

A **predator** is an animal that kills and eats other animals. The animal that is killed is the **prey.** The relationship between these two animals is known as **predation.** For example, wolves hunt and eat deer. The wolf is a predator. The deer is its prey.

Another predator is the lynx. It preys on the snowshoe hare. When the lynx population increases, the hare population decreases. What happens to the size of the lynx population when the hare population is reduced?

Figure 23–19 *Predators can be found in water as well as on land. The leopard would like to catch up with the baboon (left). The snapping turtle is about to make a meal of the fish (right). What is the animal killed by a predator called?*

Symbiotic Relationships

The word **symbiosis** means "living together." Symbiosis is a relationship in which one organism lives on, near, or even inside another organism. A symbiotic relationship may benefit either one partner or both partners in the relationship. Sometimes one partner benefits only by harming the other.

COMMENSALISM   A symbiotic relationship in which only one organism in the partnership benefits is **commensalism.** The other organism in the relationship is neither helped nor harmed.

One example of commensalism involves small birds, such as sparrows and wrens, that benefit from a large, fierce hawk called the osprey. The osprey builds a big platform-shaped nest for its eggs. The smaller birds set up their homes beneath the osprey's nest. Because the osprey eats mostly fish, these smaller birds are in no danger from the osprey. Instead, the little birds gain protection from their enemies by living close to the fierce hawk.

Tropical orchids show another example of commensalism. These beautiful plants are able to survive in dense jungles by growing high above the shade in the branches of trees. There the orchids

Figure 23–20   *Tropical orchids obtain sunlight by growing high on the branches of a tree. The relationship between the orchids and the tree is called commensalism. Is the tree harmed by the orchids?*

Figure 23–21   *The tiny, crusty-looking growths on this gray whale are animals known as barnacles. By attaching themselves to the whale, the barnacles get a free ride through the vast ocean, which increases their chances of finding food. The whale is not harmed by these passengers. What type of symbiotic relationship is this?*

Figure 23–22 *This California dodder plant is a parasite. It wraps itself around a host plant and uses the host for support and food. Is the host harmed in this relationship?*

get the sunlight they need. And because their roots are exposed, they can take water right out of the air. The tree is neither harmed nor helped by this relationship.

MUTUALISM   A symbiotic relationship that is helpful to both organisms is called **mutualism.** Food and protection are two common reasons organisms share such a partnership.

In certain areas of Africa, a small bird known as the honey guide bird lives in a mutualistic relationship with a furry little animal called a ratel. The honey guide bird loves to eat beeswax, but it is too small to break into a bee's nest easily. The ratel likes honey, but it cannot always find a supply on its own. So the two animals work together. The honey guide locates a bee's nest and chirps loudly. The ratel moves toward the sound. This chirping and following continues until the ratel reaches the bee's nest. With its sharp claws, the ratel rips the nest open and enjoys a fine feast. The honey guide bird then gets its chance to finish the beeswax.

In another example of mutualism, one animal helps feed another in return for protection. Large fish-eating birds called herons and ibises make their nests in the trees of the Florida swamps. At the base

Figure 23–23 *This furry ratel (top) works with the honey guide bird (bottom) so that both can find their favorite foods—honey and beeswax. What is the name for this symbiotic relationship?*

of these trees live poisonous snakes who gather there to catch pieces of fish the birds may drop. At the same time these snakes are getting a free meal, they are providing protection for the herons and ibises. You see, the poisonous snakes keep away raccoons and other animals that would feast on the birds' eggs and baby chicks.

PARASITISM So far you have seen that in a relationship between two organisms, both may benefit or just one may benefit. In the relationship called **parasitism,** however, one partner not only does *not* benefit but is actually harmed by the other organism. The organism that benefits is the **parasite.** The organism that the parasite lives off and harms is called the **host.** A parasite has special adaptations that help it take advantage of its host.

The sea lamprey, a very primitive fish that lacks jaws, is a blood-sucking parasite. This strange-looking fish resembles a swimming tube that has fins on its back. The lamprey's round mouth acts as a suction cup by which the lamprey attaches itself to other fish. Even as the victim tries to shake the lamprey loose, the parasite's toothed tongue carves a hole in the fish and sucks out its blood. Small fish often die from loss of blood. However, a successful parasite does not actually kill its victim. Can you explain why?

Plants, as well as animals, can be parasites. The dodder plant lives by obtaining all of its food from host plants, such as clover and alfalfa. Wrapping its pale stem around the host plant, the dodder pushes its "suckers" onto its host. Then it releases itself completely from the soil and stays attached to the host plant for support and food.

Balance in the Ecosystem

An ecosystem is a finely balanced environment in which living things successfully interact in order to survive. Disturbance in this balance in one part of the ecosystem can cause problems in another part of the ecosystem. Such disturbances can be the result of nature or of human interference.

Sometimes human interference can destroy an ecological system. Mono Lake, in eastern California,

Figure 23–24 *Poisonous snakes live at the base of the tree in which these blue herons make their nest. The relationship between these organisms is one of mutualism. The snakes eat bits of food dropped by the herons. How do the herons benefit?*

Figure 23–25 *Mono Lake in California was once home to more than 80 species of birds, as well as many plants and other organisms. What human actions caused the disruption of this ecosystem?*

Figure 23–26 *Scientists are studying the birds and other animals around Mono Lake in an effort to once again make Mono Lake a suitable habitat for such wildlife.*

is a beautiful saltwater lake fed by streams of melting snow from the Sierra Nevada Mountains. Two small islands within Mono Lake attract thousands of birds, especially sea gulls. More than 80 species of birds nest on these islands and feed on the lake's shrimp, flies, and algae. That is, until 1981. During the summer of that year, many baby sea gulls were found dead. How strange this seemed to be for an ecosystem that had long provided food, water, and shelter for the sea gull population. As people investigated the situation, they discovered that the ecosystem had been disturbed by actions taken far away from the lake many years before.

About 40 years ago, the city of Los Angeles began to use water from the major streams that feed into Mono Lake. Less and less water emptied into Mono Lake, and the lake began to dry up. Thousands of acres of dust formed where there was once water.

As the amount of water in the lake decreased, the concentration of salt dissolved in the water increased. The shrimp that sea gulls fed on could not survive in water so salty. As the shrimp died, less food was available for the sea gulls. Baby sea gulls starved to death.

To make matters worse, as the water level in Mono Lake dropped, a land bridge that connected the shore to the nesting islands formed. Coyotes crossed this bridge, killed many sea gulls, and invaded the gulls' nests.

Many people want to save Mono Lake. They realize that human actions have damaged the lake and disturbed an ecosystem. Without a water conservation plan, Mono Lake will continue to dry up. And the delicate balance between living and nonliving things in this ecosystem will be seriously altered.

The balance of an ecosystem can be upset by a natural disaster as well as by human beings. In May of 1980, Mount St. Helens, a volcano in the state of Washington, exploded. Thousands of trees, shrubs, flowers, and animals were destroyed by the eruption. Volcanic ash covered the soil as far away as 14 kilometers from the volcano. What had once been a beautiful, green forest soon looked like the barren surface of the moon.

Within a month, however, life began to return to the area. Roots of the red-flowered fireweed bush and other plants that had survived pushed growing stems up through the ash. These plants attracted insects such as aphids, which feed on the juices of the fireweed. In turn, birds came into the area to feed on insects. Spiders crawled on the ash-covered surface. When these animals died, their bodies fertilized the ash, returning important nutrients to the soil. Hoofed animals such as elk wandered through the area, breaking up the ash cover and leaving holes through which more seeds could sprout.

Once plant and animal life started again, relationships among organisms were reestablished. What

Figure 23–27 *In May 1980, the lush green forests of Mount St. Helens (bottom left) were destroyed by a volcanic eruption. Volcanic ash covered the soil and fallen trees (top). Eventually life began to appear amid the ash (center). Gradually an ecosystem is being reborn (bottom right).*

was once a bleak, lifeless landscape is now becoming an area filled with colorful flowers, green shrubs, and growing trees. Scientists are carefully studying this area for clues to the secret of how living things can turn a barren land into a lively ecosystem inhabited by interdependent organisms.

SECTION REVIEW

1. Describe five relationships that occur between organisms.
2. Define limiting factors and give three examples.
3. Compare predators and parasites.
4. How could the spraying of an insecticide interfere with the balance of an ecosystem?

23–4 Ecological Succession

As you have just learned, scientists have been studying the rebirth of the ecosystem around Mount St. Helens in order to find out about the interdependence among organisms. However, they are also learning something about the changes that occur in ecosystems over time. As time passes in a particular community, one organism follows, or succeeds, another. **The process of gradual change within a community is called ecological succession.**

The process of **ecological succession** can totally change what a place looks like. For example, what may be solid ground under your feet right now may once have been a deep pond swarming with fish. Here is how the transformation would have taken place.

Heavy rains fall, deep snow melts, or a tiny stream takes a different course. The water travels over the land and settles in low places. If the water does not drain away or evaporate, a pond forms.

At first, the pond is empty of living things. Then a water bird pays a visit. It stays for a while, finds nothing to eat, and flies off. But it may leave behind something special—fertilized fish eggs and water plant seeds washed from its feet. The fish

Figure 23–28 *As time passes, a pond will gradually fill in and become a forest. The kinds of plants and animals in the area change as does the physical appearance of the area. What is this process of gradual change called?*

eggs have hitched a ride from some distant lake. Another visitor arrives. It is the wind, which deposits the seeds of land plants around the pond.

As time passes, the fish and plants grow and reproduce. The early generations die, and their remains sink to the bottom of the pond. A fertile mud begins to form there. More time passes. Many rains fall, washing soil into the pond. The pond gets shallower and shallower. Now new organisms can get a foothold. Reeds and cattails start growing around the edges of the pond.

Meanwhile, down at the pond's bottom the mud has grown so rich that water plants can grow in ever-increasing numbers. As they grow, they use up great quantities of oxygen. Fish, which need oxygen, begin to die. The plants take over the pond, growing all across it. The last fish vanish but other animals find the new environment ideal. These new inhabitants are air-breathing frogs and turtles.

Eventually only patches of open water are left. The pond has become a marsh. Like the pond, the marsh keeps filling. It becomes dry land. Rabbits and deer roam where fish and frogs once lived. Bushes take root and then trees. What began as a pond has become a forest.

Succession takes a long time. To go from a tiny pond to a forest can take more than a century. Outside forces, however, can speed up or slow succession. Fire can burn a developing forest and set it back. Floods can fill a dying pond with water again. But, sooner or later, succession will change the face of the land.

SECTION REVIEW

1. Define ecological succession.
2. How do thickly growing water plants in ponds help cause the death of fish?
3. Why are most people unaware of ecological succession even though it may be occurring all around them?

23–5 Cycles in Nature

Earlier in this chapter, you learned how energy from the sun is transferred through the ecosystem. Some of that energy is used by plants to make food. And, each day the sun replaces that energy. Without sunlight, living things could not survive.

However, there are other things in the environment that living things need, such as food, air, and water. The chemicals that make up these substances are found in the natural world. Although great amounts of new chemicals are not added to the environment each day, chemicals can be recycled, or reused. **Nitrogen, water, oxygen, and carbon dioxide flow in cycles through the living and nonliving parts of the environment.**

The Nitrogen Cycle

About 79 percent of the atmosphere is made up of "free" nitrogen, or nitrogen that is not combined with other elements. Almost all living things need nitrogen to help build proteins and certain other body chemicals. However, in spite of being surrounded by nitrogen gas, most organisms cannot use this free nitrogen in the air. Most living things can use only nitrogen that is combined with other elements in compounds. But how are these compounds made? Among the "factories" for making nitrogen compounds are tiny, one-celled bacteria.

Figure 23–29 *Plants use energy from the sun to make food. Each day that energy is replaced.*

A few kinds of bacteria can take nitrogen directly from the air and form nitrogen compounds. This process is called **nitrogen fixation.** Some of these nitrogen-fixing bacteria live in soil. Others grow on the roots of plants known as legumes (LEHG-yoomz), including beans, peas, and peanuts. The bacteria supply the plants with usable nitrogen.

Bacteria build nitrogen compounds in another way. When plants and animals die and decay, nitrogen in the decaying matter may be combined with hydrogen, oxygen, and other elements. This manufacturing job is done by several different bacteria. One family of compounds formed in this way consists of nitrogen compounds called nitrates. Nitrates can be taken directly from the soil by plants.

The formation of nitrogen compounds by bacteria is only the first step in what is known as the nitrogen cycle. Plants use the nitrogen compounds to make food. Animals may then eat the plants or other animals that have eaten plants. When the plants and animals die, the nitrogen compounds return to the soil. Nitrogen can go back and forth between the soil and plants and animals many times before reentering the atmosphere. Eventually, however, bacteria called denitrifying bacteria break down nitrogen compounds such as nitrates. In the process, free nitrogen is released into the air. The cycle is then complete.

Figure 23–30 *The tiny, round structures on the roots of this pea plant contain nitrogen-fixing bacteria. These bacteria take nitrogen from the air and form nitrogen compounds needed by the plant. What is this process called?*

Figure 23–31 *In the nitrogen cycle, nitrogen passes from the atmosphere to living things and then back to the atmosphere. How do animals get nitrogen?*

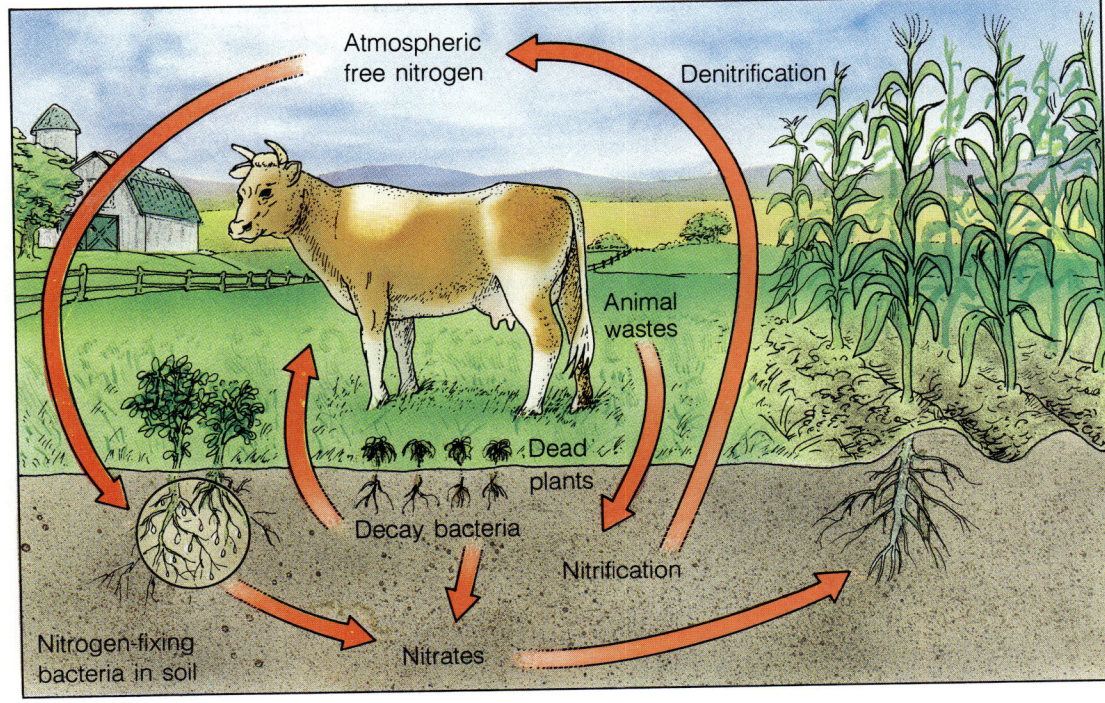

Sharpen Your Skills

Evaporation

Does water evaporate from a leaf? Here is a way to find out.

1. Cover a leaf on a house plant or a tree with a plastic sandwich bag.

2. Tie the bag to the stem with a piece of string. Be careful not to bend or twist the leaf.

3. Observe the leaf and bag after 24 hours. Record and explain your observations.

Figure 23–32 *The circulation of water from the atmosphere to the earth and back to the atmosphere is called the water cycle. How does water from the atmosphere reach the surface of the earth?*

The Water Cycle

Water circulates continually between the atmosphere and the surface of the earth. Most of the water on earth is found in lakes, streams, and especially oceans. Surface water in these lakes, streams, and oceans is heated by the sun and turns to a vapor, or gas. This process is called **evaporation.** The water vapor then rises up into the air. In the upper atmosphere, water vapor cools and changes into liquid droplets. It is these droplets that form clouds. Eventually, the droplets fall back to the surface as rain and snow, or some other form of precipitation.

Most precipitation falls directly back into the oceans, lakes, and streams. Some strikes the surface of the land and then flows into these bodies of water. In either case, water that evaporated into the air has returned to the surface of the earth, only to continue the cycle.

Not all water, of course, goes directly back into rivers and oceans. Some is taken in by living things. But here the cycle continues. Plants may take in liquid water through their roots, but they also release some water vapor through their leaves. Animals drink water, but they also return water to the environment when they breathe and when they release their wastes.

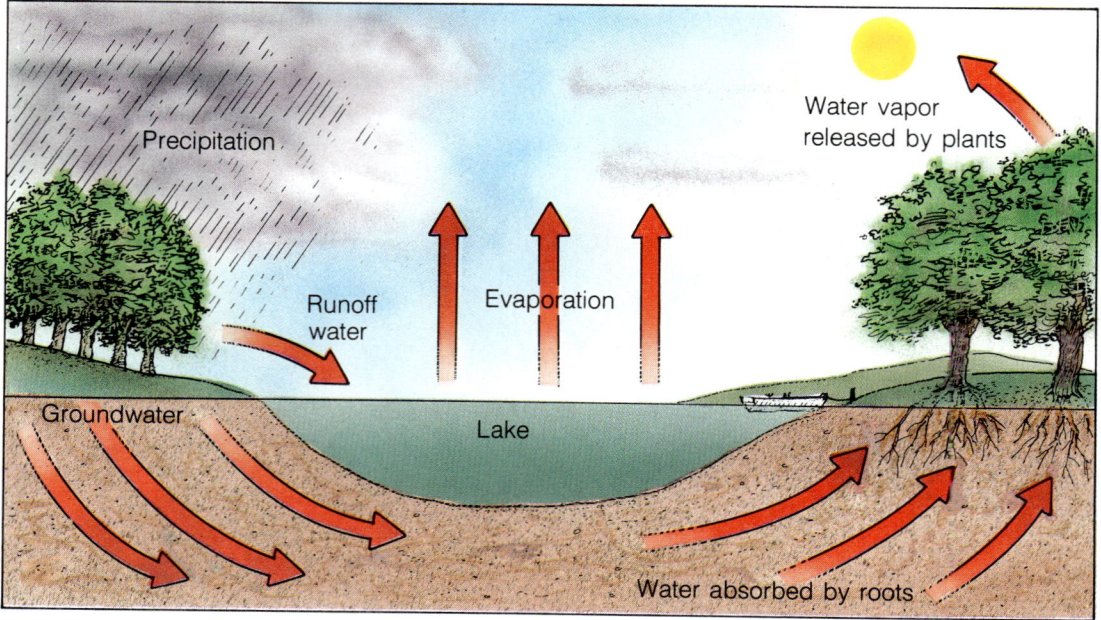

Figure 23–33 *In the oxygen/carbon dioxide cycle, photosynthesis and respiration are the major processes that circulate oxygen and carbon dioxide through the ecosystem. What organisms perform photosynthesis?*

The Oxygen/Carbon Dioxide Cycle

Like all animals, you need oxygen to live. The atmosphere, which is 20 percent oxygen, supplies you and other animals with this vital gas. Clearly, animals would have used up the available oxygen supply in the atmosphere millions of years ago if something did not return the oxygen to the atmosphere.

Consider this: When you inhale, you take in oxygen. When you exhale, you release the waste gas carbon dioxide. If something used carbon dioxide and released oxygen, it would be the perfect organism to balance your use of oxygen. That something, as you know, is green plants. In photosynthesis, green plants use carbon dioxide gas from the atmosphere, water from the soil, and the energy of sunlight to make food. During this process, plants break down water molecules into hydrogen and oxygen atoms. The oxygen escapes from plants into the atmosphere. This trade between plants and animals is the key to the oxygen/carbon dioxide cycle.

SECTION REVIEW

1. Describe the nitrogen cycle.
2. Explain how water flows between the atmosphere and the surface of the earth.
3. Describe the oxygen/carbon dioxide cycle.
4. The water you drink today may once have been part of a dinosaur. Explain.

Observing Balance in an Ecosystem

Problem

How does adding lawn fertilizer affect the balance of an aquatic ecosystem?

Materials *(per group)*

2 2-L wide-mouthed jars
pond water
8 *Elodea* (or other aquatic plant)
lawn fertilizer (or house plant food)
teaspoon

Procedure

1. Label the jars A and B.
2. Fill each jar three-fourths full with pond water.
3. Place four *Elodea* in each jar.
4. Add one-half teaspoon of lawn fertilizer to jar B.
5. Place the jars next to each other in a lighted area.
6. Observe the jars daily for three weeks. Record your observations.

Observations

Were there any differences between jars A and B? If so, when did you observe these differences?

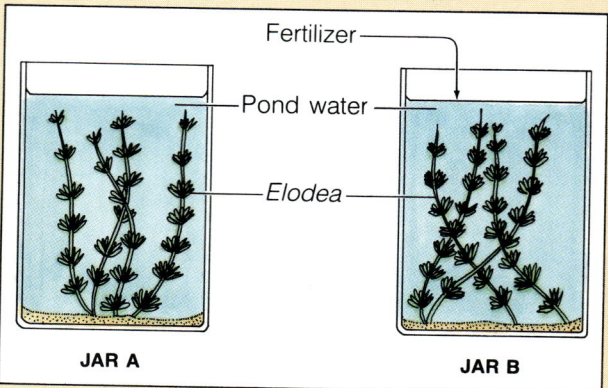

Conclusions

1. What was the control in this experiment? The variable?
2. Why did you place the jars next to each other? Why did you place them in the light?
3. What effect did the fertilizer have on the *Elodea?*
4. Lawn fertilizer contains nitrogen, phosphorus, and potassium. These nutrients are often present in sewage as well. Predict the effects of dumping untreated sewage into ponds and lakes.

CHAPTER REVIEW

SUMMARY

23–1 Living Things and Their Environment

❏ An ecosystem is a group of organisms in an area that interact with each other together with their nonliving environment.

❏ The living part of an ecosystem is called a community.

❏ A population is a group of the same type of organisms living together in the same area.

❏ The place in a community in which an organism lives is its habitat. Its niche is its role in the community.

23–2 Food and Energy in the Environment

❏ In an ecosystem, organisms are classfied as either producers, consumers, or decomposers.

❏ A food chain illustrates how groups of organisms within an ecosystem get energy. At each level in a food chain, energy is lost.

23–3 Relationships in an Ecosystem

❏ Competition among organisms for food, water, and shelter exists in all ecosystems.

❏ Symbiosis is a relationship in which two organisms live together for the benefit of either one or both of the partners.

❏ Disturbances in one part of an ecosystem can cause problems in another part.

23–4 Ecological Succession

❏ The process of gradual change within a community is called ecological succession. Ecological succession usually takes a long time, often spanning centuries.

23–5 Cycles in Nature

❏ Nitrogen, water, oxygen, and carbon dioxide flow in cycles through living and nonliving parts of the environment.

VOCABULARY

Define each term in a complete sentence.

abiotic factor	decomposer	food web	nitrogen fixation	predator
autotroph	ecological succession	habitat	omnivore	prey
biotic factor	ecology	herbivore	parasite	producer
carnivore	ecosystem	heterotroph	parasitism	scavenger
commensalism	environment	host	photosynthesis	symbiosis
community	evaporation	limiting factor	population	
competition	food chain	mutualism	predation	
consumer		niche		

CONTENT REVIEW: MULTIPLE CHOICE

On a separate sheet of paper, write the letter of the answer that best completes each statement.

1. The nonliving things in an environment are called
   a. abiotic factors.   b. biotic factors.
   c. limiting factors.   d. populations.
2. The study of the relationships and interactions of living things with one another and with their environment is called
   a. ecosystem.   b. ecology.   c. symbiosis.   d. competition.

3. The living part of any ecosystem is called a (an)
   a. environment.   b. niche.   c. community.   d. habitat.

4. An organism's role in its community is its
   a. habitat.   b. ecosystem.   c. population.   d. niche.

5. Which is *not* needed by green plants for photosynthesis?
   a. water   b. sunlight   c. carbon dioxide   d. nitrogen

6. Organisms that make their own food are called
   a. autotrophs.   b. heterotrophs.   c. consumers.   d. decomposers.

7. Omnivores eat
   a. plants only.   b. animals only.
   c. both plants and animals.   d. neither plants nor animals.

8. A symbiotic relationship in which one partner benefits and the other is harmed
   is known as
   a. mutualism.   b. commensalism.   c. parasitism.   d. predation.

9. The delicate balance of an ecosystem can be upset by
   a. human interference.   b. nature.
   c. either of these.   d. neither of these.

10. About 79 percent of the atmosphere is made up of
   a. oxygen.   b. nitrogen.   c. carbon dioxide.   d. water vapor.

CONTENT REVIEW: COMPLETION

On a separate sheet of paper, write the word or words that best complete each statement.

1. A group of organisms together with their nonliving environment is called a (an) _____.

2. A (An) _____ is a group of the same type of organism living together in the same area.

3. Plants make their own food by the process of _____.

4. An organism that feeds directly or indirectly on producers is called a (an) _____.

5. A (An) _____ is a flesh-eater.

6. A (An) _____ includes all the food chains in an ecosystem that are connected together.

7. An animal that kills and eats other animals is a (an) _____.

8. _____ is a relationship in which one organism lives on, near, or even inside another organism.

9. The process of gradual change within a community is called _____.

10. A few kinds of bacteria take nitrogen directly from the air and form nitrogen compounds during _____.

CONTENT REVIEW: TRUE OR FALSE

Determine whether each statement is true or false. Then on a separate sheet of paper, write "true" if it is true. If it is false, change the underlined word or words to make the statement true.

1. Plants and animals are <u>biotic</u> factors in the environment.

2. The place in which an organism lives is its <u>niche</u>.

552

3. Organisms in an ecosystem cannot occupy the same <u>habitat</u>.
4. Consumers are also called <u>heterotrophs</u>.
5. <u>Decomposers</u> form the base of the energy pyramid.
6. <u>Commensalism</u> is a relationship in which organisms struggle with one another and their environment.
7. A type of symbiosis that benefits both organisms is <u>mutualism</u>.
8. A <u>parasite</u> is an animal that kills and eats other animals.
9. Most chemicals on earth <u>cannot</u> be recycled.
10. Water on the earth's surface turns to gas in a process called <u>evaporation</u>.

CONCEPT REVIEW: SKILL BUILDING

Use the skills you have developed in the chapter to complete each activity.

1. **Applying definitions** The African tickbird lives on the back of the rhinoceros, where it picks bloodsucking ticks off the back of the huge animal. What type of symbiosis is this? Explain.
2. **Making diagrams** Draw a food web including a snake, mouse, wolf, rabbit, frog, grass, hawk, grasshopper, and owl. Identify each organism as a producer or a consumer.
3. **Making predictions** Suppose a new predator was introduced into an island ecosystem and reproduced successfully. What could happen to some of the other predators and the prey on the island?
4. **Expressing an opinion** Recently there has been concern that by burning fossil fuels, such as coal and oil, people are adding a lot of carbon dioxide to the atmosphere. In addition, people are destroying forests at a rapid rate. How could this affect the carbon dioxide/oxygen cycle?

5. **Developing a model** Suppose you were in charge of building a town. How could you design it so that it would be part of the existing ecosystem?
6. **Interpreting graphs** What conclusion about human population growth can you draw from the following graph?

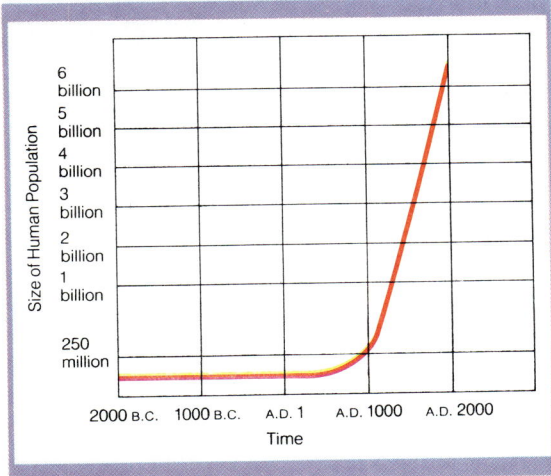

CONCEPT REVIEW: ESSAY

Discuss each of the following in a brief paragraph.

1. Define each of the following terms and explain how they are related to one another: ecosystem, community, population, habitat, niche, abiotic factor, biotic factor, environment.
2. Describe the role green plants play in an ecosystem.
3. Compare three types of symbiosis. Give an example of each.
4. Describe the process of ecological succession in which a pond gradually becomes a forest.
5. Briefly describe the nitrogen, water, and oxygen/carbon dioxide cycles.

Biomes

24

CHAPTER OBJECTIVES

After completing this chapter, you will be able to

24–1 Explain how biomes are classified.

24–1 List eight major biomes.

24–2 Describe the tundra.

24–3 Compare the three forest biomes.

24–4 Describe the grassland.

24–5 Compare hot and cold deserts.

24–6 Classify the two major water biomes.

24–6 List the factors that affect marine and freshwater life.

"At first I saw only a moving line. It seemed as if . . . countless boulders were growing out of the hills and were suddenly gliding downhill over the icy surface in slow motion." This is how a naturalist described his first view of a migrating caribou herd. Caribou are a type of reindeer. There can be hundreds of thousands of them in one herd. This particular herd spread ten kilometers across. And it took hours before the last animal passed out of sight.

Once herds of caribou were common in Maine. But hunting and other human activities drove them farther north to Canada. However, with a little luck and a lot of hard work, Maine will once again be home to these graceful animals.

It is all part of the Maine Caribou Reintroduction Project. In late 1986, game wardens captured 30 caribou in Newfoundland, a province of Canada. They weighed and measured the animals and tested them for disease. Then game wardens brought them to the University of Maine at Orono. There the caribou will be used as a "nursery herd," or breeding herd, to breed more caribou.

The caribou born to this nursery herd will be slowly introduced to their new Maine environment. Eventually, even the original nursery herd will be set free. If all goes as planned, by 1996 there may be as many as 100 caribou roaming through Maine.

Today caribou can be found in northern North America. They spend the summers in an area called the tundra. In the winter, caribou often travel to areas south of the tundra called coniferous forests. In the following pages, you will learn the difference between the tundra and a coniferous forest. You will also learn about other areas of the world and the organisms that live in them.

Caribou such as these migrate in herds in search of food.

24–1 Biomes of the World

If you were to take a trip around the world, you would quickly discover that different plants and animals live in different areas. And, you might wonder why some plants and animals survive in one area while completely different organisms survive in another area.

The type of animals that live in an area depends largely on what plants grow there. Zebras, for instance, eat grass and are found in grassy plains. Lions hunt plant eaters, such as zebras, for food. So they too are found in grassy plains.

The plant life in an area is determined mainly by **climate.** Climate is the average weather in a particular place over a long period of time. Temperature and precipitation are two important factors that determine climate. Trees, for example, grow tall and thick where the climate is warm and rainy.

Scientists have classified areas with similar climates, plants, and animals into divisions called **biomes.** Because biome divisions are merely a system to help scientists describe the natural world, not all scientists agree on the kinds and numbers of biomes. However, most scientists accept at least six land biomes and two water biomes. **The major land**

Figure 24–1 *This map shows the distribution of biomes throughout the world. In which biome do you live?*

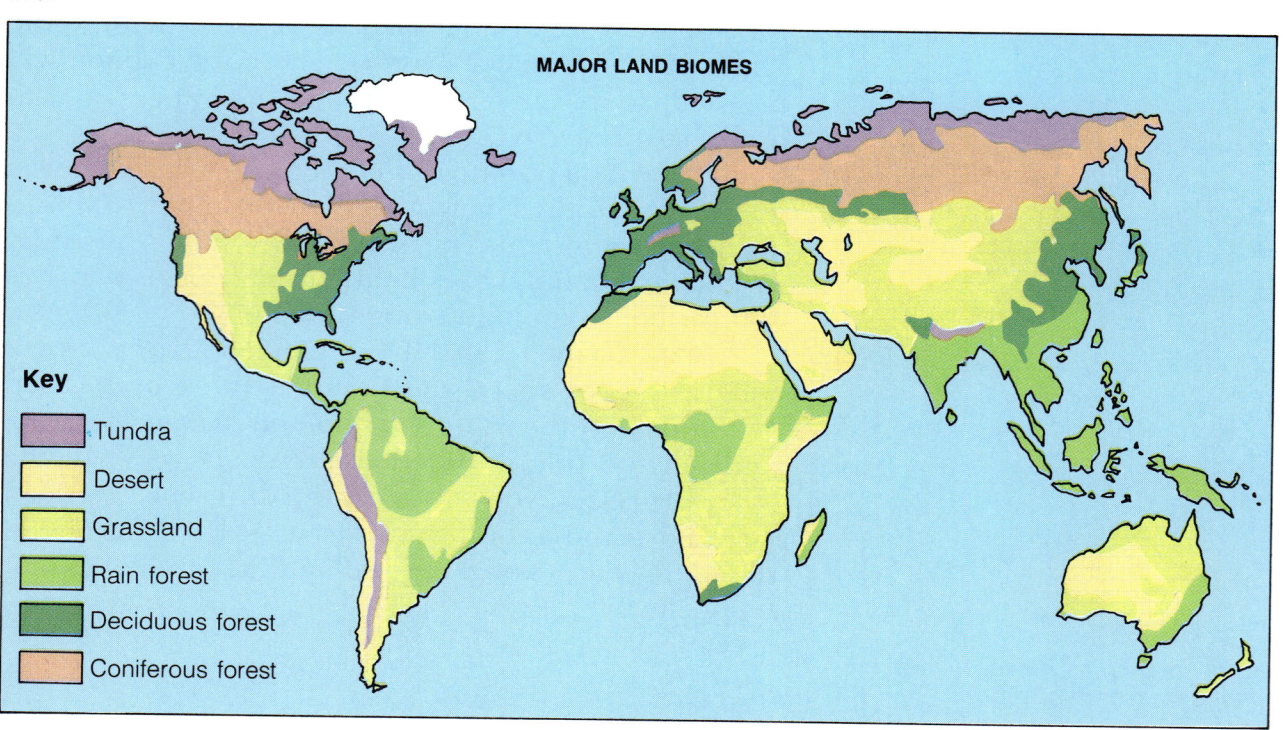

MAJOR LAND BIOMES

Key

- Tundra
- Desert
- Grassland
- Rain forest
- Deciduous forest
- Coniferous forest

biomes are tundra, coniferous forest, deciduous forest, tropical rain forest, grassland, and desert. The two water biomes are the marine, or ocean, biome and the freshwater biome. See Figure 24–1. Which biome do you live in?

SECTION REVIEW

1. What characteristics are used to divide the world into biomes?
2. List eight major biomes.
3. Why is it difficult to tell exactly where one biome ends and another begins?

24–2 The Tundra Biome

Section Objective

To describe the tundra biome

Imagine you are going on a trip through the world's biomes. Your first stop is the **tundra,** which rims the Arctic Ocean all around the North Pole. You set up camp near the ocean in Canada's Northwest Territories. It is winter and, despite your heavy clothing, the wind cuts to the bone.

The climate of the tundra is very cold and dry. It is, in fact, like a cold desert. The temperature rarely rises above freezing. And, during most years, less than 25 cm of rain and snow fall on the tundra.

Figure 24–2 *The long shaggy coats on these musk oxen (left) help them to survive on the tundra through the long, cold winter. Dwarf fireweed (right) are among the flowers that bloom on the tundra during the brief spring and summer seasons.*

Figure 24–3 *Lichens, the favorite food of caribou, cover the rocks and bare ground of the tundra. What two organisms are lichens actually composed of?*

Most water on the tundra is locked in ice within the soil. Even in the tundra's spring and summer, which lasts three months, the soil stays permanently frozen up to about a finger's length of the surface. The frozen soil is called **permafrost.** Permafrost, along with fierce tundra winds, prevents large trees from rooting. The few trees that do grow are dwarf birches and willows less than knee-high.

On the tundra, you will see lichens covering rocks and bare ground like a carpet. Each of these plantlike organisms is actually a fungus and an alga living together. Lichens are common on the tundra, where they are a main food of caribou. The caribou go where the snow is thinnest so that they can find lichens easily. Wolves often travel close behind caribou herds, preying on the old and weak.

As you leave your camp, you see great, shaggy beasts with drooping horns pawing through the snow to eat dwarf willows. They are musk oxen. Their shaggy coats insulate them from the cold.

Many small animals inhabit the tundra too. Among the most common are lemmings, which are rodents that look like meadow mice. When winter approaches, the claws of some kinds of lemmings grow thick and broad, helping them burrow in the snow, ice, and frozen soil.

You stay on the tundra until spring. As the surface soil melts, pools of water form. Clouds of mosquitoes gather about them. Unless you wear netting over your face, they make you miserable. With spring, the tough grasses and small flowers of the tundra burst into life. The sky is filled with birds.

Figure 24–4 *One of the smallest mammals on the tundra is the lemming. This lemming has dug into the snow to find some grasslike plants to feed on.*

Vast flocks of ducks, geese, and shore birds, such as sandpipers, migrate from the south to nest on the tundra. Weasels and arctic foxes hunt the young birds in their nests. Ground squirrels, which hibernate in burrows for the cold months, awaken. The days are long and sunny, but on some nights, it is frosty. The frost is a hint that, on the tundra, winter is never very far off. It is also a hint that it is time for you to pack your gear and head south.

SECTION REVIEW

1. Describe the climate of the tundra.
2. Name six organisms common to the tundra.
3. What characteristics would you expect animals that stay on the tundra all year long to have?

24–3   Forest Biomes

After leaving the tundra, you head south toward the world's forest biomes. **The three major forest biomes are coniferous forest, deciduous forest, and tropical rain forest.** The northernmost forest biome is the **coniferous forest.** Coniferous trees, or conifers, also called evergreens, produce their seeds in cones.

Coniferous Forests

The coniferous forest is a belt across Canada, Alaska, northern Asia, and northern Europe. Fingers of this forest reach south along the high slopes of mountains such as the Rockies, where the climate is colder than in the lands below. The yearly rainfall in this biome is between 50 and 125 centimeters.

The coniferous forest has fewer types of trees than forests in warmer regions. Not many kinds of trees can stand the cold northern winters as well as firs, spruces, pines, and other conifers. Their needles have a waxy covering that protects them from freezing. Because of the cold, fallen branches, needles and dead animals do not decay as fast as in warmer regions. Because the decay of plant and animal remains is one of the main factors in producing fertile soil, the soil of the coniferous forest is not

Section Objective

To compare the three major forest biomes

Figure 24–5 *A coniferous forest contains mainly coniferous trees, which produce their seeds in cones. The needles, or leaves, of these trees have a waxy covering. How does this covering help the trees survive?*

Figure 24–6 *A moose lunches on lily pads in a pond in a coniferous forest. What is another name for a coniferous forest?*

Figure 24–7 *The crossbill uses its specially adapted beak to pry the scales of a pine cone apart and get to the seeds.*

particularly rich. Poor soil is another reason that many kinds of trees are unable to grow here.

Together with poor soil, shade from the thick conifer branches keeps many plants from growing on the ground. You find that as you hike through the coniferous forest you hardly ever have to hack through underbrush.

Because it is late spring, the ground is soggy and pools of water dot the forest floor. Unlike perma-frost, soil in the coniferous forest thaws completely in spring, making some parts of it like a swamp. Indeed, these areas are often called **taiga** (TIGH-guh), a Russian word meaning "swamp forest."

Near a pond, you see a huge moose, shoulder deep in water. It dips its head and comes up with a mouthful of juicy plants from the bottom. The lake has been formed behind a dam of sticks and logs, built by beavers along a stream. Perhaps, as you move along, you see a Canadian lynx hunting a snowshoe hare, or a weasel hunting a red squirrel. Some animals found on the tundra also inhabit the coniferous forest, as you discover at night when you hear the howling of a wolf pack.

Bird songs awaken you in the morning. War-blers, which leave for the south in the autumn, chirp. Gray jays, which stay all year round can also be heard. A reddish bird with a crisscrossed bill lands on a pine cone. It is a crossbill, specially adapted to feeding on pines. In a moment, you see how. It pries the scales of the pine cone apart with its bill and takes out a seed with its tongue. Another bird of the biome, the spruce grouse, feeds on nee-dles and buds of spruce and other conifers.

Deciduous Forests

Heading south from the coniferous forest biome, you reach a **deciduous forest.** Deciduous forests contain deciduous trees, such as oaks and maples, which shed their leaves in the autumn. New leaves grow back in the spring. The deciduous forest starts around the border of the northeastern United States and Canada. It covers the eastern United States. Other deciduous forests grow throughout most of Europe and eastern Asia.

Deciduous forests grow where there is between 75 and 150 centimeters of rain a year. Summers are warm, and winters are cold, but not as cold as in the northern coniferous forests. You wander through oaks, maples, beeches, and hickories. A thick carpet of dead leaves rustles underfoot. The decaying leaves help make the forest soil richer than that of the coniferous forest. Insects, spiders, snails, and worms live on the forest floor. During the early spring, when the new leaves still are not fully grown, the forest floor is brightly lit by the sun. Wildflowers and ferns grow almost everywhere.

An occasional mouse scurries across the forest floor. Many more small mammals are out of sight under the leaves. A gray squirrel watches you, then disappears in the branches. Suddenly, up ahead, you hear a stir in the brush and see a flash of white. A white-tailed deer has spotted you and dashed away, showing the snowy underside of its tail.

Figure 24–8 *Although a deciduous forest contains mainly deciduous trees, some coniferous trees can usually be found there as well. During which season was this photograph taken?*

Figure 24–9 *Notice the mushroom amid the fallen leaves on this deciduous forest floor. When the leaves decay, nutrients that help new trees grow are returned to the soil. Why is the soil in a deciduous forest richer than the soil in a coniferous forest?*

Figure 24–10 *These baby raccoons are at home in the trunk of a tree. What other animals might you see in a deciduous forest?*

Figure 24–11 *This photograph of a tropical rain forest in Hawaii shows the different layers of the forest. What is the top layer called?*

By the side of a small stream, you spy a print in the mud. It looks almost like a small human hand. The print was made by a raccoon searching for frogs during the night. Thrushes, woodpeckers, and blue jays pass back and forth between the trees. A relative of the spruce grouse, the ruffed grouse, rests in a brush tangle and watches you pass. Under a rotting log, you find a spotted salamander, jet black with big yellow spots. A black snake slithers away as it senses your approach.

By winter, many of the birds migrate south. Snakes and frogs hibernate. Raccoons, which grow fat in autumn, spend the coldest weather sleeping in dens. However, they do come out during warm spells. The trees in winter are bare. But when spring comes, the leaves will bud and birds will return. And once again the deciduous forest will come to life.

Tropical Rain Forests

Now your travels bring you farther south, all the way to the Amazon River of South America. You camp there in the **tropical rain forest.** Rain forests get at least 200 centimeters of rain yearly. You discover that it rains for at least a short while almost every day. The climate of the tropical rain forest is warm year round, so plants grow all 12 months. This biome is also found in central Africa, southern Asia, and even a bit in Australia.

After only a few minutes, your clothes are soaking with dampness and perspiration. The air is muggy and still, although not as hot as you expected. It hardly ever gets warmer in the rain forest than it does on a scorching summer day in Chicago or New York City. Why? The answer is overhead, where the tops of trees 35 meters or taller meet to form a green roof. This top layer of the forest is called a **canopy** (KAN-uh-pee). The canopy shades the forest floor like an umbrella.

Rain forests have more varied plant life than any other land biome. The rain forest you are visiting has more than 40,000 plant species. Most plant life grows in the sunlit canopy. Vines called lianas (lee-AHN-uhz), some thicker than your leg and more than 50 meters long, twist through the branches of tall trees. Orchids and ferns perch on branches and in tree hollows.

Animal life of the rain forest is also marvelously varied. However, many of the rain forest's creatures are out of your sight. Above the canopy, harpy eagles use their keen eyes to search for monkeys and other prey. The canopy is also full of parrots, toucans, and hundreds of other kinds of birds. At night, bats move among the trees.

Boa constrictors and wild cats called ocelots climb into the lower trees that grow just under the canopy. There they hunt birds and monkeys. In the distance, you hear the roaring of a jaguar. Below, the soil is full of insects and other small creatures. As you can see, from the tops of the biggest trees to the soil, the tropical rain forest is like an animal apartment house.

Sharpen Your Skills

Forest Food Webs

On a piece of posterboard, construct a food web for each of the forest biomes. Use string to show the connections between organisms. Label each organism. List the food chains in each biome.

Figure 24–12 *Many types of animals live in a tropical rain forest. A king vulture (top left) or a slow loris (top right) might lurk in the branches of a tree. A poison arrow frog (bottom left) or a scarab beetle (bottom right) is likely found on the forest floor.*

SECTION REVIEW

1. Compare the climates of the three major forest biomes.
2. What is another name for a coniferous forest?
3. What is the difference between a coniferous tree and a deciduous tree?
4. The tropical rain forest has a greater variety of species than any other biome. What characteristics of the tropical rain forest could account for this?

Section Objective

To describe the characteristics of grasslands

24–4 Grassland Biomes

From the Amazon rain forest, you travel across the Atlantic Ocean to the **grasslands** of East Africa. **In a grassland biome, between 25 and 75 centimeters of rain falls yearly. Grasses are the main group of plants.** Africa has the largest grasslands in the world, although other large grasslands are found in western North America, central Asia, South America, and near the coasts of Australia. Another name for a grassland is **savanna.**

Your camp is in a field of grass occasionally dotted with thorny trees called acacias (uh-KAY-chuhz). There are few trees in the grassland because of the low rainfall. Fires, which often rage over grasslands, also prevent widespread tree growth.

Figure 24–13 *These lions hunt for food on the African grasslands.*

The animals that roam grasslands also keep trees from spreading by eating new shoots before they grow too large. Even large trees are not safe, as you realize when you see a herd of elephants tearing up acacias and feeding on their leaves.

Grass, on the other hand, can take trampling and low rainfall and still survive. That is why grasslands can feed the vast herds of big plant eaters, such as the zebras and antelope that you see around you. These animals in turn are the food of lions, African wild dogs, and cheetahs.

Many mice, rats, and other small animals also inhabit grasslands, eating seeds, sprouts, and insects. Snakes prowl the ground hunting these creatures. As you walk about your camp at night, you must take care not to step on a puff adder or other poisonous snake as it searches for prey.

The smaller creatures, including snakes, are the food of the keen-eyed hawks and eagles that continually sail over the savanna or perch in the acacias. Vultures circle in the sky, ready to feed on the remains of dead animals.

Different types of plant eaters and meat eaters live on grasslands in other parts of the world. For

Figure 24–14 *Bison, which once roamed American grasslands in great numbers, still exist in the nation's national parks. What do bison eat?*

CAREER   *Biogeographer*

HELP WANTED: BIOGEOGRAPHER To study the influence of certain human activities on plant and animal life. Bachelor's degree in geography and master's degree in geography or biological sciences required. Doctorial degree in biogeography highly desirable. Research results will be published.

Students who wake up and go to school in other parts of the world know the animals and plants that live in their region, but those living things may not resemble the ones in your area at all. For example, palm trees are usually found in warm areas, cactuses in deserts, and redwoods in California. Except for the ones in zoos, koalas live only in Australia and giraffes live only in Africa. Scientists who study the way in which plants and animals are scattered throughout the world are called **biogeographers.** Their research helps us to understand why organisms live where they do.

Biogeographers know about botany, zoology, meteorology, politics, and other areas of study. Since writing research papers is part of their job, good communication skills are needed. If you enjoy reading, studying, and doing research, you may consider biogeography as a career. To learn more about this diversified field, write to the Association of American Geographers, Biogeography Specialty Group, 1710 Sixteenth Street N.W., Washington, DC 20009.

example, large plant eaters such as elk and bison, were once common on North American grasslands. They were hunted by wolves and cougars. However, much of North America's grasslands have been turned to farms and ranches, and most of the big wild animals are gone. Increasingly, the same thing is happening in other grasslands of the world.

SECTION REVIEW

1. Describe the main characteristics of grasslands.
2. What are three factors that prevent trees from overrunning grasslands?
3. Predict how lions would be affected if a disease killed all the zebras and antelopes on the African grassland.

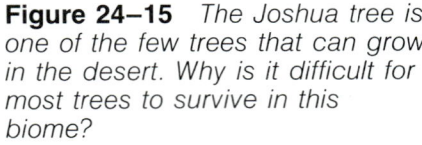
To describe the characteristics of deserts

24–5 Desert Biomes

North of the savannas, the grasslands in Africa become increasingly dry. Eventually you come to the Sahara Desert, which covers almost all of North Africa. **A desert biome is an area that receives less than 25 centimeters of rainfall a year.** Other **desert** biomes are in western North America, western Asia, the center of Australia, and along the west coast of South America.

A desert can be hot or cool. The Sahara, is a hot desert, scorching by day, chilly at night. In cool des-

Figure 24–15 *The Joshua tree is one of the few trees that can grow in the desert. Why is it difficult for most trees to survive in this biome?*

Figure 24–16 *The long ears on this jackrabbit (right) are an adaptation for desert life. The animal is able to lose heat through its ears. How is this organ pipe cactus (left) adapted for survival in the desert?*

erts, such as the Gobi Desert in northern China, there is also a great difference between daytime and nighttime temperatures. But in a cool desert, winter temperatures may drop below freezing.

As you walk through the desert, you notice that the plants in the desert are adapted to the lack of rainfall. Many have widespread roots near the surface. This enables them to absorb water quickly before the water evaporates. Plants, such as cactus, have thick fleshy stalks that help them store water.

You look hard but see few animals in the Sahara. Even so, they are there. By day, creatures such as lizards and rodents often escape the heat in underground burrows. Night brings them to the surface searching for food.

Like the plants, desert animals must live on as little water as possible. Most of the water used by desert creatures comes from seeds and other food. And they lose almost no water in their wastes. Only in these ways can they survive in a world where rain hardly ever falls.

SECTION REVIEW

1. Describe the main characteristics of deserts.
2. In what ways are desert plants adapted to the lack of rainfall?
3. Why do you think there are no tall trees in the desert?

Sharpen Your Skills

Cactus Adaptations

Now you can investigate a desert plant for yourself.

1. Obtain a cactus plant.

2. Carefully observe the outside of the plant with a hand lens. Record your observations.

3. Predict what the inside of the cactus would be like. Then, use a scissor to cut off the tip of the cactus plant. **CAUTION:** *Be careful when using a scissor. Also be careful when handling the cactus because these plants often have sharp spines.*

How did the inside of the cactus compare to your predictions? What adaptations does a cactus have that allow it to survive in a desert?

24–6 Water Biomes

Your trip around the world would not be complete without a visit to the water biomes. After all, most of the earth's surface is covered with water. **The two major water biomes are the marine biome and the freshwater biome.**

The Marine Biome

The **marine biome,** or ocean biome, covers about 70 percent of the earth. Organisms that live in this biome have adaptations that allow them to survive in water that is about 3.5 percent salt. Other factors that affect ocean organisms are sunlight, temperature, water pressure, and water movement. The ocean can be divided into different zones, or areas, based on these factors.

Most marine organisms live near the surface or near the shore. Here, algae and microscopic plants called **phytoplankton** can receive sunlight. They use the sunlight to produce food through the process of photosynthesis. Almost all of the animals in the ocean depend either directly or indirectly on these plants for food.

Marine animals come in a variety of shapes and sizes. In addition to the many types of fish that live

Figure 24–17 *The angler fish is a marine organism that inhabits the deep ocean. What purpose does the lure attached to its head serve?*

Figure 24–18 *In this marine biome, a school of blue damsel fish swim by as a clown fish hides among the tentacles of a sea anemone. What factors affect organisms that live in the oceans?*

in the ocean, there are organisms such as crabs, oysters, starfish, octopus, and whales.

Some marine organisms live in the deep ocean. Many of them have unusual adaptations that allow them to survive in this dark environment. For example, some deep-sea octopuses actually have organs that are capable of producing light.

The Freshwater Biome

The **freshwater biome** includes both still and running water. Lakes, ponds, and swamps are still water. Streams and rivers are running water.

As in the marine biome, there are a number of factors that affect freshwater life. These factors are temperature, sunlight, the amount of oxygen and food available, and the speed at which the water moves. These factors determine which organisms live in a freshwater environment.

For example, the organisms that survive in fast-moving streams have special structures that keep them from being swept away. Many plants have strong roots that anchor them to stream bottoms. Others have stems that bend easily with moving water. And the young of some insects have sucker-like structures on the bottoms of their bodies to help them attach to rocks or other objects.

If you were to visit a pond or a lake, you would see such common freshwater plants as waterlilies, duckweed, algae, and cattails. If you looked closely, you might see a fish or a water snake gliding past. Or, you might notice the bulging eyes of a frog staring out at you. You would not see the microscopic plants and animals that are part of any freshwater biome. But if you were very quiet, you might catch a glimpse of a duck or a raccoon. These animals visit freshwater biomes to feed or nest. Can you name any other animals that visit freshwater biomes?

SECTION REVIEW

1. Describe the two major water biomes.
2. Compare factors that affect marine life to factors that affect freshwater life.
3. Why are there no green plants on the bottom of the ocean?

Figure 24–19 *This bullfrog blends in with the plants in a pond. What type of biome is this?*

Figure 24–20 *An aquatic garter snake appears among the water plants, while a dragonfly rests on a stick. What factors determine which organisms can live in a freshwater environment?*

Building a Biome

Problem

How do different plants grow in each biome?

> **Materials** *(per group)*
>
> 2-L cardboard milk carton
> sandy soil or potting soil
> seeds: 5 lima beans, 30 rye grass, 10
>   impatiens
> clear plastic wrap   index card
> scissors   tape
> lamp   timer

Procedure

1. Your teacher will assign your group one of the following biomes: desert, grassland, rain forest, or deciduous forest.
2. Cut the entire front wall from a milk carton. Poke a few small holes in the uncut side for drainage. Staple the spout closed. **CAUTION:** *Be careful when handling the scissors.*
3. Fill the carton with soil to within 3 cm of the top. Note: If you have been assigned the desert biome, use sandy soil.
4. At one end of the carton, plant impatiens seeds. In the middle of the carton, plant lima bean seeds. Scatter rye seeds on the soil surface at the other end of the carton.
5. On your index card, write the names of your group, the seeds, and the type of biome. Tape the card to the carton.

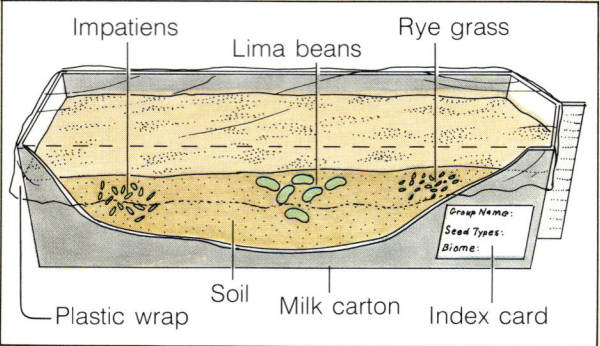

6. Water the seeds well. Cover the open part of the carton with plastic wrap.
7. Put the carton in a warm place where it will remain undisturbed. Observe the carton every day.
8. After the seeds have sprouted, and depending upon which biome your group has, give it the following amounts of light and water.
   Desert: little water, 5–6 hrs. light
   Grassland: medium water, 5–6 hrs. light
   Deciduous forest: medium water, 1–2 hrs. light
   Rain forest: much water, no direct light
   "Much water": Keep soil surface wet.
   "Medium water": Let surface dry, then add water.
   "Little water": Let soil dry to a depth of 2.5 cm.
9. Observe the development of the plants in your biome and in the biomes of the other groups. Record your observations.

Observations

1. After the seeds have sprouted and have grown for a week, describe the growth in each biome.
2. In which biome did most of the seeds grow best?
3. Where did the grass seeds grow best? The beans? The impatiens?
4. Which plants grew well in more than one biome?
5. How do beans react to little light?

Conclusions

1. Explain why the plants grew differently in each biome.
2. Why did the seeds need water when they were planted?
3. What was the variable in the experiment?
4. Predict how the impatiens, lima bean, and rye seeds would grow in the tundra and coniferous forest biomes.

CHAPTER REVIEW

SUMMARY

24–1 Biomes of the World

❑ Scientists classify biomes by their climate, plant life, and animal life.

❑ There are six major land biomes and two major water biomes.

24–2 The Tundra Biome

❑ The tundra, a northern biome, has a cold, dry climate.

❑ The frozen soil of the tundra is called permafrost.

❑ Animals that cannot survive the long tundra winter migrate farther south to warmer climates.

24–3 Forest Biomes

❑ Coniferous forests contain mainly conifers, or trees that produce their seeds in cones.

❑ The soil of the coniferous forest is not especially fertile, and many kinds of trees and plants are unable to grow there.

❑ Deciduous forests contain mainly deciduous trees, or trees that shed their leaves seasonally.

❑ The soil of the deciduous forest is more fertile than that of the coniferous forest.

❑ The tall canopy of the tropical rain forest prevents much sunlight from reaching the forest floor.

❑ Tropical rain forests contain more kinds of plants and animals than any other biome.

24–4 Grassland Biomes

❑ In a grassland biome, between 25 and 75 centimeters of rain falls yearly, and grasses are the main plants.

❑ Low rainfall, grazing animals, and fires prevent trees from overgrowing grasslands.

❑ The thick grasses in grasslands provide food for plant-eating animals. These animals, in turn, are hunted by meat-eating animals.

24–5 Desert Biomes

❑ In a desert biome, rainfall is less than 25 centimeters per year. A desert can be hot or cool.

❑ By conserving water, desert plants and animals have adapted to life in the desert.

24–6 Water Biomes

❑ Marine biomes cover 70 percent of the earth. Life in the ocean is affected by sunlight, temperature, water pressure, water movement, and the amount of salt dissolved in the ocean.

❑ Freshwater biomes include still and running water. Freshwater organisms are affected by temperature, light, the amount of food and oxygen available, and the speed at which the water moves.

VOCABULARY

Define each term in a complete sentence.

biome	desert	savanna
canopy	freshwater biome	taiga
climate	grassland	tropical rain forest
coniferous forest	marine biome	tundra
deciduous forest	permafrost	
	phytoplankton	

CONTENT REVIEW: MULTIPLE CHOICE

On a separate sheet of paper, write the letter of the answer that best completes each statement.

1. The type of plant life in a biome is determined mainly by
   a. animal life.   b. other plant life.
   c. climate.   d. other plant and animal life.
2. Two important factors for determining climate are
   a. plant and animal life.   b. temperature and rainfall.
   c. temperature and plant life.   d. temperature and animal life.
3. Which biome has the shortest summer?
   a. tundra   b. desert   c. coniferous forest   d. grassland
4. Which is the northernmost biome?
   a. tropical rain forest   b. coniferous forest
   c. deciduous forest   d. grassland
5. Which of these is *not* a deciduous tree?
   a. oak   b. maple   c. pine   d. hickory
6. The yearly rainfall in a tropical rain forest is at least
   a. 25 cm.   b. 100 cm.   c. 200 cm.   d. 500 cm.
7. Which biome has the greatest variety of plant and animal species?
   a. coniferous forest   b. tropical rain forest
   c. grassland   d. desert
8. In a grassland, the yearly rainfall is
   a. less than 25 cm.   b. between 25 and 75 cm.
   c. between 75 and 200 cm.   d. over 200 cm.
9. About what percent of the earth's surface is covered by the marine biome?
   a. 25 percent   b. 50 percent   c. 70 percent   d. 85 percent
10. Which of the following is *not* an example of still water?
    a. lake   b. pond   c. swamp   d. stream

CONTENT REVIEW: COMPLETION

On a separate sheet of paper, write the word or words that best complete each statement.

1. The average weather in a particular place over a long period of time is known as its _____.
2. The _____ biome is found near the North Pole.
3. Soil that stays permanently frozen is called _____.
4. Pine, fir, and spruce are _____ trees.
5. The top layer of a tropical rain forest is called a (an) _____.
6. A biome that receives at least 200 cm of rain yearly is the _____.
7. Zebras, elephants, and lions are found in African _____.
8. Very hot days and chilly nights are common in the _____ biome.
9. Microscopic plants found in water biomes are called _____.
10. Duckweed, cattails, and waterlilies are found in _____ biomes.

CONTENT REVIEW: TRUE OR FALSE

Determine whether each statement is true or false. Then on a separate sheet of paper, write "true" if it is true. If it is false, change the underlined word or words to make the statement true.

1. Most scientists accept at least <u>ten</u> major land biomes and two water biomes.
2. The climate of the tundra is cold and <u>wet</u>.
3. Caribou often travel from the tundra to the <u>grasslands</u> during the winter.
4. <u>Conifers</u> are trees that produce seeds in cones.
5. <u>Deciduous</u> trees shed their leaves each autumn.
6. The <u>grassland</u> biome has more than 40,000 plant species.
7. The largest grasslands in the world are in <u>the United States</u>.
8. A desert biome receives <u>less</u> than 25 cm of rainfall a year.
9. Most marine organisms live <u>deep in the ocean</u> or near the shore.
10. Duckweed and waterlilies are common <u>marine</u> plants.

CONCEPT REVIEW: SKILL BUILDING

Use the skills you have developed in the chapter to complete each activity.

1. **Classifying objects** In which biome or biomes would each of the following organisms be found?

   a. maple tree   e. cattail
   b. cactus   f. spruce tree
   c. zebra   g. musk ox
   d. caribou   h. parrot

2. **Making charts** Construct a chart in which you list each of the eight major biomes, its climate, yearly rainfall, common plants, and common animals.
3. **Making diagrams** Choose five of the eight biomes and make one food chain for each.
4. **Applying concepts** As you climb a mountain, you may walk through several biomes. Explain how this is possible.
5. **Making predictions** The Arctic tundra has come to be valued for its natural resources, such as oil and gas. It has also become a vacation spot for fishers and hunters. Predict how development might affect the tundra biome.
6. **Relating concepts** An estuary is an area where fresh water and salt water meet. Explain why it is difficult for scientists to classify estuaries as either freshwater or marine biomes.

CONCEPT REVIEW: ESSAY

Discuss each of the following in a brief paragraph.

1. Briefly describe each of the six major land biomes.
2. Explain why some people call the tundra biome a cold desert.
3. Describe how the tropical rain forest is like an animal apartment house.
4. Why is climate important in dividing the world into biomes?
5. Why do you think it is difficult for scientists to agree on the number and kinds of biomes?
6. Explain why those organisms that live in the ocean are unable to survive in freshwater biomes.

Conservation of Natural Resources

25

CHAPTER OBJECTIVES

After completing this chapter, you will be able to

25-1 Classify renewable and nonrenewable resources.

25-2 Compare methods of soil conservation.

25-2 Identify sources of land pollution.

25-3 Describe sources and effects of air pollution.

25-4 Describe sources and effects of water pollution.

25-4 Identify methods of preventing water pollution.

25-5 Compare alternative energy sources.

25-6 Describe methods used to conserve living things.

It was a typical early morning in Prince William Sound, near the remote town of Valdez on Alaska's southern coast. The weather was clear and there was barely any wind. Coast guard spokesman Rick Meidt could hardly have expected the radio call he received at 12:27 AM. The call came from the *Exxon Valdez,* a huge oil tanker loaded with over one million barrels of crude oil. "I've run aground and we've lost 150,000 barrels," said the voice over the radio. And so began the worst oil spill disaster on U.S. record.

On March 29, 1989, the *Exxon Valdez* struck Bligh reef in Prince William Sound. The reef ruptured the hull of the ship, eventually releasing over 240,000 barrels of oil into the water. As the oil slick began to grow, environmentalists, state officials, experts from Exxon and the U.S. government began to develop a plan to clean up the oil as quickly as possible. Their efforts, however, were far too slow to stop the spread of oil into one of the world's most important fisheries. All manner of sea life became threatened by the oil. Dead sea otters, birds, and many fish such as herring and salmon soon washed up on oil-slicked shores. Most experts agree that it will be many years before the final results of the spill on the environment have been determined.

Sea otters and salmon are natural resources. In the following pages, you will find out more about the earth's natural resources. And you will learn how people can destroy or protect these valuable resources.

Hundreds of thousands of barrels of oil leaked out of the Exxon Valdez after it struck a reef off the shore of Alaska.

You use **natural resources** every day of your life. Natural resources are materials taken from the environment and used by people. Some natural resources, such as forests or wildlife, are living. Other natural resources, such as water, soil, and coal, are nonliving.

Scientists classify natural resources as either renewable or nonrenewable. A **renewable resource** is one that is replaced by nature. Water, for instance, is a renewable resource because it is replaced by rain. A **nonrenewable resource** cannot be replaced by nature. Fuels such as coal, oil, and natural gas are nonrenewable resources.

People need natural resources in order to survive. But, there are now more than 5 billion people on the earth. By the year 2000, the world's population is expected to be over 6 billion. And, as the number of people on the earth increases, the rate at which people use resources also increases.

In recent years, people have become concerned about conserving natural resources. **Conservation** is the wise use of natural resources so that they will not be used up too soon or used in a way that will damage the environment. Nonrenewable resources must be conserved because they cannot be replaced. But renewable resources must also be used wisely. In some parts of the world, renewable resources,

Figure 25–1 *Like many people who fish for sport, this man is letting his catch go free again rather than killing it. How does his action help conserve a natural resource?*

such as water, are in short supply. In addition, if people are not careful, they may damage resources, such as water, so that they can no longer be used.

SECTION REVIEW

1. Compare renewable and nonrenewable resources.
2. What is conservation?
3. Name three natural resources that you use every day. Classify them as renewable or nonrenewable.

25–2 Soil

Section Objective

To compare methods of conserving soil

Soil is a renewable resource. But it must be conserved because it is difficult to replace. It can take nature 200 to 400 years to form 1 centimeter of **topsoil.** Topsoil is the rich upper layer of soil, which crops need to grow. Growing crops is one of the most important uses of soil.

Erosion and Depletion

In many areas, topsoil is being lost because of **erosion.** Erosion is the carrying off of soil by water or wind. Although erosion is a natural process, human activities have sped up its pace. Across the world, topsoil is being lost at a yearly rate of up to 10 times the rate at which new soil forms.

Figure 25–2 *In many parts of the world, forests are cut down to make room for farms, roads, or buildings. When this occurs, the roots of trees and other plants no longer prevent the forest soil from being swept away by wind or water. What is this process called?*

Sharpen Your Skills

Erosion

If the world loses 1.8 billion kilograms of soil to erosion each year, how many kilograms of soil are lost after 3.5 years?

Much of the erosion results from people clearing the land of plants. People cut forests, for example, to make room for farms or to supply themselves with timber or firewood. People also clear land to build roads and buildings. In some areas, grazing livestock, such as cattle, remove plants. Without plants, whose spreading roots help hold soil down, wind and water can easily carry soil away. When too much soil is lost, no plants can grow.

Erosion is not the only danger to soil. Nutrients needed by plants are washed away by water through a process called **depletion.** Depletion also occurs when one type of crop is grown in an area for too long. Corn, for example, removes important nitrogen compounds called nitrates from the soil. In Chapter 23, you read about the importance of nitrates in the soil. If corn is grown in the same soil for many years in a row, the nitrates in the soil will be used up.

Soil Conservation

Certain conservation measures can be taken to protect soil as a resource. One such measure simply involves careful planning. Different soils are best suited for different purposes and should be used for those purposes. For example, some soil is best for growing trees, while other soil is best for growing grass for farm animals to feed on. Areas that can produce the best crops should become farmland.

Conservation measures used to protect soil include terracing, contour farming, strip cropping, windbreaks, crop rotation, and prevention of overgrazing. In a conservation method called **terracing,** a series of level steps are made on a slope. The terracing makes the slope look like a staircase. Another method of soil conservation is **contour farming.** This involves plowing the land across a slope rather than up and down the slope. Terracing and contour farming slow down the runoff of water. The water can then soak into the soil instead of washing the soil away.

Still another method of controlling erosion is **strip cropping.** In strip cropping, farmers plant low strips of cover crops between strips of other crops.

Figure 25–3 *This photograph shows a conservation method that involves plowing the land across a slope rather than up and down the slope. What is this method called?*

Figure 25–4 *A method used by farmers to conserve soil is terracing. This method involves making a series of level steplike plots on a slope to slow water runoff.*

Cover crops are crops, such as hay and wheat, that completely cover the soil. These crops help hold down the soil between other crops, such as corn.

Farmers can prevent erosion due to wind by planting **windbreaks.** Windbreaks are rows of trees planted between fields of crops. The trees help prevent the wind from blowing the soil away.

Finally, farmers can prevent erosion by not allowing their animals to overgraze in a particular area. The grasses that the animals feed on have deep root systems that hold soil in place. In many areas, grasses prevent erosion.

Other conservation measures are used to return nutrients to the soil. One method is to add fertilizers to the soil. Fertilizers contain nutrients such as nitrogen, phosphorus, and potassium. Another method is called **crop rotation.** This involves farming alternate crops on the same land each year. Some crops, such as clover or peanuts, naturally put nitrogen back into the soil. So farmers may plant a nitrogen-using crop, such as corn, one year and a nitrogen-producing crop, such as clover, the next. This keeps soil nutrients from being used up.

Land Pollution

Erosion and depletion are not the only ways in which soil can be damaged, however. Sometimes soil is damaged by **pollution.** Pollution is the introduction of harmful or unwanted substances into the

Figure 25–5 *The people in this photograph are not dressed to go to the moon. Their suits are designed to protect them from the hazardous wastes they are handling. What are hazardous wastes?*

Figure 25–6 *This map shows some of the worst hazardous waste sites in the United States. How many sites are located in your home state?*

environment. Land pollution is caused by wastes that are disposed of on land.

One type of land pollution is caused by garbage. Garbage is produced by people. In fact, nearly 3.6 kilograms of garbage per person is thrown away every day. Garbage that is thrown out in improper places is called **litter.** If you look around your neighborhood, you may see litter such as paper, cans, bottles, or plastic materials.

Although people cannot help but produce garbage, they can reduce the amount of wasted resources in garbage. For example, people can save parts of garbage that can be used again, such as bottles. These materials can be brought to recycling plants, where they are processed to be used again.

A very serious type of land pollution is caused by industry. The chemical wastes that are produced in industrial processes are often buried on land. Often these chemical wastes are **hazardous wastes.** Hazardous wastes are wastes that burn easily, are poisonous, or react dangerously with other substances. When hazardous wastes are not buried properly, they leak out of their containers and severely damage the environment. If the wastes leak into the water supply, they can contaminate the water so that it can no longer be used for drinking.

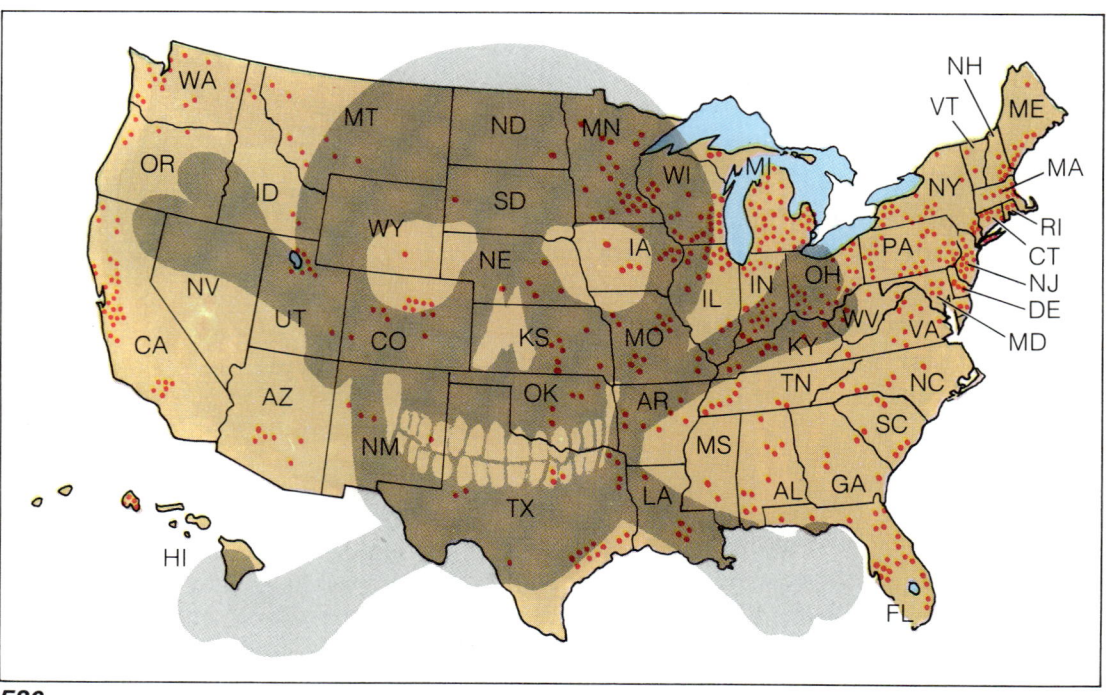

SECTION REVIEW

1. Describe five methods used by farmers to control soil erosion.
2. Compare erosion and depletion.
3. What are three sources of land pollution?
4. Why is it an advantage for a farmer to use soil conservation methods?

25–3 Air

Like soil, air is a natural resource. The air that makes up the earth's atmosphere is a mixture of several gases. The main gases include oxygen, nitrogen, carbon dioxide, and water vapor. Air can be polluted by burning **fossil fuels** such as gasoline, oil, and coal. Fossil fuels are products of decayed plants and animals that are preserved in the earth's crust over millions of years. Gases and particles given off when these fuels are burned are called emissions (ih-MIHSH-uhnz). Emissions can pollute the air. And, polluted air is dangerous to life.

Air Pollution

Air pollution is a serious problem especially in industrialized countries that burn a lot of fossil fuels. **When fossil fuels are burned, a variety of pollutants, or harmful or unwanted substances, are**

Figure 25–7 *The traffic jam on the left is in Mexico City, one of the most crowded and polluted cities in the world. The photograph on the right is a close-up of the pollutants found in automobile exhaust.*

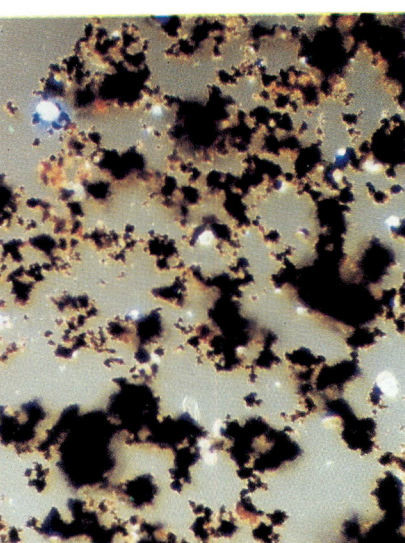

Sharpen Your Skills

Burning Fossil Fuels

No matter how clean a fossil fuel is, it produces carbon dioxide gas as it burns. The buildup of this gas in the atmosphere may produce what is called the "greenhouse effect."

Using books and reference materials in the library, find out more about the greenhouse effect. Report your findings to your class.

released into the air. These pollutants include nitrogen and sulfur gases and tiny solid particles in smoke. Much of this pollution comes from industry. But a major source of air pollution in the United States is emissions from car and truck exhausts.

In many areas, air pollutants mix with fog to form a thick smoke called **smog.** Smog and other forms of air pollution can damage people's health. Air pollutants can cause eye irritation and respiratory, or breathing, problems. People with lung diseases often must stay indoors when pollution levels are very high.

Air pollution is made worse by a flip-flop in layers of air called a **temperature inversion.** Normally, warm air near the ground rises. The warm air carries pollutants high into the atmosphere, away from people. But sometimes a layer of cool air sneaks in under the warm air. The cool layer cannot rise through the warm air above it. So the cool air is trapped, along with the pollutants that enter it.

Another serious environmental problem caused by air pollution is **acid rain.** Acid rain forms when water vapor in the air mixes with sulfur and nitro-

Figure 25–8 *Air pollution is made worse when cool air containing pollutants becomes trapped under a layer of warmer air. What is this flip-flop of air layers called?*

Cold, clean air

Warm air

Trapped cool air

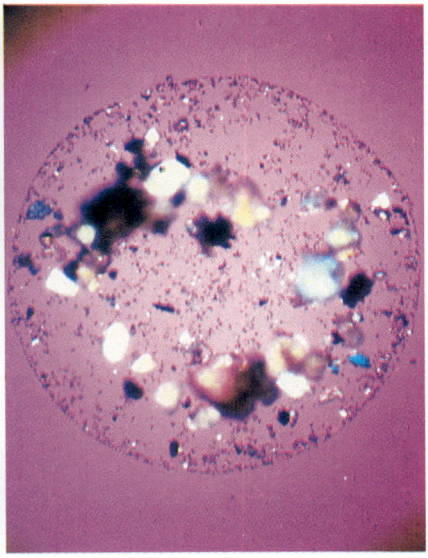

Figure 25–9 *The person in the photograph on the left is pumping lime into a lake that has been damaged by acid rain. The lime will help neutralize the acid in the lake water. The photograph on the right is a close-up of a drop of acid rain. Which pollutants cause acid rain?*

gen compounds called oxides. These compounds are released from factory smoke stacks and car exhausts. When the oxides mix with the water vapor, they form droplets of sulfuric or nitric acid. Eventually these acids fall to the earth as acid rain. Acid rain that falls on land can damage buildings, statues, plant life, and soil. Acid rain that falls into lakes can destroy both plant and animal life.

Protecting the Air

Today there are a number of laws designed to control air pollution. The Environmental Protection Agency, or EPA, sets limits on the amounts of pollutants that can be released into the air. Automobile manufacturers are required to control gas emissions from car exhausts. And, in many areas, it is against the law to burn garbage or leaves. In addition, certain devices have been developed to help reduce the amount of pollutants released from smokestacks.

SECTION REVIEW

1. What are two major sources of air pollution?
2. How does a temperature inversion make air pollution worse?
3. Describe the harmful effects of acid rain.
4. Suggest a way to reduce the amount of air pollution caused by automobile emissions.

Sharpen Your Skills

How Acidic Is Your Rain?

Find out whether the rain in your area is acidic.

1. Next time it rains, collect a sample of rainwater in a jar. Label it Sample A.
2. Place some distilled water in another jar and label it Sample B.
3. Use pH paper to determine the acidity of each of the samples. Record your findings.

Was there a difference in acidity levels between the two samples? What could have caused the difference?

25–4 Water

Water is a natural resource used for drinking, bathing, cooking, cleaning, growing crops, and manufacturing goods. In addition, both fresh water and salt water provide important sources of food, such as fish. And, people use water for recreational activities such as swimming, boating, and fishing. Even though water is a renewable resource, there is a limited amount of fresh water on earth. **People can help conserve water by careful planning and pollution prevention.**

Water Conservation

As the human population increases, more and more water is used. And while some parts of the world have adequate water supplies, other parts of the world are dry. In some areas of Africa, for example, it might not rain for as long as seven years.

Much of the fresh water used by people comes from reservoirs (REHZ-uhr-vwahrz). A reservoir is a place where fresh water is stored. Another important source of water is **groundwater,** or underground water. In the United States, more than 300 billion liters of groundwater are taken out of the ground each day. Most of this water is used for factories and farms. Groundwater is especially important in places such as the western United States and northern Africa. These areas have little surface water in lakes and rivers.

It takes a few hundred years for a large amount of groundwater to accumulate. In many areas, groundwater is being used up faster than it is being replaced. As a result, the level of groundwater drops and eventually streams and lakes may dry up.

Groundwater can be conserved by protecting **watersheds.** A watershed is a land area in which surface runoff drains into a river or system of rivers and streams. Plants in a watershed help keep the soil in place. Water is then allowed to seep back into the ground.

What else can be done about our dwindling water supplies? One solution is to replant vegetation on barren slopes. Another solution is to build dams to form reservoirs for water storage. A third solu-

Figure 25–10 *In order for these orange trees to grow, water must be pumped into the area through artificial ditches. This process is known as irrigation.*

Figure 25–11 *In some parts of the world, such as Algeria (left) and Somalia (right) in Africa, water is a scarce and precious natural resource. Is water a renewable or nonrenewable resource?*

tion to the problem of dwindling water supplies is desalination. Desalination is the removal of salt from ocean water to make it drinkable. This process is already used in some countries such as Saudi Arabia.

Because our water supply is limited, water conservation is very important. People can help conserve water in many ways. Planning for water use by industry and farms cuts down waste. Cleaning and recycling wastewater also increases the water supply. And treating sewage, wastewater from toilets, sinks, and showers, before it is released into rivers helps avoid pollution and conserve water.

Perhaps the best way to conserve water is to simply use less of it. People can take shorter showers and fix leaky faucets. In what other ways can people help conserve water?

Water Pollution

Another important way to conserve water is to prevent water pollution. Polluted water cannot be used for drinking, bathing, or swimming. And, pollution can kill organisms that live in water. Water can be polluted in many ways.

Untreated sewage causes pollution when it is dumped into rivers and streams. Bacteria in the water break down the sewage. In the process, they use up oxygen. If too much sewage is dumped, too much oxygen is used. Fish and other organisms may then die from lack of oxygen. Some pollutants such as phosphates and nitrates, are nutrients for

Figure 25–12 *In this photograph, you can see detergent suds being dumped into a stream. What substance in detergents causes uncontrolled growth of green plants?*

Figure 25–13 *Farmers often spray pesticides on their crops to kill insects or other harmful pests. However, these pesticides may contain chemicals that pollute nearby lakes and streams. How do the chemicals get into the streams?*

aquatic plants. In a lake, the addition of these nutrients will cause plants, such as algae, to increase and use up the oxygen supply of the lake. This speeds up the natural aging process, or succession, of the lake.

Many chemicals used in farming and industry also cause water pollution. Industrial chemicals are sometimes dumped from factories into water. Some of these chemicals are **toxic,** or poisonous, even in small amounts. Fertilizers or **pesticides,** chemicals used to kill harmful insects or other pests, can also pollute water. These chemicals may be carried to water by wind or by runoff.

Some industries and power plants pollute water by releasing large amounts of heat into it. The release of heat into the environment is called **thermal pollution.** Cold water from lakes and rivers is used to cool the machinery in factories and power plants. Then the warm water is returned to the lake or river. Hot water does not hold as much oxygen as cool water does. The lack of oxygen and the increased temperature can kill fish.

Protecting the Water

There are a number of ways in which water pollution can be prevented. And scientists are always searching for methods to clean up polluted water.

Many cities and towns have treatment plants that remove pollutants from sewage. Large industries also may have their own treatment plants to clean

Figure 25–14 *A source of water pollution in oceans is oil spills from tankers (left). The oil can kill wildlife, such as fish and birds (right). The oil can also damage beaches when it washes up on shore.*

waste water before it is dumped. And, to avoid thermal pollution, some power plants and factories use cooling towers to cool hot water before it is released. In addition, scientists are developing a type of bacteria that "eats" oil droplets and cleans water after an oil spill.

SECTION REVIEW

1. Describe four ways to conserve water.
2. Identify five sources of water pollution.
3. Describe four methods used to prevent water pollution.
4. Predict at least one problem that could result from dumping hazardous wastes into the ocean.

25–5   Energy Resources

Section Objective

To describe alternative energy sources

Whether you realize it or not, you use energy every day of your life. You use it to light, heat, and cool your home. You might use it to travel to and from school as well. Most of the energy you use comes from fossil fuels. Because fossil fuels cannot be replaced, they are nonrenewable resources.

Figure 25–15 *The 1300-kilometer Trans-Alaska Pipeline crosses 200 rivers, 300 streams, and 3 mountain ranges. It was built in the 1970s to carry oil and help reduce a fuel shortage in the United States.*

Fossil Fuels

There are three major types of fossil fuels. Liquid fossil fuel is called **petroleum,** or oil. Another fossil fuel is natural gas. Coal is a solid fossil fuel.

Coal, petroleum, and natural gas are the main sources of energy for industry, transportation, and homes. Since 1900, energy use in the United States has increased tenfold. In fact, the United States uses almost 30 percent of all fossil fuel energy produced in the world.

Because the amount of fossil fuels is limited, it is important that people conserve energy. This can be done by using energy more efficiently. For example, people are now trying to develop machines that can run on less energy. Another way to save energy is, of course, to use less of it. What are some ways that you can use less energy?

No matter how much people conserve fossil fuels, the earth will someday run out of these non-renewable resources. That is why scientists are now working hard to develop alternative sources of energy. **Alternative energy sources include solar energy, water and wind energy, geothermal energy, and nuclear energy.**

Alternative Energy Sources

When developing ways to replace fossil fuel energy, scientists look for energy sources that will not pollute the environment. They also try to find sources that are renewable so that we will not run out in the future.

SOLAR ENERGY Energy from the sun is called **solar energy.** Solar energy is renewable and does not cause pollution.

Solar energy can be passive or active. An example of passive use would be placing windows in a house so that the amount of sunlight absorbed would be enough to heat the house. The problem with passive solar energy is obvious. It only works when the sun is shining.

Active solar energy involves a device called a solar collector, which is used to collect the sun's energy. Solar collectors are usually located on the

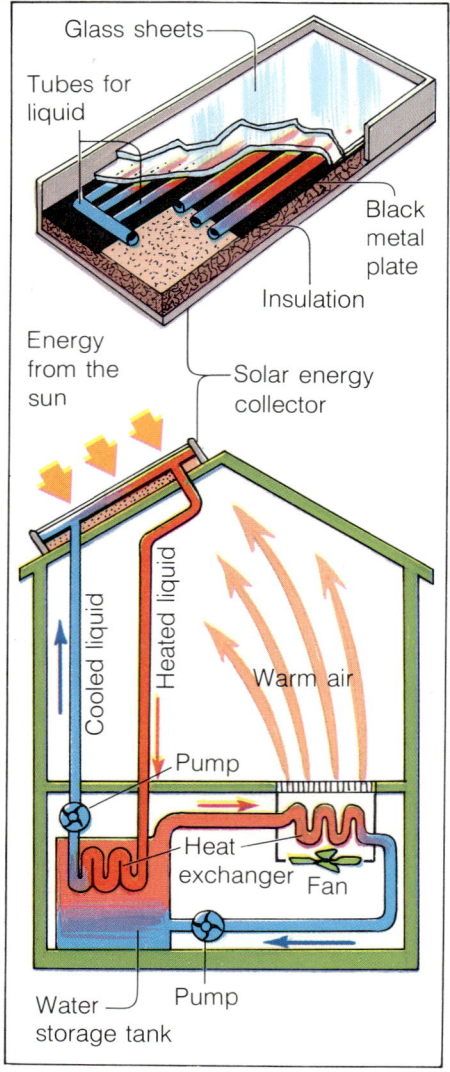

Figure 25–16 *Water in the solar panel (top) is heated by the sun. In this solar heating system, the heat is transferred at the heat exchanger (bottom) and used to provide hot water and warmth. What is the source of energy for this system?*

Glass sheets

Tubes for liquid

Black metal plate

Insulation

Energy from the sun

Solar energy collector

Cooled liquid

Heated liquid

Warm air

Pump

Heat exchanger

Fan

Water storage tank

Pump

roofs of buildings. Inside the collectors are blackened tubes that have water circulating through them. The sun heats the water as it moves through the tubes. The heated water then provides heat and hot water for the building. See Figure 25–16.

Light energy from the sun can be converted directly into electricity with a solar cell. Unfortunately, these cells are expensive and only produce a small amount of electricity. But scientists are working on improvements to make the cells more efficient.

WATER AND WIND ENERGY Energy can be produced from falling or flowing water. In fact, about one-fourth of the world's electricity is produced that way now. Dams built across rivers can hold back millions of tons of water. The water is passed through a **hydroelectric plant,** which uses energy from the moving water to produce electricity.

Figure 25–18 *For centuries, the mechanical energy of moving water has been used to turn water mills (left). Today, falling water from dams (right) is used to generate electricity in hydroelectric plants.*

Figure 25–19 *At Altamount Pass in California, thousands of windmills are used to generate electricity. Why can this method not be used in every part of the world?*

Figure 25–20 *In this power plant (left), cold water is pumped into the earth where it is heated. The hot water is then pumped up to the power plant where it is used to generate electricity. In the photograph (right), you can see the steam that results when water is heated deep within the earth. What type of energy is this?*

The oceans provide another way to use moving water to produce energy. Ocean water rises and falls with the tides. In some coastal areas, the difference between high and low tide can be as much as 15 meters. Ocean tides cause water to flow into and out of bays. A power plant built near a bay entrance can use moving water to produce electricity.

Just as moving water can be used to produce energy, so can moving air, or wind. Windmills are built in areas where winds are fairly constant. Windmills use wind energy to produce electricity.

GEOTHERMAL ENERGY Energy that comes from heat created in the earth is called **geothermal energy.** The geothermal energy heats rocks within the earth. When water comes into contact with the heated rocks, it becomes steam. The hot water and steam are released at the earth's surface through hot springs and geysers (GIGH-zuhrz). People can drill wells into underground sources of hot water and steam and feed the steam into electric power plants. In this way, geothermal energy can be used to produce electricity. See Figure 25–20. Is geothermal energy renewable or nonrenewable?

NUCLEAR ENERGY Another alternative energy source is **nuclear energy.** Nuclear energy is the energy located within the nucleus of an atom. Energy is produced in nuclear power plants through a proc-

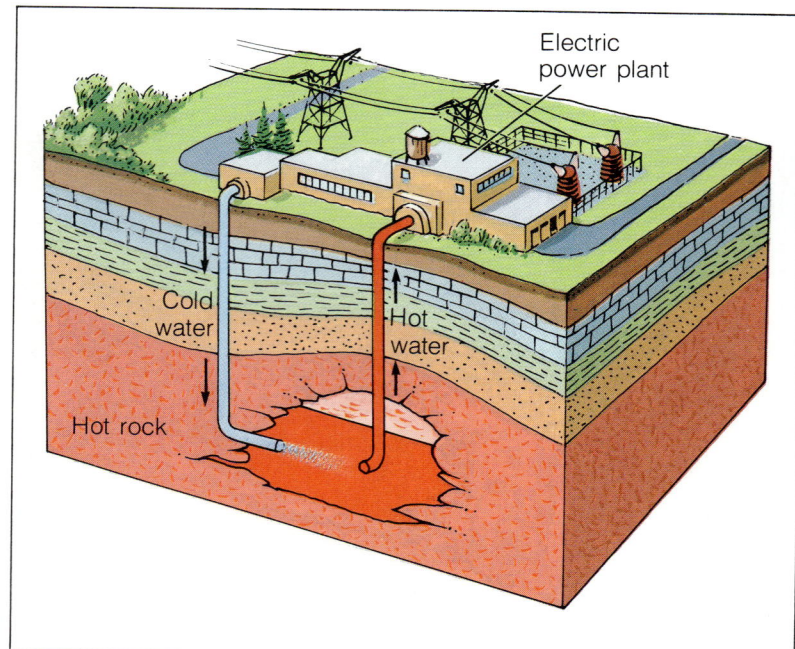

ess called **nuclear fission.** In this process, atoms of the element uranium are split to release energy. This energy is then used to heat water into steam. The steam is passed through a turbine to produce electricity.

There are several problems with nuclear fission. One problem is that **radioactive materials** are given off during the nuclear fission process. Radioactive materials give off particles and energy that are harmful to life. People have not yet developed a way to safely dispose of these radioactive wastes. And, if there is an accident at a nuclear power plant, such as the one at Chernobyl in the Soviet Union in 1986, dangerous radioactive materials can be released into the atmosphere. Another problem with nuclear fission is that it is very expensive.

Another type of nuclear energy that scientists are trying to develop is **nuclear fusion.** Nuclear fusion is the combining of atoms. Nuclear fusion produces even more energy than nuclear fission. It is fusion that produces the energy of the sun.

A nuclear fusion power plant would combine hydrogen atoms to produce energy. The hydrogen atoms could be easily obtained from water. And nuclear fusion would not produce radioactive wastes as does nuclear fission.

However, nuclear fusion power plants are only a dream for the future. Perhaps in the twenty-first century, they will become a reality.

Figure 25–21 *In the background of this photograph, you can see the cooling towers of the nuclear power plant at Three Mile Island in Pennsylvania. In 1979, an accident at the site caused concern about the safety of nuclear power plants.*

Figure 25–22 *From sugar cane grown in Hawaii (left), tiny pellets of fibrous wastes (center) are formed. These fuel pellets can then be burned in a factory (right). Is this alternative energy source renewable or nonrenewable?*

1. Why are scientists trying to develop new energy sources?
2. Briefly describe six energy sources other than fossil fuels.
3. Compare petroleum, coal, and natural gas.
4. What problems are associated with nuclear fission energy?
5. Which energy source do you think will be most common in 50 years? Explain your choice.

25–6 Living Resources

Since life began on the earth, countless species of plants and animals have appeared. Some still remain today, but untold numbers have become extinct. **Extinction,** the process by which a species passes out of existence, is a natural part of our planet's history. Species can become extinct because they cannot adjust to changes in the earth's environment. The environment is always changing. Streams shift course. Mountain chains rise. Climates become warmer or colder.

Species change too. Organisms with new traits appear. These traits may then be passed on to offspring. If new traits help a species adjust to environmental change, the species survives. If not, the

Figure 25–23 *Over the course of history, the climate on the earth has changed many times. On the left are the lush forests of Greenland many millions of years ago. On the right is Greenland as it appears today.*

species may become extinct. The race to survive often means adapting to environmental change.

Throughout most of the earth's history the race to survive occurred at a slow pace. Great natural changes, such as shifting climates, usually happen over long periods of time. So it may take hundreds, thousands, or millions of years for a species to vanish. Today, however, people and their activities have increased the pace. Some alterations people made in the environment caused it to change so rapidly that many species have not been able to keep up. These species have become extinct.

Living organisms are important natural resources. **An important part of conservation is to help protect species of plants and animals that are in danger of becoming extinct.** Because extinction is a natural process, not all species can be saved. So scientists involved in conservation often concentrate on those species that are in danger mainly due to the actions of people.

Plants

More than 300 species of plants in the United States alone have become extinct since the Europeans first came to the Americas. Worldwide, a tenth of the plants known to science are in danger of extinction. Combined, this means 20,000 different kinds of plants are close to disappearing. Plant life in general is suffering in many parts of the world. Erosion and overgrazing in the Middle East, for instance, have turned forests into bare ground.

Figure 25–24 *These oconee bell flowers grow only in a few shady woodlands in Georgia, Virginia, and the Carolinas. The plants need shade to survive. What might happen to the plants if the trees shading them are cut down?*

Figure 25–25 *The giant saguaro cactus grows only in the Sonora Desert of northwestern Mexico and the southwestern United States. Why do scientists fear that these plants may become extinct?*

The most dangerous threat to living things is the destruction of their habitat. This destruction can happen when people clear land for farms, cities, and other large communities. For example, when Europeans first arrived, forests covered most of the eastern United States. Almost all the original forest was cleared to make room for farms and towns and to harvest timber.

Today the largest forests on the earth, the tropical rain forests, are being destroyed for the same reasons. Scientists fear that half of all tropical forests will be gone by the end of this century. When the forest trees are cut, not only are the trees lost, but the thousands of other plants that live on or among the trees also vanish.

The threats to a species can come from many sources. The giant saguaro is a cactus that can grow as tall as 15 meters. It grows only in the Sonora Desert of northwestern Mexico, southwestern Arizona, and southeastern California. Cattle have trampled many young saguaros. Vehicles have crushed them as well. Other saguaros have been uprooted and taken for house plants. Resorts for tourists and housing developments have been built on the desert where the saguaros grow. People have even chopped them up for fun. For all these reasons, scientists fear saguaros will become extinct.

The threat to plants is a threat to people as well. Throughout history, about 3000 kinds of plants have been used for food. Today most food comes from about 15 kinds of plants, chiefly grains. These grains all descended from wild plants. If wild plants are destroyed, scientists wonder where the new food sources of the future will come from. Also, many medicines come from plants. Some scientists suggest that perhaps 50,000 kinds of medicines and other useful chemicals are still found in plants. Each time a species of plant vanishes, the odds of finding a new source of food or medicine go down.

Wildlife

The destruction of plants also has a direct impact on wildlife, or animals. When tropical forests are cleared, for example, wildlife living in them, such as jaguars and parrots, lose their habitat and

can quickly become extinct. Like plants, wild animals have been destroyed because people used them thoughtlessly and wastefully. In colonial times, a bird called the passenger pigeon was so numerous in eastern North America that its flocks darkened the skies. There may have been 2 billion of these birds. People felt there were so many they could shoot all they wanted for food and sport. In 1914, the last passenger pigeon died in a cage in a zoo.

Uncontrolled hunting threatens several animal species today. Although the black rhinoceros is protected by law in many African countries, people still kill it, not for food but for its horn. A powder made from the rhino horn is incorrectly considered a medicine in parts of Asia, where people pay high prices for it. Some Africans kill rhinos and sell the horns to earn money.

Human populations are growing rapidly in Africa, which causes another problem for large animals such as rhinos and elephants. People farm in areas where these animals live. Farmers do not want the animals eating crops and trampling the land. So the farmers kill the animals or ask their governments to remove the animals.

Sometimes what seem like harmless actions by people means trouble for wildlife. Farmers in southern Florida are draining marshes to grow more crops. The marshes contain snails eaten by a hawk-like bird known as the Everglades kite. As the marshes vanish, so does the kite's food. The Everglades kite is close to extinction.

Figure 25–26 *In 1988, all the world was captivated by the fate of three whales trapped in an Arctic ice pack. Rescue efforts continued for days as a path through the ice was cut. The trapped whales were forced to use a hole in the ice to obtain oxygen (left). At the same time, whale hunters continued to slaughter whales as they have done for centuries, despite the fact many species of whales are endangered or extinct (right).*

Figure 25–27 *This hawklike bird known as the Everglades kite lives in the marshes of southern Florida. How does the draining of these marshes affect the kite?*

Saving Plants and Wildlife

Just as human activities have hurt plants and wildlife, people can take action to save them. Many organizations and government agencies are devoted to conserving living things. One organization has suggested a "World Conservation Strategy." This strategy is a plan to help countries develop and use their natural resources wisely. The organization seeks ways through which development of farms and industries will not wipe out plants and animals. As part of the strategy for Indonesia, for instance, some of the money from cutting timber in tropical forests will be used to plant new trees. Programs are underway in Africa to bring more tourists to marvel at its wildlife. In this way, Africans can earn money through the tourist industry. Then they will want to leave some wildlife habitats undisturbed.

Preserving habitats is the most important step in saving plants and animals. Many countries have set up parks and preserves that cannot be developed.

Figure 25–28 *Tourists can observe wildlife from the safety of a van on this game preserve in Africa. If people can earn money from safaris, they will be more likely to leave wildlife habitats undisturbed.*

CAREER   *Conservationist*

HELP WANTED: CONSERVATIONIST
Bachelor's degree in the biological sciences or in natural resources required. To work as part of a conservation field team in the scientific management of national forest land. Involves outdoor work and research.

At one time, prairie chickens lived throughout the North American prairies. Since then, however, the prairies have been farmed with corn and wheat. The farming deprived the prairie chickens of a place to live, thus greatly reducing their numbers.

The people who see to it that the use of natural resources causes the least possible harm to the environment are called **conservationists.** Their major concern is to keep advanced technology from depleting or damaging resources such as coal, oil, water, air, and living things. As part of their conservation activities, they may seek solutions to problems caused by poisonous pesticides, insect and disease control, and illegal hunting of endangered wildlife.

There are many positions in the field of conservation. One might work as a soil scientist, wildlife manager, or researcher. A conservationist might be involved in water and land management, improving habitats, surveys, or research. Conservationists have an interest in nature and a desire to be part of a program that cares for natural resources. For more information about this rewarding field, write to the Forest Service, U.S. Department of Agriculture, P.O. Box 2417, Washington, DC 20013.

Figure 25–29 *The California condor, the largest flying land bird in North America, is an endangered species. The condor has a wingspan of 2.4 to 2.9 meters. By the mid-1980s, there were only 25 of these birds surviving.*

Another method of conservation is to limit the number of plants and animals that people use so that there are enough left to reproduce. Limitations are the reason for hunting and fishing laws that specify the size and number of the catch. However, **endangered species,** those in danger of becoming extinct, must be completely protected.

Educating people about the value of plants and wildlife also is an important part of conservation. Recently, desert tribes in Israel, for example, killed wild goats called ibexes. The government taught the tribes that if they killed too many goats, the goats would disappear. The government paid the desert people to watch after the ibexes. Within a few years, Israel had large numbers of ibexes. Moreover, leopards suddenly returned. The ibex is one of the animals on which leopards depend for food.

Figure 25–30 *Through education, the government of Israel was able to convince desert tribes that if they killed too many of these wild goats, or ibexes, the goats would disappear. Within a few years the population of the ibexes had greatly increased. The population of leopards also increased. Why?*

SECTION REVIEW

1. What are three reasons that species may become extinct?
2. Why is the extinction of plants a threat to people as well?
3. What are three ways in which endangered species can be saved?
4. How might the extinction of one species affect another species?

Classifying Litter

Problem

How can you classify the litter in your neighborhood?

Materials *(per group)*

large plastic bag
safety gloves

Procedure

1. Obtain enough litter from your teacher to fill a large plastic bag. **CAUTION:** *Wear safety gloves whenever handling litter.*
2. List all the litter that is combustible, or usually disposed of through burning.
3. List all the litter that can be recycled.
4. List all the litter that decomposes rapidly when buried.
5. Record the total number of pieces of litter.

Observations

1. Calculate the percentage of litter that is combustible. Record your calculations.

$$\% \text{ combustible} = \frac{\text{no. of combustible items}}{\text{total items}} \times 100$$

2. Calculate the percentage of litter that can be recycled. Record your calculations on a sheet of notebook paper.
3. Now calculate the percentage of litter that will decompose rapidly. Record your calculations on another sheet of notebook paper.
4. Finally, calculate the percentage of litter that will not burn, recycle, or decompose rapidly. Again, record your calculations.
5. Did you find any litter that could be reused without any recycling? List these items.

Conclusions

1. Based on your calculations, what type of litter is most common around your school or neighborhood?
2. Which litter do you consider the most serious pollution problem? Why?
3. Suggest some ways to reduce the amount of litter in your school or neighborhood.

CHAPTER REVIEW

SUMMARY

25–1 Natural Resources

❑ Natural resources are materials taken from the environment and used by people. Natural resources can be either renewable or nonrenewable.

❑ Conservation is the wise use of natural resources.

25–2 Soil

❑ Soil is a renewable resource, but it takes a very long time for it to form.

❑ In many areas, topsoil is being lost because of erosion. The loss of nutrients from soil is called depletion.

❑ Terracing, contour farming, strip cropping, windbreaks, crop rotation, and prevention of overgrazing are soil conservation measures.

❑ Litter, garbage, and hazardous waste can cause land pollution.

25–3 Air

❑ The burning of fossil fuels releases many pollutants into the air. Air pollution can be dangerous to human health.

❑ Acid rain is one serious environmental problem caused by air pollution.

25–4 Water

❑ Water is a renewable resource but there is a limited amount of fresh water on earth.

❑ Planning, treating and recycling wastewater, preventing pollution, and simply using less water are all ways to conserve water.

❑ Water pollution may be caused by untreated sewage, chemicals used in industry and farming, large amounts of heat, and oil spills.

25–5 Energy

❑ Most of the energy used in the United States comes from fossil fuels.

❑ Even if people conserve energy, the earth will someday run out of fossil fuels.

❑ Alternative energy sources include solar energy, water and wind energy, geothermal energy, and nuclear energy.

25–6 Living Resources

❑ Protecting organisms from extinction is an important part of conservation.

❑ The most dangerous threat to living things is the destruction of their habitat.

❑ Human activities have hurt plants and wildlife, but people can take action to save them.

VOCABULARY

Define each term in a complete sentence.

acid rain	geothermal energy	nuclear energy	solar energy
conservation	groundwater	nuclear fission	strip cropping
contour farming	hazardous waste	nuclear fusion	temperature inversion
crop rotation	hydroelectric plant	pesticide	terracing
depletion	litter	petroleum	thermal pollution
endangered species	natural resource	pollution	topsoil
erosion	nonrenewable resource	radioactive materials	toxic
extinction		renewable resource	watershed
fossil fuel		smog	windbreak

CONTENT REVIEW: MULTIPLE CHOICE

On a separate sheet of paper, write the letter of the answer that best completes each statement.

1. Which of the following is *not* a nonrenewable resource?
   a. coal   b. oil   c. water   d. natural gas
2. About how long does it take for a layer of topsoil to form?
   a. 10 to 20 years   b. 50 to 100 years
   c. 100 to 200 years   d. 200 to 400 years
3. The carrying off of soil by water or wind is called
   a. erosion.   b. depletion.   c. pollution.   d. conservation.
4. Which conservation measure prevents soil nutrients from being used up?
   a. crop rotation   b. windbreaks
   c. strip cropping   d. terracing
5. Acid rain can damage
   a. lakes.   b. soil.   c. statues.   d. all of these.
6. Which of the following pollutants can cause uncontrolled growth of green plants in lakes?
   a. pesticides   b. acid rain   c. phosphates   d. oil
7. Which is *not* a fossil fuel?
   a. uranium   b. petroleum   c. coal   d. natural gas
8. Energy that is created from the splitting of atoms is called
   a. solar energy.   b. nuclear fusion energy.
   c. nuclear fission energy.   d. geothermal energy.
9. Which of the following energy sources is nonrenewable?
   a. solar energy   b. geothermal energy
   c. wind energy   d. fossil fuel energy
10. The extinction of a plant species can affect
    a. people.   b. wildlife.
    c. both people and wildlife.   d. neither people nor wildlife.

CONTENT REVIEW: COMPLETION

On a separate sheet of paper, write the word or words that best complete each statement.

1. _____ are materials taken from the environment and used by people.
2. A resource that can be replaced by nature is _____.
3. The farming of alternate crops on the same land each year is called _____.
4. The introduction of harmful or unwanted substances into the environment is known as _____.
5. Wastes that burn easily, are poisonous, or react dangerously with other substances are _____.
6. A pollution problem caused by sulfur and nitrogen oxides is called _____.
7. Wastewater from toilets, sinks, and showers is called _____.
8. A (An) _____ plant uses energy from moving water to produce electricity.
9. Energy that comes from heat created in the earth is _____ energy.
10. The process by which a species passes out of existence is _____.

CONTENT REVIEW: TRUE OR FALSE

Determine whether each statement is true or false. Then on a separate sheet of paper, write "true" if it is true. If it is false, change the underlined word or words to make the statement true.

1. Fossil fuels are <u>nonrenewable</u>.
2. <u>Depletion</u> is the carrying off of soil by wind or water.
3. <u>Contour farming</u> is making a slope into a series of level steps for planting.
4. <u>Fossil fuels</u> are products of decayed plants and animals that are preserved in the earth's crust over millions of years.
5. Air pollution is made worse by a flip-flop in layers of air called <u>smog</u>.
6. There is a <u>limited</u> amount of fresh water on earth.
7. In a <u>hydroelectric plant</u>, solar energy from the sun can be converted into electricity.
8. Solar collectors are used in <u>passive</u> solar energy.
9. When <u>nuclear fusion</u> is used to produce energy, radioactive wastes are produced.
10. The passenger pigeon is an <u>endangered</u> species.

CONCEPT REVIEW: SKILL BUILDING

Use the skills you have developed in the chapter to complete each activity.

1. **Relating cause and effect** List as many causes and effects of pollution in your community as you can. Then make suggestions as to how each source of pollution can be lessened or eliminated.
2. **Making calculations** If a leaky faucet lost 53 drops of water every minute, how many drops would be lost in an hour? In a day? In a year?
3. **Relating cause and effect** For each of the following, identify the natural resource that is being conserved.
   a. national parks
   b. ban on burning leaves
   c. recycling newspapers
   d. reduced-flow shower heads
4. **Making charts** Construct a chart with three columns. In the first column, list each energy source discussed in the chapter. Use the other two columns to list advantages and disadvantages of each source.
5. **Applying concepts** Why do you think alternative energy sources such as tidal energy, wind energy, and geothermal energy are not used in all parts of the world?
6. **Making inferences** Air pollution problems, such as acid rain, often exist in areas far from the actual pollution source. How do you think this is possible?
7. **Relating concepts** Why do lakes that are damaged by acid rain often look beautiful and crystal clear?

CONCEPT REVIEW: ESSAY

Discuss each of the following in a brief paragraph.

1. List the natural resources that you use either directly or indirectly in a typical day. Then, describe a way that you might be able to conserve each resource.
2. Explain how extinction can be a natural process as well as a process brought on by human beings.
3. In what ways does the increasing human population affect natural resources?
4. What are hazardous wastes and how might they pollute air, water, and soil?
5. Explain why the conservation of nonliving resources is important in the conservation of living resources.

Biosphere II: A World Under Glass

In 1989, in the dry foothills of Arizona's Santa Catalina mountains, the doors to a miniature world will be sealed. At that time, eight people will begin a two-year adventure. These eight people will live and work inside a glass and steel "terrarium" called Biosphere II. Biosphere II will contain models of some of the biomes found on earth.

The scientists working on Biosphere II hope to develop a system that supports life as the earth does. At this point, you may well wonder why scientists would want to live and work in a sealed structure for two years when they can study the same biomes outdoors. The answer to this question can be found out in space.

Scientists hope that Biosphere II will lead to the development of future space settlements. Such settlements will be self-contained and provide food, shelter, and water for the people that live there. This kind of space station would not need to be supplied constantly with material from earth. As you can see from this example, research on Biosphere II may have applications that will prove useful far in the future.

Biosphere II is a cooperative effort of many people who might ordinarily work alone. These people, experts in many fields of study, are contributing their talents to make Biosphere II a success.

Now let's take a tour of Biosphere II. A four-story white domed building stands at one corner of Biosphere II. This building will be home for the eight "Biospherians" for two years. The building will have apartments, laboratories, a gymnasium, a library, a lounge, and a computer center.

Next to this building is a small farm. Some grains, garden vegetables, and tropical fruits such as pineapples, papayas, and bananas will be grown. Small goats, chickens, and freshwater fish will be raised here.

A rain forest biome stands next to the farm. Dr. Ghillean Prance, who works at the New York Botanical Garden, is selecting the plants that will be grown in this model biome.

Water from the rain forest will flow down a miniature mountain and collect in a stream that will run through a flat savanna biome that lies at the edge of the rain forest. The savanna section of Biosphere II is being developed by Dr. Peter Warshall, an ecologist from Tucson. He will select the bushes, small shrubs, and other savanna plants that will be grown here.

Next, the stream flows into a small freshwater marsh, complete with reeds and marsh grasses. The water then becomes part of a saltwater marsh. On earth, many tiny marine plants and animals spend part of their life in saltwater marshes such as this. Dr. Walter Adey of the Smithsonian Institution is working to develop the marshy areas of Biosphere II.

Dr. Ghillean Prance is part of the team that selected the plants and animals for Biosphere II.

Dr. Adey also helped design Biosphere II's lagoon, coral reef, and ocean areas, the next stops on our brief tour. For at least two years, these areas will be home to fish, shellfish, coral organisms, and various sea plants. There will even be a small model of a deep ocean environment in Biosphere II.

A small desert will also be contained within the dome. Tony Burgess, a desert botanist, will help select suitable plants and animals for this area. In the desert area, cactuses and mesquite bushes may share the sandy area with kangaroo rats, insects, and maybe even a scorpion or two.

It is the dream of the scientists working on Biosphere II that their work will provide the information necessary to develop a space station. In the future, you may well find yourself living in an artificial "Biosphere" far off in space. Perhaps you will even be living on a distant planet!

Biosphere II will contain organisms that live in several biomes. Data gathered from this investigation may make life in future space colonies possible.

ARE WE DESTROYING

People enjoy the scenic beauty of their national parks. But, in the United States, it seems, people may be loving their parks to death. In a recent year, more than 280 million people flocked to the nation's parks and monuments to experience nature and their own national heritage. But the experience is leaving the 335 units of the national parks system bruised and battered.

In many parks, backpacking and hiking trails that once were narrow have been widened by the footsteps of millions of hikers. Nature lovers often climb a remote mountain peak and look down from the top to view packed parking lots, smog over distant cities, and areas littered with trash. And the whine and growl of motorbikes, snowmobiles and four-wheel-drive vehicles often fill the air with fumes and noise.

The word *wilderness* seems to have changed since the first national park—Yellowstone National Park in Wyoming—was established in 1872. Before that, people often needed to be protected from the wilderness. Today, what is left of the wilderness needs protection from an ever increasing number of visitors. People visit the parks to see wildlife, and plants and to enjoy the wonders of nature. But multiply the impact on nature from one vacationer by 280 million. Add environmental problems such as air, land, and water pollution. The end result is too much for the parks to absorb.

Yellowstone National Park is one park that reflects problems facing our national park system. Congress showed great foresight when it established this national park.

Yellowstone is a huge mountain wonderland of hot springs, geysers, mudpots, forests, and magnificent river valleys.

The original intention of Congress was to protect this national treasure for Americans to enjoy forever. What Congress did not realize in 1872 was that Yellowstone Park is only a part of a much larger ecosystem. It cannot survive naturally without the meadows, rivers, lakes, and mountains that surround it.

The near extinction of Yellowstone's grizzly bears is one result of this oversight. Grizzlies live in meadows and open hills and valleys. These are located outside of the boundaries of Yellowstone Park. Years ago, the grizzlies were driven into the protected parklands by people who brought logging, mining, and homesteading to the area.

For many years, the bears thrived in the park by scavenging for food in the plentiful supply of trash in the park dumps. But in 1963, the National Park Service closed the park's dumps to grizzlies. The grizzlies responded as planned and went elsewhere for food and shelter. When the bears wandered into the areas outside of the park, they found towns and resorts. True, there was plenty of garbage to eat. But the people were not so happy to see bears near their homes. Since 1968, hundreds of grizzlies have been shot. Today, only about 200 bears survive in the Yellowstone area.

Other populations of animals have also been affected by new park policies. In the past, wolves and mountain lions in the Yellowstone area were shot. Wolves and mountain lions eat elk. Without these

OUR NATIONAL PARKS?

natural predators, the elk population has increased dramatically. Elk often eat plants down to the ground, leaving very little food for other animals such as the bighorn sheep. The increasing elk population has reduced the number of plants in the park. It is only a matter of time before a harsh winter causes many of the park's elk to starve.

One possible solution is to reintroduce wolves and other predators into the area. These animals would reduce the elk population to a number that the park's resources could reasonably support.

However, some people feel such predators might pose a danger to visitors in the park. They argue that people should be able to use the park in safety. After all, their tax dollars and user fees support the park. They feel that the parks have been set up for the enjoyment of people, and that people should be able to have free safe access to the parks. If the people of this country withdrew their support of the national park system, they argue, no wildlife would be protected from commercial development.

Similar problems are being felt in the other national parks. Should the number of people allowed to visit the national parks be reduced? Should construction be limited in areas near national parks in an effort to preserve the ecosystems near the park? What do you think?

People enjoy our national parks. However, the enormous crowds sometimes force animals from the safety of parklands.

OLD FAITHFUL GEYSER

For Further Reading

If you have an interest in a specific area of Life Science or simply want to know more about the topics you are studying, one of the following books may open the door to an exciting learning adventure.

Chapter 1: Exploring Living Things

Asimov, Isaac. *Great Ideas of Science.* Boston: Houghton Mifflin.

Land, B. *The New Explorers: Women in Antarctica.* New York: Dodd Mead.

Moulton, R. *First to Fly.* Minneapolis: Lerner.

Chapter 2: The Processes of Life

Horowitz, N. *To Utopia and Back: The Search for Life in the Solar System.* New York: W.H. Freeman & Company Publishers.

Nilsson, L. *Close to Nature.* New York: Pantheon.

Riedman, S. *Biological Clocks.* New York: Thomas Y. Crowell.

Chapter 3: Cells, Tissues, and Organ Systems

Cairns-Smith, A. *Seven Clues to the Origins of Life: A Scientific Detective Story.* New York: Cambridge University Press.

Johnson, S. *Inside an Egg.* Minneapolis: Lerner.

Prince, J. *How Animals Move.* New York: Elsevier/Nelson Books.

Chapter 4: Classification of Living Things

Linsenmaier, W. *Insects of the World.* New York: McGraw-Hill.

Silver, D. *Animal World.* New York: Random House.

Whitfield, P. *Macmillan Illustrated Animal Encyclopedia.* New York: Macmillan.

Chapter 5: Viruses and Bacteria

Nourse, A. *Viruses.* New York: Franklin Watts.

Patent, D. *Bacteria: How They Affect Other Living Things.* New York: Holiday House.

Sabin, F. *Microbes and Bacteria.* Mahwah, N.J.: Troll.

Chapter 6: Protozoans

Amos, W. *Life in Ponds and Streams.* Wash., D.C.: National Geographic Society.

Curtis, H. *The Marvelous Animals: An Introduction to the Protozoa.* New York: Natural History Press.

Teasdale, J. *Microbes.* Morristown, N.J.: Silver Burdett.

Chapter 7: Nonvascular Plants and Plantlike Organisms

Jacobs, F. *Breakthrough: The True Story of Penicillin.* New York: Dodd Mead.

Johnson, S. *Mushrooms.* Minneapolis: Lerner.

Kavaler, L. *Green Magic: Algae Rediscovered.* New York: Thomas Y. Crowell.

Chapter 8: Vascular Plants

Overbeck, Cynthia. *Carnivorous Plants.* Minneapolis: Lerner.

Selsam, M. *Tree Flowers.* New York: Morrow.

Wexler, J. *From Spore to Spore: Ferns and How They Grow.* New York: Dodd Mead.

Chapter 9: Invertebrates

Arnett, R. *Simon & Schuster's Guide to Insects.* New York: Simon & Schuster.

Horton, C. *Insects.* New York: Gloucester.

Johnson, S.A. *Snails.* Minneapolis: Lerner.

Chapter 10: Coldblooded Vertebrates

Blassingame, W. *Wonders of Sharks.* New York: Dodd Mead.

Chase, G. *The World of Lizards.* New York: Dodd Mead.

Gibbons, W. *Their Blood Runs Cold: Adventures with Reptiles and Amphibians.* University of Alabama Press.

Chapter 11: Warmblooded Vertebrates

Boitani, L. *Simon and Schuster's Guide to Mammals.* New York: Simon & Schuster.

Hosking, E. *Seabirds of the World.* New York: Facts on File.

Patent, D. *Whales: Giants of the Deep.* New York: Holiday House.

Chapter 12: Skeletal and Muscular Systems

Silverstein, A. *The Skeletal System: Frameworks for Life.* Englewood Cliffs, N.J.: Prentice-Hall.

Ward, Brian R. *The Skeleton and Movement.* New York: Franklin Watts.

Vevers, G. *Your Body: Muscles and Movement.* New York: Lothrop, Lee & Shepard.

Chapter 13: Digestive System

Fetz, S. *Going Vegetarian: A Guide for Teen-Agers.* New York: Morrow.

Peavy, L. *Food, Nutrition, and You.* New York: Scribner's.

Ward, B. *Food and Digestion.* New York: Franklin Watts.

Chapter 14: Circulatory System

Dunbar, R. *The Heart and Circulatory System.* New York: Franklin Watts.

Meredith, S. *Book of the Human Body.* Tulsa, OK: Educational Development.

Ward, B. *The Heart and Blood.* New York: Franklin Watts.

Chapter 15: Respiratory and Excretory Systems

Bruun, R. *The Human Body.* New York: Random House.

Vevers, G. *Your Body: Blood and Lungs.* New York: Lothrop, Lee & Shepard.

Ward, B. *The Lungs and Breathing.* New York: Franklin Watts.

Chapter 16: Nervous and Endocrine Systems

Berger, M. *Exploring the Mind and Brain.* New York: Thomas Y. Crowell.

Simon, H. *Sight and Seeing: A World of Light and Color.* New York: Philomel Books.

Ward, B. *The Brain and Nervous System.* New York: Franklin Watts.

Chapter 17: Reproduction and Development

Madras, L. *What's Happening to My Body?: A Growing Up Guide for Mothers and Daughters.* New York: Newmarket Press.

Miller, J. *The Facts of Life.* New York: Viking.

Ward, B. *Birth and Growth.* New York: Franklin Watts.

Chapter 18: Infectious Disease and Chronic Disorders

Fromer, M. *AIDS: Acquired Immune Deficiency Syndrome.* New York: Pinnacle Press.

Hyde, M. *VD-STD: The Silent Epidemic.* New York: McGraw-Hill.

Patent, D. *Germs!* New York: Holiday House.

Chapter 19: Drugs, Alcohol, and Tobacco

Berger, G. *Addiction: Its Causes, Problems, and Treatments.* New York: Franklin Watts.

Claypool, J. *Alcohol and You.* New York: Franklin Watts.

Hyde, M. *Know About Smoking.* New York: McGraw-Hill.

Chapter 20: Genetics

Asimov, A. *How Did We Find Out About DNA?* New York: Walker.

Asimov, A. *How Did We Find Out About Genes?* New York: Walker.

Gonick, L. *Cartoon Guide to Genetics.* New York: Harper and Row.

Chapter 21: Applied Genetics

Bernstein, S. *New Frontiers in Genetics.* New York: Messner.

Silverman, A. *The Genetics Explosion.* New York: Macmillan.

Stwertka, E. *Genetics Engineering.* New York: Franklin Watts.

Chapter 22: Changes in Living Things

Branley, F. *Dinosaurs, Asteroids, and Superstars: Why the Dinosaurs Disappeared.* New York: Thomas Y. Crowell.

Cork, B. *The Young Scientist Book of Evolution.* Tulsa, OK: EDC Publishing.

Smith, M. *Living Fossils.* New York: Dodd Mead.

Chapter 23: Interactions Among Living Things

Carson, R. *The Edge of the Sea.* Boston: Houghton Mifflin.

Hughey, P. *Scavengers and Decomposers: The Cleanup Crew.* New York: Atheneum.

Sacket, R. *Edge of the Sea.* Alexandria, Va.: Time-Life Books.

Chapter 24: Biomes

Milne, L. *The Mystery of the Bog Forest.* New York: Dodd.

Page, J. *Arid Lands.* Alexandria, Va.: Time-Life Books.

The Editors of Time-Life Books. *Grasslands and Tundra.* Alexandria, Va.: Time-Life Books.

Chapter 25: Conservation of Natural Resources

Carson, J. *Deserts and People.* Morristown, N.J.: Silver Burdett.

Grove, N. *Wild Lands for Wildlife: America's National Refuges.* Wash., D.C.: National Geographic Society.

Polking, K. *Oceans of the World: Our Essential Resource.* New York: Philomel Books.

The metric system of measurement is used by scientists throughout the world. It is based on units of ten. Each unit is ten times larger or ten times smaller than the next unit. The most commonly used units of the metric system are given below. After you have finished reading about the metric system, try to put it to use. How tall are you in metrics? What is your mass? What is your normal body temperature in degrees Celsius?

COMMONLY USED METRIC UNITS

Length The distance from one point to another

meter (m)   1 meter = 1000 millimeters (mm)

(a meter is   1 meter = 100 centimeters (cm)
slightly
longer than   1000 meters = 1 kilometer (km)
a yard)

Volume The amount of space an object takes up

liter (L)   1 liter = 1000 milliliters (mL)

(a liter is slightly
larger than a quart)

Mass The amount of matter in an object

gram (g)   1000 grams = 1 kilogram (kg)

(a gram has a mass
equal to about one
paper clip)

Temperature The measure of hotness or coldness

degrees   0°C = freezing point of water
Celsius (°C)
                   100°C = boiling point of water

METRIC – ENGLISH EQUIVALENTS

2.54 centimeters (cm) = 1 inch (in.)
1 meter (m) = 39.37 inches (in.)
1 kilometer (km) = 0.62 miles (mi)
1 liter (L) = 1.06 quarts (qt)
250 milliliters (mL) = 1 cup (c)
1 kilogram (kg) = 2.2 pounds (lb)
28.3 grams (g) = 1 ounce (oz)
$°C = 5/9 \times (°F - 32)$

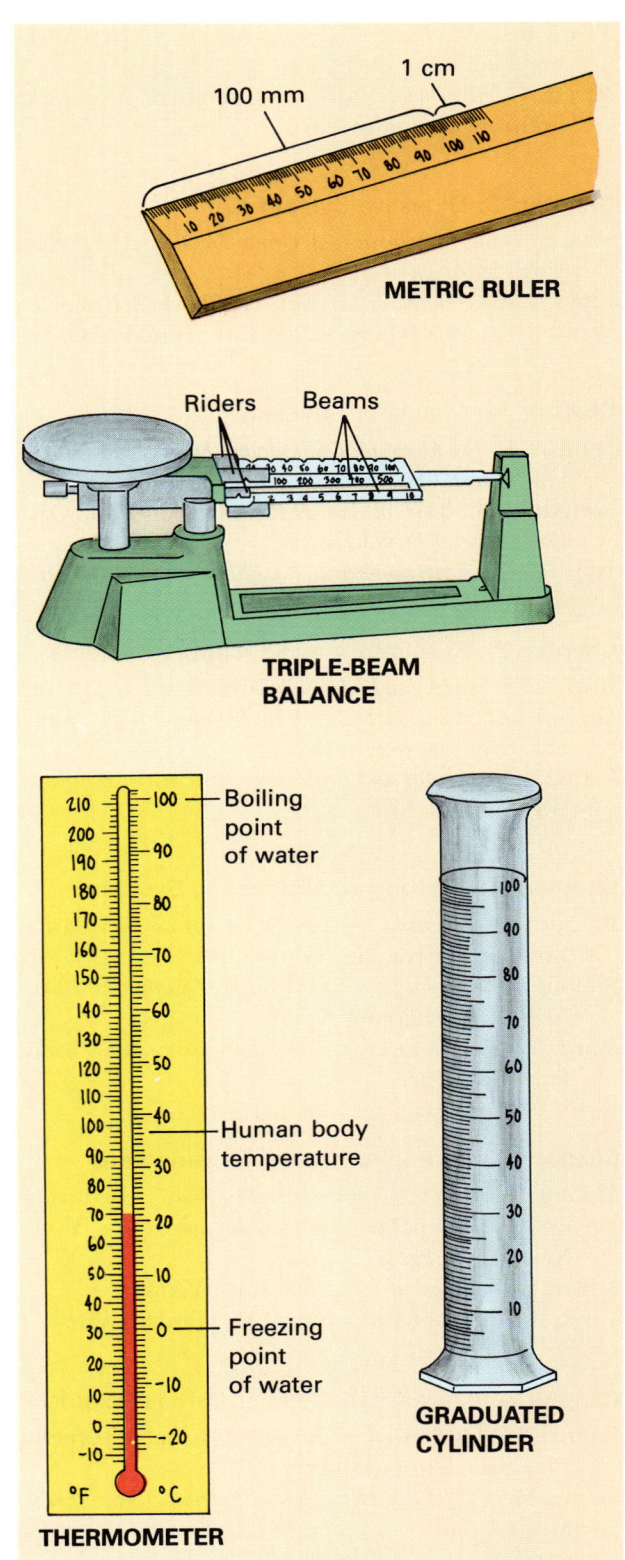

METRIC RULER

TRIPLE-BEAM BALANCE

THERMOMETER

GRADUATED CYLINDER

The microscope is an essential tool in the study of life science. It enables you to see things that are too small to be seen with the unaided eye. It also allows you to look more closely at the fine details of larger things.

The microscope you will use in your science class is probably similar to the one illustrated on the following page. This is a compound microscope. It is called compound because it has more than one lens. A simple microscope would only contain one lens. The lenses of the compound microscope are the parts that magnify the object being viewed.

Typically, a compound microscope has one lens in the eyepiece, the part you look through. The eyepiece lens usually has a magnification power of 10 ×. That is, if you were to look through the eyepiece alone, the object you were viewing would appear 10 times larger than it is.

The compound microscope may contain one or two other lenses. These two lenses are called the low- and high-power objective lenses. The low-power objective lens usually has a magnification of 10 ×. The high-power objective lens usually has a magnification of 40 ×. To figure out what the total magnification of your microscope is when using the eyepiece and an objective lens, multiply the powers of the lenses you are using. For example, eyepiece magnification (10 ×) multiplied by low-power objective lens magnification (10 ×) = 100 × total magnification. What is the total magnification of your microscope using the eyepiece and the high-power objective lens?

To use the microscope properly, it is important to learn the name of each part, its function, and its location on your microscope. Keep the following procedures in mind when using the microscope:

1. Always carry the microscope with both hands. One hand should grasp the arm, and the other should support the base.
2. Place the microscope on the table with the arm toward you. The stage should be facing a light source.
3. Raise the body tube by turning the coarse adjustment knob.
4. Revolve the nosepiece so that the low-power objective lens (10 ×) is directly in line with the body tube. Click it into place. The low-power lens should be directly over the opening in the stage.
5. While looking through the eyepiece, adjust the diaphragm and the mirror so that the greatest amount of light is coming through the opening in the stage.
6. Place the slide to be viewed on the stage. Center the specimen to be viewed over the hole in the stage. Use the stage clips to hold the slide in position.
7. Look at the microscope from the side rather than through the eyepiece. In this way, you can watch as you use the coarse adjustment knob to lower the body tube until the low-power objective *almost* touches the slide. Do this slowly so you do not break the slide or damage the lens.
8. Now, looking through the eyepiece, observe the specimen. Use the coarse adjustment knob to *raise* the body tube, thus raising the low-power objective away from the slide. Continue to raise the body tube until the specimen comes into focus.
9. When viewing a specimen, be sure to keep both eyes open. Though this may seem strange at first, it is really much easier on your eyes. Keeping one eye closed may create a strain, and you might get a headache. Also, if you keep both eyes open, it is easier to draw diagrams of what you are observing. In this way, you do not have to turn your head away from the microscope as you draw.
10. To switch to the high-power objective lens (40 ×), look at the microscope from the side. Now, revolve the nosepiece so that the high-power objective lens clicks into place. Make sure the lens does not hit the slide.
11. Looking through the eyepiece, use only the fine adjustment knob to bring the specimen into focus. Why should you not use the coarse adjustment knob with the high-power objective?
12. Clean the microscope stage and lens when you are finished. To clean the lenses, use lens paper only. Other types of paper may scratch the lenses.

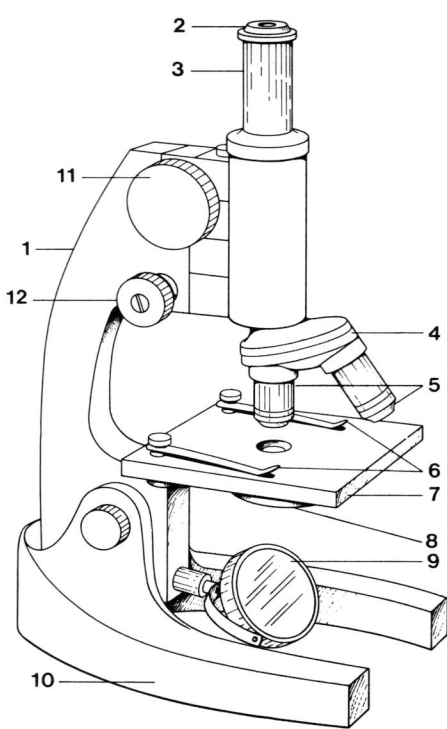

Microscope Parts and Their Functions

1. **Arm** Supports the body tube
2. **Eyepiece** Contains the magnifying lens you look through
3. **Body tube** Maintains the proper distance between the eyepiece and objective lenses
4. **Nosepiece** Holds high- and low-power objective lenses and can be rotated to change magnification
5. **Objective lenses** A low-power lens which usually provides 10X magnification, and a high-power lens which usually provides 40X magnification
6. **Stage clips** Hold the slide in place
7. **Stage** Supports the slide being viewed
8. **Diaphragm** Regulates the amount of light let into the body tube
9. **Mirror** Reflects the light upward through the diaphragm, the specimen, and the lenses
10. **Base** Supports the microscope
11. **Coarse adjustment knob** Moves the body tube up and down for focusing
12. **Fine adjustment knob** Moves the body tube slightly to sharpen the image

One of the first things a scientist learns is that working in the laboratory can be an exciting experience. But the laboratory can also be quite dangerous if proper safety rules are not followed at all times. To prepare yourself for a safe year in the laboratory, read over the following safety rules. Then read them a second time. Make sure you understand each rule. If you do not, ask your teacher to explain any rules you are unsure of.

Dress Code

1. Many materials in the laboratory can cause eye injury. To protect yourself from possible injury, wear safety goggles whenever you are working with chemicals, burners, or any substance that might get into your eyes. Never wear contact lenses in the laboratory.

2. Wear a laboratory apron or coat whenever you are working with chemicals or heated substances.

3. Tie back long hair to keep it away from any chemicals, burners and candles, or other laboratory equipment.

4. Remove or tie back any article of clothing or jewelry that can hang down and touch chemicals and flames before working in the laboratory.

General Safety Rules

5. Read all directions for an experiment several times. Follow the directions exactly as they are written. If you are in doubt about any part of the experiment, ask your teacher for assistance.

6. Never perform activities that are not authorized by your teacher. Obtain permission before "experimenting" on your own.

7. Never handle any equipment unless you have specific permission.

8. Take extreme care not to spill any material in the laboratory. If spills occur, ask your teacher immediately about the proper cleanup procedure. Never simply pour chemicals or other substances into the sink or trash container.

9. Never eat in the laboratory. Wash your hands before and after each experiment.

First Aid

10. Report all accidents, no matter how minor, to your teacher immediately.

11. Learn what to do in case of specific accidents, such as getting acid in your eyes or on your skin. (Rinse acids on your body with lots of water.)

12. Become aware of the location of the first-aid kit. Your teacher should administer any required first aid due to injury. Or your teacher may send you to the school nurse or call a physician.

13. Know where and how to report an accident or fire. Find out the location of the fire extinguisher, phone, and fire alarm. Keep a list of important phone numbers such as the fire department and school nurse near the phone. Report any fires to your teacher at once.

Heating and Fire Safety

14. Again, never use a heat source such as a candle or burner without wearing safety goggles.

15. Never heat a chemical you are not instructed to heat. A chemical that is harmless when cool can be dangerous when heated.

16. Maintain a clean work area and keep all materials away from flames.

17. Never reach across a flame.

18. Make sure you know how to light a Bunsen burner. (Your teacher will demonstrate the proper procedure for lighting a burner.) If the flame leaps out of a burner toward you, turn the gas off immediately. Do not touch the burner. It may be hot. And never leave a lighted burner unattended!

19. Point a test tube or bottle that is being heated away from you and others. Chemicals can splash or boil out of a heated test tube.

20. Never heat a liquid in a closed container. The expanding gases produced may blow the container apart, injuring you or others.

21. Never pick up a container that has been heated without first holding the back of your hand near it. If you can feel the heat on the back of your hand, the container may be too hot to handle. Use a clamp or tongs when handling hot containers.

Using Chemicals Safely

22. Never mix chemicals for the "fun of it." You might produce a dangerous, possibly explosive substance.

23. Never touch, taste, or smell a chemical that you do not know for a fact is harmless. Many chemicals are poisonous. If you are instructed to note the fumes in an experiment, gently wave your hand over the opening of a container and direct the fumes toward your nose. Do not inhale the fumes directly from the container.

24. Use only those chemicals needed in the activity. Keep all lids closed when a chemical is not being used. Notify your teacher whenever chemicals are spilled.

25. Dispose of all chemicals as instructed by your teacher. To avoid contamination, never return chemicals to their original containers.

26. Be extra careful when working with acids or bases. Pour such chemicals over the sink, not over your workbench.

27. When diluting an acid, pour the acid into water. Never pour water into the acid.

28. Rinse any acids off your skin or clothing with water. Immediately notify your teacher of any acid spill.

Using Glassware Safely

29. Never force glass tubing into a rubber stopper. A turning motion and lubricant will be helpful when inserting glass tubing into rubber stoppers or rubber tubing. Your teacher will demonstrate the proper way to insert glass tubing.

30. Never heat glassware that is not thoroughly dry. Use a wire screen to protect glassware from any flame.

31. Keep in mind that hot glassware will not appear hot. Never pick up glassware without first checking to see if it is hot.

32. If you are instructed to cut glass tubing, fire-polish the ends immediately to remove sharp edges.

33. Never use broken or chipped glassware. If glassware breaks, notify your teacher and dispose of the glassware in the proper trash container.

34. Never eat or drink from laboratory glassware. Thoroughly clean glassware before putting it away.

Using Sharp Instruments

35. Handle scalpels or razor blades with extreme care. Never cut material toward you; cut away from you.

36. Notify your teacher immediately if you are cut in the laboratory.

Animal Safety

37. No experiments that will cause pain, discomfort, or harm to mammals, birds, reptiles, fish, and amphibians should be done in the classroom or at home.

38. Animals should be handled only if necessary. If an animal is excited or frightened, pregnant, feeding, or with its young, special handling is required.

39. Your teacher will instruct you as to how to handle each animal species that may be brought into the classroom.

40. Clean your hands thoroughly after handling animals or the cage containing animals.

End-of-Experiment Rules

41. When an experiment is completed, clean up your work area and return all equipment to its proper place.

42. Wash your hands after every experiment.

43. Turn off all burners before leaving the laboratory. Check that the gas line leading to the burner is off as well.

Glossary

Pronunciation Key

When difficult names or terms first appear in the text, they are respelled to aid pronunciation. A syllable in SMALL CAPITAL LETTERS receives the most stress. The key below lists the letters used for respelling. It includes examples of words using each sound and shows how the words would be respelled.

Symbol	Example	Respelling
a	hat	(hat)
ay	pay, late	(pay), (layt)
ah	star, hot	(stahr), (haht)
ai	air, dare	(air), (dair)
aw	law, all	(law), (awl)
eh	met	(meht)
ee	bee, eat	(bee), (eet)
er	learn, sir, fir	(lern), (ser), (fer)
ih	fit	(fiht)
igh	mile, sigh	(mighl), (sigh)
oh	no	(noh)
oi	soil, boy	(soil), (boi)
oo	root, rule	(root), (rool)
or	born, door	(born), (dor)
ow	plow, out	(plow), (owt)

Symbol	Example	Respelling
u	put, book	(put), (buk)
uh	fun	(fuhn)
yoo	few, use	(fyoo), (yooz)
ch	chill, reach	(chihl), (reech)
g	go, dig	(goh), (dihg)
j	jet, gently, bridge	(jeht), (JEHNT lee), (brihj)
k	kite, cup	(kight), (kuhp)
ks	mix	(mihks)
kw	quick	(kwihk)
ng	bring	(brihng)
s	say, cent	(say), (sehnt)
sh	she, crash	(shee), (krash)
th	three	(three)
y	yet, onion	(yeht), (UHN yuhn)
z	zip, always	(zihp), (AWL wayz)
zh	treasure	(TREH zher)

abiotic factor: nonliving things in an environment

acid rain: rain containing nitric acid and sulfuric acid

acquired immunity: immunity that develops during a person's lifetime

adaptation: change that increases an organism's chances of survival

adrenal (uh-DREE-nuhl): endocrine gland on top of each kidney that produces the hormone adrenaline

agar (AG-uhr): jellylike material produced by some red algae on which scientists grow bacteria

air bladder: tiny grape-shaped structure that acts like an inflatable life-preserver in brown algae

air sac: hollow structure connected to a bird's lungs that increases the amount of air to the lungs

alcoholism: disease caused by drinking large amounts of alcohol daily

alga (AL-guh; plural: algae): nonvascular plant that contains chlorophyll

allele (uh-LEEL): each member of a gene pair

allergy: reaction that occurs when the body is especially sensitive to certain substances called allergens

alveolus (al-VEE-uh-luhs; plural: alveoli): grapelike cluster of tiny balloons in the lungs

amino acid: building block of protein

amniocentesis (am-nee-oh-sehn-TEE-sihs): process of removing fluid from the sac surrounding a developing baby

amnion: clear membrane that forms a fluid-filled sac around the embryo in the uterus

amniotic sac: fluid-filled sac that cushions and protects the developing baby in the uterus

amoeba: sarcodine, a type of protozoan, that lives in fresh water and moves by means of pseudopods

anal pore: structure in a paramecium through which undigested food is eliminated

anaphase: third stage of mitosis during which the chromosomes split apart

anatomy: study of the structure of living things

angiosperm (AN-jee-uh-sperm): type of seed plant whose seeds are covered by a protective wall

annual ring: one year's growth of xylem cells

antibiotic: chemical that destroys or weakens disease-causing bacteria

antibody: protein produced by certain kinds of white blood cells in response to an invasion by a particular organism or substance

antigen: invading organism or substance

aorta (ay-OR-tuh): largest blood vessel in the body

artery: blood vessel that carries blood away from the heart

arthropod (AHR-thruh-pahd): invertebrate that has jointed legs and an exoskeleton

asexual reproduction: reproduction requiring only one parent

atherosclerosis (ath-uhr-oh-skluh-ROH-sihs): thickening of the inner lining of the arteries

atom: tiny particle of matter with a nucleus containing protons and neutrons and an electron cloud containing electrons

atrium (plural: atria): upper heart chamber

autonomic nervous system: division of the peripheral nervous system that controls all involuntary body processes

autotroph (AWT-uh-trohf): organism that can make its own food from simple substances

axon: fiber that carries messages away from a body cell

bacillus (buh-SIHL-uhs; plural: bacilli): rod-shaped bacterium

bacteriophage (bak-TEER-ee-uh-fayj): virus that infects bacteria

bacterium (plural: bacteria): unicellular microorganism that does not have a nucleus

benign: harmless

bile: substance produced by the liver that aids in digestion

binary fission: reproductive process in which a cell divides into two cells

binomial nomenclature (bigh-NOH-mee-uhl NOH-muhn-klay-cher): naming system in which organisms are given two names: a genus and a species

bioluminescence (bigh-oh-loo-muh-NEHS-uhns): glow produced by some fire algae

biome: division of area with similar climate, plants, and animals

biotechnology (bigh-oh-tehk-NAHL-uh-jee): application of technology to the study and solution of problems involving living things

biotic factor: living things in an environment

botany: study of plants

bronchus (BRAHNG-kuhs; plural: bronchi): tube that branches off from the trachea

budding: reproductive process in yeast, in which a new yeast cell is formed from a tiny bud

Calorie: amount of heat energy needed to raise the temperature of 1 kilogram of water 1°C

cambium (KAM-bee-uhm): growth tissue of the stem where xylem and phloem cells are produced

camouflage (KAM-uh-flahzh): hiding from enemies by blending in with the surroundings

cancer: abnormal and uncontrolled cell reproduction

canine (KAY-nighn): sharp pointed tooth used for tearing and shredding meat

canopy (KAN-uh-pee): roof formed by tall trees in the forest

cap: umbrella-shaped part of a mushroom, which is part of the mushroom's fruiting body

capillary: tiny, thin-walled blood vessel

capsule: cup-shaped part of the nephron

carbohydrate: energy-rich substance found in foods such as vegetables, cereal grains, and breads

carcinogen (kahr-SIHN-uh-juhn): cancer-causing substance

cardiac muscle: muscle found only in the heart

cardiovascular (kahr-dee-oh-VAS-kyoo-luhr) **disease:** disease that affects the heart and blood vessels

carnivore (KAHR-nuh-vor): flesh-eating mammal

cartilage: flexible tissue that gives support and shape to body parts

cell: basic unit of structure and function of a living thing

cell membrane: thin, flexible envelope that surrounds a cell

cellulose: long chain of sugar molecules manufactured by a cell that makes up the cell wall

cell wall: outermost boundary of plant and bacterial cells that is made of cellulose

Celsius: temperature scale used in the metric system in which water freezes at 0° and boils at 100°

central nervous system: part of the nervous system made up of the brain and spinal cord

centriole (SEHN-tree-ohl): structure outside the nucleus in animal cells that plays a part in cell division

cerebellum (sair-uh-BEHL-uhm): part of the brain that controls balance and posture

cerebrum ((SAIR-uh-bruhm): part of the brain that controls the senses, thought, and conscious activities

chemical digestion: breaking down of food by enzymes

chlorophyll: green substance, needed for photosynthesis, found in green plant cells

chloroplast: large, irregularly shaped structure that contains the green pigment chlorophyll; food-making site in green plants

cholesterol: fatty substance found in animal fats, meats, and dairy products

chordate (KOR-dayt): phylum of vertebrates

chromatin: threadlike coils of chromosomes

chromosome: rod-shaped cell structure that directs the activities of a cell and passes on the traits of a cell to new cells

chromosome theory: theory that states that genes are found on chromosomes and that genes are carried from the parental generation to the next generation on chromosomes

chronic disorder: lingering, or lasting, illness

cilium (SIHL-ee-uhm; plural: cilia): small, hairlike projection on the outside of a ciliate that acts like a tiny oar and helps the organism move

ciliate (SIHL-ee-ayt): protozoan that moves by means of cilia

circulatory system: body system that delivers food and oxygen to body cells and carries carbon dioxide and other waste products away from body cells

cirrhosis (suh-ROH-suhs): loss of liver function due to alcohol abuse

class: classification grouping between phylum and order

classification: grouping of living things according to similar characteristics

climate: average weather in a particular place over a long period of time

club fungus: fungus that produces spores in a club-shaped structure

coccus (KAHK-uhs; plural: cocci): sphere-shaped bacterium

coelenterate (sih-LEHN-tuh-rayt): phylum of invertebrates that contain a central cavity with only one opening

cochlea (KAHK-lee-uh): spiraling tube in the inner ear from which nerve impulses are carried to the brain

coldblooded: having a body temperature that can change somewhat with changes in the temperature of the environment

colorblindness: sex-linked trait that causes the inability to distinguish between certain colors

commensalism: symbiotic relationship in which only one partner in the relationship benefits

communicable (kuh-MYOO-nih-kuh-buhl) **disease:** disease transmitted among people by harmful organisms such as viruses and bacteria; infectious disease

community: living part of any ecosystem

competition: struggle among living things to get the proper amount of food, water, and energy

compound: two or more elements chemically combined

compound light microscope: microscope having more than one lens and that uses a beam of light to magnify objects

coniferous forest: northernmost forest biome, which contains conifers, or cone-bearing trees

conjugation (kahn-juh-GAY-shuhn): type of sexual reproduction in which hereditary material is exchanged

connective tissue: type of tissue that provides support for the body and unites its parts

conservation: wise use of natural resources so that they will not be used up too soon or used in a way that will damage the environment

consumer: organism that feeds directly or indirectly on producers

contour farming: farming method in which a slope is plowed horizontally across its face to avoid erosion

contour feather: large feather used for flight that is found on a bird's wing and on most of the bird's body

contractile vacuole: structure in protozoans through which excess water is pumped out

control: experiment done in exactly the same way as another experiment, but without the variable

cornea: transparent protective covering of the eye

cotyledon (kaht-uhl-EED-uhn): leaflike structure of a young plant that stores food

crop: saclike organ that stores food in an earthworm

crop rotation: farming method of alternating the growth of different crops each year on the same land

cross-pollination: process in which pollen is transferred from the male part of one flower to the female part of another flower

cuticle (KYOOT-ih-kuhl): waxy covering of the epidermis that prevents the loss of too much water from a leaf

cytoplasm: all the protoplasm, or living material, outside the nucleus of a cell

data: recorded observations and measurements

deciduous forest: forest biome that contains deciduous trees, which shed their leaves in the autumn

decomposer: organism that feeds on dead organic matter and breaks it down into simpler substances

dendrite: fiber that carries messages from a neuron toward the cell body

density: measure of how much mass is contained in a given volume of an object

depletion: process in which nutrients are washed away from the soil by water

depressant: drug that slows down the actions of the nervous system

dermis: inner layer of the skin

desert: biome that receives less than 25 centimeters of rainfall a year

diaphragm (DIGH-uh-fram): muscle at the bottom of the chest that aids in breathing

diatom (DIGH-uh-tahm): type of golden algae that is made of a tough glasslike material

diffusion: process by which food molecules, oxygen, water, and other materials enter and leave a cell through the cell membrane

digest: break down

digestion: process by which food is broken down into simpler substances

digestive system: body system in which food is broken down into simpler substances for use by the body

dinoflagellate (digh-nuh-FLAJ-uh-liht): type of fire algae that has two flagella

division of labor: division of the work among the different parts of an organism's body that keeps an organism alive

DNA (deoxyribonucleic acid): nucleic acid that stores the information needed to build proteins and carries genetic information about an organism

dominant: stronger trait in genetics

down feather: short, fuzzy type of feather used for insulation

Down's syndrome: condition that results when the twenty-first pair of chromosomes has an extra chromosome

drug: substance that has an effect on the body

drug abuse: using too much of a drug or using a drug in a way doctors would not approve

eardrum: membrane in the ear that vibrates when struck by sound waves

echinoderm (ih-KEE-nuh-derm): invertebrate with rough, spiny skin

ecological succession: process of gradual change within a community

ecology: study of relationships and interactions of living things with one another and with their environment

ecosystem: group of organisms in an area that interact with one another, together with their nonliving environment

effector: part of the body that carries out the instructions of the nervous system

egg: female sex cell; ovum

electron: negatively charged particle of an atom

electron microscope: microscope that uses a beam of electrons to magnify objects

element: pure substance that cannot be separated into simpler substances by ordinary chemical processes

embryo (EHM-bree-oh): newly formed organism that is the product of fertilization

embryology: study of developing organisms

endangered species: species in danger of becoming extinct

endocrine (EHN-duh-krihn) **gland:** gland that releases hormones directly into the bloodstream

endocrine system: body system that produces chemicals that control the body

endoplasmic reticulum (ehn-duh-PLAZ-mihk rih-TIHK-yuh-luhm): clear tubular passageways leading out from the nuclear membrane that are involved in the transport of proteins

endoskeleton: internal skeleton

endospore: spherical structure that protects a bacterium when environmental conditions are not suitable for bacterial growth

environment: all the living and nonliving things with which an organism interacts

enzyme: chemical substance that helps control chemical reactions

epidermis (ehp-uh-DER-muhs): outer, protective layer of a leaf; outermost layer of the skin

epiglottis (ehp-uh-GLAHT-ihs): small flap of tissue that closes over the windpipe

epithelial tissue: type of tissue that forms a protective surface for the body and lines the cavities and other body parts

erosion: process in which soil is carried off by water or wind

esophagus (ih-SAHF-uh-guhs): pipe-shaped tube that transports food to the stomach

estrogen: hormone that triggers the broadening of the hips in females and starts the maturation of egg cells in the ovaries

evaporation: process in which radiant heat from the sun turns liquid water into a gas

evolution: change in a species over time

excretion: process of getting rid of waste materials

excretory (EHKS-kruh-tor-ee) **system:** body system that removes body wastes

exhale: breathe out

exocrine (EHK-suh-krihn) **gland:** gland that releases its chemicals through ducts, or tubes, into a nearby organ

exoskeleton: rigid, outer covering of an organism

external fertilization: fertilization that occurs outside the body of the female

extinct: having died out

extinction: process by which a species passes out of existence

eyespot: reddish structure in *Euglena* that is sensitive to light

Fallopian (fuh-LOH-pee-uhn) **tube:** oviduct; tube through which an egg travels from the ovary

family: classification grouping between order and genus

fang: special tooth in snakes used to inject venom into their prey

fat: substance that supplies the body with energy and also helps support and cushion the vital organs in the body

feces (FEE-seez): solid waste that is eliminated from the body

feedback mechanism: process in the endocrine system in which the production of a hormone is controlled by the concentration of another hormone

fermentation: energy-releasing process in which sugars and starches are changed into alcohol and carbon dioxide; process by which yeasts obtain their energy

fertilization: joining of the egg and the sperm

fetus (FEET-uhs): developing baby from the eighth week until birth

fibrin (FIGH-bruhn): substance that traps blood cells and plasma, forming a scab

flagellate (FLAJ-uh-layt): protozoan that moves by means of flagella

flagellum (fluh-JEHL-uhm; plural: flagella): long, thin whiplike structure that propels an organism

flower: structure containing the reproductive organs of an angiosperm

food chain: food and energy links between the different plants and animals in an ecosystem

food vacuole (VAK-yoo-wohl): spherical structure in protozoans that digests food particles

food web: all the food chains in an ecosystem that are connected

formula: combination of chemical symbols that shows the elements that make up a compound and the number of atoms of each element in a molecule of that compound

fossil: imprint or remains of plants or animals that existed in the past

fossil fuel: product of decayed plants and animals that is preserved in the earth's crust over millions of years

fossil record: most complete biological record of life on earth

freshwater biome: biome that contains freshwater lakes, ponds, swamps, streams, and rivers

frond: leaf of a fern plant

fruit: ripened ovary of an angiosperm

fruiting body: spore-containing structure in a fungus

fungus (FUHNG-uhs; plural: fungi): nonvascular plantlike organism that has no chlorophyll

gall bladder: organ that stores bile

gene: basic unit of heredity

genetic engineering: process in which genes, or parts of DNA, are transferred from one organism into another organism

genetics: study of heredity

genotype (JEE-nuh-tighp): genetic makeup of an organism

genus (plural: genera): group of organisms that are closely related; classification group between family and species

geothermal energy: energy that comes from heat created in the earth

germination: early growth stage, or the "sprouting," of a young plant

gill: spore-producing structure in a mushroom; structure through which water-dwelling animals obtain their oxygen

gizzard: structure in an earthworm that grinds up food

gradualism (GRAJ-oo-wuhl-ihz-uhm): belief that evolution is a slow and steady process

grassland: biome made up mainly of grasses that receive between 25 and 75 centimeters of rainfall yearly

gravity: force of attraction between objects

groundwater: underground water

guard cell: sausage-shaped cell that regulates the opening and closing of stomata

gullet: funnel-shaped structure in a paramecium through which food passes from the oral groove to the food vacuole

gymnosperm (JIHM-nuh-sperm): type of seed plant whose seeds are not covered by a protective wall

habitat: place in which an organism lives

half-life: time it takes for half of a radioactive element to decay

hallucinogen (huh-LOO-suh-nuh-jehn): drug that produces powerful hallucinations

Haversian (huh-VER-shuhn) **canal:** passageway running through the thick bone, containing blood vessels and nerves

hazardous waste: waste that burns easily, is poisonous, or reacts dangerously with other substances

hemoglobin (hee-muh-GLOH-bihn): iron-containing protein in red blood cells

hemophilia (hee-muh-FIHL-ee-uh): inherited disease that causes the blood to clot slowly or not at all

herbaceous (huhr-BAY-shuhs) **stem:** soft, green plant stem

herbivore (HER-buh-vor): organism that eats only plants

heterotroph (HEHT-uh-roh-trohf): organism unable to make its own food

hibernation: winter sleep during which all body activities slow down

holdfast: rootlike structure that attaches an alga to a rock or other object on the ocean floor

homeostasis (hoh-mee-oh-STAY-sihs): ability of an organism to keep conditions inside its body the same even though conditions in its external environment change

homologous (hoh-MAHL-uh-guhs): similar in structure

hormone: chemical messenger that travels through the blood

host: organism in which another organism lives

hybrid: organism produced through hybridization; organism with two different genes for a particular trait

hybridization (high-bruh-duh-ZAY-shuhn): crossing of two genetically different but related species of an organism

hybrid vigor: ability of hybrids to grow faster or larger than their parents

hydroelectric plant: plant that uses energy from moving water to produce electricity

hypertension: high blood pressure

hypha (HIGH-fuh; plural: hyphae: HIGH-fee): thread-like structure in fungi that produces enzymes to break down living or dead organisms

hypothalamus (high-puh-THAL-uh-muhs): endocrine gland at the base of the brain that controls body temperature, water balance, appetite, and sleep

hypothesis: suggested solution to a problem

immune (ih-MYOON) **system:** body's defense system against disease

immunity (ih-MYOON-uh-tee): body's ability to fight off disease without becoming sick

inbreeding: breeding that involves crossing plants or animals that have the same or very similar sets of genes

incisor (ihn-SIGH-zuhr): front tooth used for biting

incomplete dominance: condition that occurs when a gene is neither dominant nor recessive

incubate: warm eggs by sitting on them until they hatch

infection: state caused when the body is invaded by disease-causing organisms

infectious disease: disease that is transmitted among people by harmful organisms such as viruses and bacteria; communicable disease

inflammation (ihn-fluh-MAY-shuhn): body's response to an attack by disease-causing organisms

ingestion: taking in food; eating

inhale: breathe in

interferon: substance produced by a body cell when invaded by a virus

internal fertilization: fertilization that takes place inside the body of the female

interneuron: type of neuron that connects sensory and motor neurons

interphase: period between one mitosis and the next

invertebrate: animal without a backbone

iris: circular, colored portion of the eye that regulates the amount of light entering the eye

joint: place where two bones meet

kidney: major excretory organ

kilogram (kg): basic unit of mass in the metric system

kingdom: largest classification grouping

large intestine: organ in the digestive system in which water is absorbed and undigested food is stored

larva: stage of insect that develops from an egg

larynx (LAR-ihngks): voice box

law: scientific theory that has been tested many times and is generally accepted as true

lens: curved piece of glass that bends light rays as they pass through it; part of the eye that focuses the light ray coming into the eye

lichen (LIGH-kuhn): organism made up of a fungus and an alga that live together in a symbiotic relationship

life span: maximum length of time an organism can be expected to live

ligament: stringy connective tissue that holds the bones together

limiting factor: living or nonliving factor in an environment that can stop a population from increasing in size

liter (L): basic unit of volume in the metric system

litter: material that is disposed of in improper places

liver: organ that produces bile

liverwort: small, green nonvascular plant with flat leaflike parts

lung: main respiratory organ

lymph: plasma that leaks out of the blood and surrounds and bathes body cells

lysosome (LIGH-suh-sohm): small, round structure involved in the digestive activities of a cell

macronucleus: large nucleus in a paramecium that controls all life functions except reproduction

malignant: life-threatening

mammary gland: structure in a female mammal that produces milk

mantle: part of a mollusk that produces material that makes up the hard shell

marine biome: ocean biome

marrow: soft material inside a bone

marsupial (mahr-SOO-pee-uhl): pouched mammal

mass: measure of the amount of matter in an object

matter: anything that takes up space and has mass

mechanical digestion: physical action of breaking down food into smaller pieces

medulla (mih-DUHL-uh): part of the brain located at the base of the brain stem that controls involuntary body processes

meiosis (migh-OH-sihs): process that results in cells with only half the normal number of chromosomes

menopause: physical change in females after which menstruation and ovulation stop

menstrual (MEHN-struhl) **cycle:** monthly cycle of change that occurs in the female reproductive system

menstruation (mehn-STRAY-shuhn): process in which the blood and tissue from the thickened lining of the uterus pass out of a female's body through the vagina

metabolism (muh-TAB-uh-lihz-uhm): all chemical activities in an organism

metamorphosis (meht-uh-MOR-fuh-sihs): change in appearance due to development

metaphase: second stage of mitosis

meter (m): basic unit of length in the metric system

metric system: universal system of measurement

microbiology: study of microorganisms

micronucleus: small nucleus that controls reproduction in a paramecium

microorganism: microscopic organisms

microscope: instrument that produces an enlarged image of an object

migrate: move to a new environment during the course of a year

mineral: simple substance found in nature that helps maintain the normal functioning of the body

mitochondrion (might-uh-KAHN-dree-uhn; plural: mitochondria): rod-shaped structure that is referred to as the powerhouse of a cell

mitosis (migh-TOH-sihs): duplication and division of the nucleus and of the chromosomes during cell reproduction

molar: back tooth that grinds and crushes food

molecular clock: scale used to estimate the rate of change in proteins over time

molecule (MAHL-uh-kyool): smallest particle of a compound having all the properties of that compound

mollusk (MAHL-uhsk): invertebrate with a soft, fleshy body that is often covered by a hard shell

molt: periodically shed one's skin

moneran: member of the Monera kingdom

monoclonal (MAHN-uh-kloh-nuhl) **antibody:** substance produced by the joining of cancer cells with antibody-producing white blood cells

monotreme (MAHN-uh-treem): egg-laying mammal

moss: small, green nonvascular plant that has stemlike and leaflike parts

motor neuron: type of neuron that carries messages from the central nervous system to effectors

multiple allele: more than two alleles that combine to determine a certain characteristic

multicellular: having many cells

muscle tissue: type of tissue that has the ability to contract and make the body move

mutation: change in genes or chromosomes that causes a new trait to be inherited

mutualism: symbiotic relationship that is helpful to both organisms

natural immunity: immunity that is present at birth and protects people from some diseases that infect other types of organisms

natural resource: material produced by the environment and used by people

natural selection: survival and reproduction of those organisms best adapted to their surroundings

nematocyst (NEHM-uh-toh-sihst): stinging cell that is found in the mouth of a coelenterate

nephron (NEHF-rahn): microscopic chemical filtering factory in the kidneys

nerve impulse: message carried throughout the body by nerves

nerve tissue: type of tissue that carries messages back and forth between the brain and spinal cord and to every part of the body

nervous system: body system that controls all of the activities of the body

neuron: nerve cell

neutron: neutral particle in the nucleus of an atom

niche: role of an organism in its community or environment

nitrogen base: substance in DNA that contains the element nitrogen

nitrogen fixation: process by which some kinds of bacteria take nitrogen directly from the air and form nitrogen compounds

nondisjunction (nahn-dihs-JUHNGK-shuhn): failure of chromosomes to separate during meiosis

nonrenewable resource: natural resource that cannot be replaced by nature

nonvascular (nahn-VA-skyuh-luhr) **plant:** plant lacking transportation tubes that carry water and food throughout the plant

nostril: opening in the nose

nuclear energy: energy located within the nucleus of an atom

nuclear fission: process in which atoms of the element uranium are split to release energy

nuclear fusion: process by which atoms are combined and energy is released

nuclear membrane: thin membrane that separates the nucleus from the rest of a cell

nucleic (noo-KLEE-ihk) **acid:** large organic compound that stores information that helps the body make the proteins it needs

nucleolus (noo-KLEE-uh-luhs): cell structure located in the nucleus and made up of RNA and protein

nucleus (NOO-klee-uhs): cell structure that directs all the activities of the cell

nutrient: usable portion of food

oil: energy-rich substance

omnivore (AHM-nuh-vor): organism that eats both plants and animals

opiate: pain-killing drug produced from the opium poppy

oral groove: indentation in a paramecium through which food particles enter

order: classification grouping between class and family

organ: group of different tissues working together; third level of organization in an organism

organelle (or-guh-NEHL): tiny cell structure

organism: entire living thing that carries out *all* the basic life functions

organ system: group of organs that work together to perform certain functions; fourth level of organization in an organism

osmosis: special type of diffusion by which water passes into and out of the cell

ossification (ahs-uh-fuh-KAY-shuhn): process in which cartilage disappears and is replaced by bone

ovary (OH-vuhr-ee): hollow structure that contains the egg cells of a flower; female sex gland; endocrine gland that produces female hormones

over-the-counter drug: drug that can be bought without a prescription

oviduct: Fallopian tube; tube through which an egg travels from the ovary

ovulation: process in which an egg is released from the ovary into the Fallopian tube

ovule (OH-vyool): structure that contains the female sex cells of a seed plant

pacemaker: special tissue in the heart that controls the pace at which the heart beats

palisade (pal-uh-SAYD) **layer:** long, cylindrical food-making cells of the upper mesophyll in a leaf

pancreas (PAN-kree-uhs): organ that produces pancreatic juice and insulin

paramecium (par-uh-MEE-see-uhm; plural: paramecia): type of protozoan that moves by means of cilia

parasite (PAR-uh-sight): organism that feeds on other living organisms

parasitism: symbiotic relationship in which one organism is harmed by the other organism

parathyroid: endocrine gland producing a hormone that controls the level of calcium in the blood

pasteurization: process in which milk is heated to destroy the bacteria that would cause the milk to spoil quickly

pellicle (PEHL-ih-kuhl): hard membrane that covers the outer surface of a paramecium

pepsin: enzyme produced by the stomach that digests protein

periosteum (pehr-ih-AHS-tee-uhm): tough membrane containing bone-forming cells and blood vessels that surrounds the solid bone

peripheral (puh-RIHF-uh-ruhl) **nervous system:** part of the nervous system that branches out from the central nervous system and includes a network of nerves and sense organs

peristalsis (pair-uh-STAHL-sihs): powerful wave of muscle contractions that pushes food through the digestive system

permafrost: permanently frozen tundra soil

pesticide: chemical used to kill harmful insects or other pests

petal: colorful leaflike structure that surrounds the male and female reproductive organs in a flower

petroleum: liquid fossil fuel

phenotype (FEE-nuh-tighp): visible characteristic of an organism

pheromone (FAIR-uh-mohn): chemical substance given off by insects and other animals to attract a mate

phloem (FLOH-uhm): tubelike plant tissue that carries food down the plant

photosynthesis (foh-tuh-SIHN-thuh-sihs): process by which organisms use energy from sunlight to make their own food

phylum (FIGH-luhm; plural: phyla): second largest classification grouping

physical dependence: effect of drug abuse in which the body cannot function properly without the drug

phytoplankton: microscopic plants that live on the surface of the ocean

pistil: female reproductive organ of a flower

pituitary (pih-TOO-uh-tair-ee): endocrine gland located below the hypothalamus that produces hormones that control many body processes

placenta (pluh-SEHN-tuh): structure through which developing mammals receive food and oxygen while in the mother

plankton: small organisms that float or swim near the surface of water

plasma: yellowish liquid portion of blood

plasmid: bacterial DNA in the form of a ring

platelet (PLAYT-liht): blood cell fragment that aids in blood clotting

pollen: contains the male sex cells of a seed plant

pollination: transfer of pollen from the male part to the female part of a flower

pollution: introduction of harmful or unwanted substances into the environment

population: group of the same type of organism living together in the same area

pore: opening on the outer surface of an animal through which materials enter or leave

poriferan (po-RIHF-uhr-uhn): member of the phylum porifera

predator: animal that kills and eats other animals

predation: relationship that exists between a predator and its prey

prescription drug: drug that requires a doctor's prescription

prey: animal that is killed and eaten by a predator

primate: order of mammals that includes humans, apes, and monkeys

probability: possibility that an event may or may not take place

producer: organism that can make its own food

prophase: first stage of mitosis during which the nuclear membrane begins to disappear

protein: substance used to build and repair cells; made up of amino acids

proton: positively charged particle in the nucleus of an atom

protoplasm: all the living materials in a cell

protozoan: member of the Protista kingdom; unicellular, animallike organism with a nucleus

pseudopod (SOO-duh-pahd): extension of the cytoplasm of a sarcodine that is used in moving and getting food

psychological dependence: emotional need for a drug

ptyalin (TIGH-uh-lihn): enzyme in saliva that breaks down some starches into sugars

puberty (PYOO-buhr-tee): beginning of adolescence

punctuated equilibrium (PUHNGK-choo-wayt-uhd ee-kwuh-LIHB-ree-uhm): theory that evolution occurs in rapid and sudden changes in a species after a long period of little or no change

pupa (PYOO-puh): stage in an insect's life that follows the larva stage

pupil: circular opening at the center of the iris

pus: white substance in an infected area made up of dead bacteria, destroyed body cells, and dead white blood cells

radioactive dating: method based on radioactive elements used by scientists to measure the age of fossils or the age of the rocks in which fossils are found

radioactive material: material given off during the nuclear fission process that is harmful to life

radula: filelike structure in the mouth of a univalve used to scrape food from an object

receptor: part of the nervous system that responds to stimuli

recessive: weaker trait in genetics

recombinant DNA: new piece of DNA produced by combining parts of separate DNA strands

red blood cell: cell that carries oxygen throughout the body

reflex: automatic reaction to a stimulus

regeneration: ability of an organism to regrow lost parts

relative dating: method of determining the age of fossils that involves comparing the rock layers in which the fossils were formed

renewable resource: natural resource that is replaced by nature

replication (rehp-luh-KAY-shuhn): process in which DNA molecules form exact duplicates

reproduction: process by which living things give rise to the same type of living thing

reproductive system: body system in which the male and female sex cells are produced and which enables an organism to produce offspring

respiration: process by which living organisms take in oxygen and use it to produce energy

respiratory system: body system that gets oxygen into the body and removes carbon dioxide and water from the body

response: some action or movement of an organism brought on by a stimulus

retina (REHT-uhn-uh): inner eye layer on which an image is focused

rhizoid (RIGH-zoid): rootlike structure through which mosses absorb water

rhizome (RIGH-zohm): underground stem of a fern

ribosome: grainlike body made up of RNA and attached to the inner surface of an endoplasmic passageway; a protein-making site of the cell

RNA (ribonucleic acid): nucleic acid that "reads" the genetic information carried by DNA and guides the protein-making process

root hair: microscopic extension of an individual cell that greatly increases the root's surface area and helps the plant absorb water and minerals from the soil

sac fungus: fungus that produces spores in a saclike structure

saliva: liquid produced by the salivary glands that moistens food and contains an enzyme that breaks down starches

saprophyte (SAP-ruh-fight): organism that feeds on dead organisms

sarcodine (SAHR-kuh-deen): protozoan that moves by means of pseudopods

savanna: grassland

scavenger: organism that feeds on dead animals

scientific method: systematic approach to problem solving

scrotum (SKROHT-uhm): external sac in males that contains the testes

sedimentary rock: type of rock formed from layers of mud and sand that harden slowly over time

seed: structure from which a plant grows; contains a young plant, stored food, and a seed coat

selective breeding: crossing of animals or plants that have desirable characteristics to produce offspring with desirable characteristics

self-pollination: process in which a plant pollinates itself

semicircular canal: curved tube in the inner ear that is responsible for balance

sensory neuron: type of neuron that carries messages from special receptors to the central nervous system

sepal (SEE-puhl): leaflike structure enclosing a flower when it is still a bud

septum: thick wall of tissue that separates the heart into right and left sides

seta (plural: setae): bristle on the segment of an earthworm that helps it pull itself along the ground

sex chromosome: chromosome that determines the sex of an organism

sex-linked trait: characteristic passed from parent to child on a sex chromosome

sexual reproduction: reproduction usually requiring two parents

sickle-cell anemia: inherited blood disease that causes red blood cells to become sickle shaped

skeletal muscle: muscle that is attached to bone and moves the skeleton

skin: outer covering of the body

slime mold: bloblike organism resembling a protozoan and a fungus during the two stages of its life cycle

small intestine: organ in the digestive system in which most digestion takes place

smog: thick cloud of pollutants

smooth muscle: muscle responsible for involuntary movement

solar energy: energy from the sun

sorus (SOH-ruhs; plural: sori): spore case on the underside of a fern frond

species (SPEE-sheez): group of organisms that are able to interbreed and produce young

sperm: male sex cell

sphygmomanometer (sfihg-moh-muh-NAHM-uh-tuhr): instrument for measuring blood pressure

spinal cord: part of the nervous system that connects the brain with the rest of the nervous system

spirillum (spigh-RIHL-uhm; plural: spirilla): spiral-shaped bacterium

spongy layer: irregularly shaped food-making cells of the lower mesophyll of a leaf

spontaneous generation: theory that states that life can spring from nonliving matter

spore: tiny reproductive cell

sporozoan (SPOR-uh-zoh-uhn): protozoan that has no means of movement

stalk: stemlike structure in a mushroom that supports the cap

stamen (STAY-muhn): male reproductive organ of a flower

stigma (STIHG-muh): structure at the top of pistil

stimulant: drug that speeds up the activities of the nervous system

stimulus (plural: stimuli): signal to which an organism reacts; change in the environment

stoma (STOH-muh; plural: stomata): opening in the lower surface of the epidermis

stomach: J-shaped, muscular organ connected to the end of the esophagus in which foods are physically and chemically digested

strip cropping: farming method in which strips of cover crops are grown between strips of other crops to hold down the soil

style: slender tube that connects the ovary to the stigma

swim bladder: sac filled with air that enables bony fish to rise or sink in water

symbiosis (sihm-bigh-OH-sihs): relationship in which an organism lives on, near, or in another organism

symptom: sign of disease

synapse (sih-NAPS): tiny gap between an axon and a dendrite

taiga (TIGH-guh): northernmost area of a coniferous forest biome

talon: sharp claw of a bird

taxonomy (tak-SAHN-uh-mee): science of classification

telophase: fourth stage of mitosis resulting in the formation of two individual cells

temperature inversion: atmospheric condition in which a layer of cool air containing pollutants is trapped near the ground under a layer of warm air

tendon: connective tissue that connects muscle to bone

terracing: farming method in which a slope is made into a series of level plots in steplike fashion to avoid erosion

territory: area where an animal lives

testis (TEHS-tihs; plural: testes): male sex gland; endocrine gland that produces male hormones

testosterone (tehs-TAHS-tuh-rohn): hormone responsible for the growth of facial and body hair, broadening of the shoulders, and deepening of the voice in males

theory: most logical explanation of events that happen in nature

thermal pollution: release of heat into the environment

thyroid (THIGH-roid): endocrine gland that produces a hormone that controls metabolism

tissue: group of cells that are similar in structure and perform a special function; second level of organization in an organism

tolerance: effect of drug abuse in which a person must take more and more of a drug each time to get the same effect

topsoil: rich upper layer of soil

toxic (TAHK-sihk): poisonous

toxin: poison

trachea (TRAY-kee-uh): windpipe; carries air to lungs

trait: characteristic of an organism

transfusion: process of transferring blood from one body to another

transpiration: process for regulating water loss through the leaves of a plant

tropical rain forest: forest biome that receives at least 200 centimeters of rain yearly

tropism (TROH-pihz-uhm): movement of a plant toward or away from a stimulus

tuber: underground stem of a plant

tumor: swelling of tissue that develops separately from the tissue surrounding it

tundra: biome that rims the Arctic Ocean around the North Pole and has a cold, dry climate

umbilical (uhm-BIHL-ih-kuhl) **cord:** structure that connects an embryo to its mother and transports food, oxygen, and wastes

unicellular: one celled

urea: (yoo-REE-uh): nitrogen waste formed in the liver

ureter (yoo-REET-uhr): tube that conducts urine to the urinary bladder

urethra (yoo-REE-thruh): tube through which urine passes out of the body

urinary bladder: sac of tissue that stores urine

uterus (YOOT-uhr-uhs): pear-shaped structure in which the early development of a baby takes place

vaccine: substance that increases immunity

vacuole (VA-kyoo-wohl): large, round sac in the cytoplasm of a cell that stores water, food, enzymes, and other materials

valve: small flap of tissue between the upper and lower chambers of the heart

variable: factor being tested in an experiment

variation: difference in members of the same species

vascular plant: plant that contains transporting tubes that carry materials throughout the plant

vein: blood vessel that carries blood to the heart

venom: poison produced in special glands by snakes

ventricle: lower chamber of the heart

vertebra (VER-tuh-bruh; plural: vertebrae): bone that makes up a vertebrate's backbone

vertebrate (VER-tuh-briht): animal with a backbone

villus (VIHL-uhs; plural: villi): hairlike projection in the small intestine through which food is absorbed into the bloodstream

virus: tiny particle that contains hereditary material

vitamin: nutrient that helps regulate growth and normal body functioning

vocal cord: tissue in the larynx that vibrates with the passage of air to form sounds

warmblooded: having a constant body temperature

watershed: land area in which surface runoff drains into a river or system of rivers and streams

weight: measure of the force of attraction between objects due to gravity

white blood cell: blood cell that acts as a defense system against disease

windbreak: row of trees planted between fields of crops to prevent erosion due to wind

withdrawal: effect of drug abuse that occurs when a person who is physically dependent on a drug is taken off that drug

woody stem: rigid plant stem

xylem (ZIGH-luhm): tubelike plant tissue that carries water and minerals through the plant

zoology: study of animals

zygote: fertilized egg

Index